Methods in Molecular Biology

Series Editor
John M. Walker
School of Life and Medical Sciences
University of Hertfordshire
Hatfield, Hertfordshire, AL10 9AB, UK

For further volumes:
http://www.springer.com/series/7651

Cell Migration

Methods and Protocols

Edited by

Alexis Gautreau

Laboratory of Biochemistry, Ecole Polytechnique, Palaiseau, France

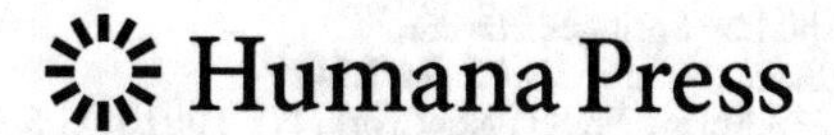

Editor
Alexis Gautreau
Laboratory of Biochemistry
Ecole Polytechnique
Palaiseau, France

ISSN 1064-3745 ISSN 1940-6029 (electronic)
Methods in Molecular Biology
ISBN 978-1-4939-9261-4 ISBN 978-1-4939-7701-7 (eBook)
https://doi.org/10.1007/978-1-4939-7701-7

Cover Illustration: Montage of time-lapse, pseudo-colored, fluorescent images of neutrophils emerging from a drop of blood to migrate through channels in response to chemokine gradients (Courtesy of Xiao Wang and Daniel Irimia, Harvard Medical School)

Printed on acid-free paper

This Humana Press imprint is published by Springer Nature
The registered company is Springer Science+Business Media, LLC
The registered company address is: 233 Spring Street, New York, NY 10013, U.S.A.

Preface

Cell migration is a characteristic of animal cells. Cell migration is highly regulated as it is critical to the development of embryos, to the patrolling of immune cells, and to the correct establishment of brain layers in the brain cortex. Cell migration is also a process of utmost importance in pathologies. We need to better learn how to control cell migration to enhance it in order to promote wound healing, for example, or to block it in order to prevent metastasis formation during cancer progression. This is a major challenge, as the migration of tumor cells is particularly plastic, since it can transit from one type to another type of cell migration.

The recent years have witnessed an explosion of new concepts and techniques applied to cell migration. Numerous assays have thus been developed to characterize cell migration in vitro, ex vivo, and in vivo. This field of cell biology has rapidly become quantitative with the import of concepts and techniques from physics. The imaging of molecular machines powering cell migration has been greatly improved and has now reached the resolution of single molecules. Furthermore, cell migration can be guided by light and through various microfabricated devices.

Cutting edge and comprehensive, *Cell Migration: Methods and Protocols* encompasses various aspects through clear and practical chapters written by experts in their field. These chapters provide specialists and nonspecialists with the latest rotocols to observe, quantify, and control cell migration. Each chapter introduces its topic, lists the necessary materials and reagents, and provides a step-by-step protocol, together with tips on troubleshooting and avoiding known pitfalls.

Palaiseau, France *Alexis Gautreau*

Contents

Contributors

ANTONINA Y. ALEXANDROVA · *Laboratory of Mechanisms of Carcinogenesis, N.N. Blokhin Russian Cancer Research Center, Moscow, Russian Federation*

GUILLAUME ALLIO · *INSERM UMR_S1109, Tumor Biomechanics, Institut d'Hématologie et d'Immunologie, Strasbourg Cedex, France; Université de Strasbourg, Strasbourg, France; LabEx Medalis, Université de Strasbourg, Strasbourg, France; Fédération de Médecine Translationnelle de Strasbourg (FMTS), Strasbourg, France; Centre de Biologie du Développement, UMR 5547 CNRS/Université Paul Sabatier, Toulouse, France*

CHRISTOPHE AMPE · *Department of Biochemistry, Faculty of Medicine and Health Care, Ghent University, Ghent, Belgium; Cancer Research Institute Ghent (CRIG), Ghent, Belgium*

FLORA ASCIONE · *Laboratoire PhysicoChimie Curie, Institut Curie, PSL Research University—Sorbonne Universités, UPMC—CNRS—Equipe labellisée Ligue Contre le Cancer, Paris, France*

RAFAELE ATTIA · *Institut Curie, PSL Research University, CNRS, UMR 144, Paris, France; Institut Pierre-Gilles de Gennes, PSL Research University, Paris, France*

KARIMA BAKKALI · *Department of Biochemistry, Faculty of Medicine and Health Care, Ghent University, Ghent, Belgium*

JORGE BARBAZAN · *Institut Curie, PSL Research University, CNRS, UMR 144, Paris, France*

LUCIE BARBIER · *Institut Curie, PSL Research University, CNRS, UMR 144, Paris, France; Institut Pierre-Gilles de Gennes, PSL Research University, Paris, France; Université Paris Sud, Université Paris-Saclay, Orsay, France*

BATTUYA BAYARMAGNAI · *Department of Bioengineering, Temple University, Philadelphia, PA, USA*

SVEN BOGDAN · *Institut für Physiologie und Pathophysiologie, Abteilung Molekulare Zellphysiologie, Phillips-Universität Marburg, Marburg, Germany*

ISABELLE BONNET · *Laboratoire PhysicoChimie Curie, Institut Curie, PSL Research University—Sorbonne Universités, UPMC—CNRS—Equipe labellisée Ligue Contre le Cancer, Paris, France*

ARTHUR BOUTILLON · *Laboratory for Optics and Biosciences, Ecole Polytechnique, CNRS, INSERM, Université Paris-Saclay, Palaiseau Cedex, France*

STEFAN BRÜHMANN · *Institute for Biophysical Chemistry, Hannover Medical School, Hannover, Germany*

AXEL BUGUIN · *Laboratoire PhysicoChimie Curie, Institut Curie, PSL Research University—Sorbonne Universités, UPMC—CNRS—Equipe labellisée Ligue Contre le Cancer, Paris, France*

YONGPING CHAI · *Tsinghua-Peking Center for Life Sciences, School of Life Sciences and MOE Key Laboratory for Protein Science, Tsinghua University, Beijing, China*

GUILLAUME CHARRAS · *London Centre for Nanotechnology, University College London, London, UK; Institute for the Physics of Living Systems, University College London, London, UK; Department of Cell and Developmental Biology, University College London, London, UK*

YANG CHEN · *State Key Laboratory of Membrane Biology, Tsinghua University-Peking University Joint Center for Life Sciences, School of Life Sciences, Tsinghua University, Beijing, China*

ALEKSANDRA S. CHIKINA · *Laboratory of Mechanisms of Carcinogenesis, N.N. Blokhin Russian Cancer Research Center, Moscow, Russian Federation; CNRS UMR144/Institut Curie, Paris, France*

FU-SHENG CHOU · *Department of Pediatrics, University of Kansas Medical Center, Kansas City, KS, USA; Children's Mercy-Kansas City, Kansas City, MO, USA*

OLIVIER COCHET-ESCARTIN · *Laboratoire PhysicoChimie Curie, Institut Curie, PSL Research University—Sorbonne Universités, UPMC—CNRS—Equipe labellisée Ligue Contre le Cancer, Paris, France*

JULIA DAMIANO-GUERCIO · *Institute for Biophysical Chemistry, Hannover Medical School, Hannover, Germany*

IRENE DANG · *Department of Systems Pharmacology and Translational Therapeutics, University of Pennsylvania, Philadelphia, PA, USA; Ecole Polytechnique, Université Paris-Saclay, BIOC-CNRS UMR7654, Palaiseau, France*

NICOLAS B. DAVID · *Laboratory for Optics and Biosciences, Ecole Polytechnique, CNRS, INSERM, Université Paris-Saclay, Palaiseau Cedex, France*

MAXIME DEFORET · *Laboratoire PhysicoChimie Curie, Institut Curie, PSL Research University—Sorbonne Universités, UPMC—CNRS—Equipe labellisée Ligue Contre le Cancer, Paris, France*

GUILLAUME DUCLOS · *Laboratoire PhysicoChimie Curie, Institut Curie, PSL Research University—Sorbonne Universités, UPMC—CNRS—Equipe labellisée Ligue Contre le Cancer, Paris, France*

JAN FAIX · *Institute for Biophysical Chemistry, Hannover Medical School, Hannover, Germany*

GAUTIER FOLLAIN · *INSERM UMR_S1109, Tumor Biomechanics, Institut d'Hématologie et d'Immunologie, Strasbourg Cedex, France; Université de Strasbourg, Strasbourg, France; LabEx Medalis, Université de Strasbourg, Strasbourg, France; Fédération de Médecine Translationnelle de Strasbourg (FMTS), Strasbourg, France*

CÉDRIC FUCHS · *INSERM UMR_S1109, Tumor Biomechanics, Institut d'Hématologie et d'Immunologie, Strasbourg Cedex, France; Université de Strasbourg, Strasbourg, France; LabEx Medalis, Université de Strasbourg, Strasbourg, France; Fédération de Médecine Translationnelle de Strasbourg (FMTS), Strasbourg, France*

N. GAUTAM · *Department of Anesthesiology, Washington University School of Medicine, St. Louis, MO, USA; Department of Genetics, Washington University School of Medicine, St. Louis, MO, USA*

ALEXIS GAUTREAU · *Ecole Polytechnique, Université Paris-Saclay, BIOC-CNRS UMR7654, Palaiseau, France; School of Biological and Medical Physics, Moscow Institute of Physics and Technology, Dolgoprudny, Moscow Region, Russian Federation*

JULIE GAVARD · *Team SOAP, CRCINA, INSERM, CNRS, Université de Nantes, Nantes, France*

GRÉGORY GIANNONE · *UMR 5297, Interdisciplinary Institute for Neuroscience, University of Bordeaux, Bordeaux, France; UMR 5297, CNRS, Interdisciplinary Institute for Neuroscience, Bordeaux, France*

FLORENCE A. GIGER · *Centre for Developmental Neurobiology, MRC CNDD, IoPPN, King's College London, London, UK*

BOJANA GLIGORIJEVIC · *Department of Bioengineering, Temple University, Philadelphia, PA, USA; Cancer Biology Program, Fox Chase Cancer Center, Philadelphia, PA, USA*
NATALYA A. GLOUSHANKOVA · *N.N. Blokhin Russian Cancer Research Center, Moscow, Russian Federation*
JACKY G. GOETZ · *INSERM UMR_S1109, Tumor Biomechanics, Institut d'Hématologie et d'Immunologie, Strasbourg Cedex, France; Université de Strasbourg, Strasbourg, France; LabEx Medalis, Université de Strasbourg, Strasbourg, France; Fédération de Médecine Translationnelle de Strasbourg (FMTS), Strasbourg, France*
SÉBASTIEN HARLEPP · *Université de Strasbourg, Strasbourg, France; LabEx Medalis, Université de Strasbourg, Strasbourg, France; Fédération de Médecine Translationnelle de Strasbourg (FMTS), Strasbourg, France; IPCMS, UMR7504, Strasbourg, France; LabEx NIE, Université de Strasbourg, Strasbourg, France*
LYNN HUYCK · *Department of Biochemistry, Faculty of Medicine and Health Care, Ghent University, Ghent, Belgium; UZ Gent VINRAD, Ghent, Belgium*
DANIEL IRIMIA · *Division of Surgery, Department of Surgery, BioMEMS Resource Center, Innovation and Bioengineering, Massachusetts General Hospital, Harvard Medical School, Boston, MA, USA; Shriners Burns Hospital, Boston, MA, USA*
KATHRYN A. JACOBS · *Team SOAP, CRCINA, INSERM, CNRS, Université de Nantes, Nantes, France*
ASIER JAYO · *Universidad CEU San Pablo, Madrid, Spain*
FRIEDA KAGE · *Department of Cell Biology, Helmholtz Centre for Infection Research, Braunschweig, Germany; Division of Molecular Cell Biology, Zoological Institute, Technische Universität Braunschweig, Braunschweig, Germany*
WLODZIMIERZ KOROHODA · *Department of Cell Biology, Faculty of Biochemistry, Biophysics and Biotechnology, Jagiellonian University, Krakow, Poland*
DENIS KRNDIJA · *Institut Curie, PSL Research University, CNRS, UMR 144, Paris, France*
SLAWOMIR LASOTA · *Department of Cell Biology, Faculty of Biochemistry, Biophysics and Biotechnology, Jagiellonian University, Krakow, Poland*
CLAIRE LECLECH · *INSERM, UMRS-839, Institut du Fer à Moulin, Paris, France; Sorbonne Université, UPMC University Paris 6, UMRS-839, Paris, France; Institut du Fer à Moulin, Paris, France*
YING LI · *State Key Laboratory of Membrane Biology, Tsinghua University-Peking University Joint Center for Life Sciences, School of Life Sciences, Tsinghua University, Beijing, China*
CHRISTOF LITSCHKO · *Institute for Biophysical Chemistry, Hannover Medical School, Hannover, Germany*
JOHN G. LOCK · *EMBL Australia Node in Single Molecule Science, School of Medical Sciences, and ARC Centre of Excellence in Advanced Molecular Imaging, University of New South Wales, Sydney, Australia*
LIANG MA · *State Key Laboratory of Membrane Biology, Tsinghua University-Peking University Joint Center for Life Sciences, School of Life Sciences, Tsinghua University, Beijing, China*
SELMA MAACHA · *Department of Surgery, Vanderbilt University Medical Center, Nashville, TN, USA*
ZBIGNIEW MADEJA · *Department of Cell Biology, Faculty of Biochemistry, Biophysics and Biotechnology, Jagiellonian University, Krakow, Poland*

MAJID MALBOUBI · *Department of Engineering Science, University of Oxford, Oxford, UK; London Centre for Nanotechnology, University College London, London, UK*

LENNART MARTENS · *Department of Biochemistry, Faculty of Medicine and Health Care, Ghent University, Ghent, Belgium; Cancer Research Institute Ghent (CRIG), Ghent, Belgium; Medical Biotechnology Center, VIB, Ghent, Belgium*

SOPHIE MASSOU · *UMR 5297, Interdisciplinary Institute for Neuroscience, University of Bordeaux, Bordeaux, France; UMR 5297, CNRS, Interdisciplinary Institute for Neuroscience, Bordeaux, France*

PAOLA MASUZZO · *Department of Biochemistry, Faculty of Medicine and Health Care, Ghent University, Ghent, Belgium; Cancer Research Institute Ghent (CRIG), Ghent, Belgium; Medical Biotechnology Center, VIB, Ghent, Belgium*

AMINE MEHIDI · *UMR 5297, Interdisciplinary Institute for Neuroscience, University of Bordeaux, Bordeaux, France; UMR 5297, CNRS, Interdisciplinary Institute for Neuroscience, Bordeaux, France*

XENIA MESHIK · *Department of Anesthesiology, Washington University School of Medicine, St. Louis, MO, USA*

CHRISTINE MÉTIN · *INSERM, UMRS-839, Institut du Fer à Moulin, Paris, France; Sorbonne Université, UPMC University Paris 6, UMRS-839, Paris, France; Institut du Fer à Moulin, Paris, France*

SARAH MOITRIER · *Laboratoire PhysicoChimie Curie, Institut Curie, PSL Research University—Sorbonne Universités, UPMC—CNRS—Equipe labellisée Ligue Contre le Cancer, Paris, France*

NICOLAS MOLINIE · *Ecole Polytechnique, Université Paris-Saclay, BIOC-CNRS UMR7654, Palaiseau, France*

PENNY E. MORTON · *Randall Division of Cell and Molecular Biophysics, New Hunts House, Guys Campus, King's College London, London, UK*

BENEDIKT M. NAGEL · *Institut für Physiologie und Pathophysiologie, Abteilung Molekulare Zellphysiologie, Phillips-Universität Marburg, Marburg, Germany*

PATRICK R. O'NEILL · *Department of Anesthesiology, Washington University School of Medicine, St. Louis, MO, USA; Department of Genetics, Washington University School of Medicine, St. Louis, MO, USA*

THOMAS ORRÉ · *UMR 5297, Interdisciplinary Institute for Neuroscience, University of Bordeaux, Bordeaux, France; UMR 5297, CNRS, Interdisciplinary Institute for Neuroscience, Bordeaux, France*

NAËL OSMANI · *INSERM UMR_S1109, Tumor Biomechanics, Institut d'Hématologie et d'Immunologie, Strasbourg Cedex, France; Université de Strasbourg, Strasbourg, France; LabEx Medalis, Université de Strasbourg, Strasbourg, France; Fédération de Médecine Translationnelle de Strasbourg (FMTS), Strasbourg, France*

GUANGSHUO OU · *Tsinghua-Peking Center for Life Sciences, School of Life Sciences and MOE Key Laboratory for Protein Science, Tsinghua University, Beijing, China*

MARIA CARLA PARRINI · *Institut Curie, Centre de Recherche, Paris Sciences et Lettres Research University, Paris, France; ART Group, Inserm U830, Paris, France*

MADDY PARSONS · *Randall Division of Cell and Molecular Biophysics, King's College London, London, UK*

PERRINE PAUL-GILLOTEAUX · *Institut Curie, Centre de Recherche, Paris Sciences et Lettres Research University, Paris, France; Cell and Tissue Imaging Facility (PICT-IBiSA), CNRS*

UMR144, Paris, France; SFR Santé Francois Bonamy CNRS INSERM Université de Nantes, Nantes, France

LOUISIANE PERRIN · *Department of Bioengineering, Temple University, Philadelphia, PA, USA*

MATTHIEU PIEL · *Institut Curie, PSL Research University, CNRS, UMR 144, Paris, France; Institut Pierre-Gilles de Gennes, PSL Research University, Paris, France*

KAMYAR ESMAEILI POURFARHANGI · *Department of Bioengineering, Temple University, Philadelphia, PA, USA*

OLIVIER ROSSIER · *UMR 5297, Interdisciplinary Institute for Neuroscience, University of Bordeaux, Bordeaux, France; UMR 5297, Interdisciplinary Institute for Neuroscience, CNRS, Bordeaux, France*

KLEMENS ROTTNER · *Department of Cell Biology, Helmholtz Centre for Infection Research, Braunschweig, Germany; Division of Molecular Cell Biology, Zoological Institute, Technische Universität Braunschweig, Braunschweig, Germany*

SVETLANA N. RUBTSOVA · *N.N. Blokhin Russian Cancer Research Center, Moscow, Russian Federation*

MARIKE RÜDER · *Institut für Physiologie und Pathophysiologie, Abteilung Molekulare Zellphysiologie, Phillips-Universität Marburg, Marburg, Germany*

PABLO J. SÁEZ · *Institut Curie, PSL Research University, CNRS, UMR 144, Paris, France; Institut Pierre-Gilles de Gennes, PSL Research University, Paris, France*

TRINISH SARKAR · *Laboratoire PhysicoChimie Curie, Institut Curie, PSL Research University—Sorbonne Universités, UPMC—CNRS—Equipe labellisée Ligue Contre le Cancer, Paris, France*

SIMON SAULE · *Unité Mixte de Recherche 3347 (UMR), Unité 1021, Centre National de La Recherche Scientifique (CNRS), Institut National de la Santé Et de la Recherche Médicale (INSERM), Institut Curie, PSL Research University, Orsay, France; Unité Mixte de Recherche 3347, Unité 1021, Centre National de La Recherche Scientifique, Université Paris Sud, Université Paris-Saclay, Orsay, France*

PASCAL SILBERZAN · *Laboratoire PhysicoChimie Curie, Institut Curie, PSL Research University—Sorbonne Universités, UPMC—CNRS—Equipe labellisée Ligue Contre le Cancer, Paris, France*

ANTHONY SIMON · *Institut Curie, PSL Research University, CNRS, UMR 144, Paris, France*

MANISH KUMAR SINGH · *Institut Curie, Centre de Recherche, Paris Sciences et Lettres Research University, Paris, France; ART Group, Inserm U830, Paris, France*

JOLANTA SROKA · *Department of Cell Biology, Faculty of Biochemistry, Biophysics and Biotechnology, Jagiellonian University, Krakow, Poland*

RALITZA STANEVA · *Institut Curie, PSL Research University, CNRS, UMR 144, Paris, France; University Paris Descartes, Paris, France*

ANIKA STEFFEN · *Department of Cell Biology, Helmholtz Centre for Infection Research, Braunschweig, Germany*

STAFFAN STRÖMBLAD · *Department of Biosciences and Nutrition, Karolinska Institutet, Huddinge, Sweden*

HAWA-RACINE THIAM · *Institut Curie, PSL Research University, CNRS, UMR 144, Paris, France; Institut Pierre-Gilles de Gennes, PSL Research University, Paris, France*

MARLEEN VAN TROYS · *Department of Biochemistry, Faculty of Medicine and Health Care, Ghent University, Ghent, Belgium; Cancer Research Institute Ghent (CRIG), Ghent, Belgium*

PABLO VARGAS · *Institut Curie, PSL Research University, CNRS, UMR 144, Paris, France; Institut Pierre-Gilles de Gennes, PSL Research University, Paris, France*
DANIJELA MATIC VIGNJEVIC · *Institut Curie, PSL Research University, CNRS, UMR 144, Paris, France*
FRANÇOIS WAHARTE · *Institut Curie, Centre de Recherche, Paris Sciences et Lettres Research University, Paris, France; Cell and Tissue Imaging Facility (PICT-IBiSA), CNRS UMR144, Paris, France*
PEI-SHAN WANG · *Department of Pediatrics, University of Kansas Medical Center, Kansas City, KS, USA*
XIAO WANG · *Division of Surgery, Department of Surgery, BioMEMS Resource Center, Innovation and Bioengineering, Massachusetts General Hospital, Harvard Medical School, Boston, MA, USA; Shriners Burns Hospital, Boston, MA, USA*
DAVY WATERSCHOOT · *Department of Biochemistry, Faculty of Medicine and Health Care, Ghent University, Ghent, Belgium*
VICTOR YASHUNSKY · *Laboratoire PhysicoChimie Curie, Institut Curie, PSL Research University—Sorbonne Universités, UPMC—CNRS—Equipe labellisée Ligue Contre le Cancer, Paris, France*
HANNAH G. YEVICK · *Laboratoire PhysicoChimie Curie, Institut Curie, PSL Research University—Sorbonne Universités, UPMC—CNRS—Equipe labellisée Ligue Contre le Cancer, Paris, France*
LI YU · *State Key Laboratory of Membrane Biology, Tsinghua University-Peking University Joint Center for Life Sciences, School of Life Sciences, Tsinghua University, Beijing, China*
IRINA Y. ZHITNYAK · *N.N. Blokhin Russian Cancer Research Center, Moscow, Russian Federation*
ZHIWEN ZHU · *Tsinghua-Peking Center for Life Sciences, School of Life Sciences and MOE Key Laboratory for Protein Science, Tsinghua University, Beijing, China*
ELIZA ZIMOLAG · *Department of Cell Biology, Faculty of Biochemistry, Biophysics and Biotechnology, Jagiellonian University, Krakow, Poland*

Check for updates

Chapter 1

Random Migration Assays of Mammalian Cells and Quantitative Analyses of Single Cell Trajectories

Irene Dang and Alexis Gautreau

Abstract

Cell migration is essential to many biological processes such as embryonic development, immune surveillance and wound healing. Random cell migration refers to the intrinsic ability of cells to migrate, often called cell motility. This basal condition contrasts with directed cell migration, where cells migrate toward a chemical or physical cue. Unlike Brownian particles, however, randomly migrating cells exhibit a directional persistence, i.e., they are more likely to sustain the movement in the direction they previously took than to change, even if this direction is randomly chosen in an isotropic environment. Here we describe how to set up time-lapse recording of mammalian cells freely moving on a two-dimensional surface coated with extracellular matrix proteins, how to acquire single cell trajectories from movies and how to extract key parameters that characterize cell motility, such as cell speed, directionality, mean square displacement, and directional persistence.

Key words Cell tracking, Speed, Directionality, MSD, Persistence, Autocorrelation

1 Introduction

Cell migration is a tightly regulated process. Some cells are motile, some others are not, and the ones that are motile can move through a variety of mechanisms. However, there are two main modes of cell motility: one is called mesenchymal and is based on a lamellipodial protrusion, which is powered by actin polymerization, a second mode is called amoeboid and is based on a fast protrusion, the bleb, which corresponds to a bulge of the cell cortex [1, 2]. Physical constraints can modify the migration regime. For example, confinement can trigger fast migration by inducing a "constant bleb" at the front edge [3]. Tumor cells, in particular, exhibit a wide variety of migration mechanisms. For example, microtubule based protrusions have been reported to drive glioblastoma cell migration [4].

The aim of this chapter is to provide a basic—yet essential—characterization of cell motility. The protocol describes how to characterize cell motility in the most standard conditions, corresponding to single cells exhibiting random cell migration when

Alexis Gautreau (ed.), *Cell Migration: Methods and Protocols*, Methods in Molecular Biology, vol. 1749,
https://doi.org/10.1007/978-1-4939-7701-7_1,

plated on a two-dimensional substrate coated with extracellular matrix proteins. The main mode of motility in this situation is mesenchymal migration. The protocol we describe can easily be adapted to various cell types of interest and culture conditions. Any displacement is characterized by a vector, which has two basic features: a length corresponding to speed and an angle indicating the direction of displacement. High speed and high directionality allow cells to explore a large territory over time. This behavior is captured by the mean square displacement, whose unit is a surface area over time.

Lamellipodia depend on the sustained polymerization of branched actin networks by the Arp2/3 complex. The main characteristic of Arp2/3 based migration is to drive persistent cell motility [5, 6]. The maintenance of the leading edge at the previous front is critical for the persistence of migration [7] and was shown to depend on numerous cytoskeletal feedback loops [8]. It is therefore important to assess directional persistence, in addition to previous parameters, in order to rigorously dissect the molecular machinery and signaling pathways governing cell migration. Directional persistence is not easily computed, because this parameter is the output of an autocorrelation function. The here given protocol describes the whole procedure, from setting up the random migration assay, to analyzing single cell trajectories through the use of custom-made codes to calculate with Excel all these parameters of cell motility.

2 Materials

1. Cell culture treated dishes: 8-well chamber slides (μ-Slide 8 Wells, Ibidi®, Cat No. 80826) that allow multiple comparisons.
2. Cells: human MDA-MB-231 cells.
3. Medium: RPMI-1640 with phenol red supplemented with 10% fetal bovine serum (*see* **Note 1**).
4. 1× D-phosphate buffered saline, no calcium, no magnesium.
5. 1× Trypsin–EDTA 0.25%, with phenol red.
6. Fibronectin from bovine plasma, sterile-filtered, suitable for cell culture (Sigma-Aldrich, F1141).
7. Microscope: Axio Observer microscope (Zeiss) equipped with a Plan-Apochromat 20×/0.80 air objective, a Hamamatsu camera C10600 Orca-R2 and a Pecon Zeiss incubator XL multi S1 RED LS with an XL S heating Unit, a temperature module, a CO_2 module, a PS heating Insert, and a CO_2 cover (Fig. 1) (*see* **Note 2**).
8. ImageJ, a free analysis software (https://imagej.nih.gov/ij/) equipped with the MTrackJ plugin (http://www.imagescience.org/meijering/software/mtrackj/) [9].
9. Microsoft Excel (version 2007 or later).

3 Methods

3.1 Coating with Extracellular Matrix

1. Under sterile conditions, dilute fibronectin solution to 10 μg/mL in PBS.
2. Add 100 μL of the diluted ligand solution in each well to cover the whole surface and incubate for 1 h at 37 °C in a cell culture incubator or overnight at 4 °C.
3. Wash twice with 1× PBS, then add 250 μL of RPMI-1640 supplemented with 10% FBS and cover the slide with its sterile lid.
4. Incubate your slide in the cell culture incubator at 37 °C during cell trypsinization.

3.2 Seeding of Cells

1. In the cell culture hood under sterile conditions, trypsinize cells from healthy subconfluent cultures and make sure that cells are detached from one another by pipetting back and forth (Fig. 1a).
2. Count cells and dilute to 2×10^4 cells/mL in RPMI-1640 completed with 10% FBS.

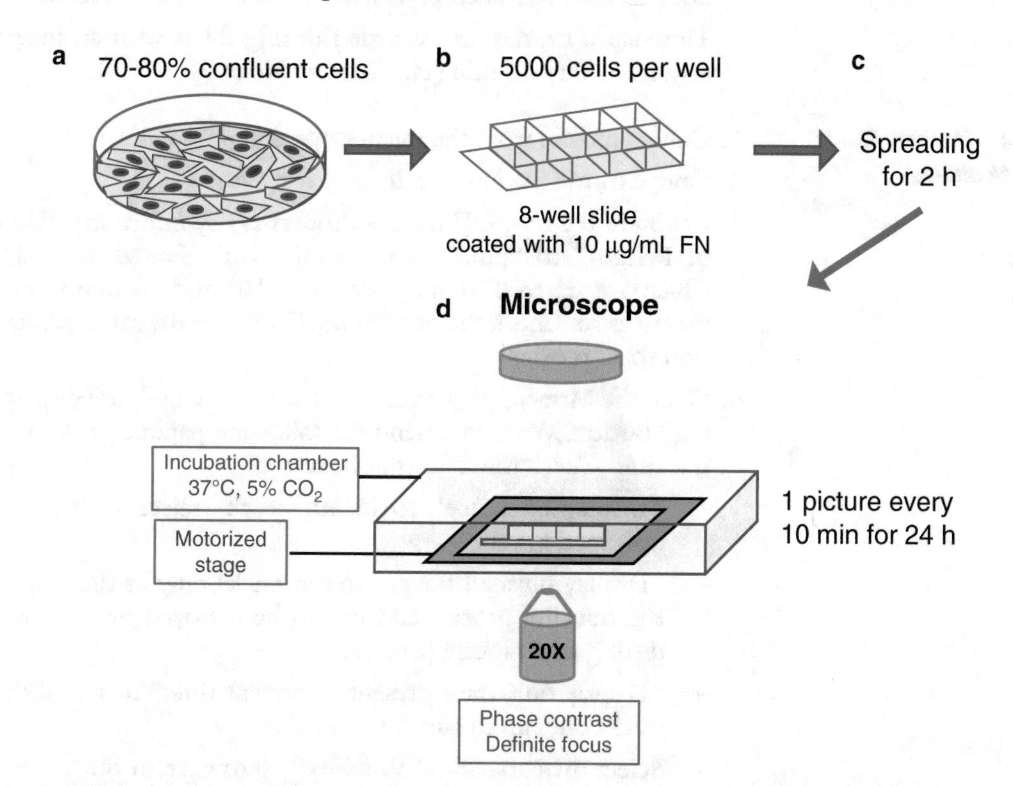

Fig. 1 Setting up the random cell migration assay. (**a**) Healthy cells from a subconfluent cell culture are trypsinized and (**b**) plated on an 8-well slide coated with 10 μg/mL fibronectin (FN). (**c**) Cells are incubated for 2 h at 37 °C, 5% CO_2 to let them adhere and spread. (**d**) The slide is placed within an incubation chamber under the inverted microscope. Pictures are acquired with a 20× phase contrast objective every 10 min for 24 h. At least five fields of view are recorded per well

3. Plate 250 μL of the cell dilution in each well (Fig. 1b) (*see* **Notes 3** and **4**). Seed two or three technical replicates for downstream statistical analysis. Put the lid on the slide.
4. Allow cells to spread for 2 h in the incubator at 37 °C under 5% CO_2 (Fig. 1c).
5. Check under the microscope that cells are spread enough to start migrating. Cells should be isolated, with a confluency around 10–15% (*see* **Note 5**).

3.3 Setting Up Image Acquisition

1. During cell spreading, turn on the microscope heating chamber and set it at 37 °C. Let it warm up for at least 2 h.
2. Turn on the CO_2 module and set it at 5%.
3. Place the 8-well slide under the microscope insert and put the CO_2 cover (*see* **Notes 6** and 7).
4. Turn on phase contrast illumination and set the objective to the 20× magnification.
5. Choose fields containing about 10–15 isolated cells (*see* **Note 8**). For statistical relevance, at least five fields per well are recorded.
6. Time-lapse movies are recorded during 24 h with an image acquired every 10 min (Fig. 1d) (*see* **Note 9**).

3.4 Tracking Cell Trajectories

1. Collect movies from the microscope computer.
2. Drag a movie file into the ImageJ software.
3. Calibrate the movie (Image > Properties) by informing "Unit of Length" to "μm," "Pixel width" and "Frame interval," which are set to 0.31 μm, (*see* **Note 10**) and 10 min in our precise case. Check the box "Global" to keep the same calibration to all movies.
4. Open the MtrackJ plug-in and click on the "Configure displaying" button. We recommend the following parameters for cell tracking. Check the following boxes:
   - "Display reference" to identify each trajectory with an assigsned number.
   - "Display finished tracks" to see tracks only at the end of the tracking process and not to be annoyed by the track during the tracking process.
   - "Display only track present at current time" to see all the tracks only up to current time.
   - Select the parameter "Visibility" up to current time.
5. To start cell tracking, click on "Add" (the letters turn to red indicating that the tracking process is recording, Fig. 2a). Track one cell by clicking on the nucleus center. Each click allows moving to the next frame. Cell tracking can be stopped

by pressing again on "Add" (the letters return black to indicate that the tracking process is ended).

6. When "Add tracks" is pressed again, a new trajectory is recorded (*see* **Note 11**). Track at least 50 cells per condition.
7. When cell tracking is finished, click on the "Measure tracks" button and check the boxes "Display point measurements" and "Display track measurement," then press OK.

Two new windows open. (1) Tracks: displays the mean parameters of each trajectory, (2) Points: displays the detail tracks point by point. Copy and paste all the data into two separate Excel files. When several movies are recorded for the same condition, it is important to paste data one following the other in the same Excel sheet. All the parameters calculated by the MTrackJ plugin are explained in the following manual (http://www.imagescience.org/meijering/software/mtrackj/manual/).

3.5 Analyzing Cell Trajectories

Here we briefly describe how to use the DiPer suite of custom-made programs [10]. If cells exhibit clearly active phases of migration interspersed by idling phases, it is worth considering a bimodal analysis, as we recently described [6].

1. Arrange your data as follows: copy the column "PID," "x," and "y" and paste the data in a new Excel file in columns D, E, and F respectively (Fig. 2b) (*see* **Notes 12** and **13**). To compare simultaneously different conditions, each worksheet of your workbook must correspond to one condition (*see* **Note 14**). To use the Virtual Basic for Application (VBA) developer, go to Excel options, check the mark "Show Developer tab in the ribbon," click on "OK." The developer appears in Excel as a ribbon. To open VBA, click on Visual Basic (up left in the VBA ribbon), a new window opens. Click on "Insert," "Module" and paste the codes found in reference [10] (Fig. 2b). Click on the small arrow "play" to run the program (or press F5).
2. To generate a plot of trajectories, launch the code "Plot_At_Origin" found in reference [10], as explained in **step 1**. A plot appears in each worksheet and combines all the cells trajectories from the given condition. Relative "x" and "y," for trajectories translated to the origin (0:0), are displayed in columns 8 and 9.
3. Cell speed is defined as the average of all instantaneous speed for all cells. To calculate cell speed, arrange your data as in **step 1**, in a new Excel file and paste the code "Speed" found in reference [10] in VBA. Run the program and enter frame interval to "10" minutes (*see* **Note 15**). Results appear in a new worksheet and show a plot computing the average of instantaneous speed by conditions with the standard errors of means (s.e.m.). Column 9 of the original worksheet contains the instantaneous speed values.

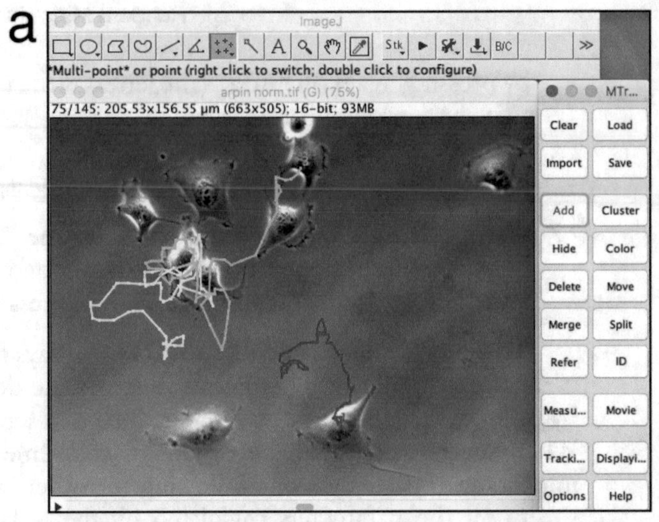

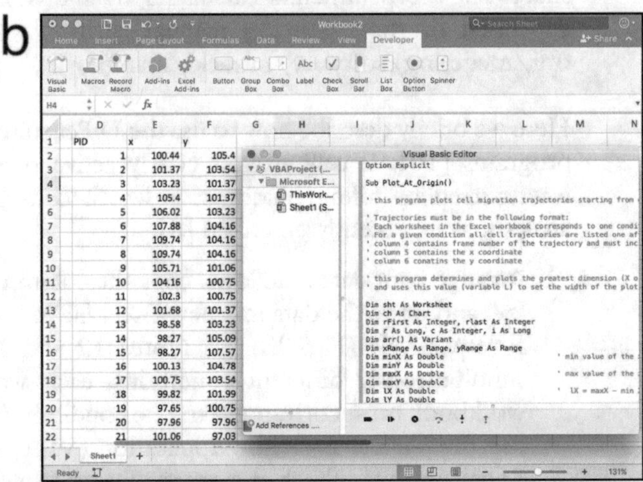

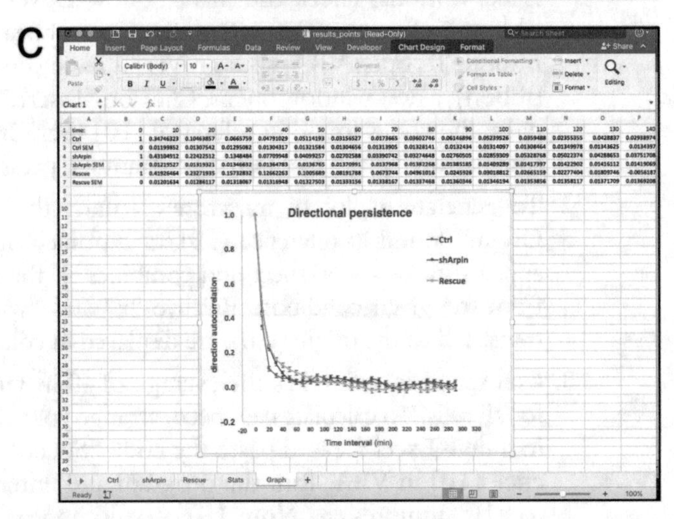

Fig. 2 Cell tracking and trajectory analysis. (**a**) Single cells are manually tracked using the MTrackJ plugin in ImageJ. The "Add" letters turn red when the tracking process is going on. (**b**) Trajectory coordinates are obtained from the MTrackJ plugin and pasted in a new Excel file worksheet as "PID" (frame number), "x," and "y" values arranged in column D, E, and F. VBA codes are launched in the Excel VBA developer and run. (**c**) For "Directional persistence," a chart appears in a new worksheet named "Graph" and displays the average autocorrelation coefficient curves for each condition. The "Stats" worksheet displays raw autocorrelation coefficients for each condition per time interval

4. Directionality (d/D) is represented by the ratio of the shortest linear distance from the start to the endpoint (d) divided by the total track distance migrated by an individual cell (D). The ratio is equal to one if the migration is straight and directional. To calculate this parameter, arrange your data as in **step 1**, and paste the code "Dir_ratio" found in reference [10]. Run the program and enter time interval to "10" minutes (*see* **Note 15**). The results are shown in the Excel file and new worksheets appear. "Last Point" worksheet gives the mean directionality and s.e.m. of cell trajectories and "Overtime" worksheet represents the directionality values calculated over time.
5. Directional persistence is an autocorrelation function based on the angle that the displacement vectors make between each other. We found this parameter to be correlated with the Arp2/3 activity [5] and to be less biased than directionality [10]. In particular this value is fully independent from cell speed. To calculate this parameter, arrange your data as described in **step 1**. Paste and launch the VBA code named "Autocorrel" found in reference [10] (*see* **Note 16**). Users must input time interval between movie frames and number of intervals desired (we recommend 30 as a starting point). A new worksheet called "Graph" appears and displays a plot, where each time point represents the average autocorrelation coefficient for the given time interval with error bars (s.e.m.) (Fig. 2c). The "Stats" worksheet displays autocorrelation coefficients for different conditions that are grouped per time intervals.
6. Mean Square Displacement can be calculated to reflect the surface area explored by the cell as a function of time. This parameter combines the influence of both speed and directionality. To calculate the MSD, arrange your data as described in **step 1**, paste and launch the VBA code named "MSD" found in reference [10] (*see* **Note 16**). Results are displayed in a new worksheet and show a log-log plot with log (MSD) on the y-axis and log (time interval) on the x-axis. Averaged MSD coefficient and s.e.m. are displayed for each condition. The slope of the curve, often called the α-value, tends to 1 for a random migration and to 2 for a directional migration. It can be easily calculated using the Excel Log and slope functions on the averaged MSD coefficient values.

4 Notes

1. Cell line and corresponding medium might be adapted to the user-specific purpose.
2. Other configurations can perform as well. We recommend, however, working with a microscope equipped with at least the following: phase contrast, a temperature and CO_2 controller, a motorized stage, an automatically controlled camera able to

acquire images at defined time intervals, and a Definite Focus module to avoid focus drift over time.

3. Eight-well slide chambers contain 1 cm^2 of growth area per well. This will plate 5×10^3 cells per well, i.e., 5×10^3 cells per cm^2. Keep a roughly similar number if using different dishes or cells.
4. The homogeneity of cell seeding is important. If you agitate the slide, cells will accumulate in the center of the well. If you mix by pipetting back and forth, cells will accumulate into the corners of the well. We found that the best homogeneity was obtained with the following procedure: we dilute the exact cell number in 250 μL of complete medium, we homogenize the cell suspension in the eppendorf tube and then we add 250 μL of the suspension into the well. Carefully, place your slide in the incubator immediately after cell seeding, paying attention not to shake the slide.
5. This procedure of spreading cells on extracellular matrix and recording immediately ensures that cells migrate on the matrix used for coating and not on the extracellular matrix proteins that cells secrete and assemble over time.
6. Pay attention that the medium does not touch the lid of the 8 well slide. Carefully carry your slide to the microscope and keep it in a horizontal position. If the medium accidentally touches the lid, this can lead to contamination and evaporation. Indeed, the drop that touches the lid evaporates fast and induces the rest of the medium to evaporate as well by capillarity.
7. Check that the slide does not move before starting the acquisition. If the slide is not properly held to the stage, the acquisition will invariably drift out of focus.
8. Cell density in the field is important. If too many cells are in the field, increasing cell-cell interactions will bias random cell motility.
9. There is an upper limit to the number of positions you can choose and this limit depends on the frame rate. The stage has to move from the previous field of view, the camera has to acquire the image before moving to the next position, and the whole cycle must be completed within the frame rate. Therefore always perform a test acquisition to make sure that there is no conflict between the different set parameters.
10. Usually, pixel size is given by the software used for image acquisition. If not, use a calibration slide, or any object you know the size of, such as a hemacytometer. Take a picture at the same magnification, open the image in ImageJ, drag a line along the object to determine the pixel size.
11. If the tracked cell divide, end your tracking before cell division, during which cells do not move.

12. Make sure that Excel recognizes your data as numbers (numbers are by default adjusted to the left of the cells). If not, pay attention to the decimal separator that your Excel configuration recognizes as numbers. You may need to replace comma by period or vice versa.
13. There is no need to convert (x, y) coordinates into microns.
14. The arrangement of data is imposed by VBA codes from reference [10]. VBA codes can be launched only once for an Excel file. Be careful to copy your data to a new Excel file prior to launching a program. If you encounter any problem with a VBA code, try first to close the Excel software, to relaunch it, create a new Excel file and paste again your data in the exact format described in Subheading 3.5, **step 1**. In our experience, this simple procedure solves most issues.
15. It is not mandatory to convert time interval to minutes. The unit of time is arbitrary.
16. MSD or autocorrelation calculations can take up to 2–3 min before the results appear. Do not perturb the calculation process by browsing on your Excel file.

References

1. Sanz-Moreno V, Gadea G, Ahn J et al (2008) Rac activation and inactivation control plasticity of tumor cell movement. Cell 135:510–523. https://doi.org/10.1016/j.cell.2008.09.043
2. Sanz-Moreno V, Marshall CJ (2010) The plasticity of cytoskeletal dynamics underlying neoplastic cell migration. Curr Opin Cell Biol 22:690–696. https://doi.org/10.1016/j.ceb.2010.08.020
3. Liu YJ, Le Berre M, Lautenschlaeger F et al (2015) Confinement and low adhesion induce fast amoeboid migration of slow mesenchymal cells. Cell 160:659–672. https://doi.org/10.1016/j.cell.2015.01.007
4. Panopoulos A, Howell M, Fotedar R, Margolis RL (2011) Glioblastoma motility occurs in the absence of actin polymer. Mol Biol Cell 22:2212–2220. https://doi.org/10.1091/mbc.E10-10-0849
5. Dang I, Gorelik R, Sousa-Blin C et al (2013) Inhibitory signalling to the Arp2/3 complex steers cell migration. Nature 503:281–284. https://doi.org/10.1038/nature12611
6. Gorelik R, Gautreau A (2015) The Arp2/3 inhibitory protein arpin induces cell turning by pausing cell migration. Cytoskeleton 72:362–371. https://doi.org/10.1002/cm.21233
7. Petrie RJ, Doyle AD, Yamada KM (2009) Random versus directionally persistent cell migration. Nat Rev Mol Cell Biol 10:538–549. https://doi.org/10.1038/nrm2729
8. Krause M, Gautreau A (2014) Steering cell migration: lamellipodium dynamics and the regulation of directional persistence. Nat Rev Mol Cell Biol 15:577–590. https://doi.org/10.1038/nrm3861
9. Meijering E, Dzyubachyk O, Smal I (2012) Methods for cell and particle tracking. Methods Enzymol 504:183–200. https://doi.org/10.1016/B978-0-12-391857-4.00009-4
10. Gorelik R, Gautreau A (2014) Quantitative and unbiased analysis of directional persistence in cell migration. Nat Protoc 9:1931–1943. https://doi.org/10.1038/nprot.2014.131

Chapter 2

Directional Collective Migration in Wound Healing Assays

Nicolas Molinie and Alexis Gautreau

Abstract

Cell migration is suppressed by confluence in a process called contact inhibition. Relieving contact inhibition upon scratching is one of the simplest ways to induce cell migration in a variety of cell types. Wound healing is probably most relevant to epithelial monolayers, because epithelial cells generally assume a barrier function, which must be restored as fast as possible by the healing process. This versatile assay, however, can also be applied to fibroblasts and to tumor cell types. Furthermore, assessing the cell response to scratch wounding requires no special equipment or reagents. It is one of the few cell migration assays, which can even be performed without videomicroscopy, since the closure of the wound can be estimated at fixed time points. Several hours after wounding, directional collective migration is easily assessed and quantified. However, cell proliferation, which is also induced by the relief of contact inhibition, is one of the confounding factors of wound healing assays that must be taken into account. A recent alternative to the scratch-induced wound is to use special inserts to seed cells into closely spaced chambers. When the insert is removed, contact inhibition is relieved, similar to the scratch-induced wound. In this chapter, we provide the protocol of the two methods and compare their advantages and disadvantages. We also provide a protocol to estimate cell proliferation upon wound healing based on the incorporation of the nucleotide analog EdU.

Key words Migration, Proliferation, Scratch, Insert, Videomicroscopy

1 Introduction

Contact inhibition is a tight control exerted on both cell migration and cell proliferation. Because relieving contact inhibition is as easy as to perform a scratch in a cell monolayer, this system has been widely used in vitro even for cells which are unlikely to be exposed to scratches in the organism. A slightly more elaborate technique requires the use of inserts, which allows to grow two cell populations separated by a thin wall of 500 μm. The simplicity of these assays relies from the facts that no special equipment or reagent is required and that the quantitative scoring of healing the wound is as simple as reporting gap closure. Because cell migration and proliferation both contribute to wound healing (Fig. 1), the ability to conclude that a treatment or a genetic manipulation affects cell

Alexis Gautreau (ed.), *Cell Migration: Methods and Protocols*, Methods in Molecular Biology, vol. 1749,
https://doi.org/10.1007/978-1-4939-7701-7_2,

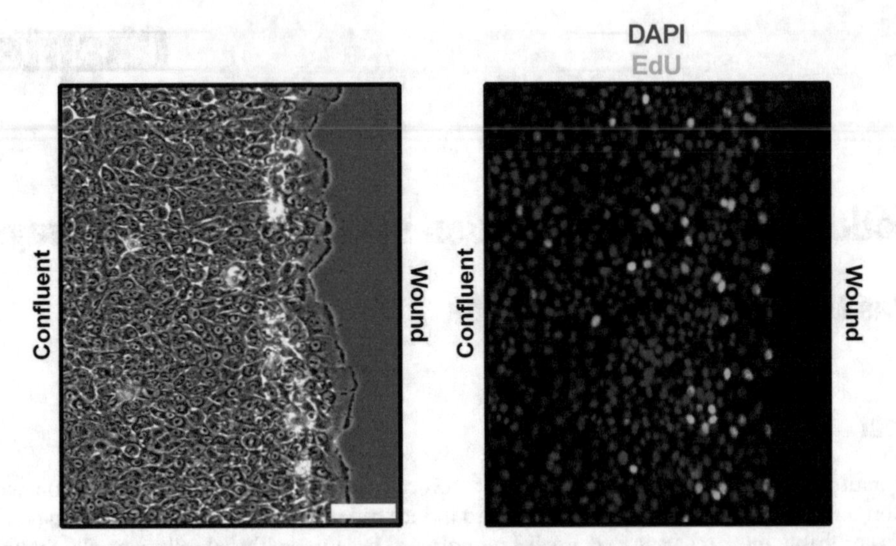

Fig. 1 A wound induces lamellipodium protrusion and cell proliferation. A MCF10A monolayer was wounded and observed by phase contrast for healing (left panel). After 4 h, the culture was incubated with EdU. At 5 h postwounding, the wound was fixed, nuclei were stained with DAPI and EdU incorporation was revealed (right panel)

migration requires to control that the effect is not mediated by cell proliferation. This is why we also provide in this chapter a simple protocol based on EdU incorporation to estimate proliferation on fixed cultures [1, 2].

Here we provide two protocols of wound healing, the historical one based on a scratch and a more recent one using an insert (Fig. 2). The scratch technique is the more basic one. It physically damages cells. Cell damage is, of course, variable and so unwanted for a reliable assay. But this is also the pathophysiology underlying the repair of a real wound. As a consequence, the width of the scratch is hard to control. But the scratch can also displace cells, which can potentially reattach in the wounded area and thus biasing analyses of gap repair (Table 1). In comparison, removal of the insert provides two cell monolayers with smooth edges separated by a well-defined gap.

Wound healing is a type of collective cell migration. Moreover it is directional, since cells develop protrusions, called lamellipodia, toward the wound. One of the limitations of these assays is that cells can be pushed passively by the others to fill in the gap, even if the treatment or genetic manipulation affects the coordination between cells of the monolayer and the resulting polarity. Indeed, during repair, the microtubule organizing center and the Golgi apparatus are both reoriented toward the wound in this assay. The reorientation of these organelles can easily be observed by immunofluorescence and was shown to require both microtubules and actin [3, 4].

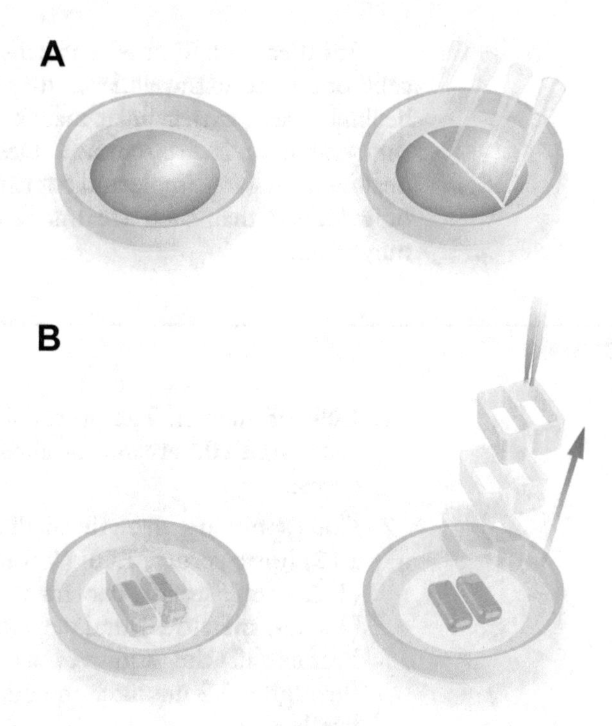

Fig. 2 Two ways to perform a wound healing assay. (**a**) Scratch a confluent cell monolayer with a pipetman tip. (**b**) Create a gap between two groups of cells by removing an insert. Cells are in the pink areas

Table 1
Pros and cons of wound healing to assess cell migration

	Advantages	**Disadvantages**
Wound healing in general	Simple and fast	Uncertainty concerning the extracellular matrix
	No experimental optimization	Influence of neighboring cells
	Directional response	No control on the directional cue
	Little equipment needed	Influence of cell proliferation
	Easy analysis	
Scratch	A physiological response	Cell damage
		Variations of the wound width
		Cell displacement
Insert	Clean controlled wound	
	Easy comparison of two cell types	
	Easy comparison of two treatments	

Another limitation is that cells usually migrate when specifically bound to extracellular matrix molecules, and that in wound healing assays, extracellular matrix molecules may be displaced by the scratch or insert removal. Despite these limitations, wound healing assays are simple and robust and provide an easy first piece of evidence that such treatment or such molecule affects cell migration.

2 Materials

1. Cells of interest. The protocol below is adapted to MCF10A cells. MCF10A are immortalized cells derived from the human breast.
2. Corresponding growth medium. For MCF10A: DMEM/F12, horse serum, EGF (0.5 mg/mL stock), hydrocortisone (1 mg/mL stock), cholera toxin (1 mg/mL stock), insulin (10 mg/mL stock), and penicillin–streptomycin (100× stock). Premix all the additives as indicated below, sterile-filter through a 0.2 μm filter and add to the DMEM/F12 medium bottle.

   25 mL of horse serum, 20 μL of EGF, 250 μL of hydrocortisone, 50 μL of cholera toxin, 500 μL of insulin, and 5 mL of penicillin–streptomycin.
3. Cell culture dish of 35 mm (*see* **Note 1**).
4. 20–200 μL pipetman tip for the scratch induced wound healing assay.
5. Culture insert for the insert removal assay (Ibidi #80209).
6. Microscope: any epifluorescent microscope can suffice for a fixed time point analysis. A videomicroscope equipped with an incubation chamber (temperature and CO_2) permits to image wound healing live. A motorized stage allows to image simultaneously different conditions (growth factor treatments, drug treatments, genetically manipulated clones compared to control ones, …).
7. Computer equipped with the free software Image J.
8. EdU: resuspend EdU (Thermofisher) at 10 mM in DMSO. Store at −20 °C as aliquots.
9. Paraformaldehyde.
10. Triton X-100.
11. 50 mM $CuSO_4$.
12. 0.5 M Tris–HCl, pH 8.5.
13. Alexa Fluor™ 488 Azide at 0.5 mM in DMSO (Thermofisher). Store at −20 °C as aliquots.

14. 1 M ascorbic acid.
15. 1× Click-iT EdU buffer: for 1 mL, add in the following order, 720 μL ddH_2O, 20 μL $CuSO_4$, 200 μL Tris pH 8.5, 10 μL Alexa Fluor® azide, 50 μL ascorbic acid.
16. DAPI.
17. Coverslips.
18. ProLong™ Diamond Antifade Mountant (Thermofisher).

3 Methods

First, coat the dish or the chamber slide with extracellular matrix (Fibronectin at 10 μg/mL) for 30 min at 37 °C or overnight at 4 °C in PBS. Rinse twice for 2 min with PBS without touching the surface.

3.1 Scratch Induced Wound Healing Assay

1. Seed cells to be studied in the right medium (*see* **Note 2**) in a dish or a chamber slide compatible with live cell imaging. The most suitable dish is a 35 mm dish. Seed 1,500,000 cells in 2 mL of growth medium to obtain a confluent monolayer after 24 h of incubation at 37 °C (*see* **Note 3**).
2. With a sterile 20–200 μL pipetman tip, scrape the confluent monolayer in the center of the dish to create a scratch. Perform the scratch in one uniform and straight movement.
3. Rinse with complete medium to remove displaced cells.

3.2 Wound Healing After Insert Removal

1. Stick the culture insert on the coated surface. Make sure that the insert is well attached to the surface to avoid cell invasion under the insert.
2. Plate 4000 cells to be studied in 70 μL into each well of the insert and incubate cells for 24 h at 37 °C (*see* **Notes 2** and **3**).
3. Gently remove the culture insert by grabbing it with sterile tweezers (Fig. 2). Wash with PBS. Fill the dish with complete growth medium (*see* **Notes 4** and **5**).

3.3 Acquisition of Microscopy Pictures and Migration Analysis

1. Preheat the microscope chamber 1 h before you place the cell culture on the stage.
2. Place the dish under a phase-contrast microscope (*see* **Note 6**). Use the 10 or 20× objective and move the frame so that the wound and the two cell fronts are simultaneously visible (*see* **Note 7**).
3. Choose several fields and record for each field the wound repair. To determine the characteristics of cell migration,

perform a time-lapse with 1 frame every 10 min. The wound repair usually last for less than 16 h. If you want to measure lamellipodial characteristics, take, at least, 1 frame every minute.

4. There are a lot of ways to measure cell migration and determine its characteristics. The most simple way is to manually measure the proportion of wound area that has been repaired at different time points ([5], *see* **Note 8**). By manually tracking each cell with the MTrackJ pluging in ImageJ, a single cell analysis can also be performed as described in ref. 6.

3.4 EdU Staining

1. Incubate the cells for 1 h at 37 °C with a prewarmed 10 μM solution of EdU diluted in complete medium to stain the cells that are in S-phase.
2. Remove medium, wash with warm PBS and fix the cells with 4% paraformaldehyde in PBS for 15 min.
3. Wash twice with 2% BSA in PBS and permeabilize the cells with 0.5% Triton X-100 in PBS for 5 min.
4. Incubate the cells with the 1× Click-iT EdU buffer for 30 min at room temperature, in the dark.
5. Wash twice with 2% BSA in PBS and stain the nuclei for 10 min with 0.1% DAPI in PBS.
6. Wash twice with 2% BSA in PBS. Remove excess liquid from the dish or the chamber slide. Add 1 drop of mounting medium and place the coverslip sample-side down onto mounting medium.
7. Acquire microscopy pictures of Alexa Fluor dye and DAPI staining
8. Compare the total number of cells (i.e., number of nuclei) with the number of S-phase cells stained by Alexa Fluor 488. Rather than counting the number of cells manually, we created a macro that automatically counts the number of nuclei (*see* **Note 9**) or the number of Alexa Fluor™ 488 positive cells (*see* **Note 10**). Copy the text of **Note 9** or **10** and paste it as a new Macro in Image J: Plugins > New > Macro. Save it with the corresponding name (DAPI or EdU). To use the macro, open the picture and run the macro: Plugins > Macro > Run. The count is given in the summary window.

4 Notes

1. One can use 2-well chamber slides as well. But use big enough wells to avoid edge effects, which degrade phase-contrast imaging and limit the field of view to the very center.

2. A variant of medium with no phenol red often exists and might provide a clearer image. If you cannot set the CO_2 concentration in your microscope, use a CO_2 independent medium such as Leibovitz's L-15. You can also supplement the regular medium with 25 mM of HEPES to maintain the pH into the physiological range (7.4–7.6).
3. The wound assay can be performed on either naive cells or transfected cells to study the effect of specific proteins on cell migration. Transient transfections of siRNAs usually give homogeneously depleted cell populations. Cells are usually transfected 2–3 days before wounding. In contrast, transiently transfected cells with plasmids give rise to very inhomogeneous cell populations and therefore we prefer to compare stable cell lines in wound healing assays.
4. We recommend waiting for 24 h. However the assay can be performed as early as 8 h after seeding if cells are properly adhered to the substratum and to each other (There should be no visible gaps before wounding) and as late as 48 h. Cell monolayers tend to detach as a sheet from the substratum if overgrown.
5. If the edge of the scratch is not smooth enough, try to increase the speed of scraping, keeping the pipette tip under an angle of 30 degrees to keep the scratch width limited.
6. Instead of following the characteristics of cell migration, you can also be interested in cell morphology or protein localization in the cell during cell migration. These parameters can be easily studied in fixed cells.
7. A microscope with motorized stages allows you to capture several fields for each condition, thus increasing the number of conditions one can analyze in a single experiment.
8. Measure the distance traveled by leader cells at each time point (every 2 h for example) and compare it to the initial width of the wound (500 μm for the Ibidi insert). To determine the value of a pixel, refer to the manufacturer's instructions of your microscope.
9.
```
setAutoThreshold("Default");
//run("Threshold...");
setAutoThreshold("Default dark");
setAutoThreshold("Default dark");
```

```
run("Convert to Mask");
run("Fill Holes");
run("Close-");
run("Fill Holes");
run("Smooth");
run("Sharpen");
setAutoThreshold("Default");
//run("Threshold...");
setThreshold(128, 255);
run("Convert to Mask");
run("Watershed");
run("Analyze Particles...", "size=200-Infin-
ity circularity=0.00-1.00 show=Nothing
include summarize add");
```

10.
```
setAutoThreshold("Default");
//run("Threshold...");
setAutoThreshold("Default dark");
setAutoThreshold("Yen dark");
run("Convert to Mask");
run("Fill Holes");
run("Close-");
run("Fill Holes");
run("Smooth");
run("Sharpen");
//run("Threshold...");
setThreshold(255, 255);
run("Convert to Mask");
run("Watershed");
run("Analyze Particles...", "size=200-Infin-
ity circularity=0.00-1.00 show=Nothing
include summarize add");
```

Acknowledgments

The authors thank Sebastien Coste for drawing Fig. 2.

References

1. Salic A, Mitchison TJ (2008) A chemical method for fast and sensitive detection of DNA synthesis in vivo. Proc Natl Acad Sci U S A 105:2415–2420
2. Buck S, Bradford JS, Gee K, Agnew B, Clarke S, Salic A (2008) Detection of S-phase cell cycle progression using 5-ethynyl-2′-deoxyuridine incorporation with click chemistry, an alternative to using 5-bromo-2′-deoxyuridine antibodies. Biotech 44:927–929
3. Magdalena J, Millard TH, Machesky LM (2003) Microtubule involvement in NIH 3T3 Golgi and MTOC polarity establishment. J Cell Sci 116:743–756
4. Magdalena J, Millard TH, Etienne-Manneville S, Launay S, Warwick HK, Machesky LM (2003) Involvement of the Arp2/3 complex and Scar2 in Golgi polarity in scratch wound models. Mol Biol Cell 14:670–684
5. Lomakina ME, Lallemand F, Vacher S, Molinie N, Dang I, Cacheux W et al (2016) Arpin downregulation in breast cancer is associated with poor prognosis. Br J Cancer 114: 545–553
6. Gorelik R, Gautreau A (2014) Quantitative and unbiased analysis of directional persistence in cell migration. Nat Protoc 9: 1931–1943

Chapter 3

An In Vitro System to Study the Mesenchymal-to-Amoeboid Transition

Aleksandra S. Chikina and Antonina Y. Alexandrova

Abstract

During the last few years, significant attention has been given to the plasticity of cell migration, i.e., the ability of individual cell to switch between different motility modes, in particular between mesenchymal and amoeboid motilities. This phenomenon is called the mesenchymal-to-amoeboid transition (MAT). Such a plasticity of cell migration is a mechanism, by which cancer cells can adapt their migration mode to different microenvironments and thus it may promote tumor dissemination. It was shown that interventions at certain regulatory points of mesenchymal motility as well as alterations of environmental conditions can trigger MAT. One of the approaches to induce MAT is to mechanically confine cells and one of the simplest ways to achieve this is to cultivate cells under agarose. This method does not require any special tool, is easily reproducible and allows cell tracking by videomicroscopy. We describe here a protocol, where MAT is associated with chemotaxis.

Key words Mesenchymal-to-amoeboid transition, Lamellipodia, Bleb, Confined conditions, Ultrapure agarose

Abbreviations

ECM Extracellular matrix
MAT Mesenchymal-to-amoeboid transition
MMPs Matrix metalloproteinases.

1 Introduction

During the last years, significant attention was focused on the study of the transition from mesenchymal to amoeboid modes of cell motility, the so-called mesenchymal-to-amoeboid transition (MAT). Both amoeboid and mesenchymal motility are features of individually migrating cells but they are based on different molecular mechanisms. Mesenchymal motility is based on membrane

Alexis Gautreau (ed.), *Cell Migration: Methods and Protocols*, Methods in Molecular Biology, vol. 1749, https://doi.org/10.1007/978-1-4939-7701-7_3,

protrusions in the form of lamellipodia and/or filopodia driven by actin polymerization. It requires integrin-based cell-matrix adhesion and depends on activity of matrix metalloproteinases (MMPs) for migration in 3D substrates. Molecular machineries which regulate cytoskeleton reorganizations during mesenchymal motility are fairly well characterized. They are activated by signals that impinge on the small GTPase Rac, which triggers actin polymerization [1–4]. Studies of the amoeboid motility of cancer cells began years ago [5], but for a long time researchers did not pay enough attention to its cellular mechanisms. Amoeboid motility was considered as an artifact or a phenomenon limited to certain cells types such as free living amoeba *Dictyostelium discoideum* or cells of leukocytic lineage of mammals. The essential role of the ability of cells to switch between motility modes was noticed when scientists began to study the effect of MMPs inhibition on the dissemination of cancer cells and showed that cancer cells which usually migrate by mesenchymal mode can switch to amoeboid motility under inhibition of MMPs. It was revealed that transition from proteolytic "mesenchymal" to protease independent "amoeboid" movement could be considered as a novel cellular and molecular adaptation pathway in sustaining cell migration [6]. That is why more recent studies have focused on the alternative—amoeboid migration mode [7, 8] and the conditions triggering migration plasticity and particularly MAT [9]. It has been shown that even though the morphological term "amoeboid" migration includes a number of rather distinct biophysical modes of cellular locomotion, from bleb-based motility to gliding [10, 11], all of them demonstrate similar characteristics including independence from proteolytic matrix degradation, failure of adhesion to the ECM and enhanced cell contractility. Amoeboid motility requires activation of the small GTPase Rho A which leads to ROCK-dependent myosin light chain phosphorylation and myosin II activation [12]. The migration of cells based on the formation of blebs as main protrusion type, meets all these characteristics, therefore blebbing associated with cell migration could be estimated as sign of amoeboid motility. Accordingly transition from lamellipodial to nonapoptotic blebbing protrusions can be considered as a sign of MAT.

For the analysis of the mesenchymal motility, 2D migration assays such as scratch wound healing or analysis of single cell motility on flat substrates have been widely used, but these tests are inappropriate for studying the amoeboid motility, because MAT leads to decrease of substrate adhesions and amoeboid cells tend to detach from the substrate. To study amoeboid motility, it is necessary to protect cells from floating, for example by developing 3D condition assays. It was shown that treatments that suppress mesenchymal motility trigger the transition of some cells to amoeboid motility. Thus, MAT can be induced in tumor cells by disrupting cell-substrate adhesions using different approaches [13–18], by inhibiting either Arp2/3 complex-dependent actin polymerization in lamellipodia [16, 19, 20], or MMPs activity [6, 21] and also by shifting the balance from actin polymerization-driven protrusions to actomyosin-dependent contractility [12, 22–24]. MAT can also be triggered by changing characteristics of the ECM. Particularly it was shown that MAT can be induced by different kinds of confinement, for example plating cells under agarose gel [16], under the glass surface of PEG-treated slides containing microspacers [17] or plating cells into glass tubes with limited diameter [25]. Finally, MAT can be triggered by embedding cells into substrate with large pores. The example of such substrate is collagen gels polymerized at low temperatures [26]. In this system, the switch to amoeboid motility depends on the pore size of the collagen gel, which itself depends on the temperature of polymerization, but does not depend on matrix stiffness, that is determined by collagen concentration. Therefore it seems that the special conditions that trigger amoeboid motility are the combination of a confined environment and of "empty" space where cells can protrude blebs. Thus the most suitable condition to study MAT is experimental confinement. One of the simplest approaches is to squeeze cells under agarose, as described by Bergert and coauthors [16]. The assay of under-agarose migration was originally developed for studying chemotaxis and chemokinesis [27] and it gave the opportunity to investigate both migration of entire cell population and dynamics of protrusion and morphology of individual cells. This method allowed to study the influence of different factors on motility effectiveness, such as substrate adhesiveness and intercellular regulations. Later it was shown that under agarose conditions can trigger the transition between motility modes. We combined observations by Bergert and coauthors, which showed that Walker 256 carcinosarcoma could switch from lamellipodia to blebs in the under-agarose conditions [16], with the original protocol by Heit [27] and stimulated cell motility toward serum gradient under confinement by agarose gel. It is a simple and reproducible method to trigger MAT in laboratory conditions.

2 Materials

1. Culture dishes with glass bottom 35 mm diameter.
2. HT1080 cell line.
3. Fetal calf serum (FCS).
4. 1 N HCl. Store at room temperature.
5. Millipore filters type GS, pore size 0.22 μm for sterilization.
6. Agarose stock solution: dissolve 4.8 g of ultrapure agarose of low gelling temperature in 100 mL ddH_2O (*see* **Note 1**). To

increase the solubility of agarose, preheat H_2O using a 95 °C water bath (or microwave). Then add part of the agarose, vortex, and heat again on the water bath. Repeat these steps until the whole volume of agarose is dissolved. Sterilize solution using autoclave. Stock solution can be stored at +4 °C up to 1 month.

7. Dulbecco's Modified Eagle Medium (DMEM) powder.
8. Sodium bicarbonate: 7.5% solution $NaHCO_3$. Sterile. Keep at 4 °C
9. 2× DMEM: prepare 100 mL by following the protocol provided by the company and using the sodium bicarbonate, but double the amount of Dulbecco's Modified Eagle Medium (DMEM) powder, adjust the pH to 7 with 1 N HCl. Sterilize the solution using a 0.22 μm filter.
10. Phosphate Buffer Solution (PBS): 137 mM NaCl, 2.7 mM KCl, 10 mM Na_2HPO_4, 1.76 mM KH_2PO_4.

For 1 L of PBS, take 800 mL of ddH_2O, add each ingredient and dissolve it (8 g NaCl, 0.2 g KCl, 1.44 g Na_2HPO_4, 0.24 g KH_2PO_4). Adjust pH to 7.4 with 1 M HCl or 1 M NaOH, add ddH_2O to 1 L. Store at room temperature. The solution is stable for 1 month.

3 Methods

1. By the day of experiment be sure that cultured cells are in the exponential phase of growth (around 70–80% of confluence).
2. Melt the agarose stock solution using water bath (*see* **Note 2**).
3. Preheat 2× DMEM at 37 °C.
4. For one dish preparation, mix 1.25 mL of agarose stock solution and 1.25 mL of 2× DMEM in a 15 mL tube. Avoid making bubbles, gently mix by flipping the tube, but not by pipetting.
5. Place the mix on the dish and let it be gelled at +4 °C (*see* **Note 3**).
6. When the agarose is gelled, make holes in it according to the design of your experiment. In our chemotactic experiment, we make a central hole of 4 mm in diameter using the wide opening of a 200 μL pipet tip and a peripheral hole of about 2 mm of diameter using the sharp end of a 200 μL pipet tip (Fig. 1; *see* **Note 4**).
7. After holes are punched, remove as much agarose from holes as possible, because it may prevent cells from adhering (*see* **Note 5**).
8. Trypsinize cells and resuspend them at 600,000/mL concentration in DMEM with low FCS (1%). Load 100–150 μL of cell

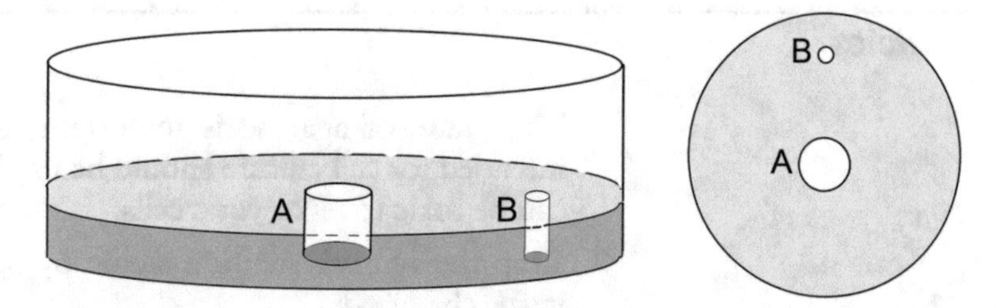

Fig. 1 Experimental design for under agarose confinement assay. A Petri dish contains an agarose gel with punched holes, a large hole for cell loading (**A**) and a small hole for chemoattractant loading (**B**)

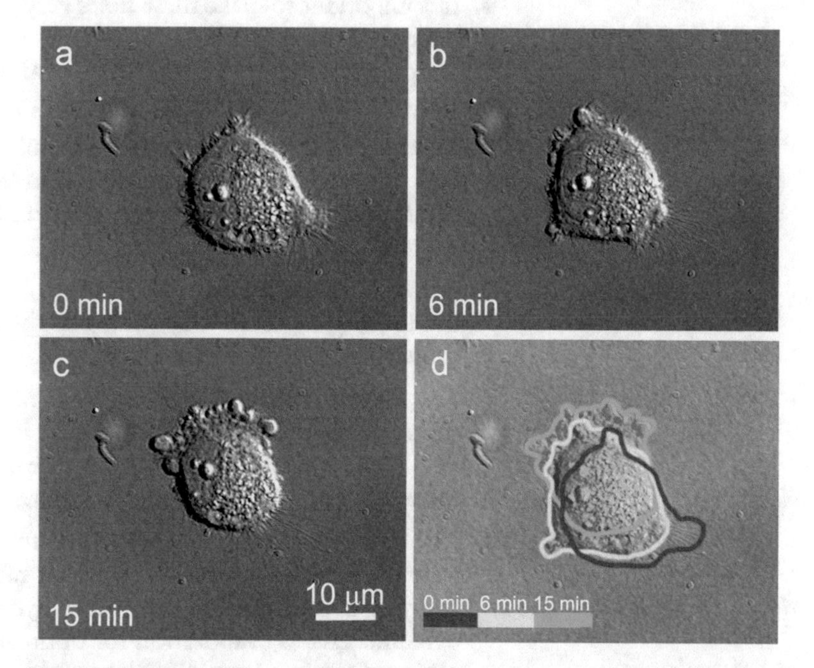

Fig. 2 Mesenchymal-to-amoeboid transition in motile HT1080 cell under agarose. Cells were imaged by DIC microscopy (**a–c**) and migration is represented by a color overlay (**d**)

suspension into 4 mm hole. After cell spreading, medium is replaced with FCS-free DMEM to starve cells and increase their response to the FCS gradient (*see* **Note 6**).

9. Load 20 μL of DMEM with 20% FCS into the 2 mm hole and incubate the dish overnight at 37 °C in a 5% CO_2 incubator (*see* **Note 7**).
10. Perform videomicroscopy using DIC imaging the next day (Fig. 2; *see* **Note 8**).
11. Determine parameters of membrane protrusion and cell migration using the ImageJ software (*see* Chapter 2).

4 Notes

1. The grade of agarose is important, only ultrapure agarose intended for cell culture should be used, because lower grades can be toxic to eukaryotic cells.
2. Keep the agarose solution sterile by opening the bottle only inside the hood.
3. In the initial protocol room temperature gelation was described [27], but in our hands, gelation at +4 °C works better. The gel is more uniform.
4. It is important to make these holes very gently and to pay special attention not to detach the agarose from the glass at the periphery of the hole. Otherwise, confinement is not sufficient and cells migrate on the substratum by mesenchymal motility.
5. To improve the adhesion of cells to substrate for better formation of initial monolayer inside 4 mm hole, one may consider coating holes with fibronectin solution (5 mg/mL) for 1 h at 37 °C.
6. For different cell lines, the optimal concentration of FCS in the loading medium may vary, but the final aim is to achieve healthy monolayer in the 4 mm hole.
7. The concentration of FCS in DMEM may need to be optimized if another cell line is used. It is important to load the FCS containing DMEM into the hole gently, to avoid pushing the attractant under the agarose layer.
8. The video frame rate is about one image per minute for cell migration analysis. The video frame rate is about one image every 5 s to capture protrusion dynamics. Blebs are very dynamic. Avoid evaporation for optimal imaging. This can be achieved using a humidified chamber or by covering the agarose with mineral oil. Videomicroscopy can be performed using different techniques, such as phase contrast or fluorescence provided that the cells express fluorescent proteins.

Acknowledgment

This work was supported by the Russian Science Foundation (Grant 16-15-10288).

References

1. Parsons JT, Horwitz AR, Schwartz MA (2010) Cell adhesion: integrating cytoskeletal dynamics and cellular tension. Nat Rev Mol Cell Biol 11:633–643
2. Pollard TD, Borisy GG (2003) Cellular motility driven by assembly and disassembly of actin filaments. Cell 112:453–465

3. Svitkina TM (2013) Ultrastructure of protrusive actin filament arrays. Curr Opin Cell Biol 25:574–581
4. Vicente-Manzanares M, Choi C, Horwitz AR (2009) Integrins in cell migration – the actin connection. J Cell Sci 122:199–206
5. Enterline HT, Cohen DR (1950) The ameboid motility of human and animal neoplastic cells. Cancer 3:1033–1038
6. Wolf K, Mazo I, Leung H et al (2003) Compensation mechanism in tumor cell migration: mesenchymal-amoeboid transition after blocking of pericellular proteolysis. J Cell Biol 160:267–277
7. Friedl P, Wolf K (2003) Tumour-cell invasion and migration: diversity and escape mechanisms. Nat Rev Cancer 3:362–374
8. Fackler OT, Grosse R (2008) Cell motility through plasma membrane blebbing. J Cell Biol 181:879–884
9. Panková K, Rösel D, Novotný M, Brábek J (2010) The molecular mechanisms of transition between mesenchymal and amoeboid invasiveness in tumor cells. Cell Mol Life Sci 67:63–71
10. Paluch EK, Raz E (2013) The role and regulation of blebs in cell migration. Curr Opin Cell Biol 25:582–590
11. Lämmermann T, Sixt M (2009) Mechanical modes of "amoeboid" cell migration. Curr Opin Cell Biol 21:636–644
12. Sahai E, Marshall CJ (2003) Differing modes of tumour cell invasion have distinct requirements for Rho/ROCK signalling and extracellular proteolysis. Nat Cell Biol 5:711–719
13. Carragher NO, Walker SM, Scott Carragher LA et al (2006) Calpain 2 and Src dependence distinguishes mesenchymal and amoeboid modes of tumour cell invasion: a link to integrin function. Oncogene 25:5726–5740
14. Van Goethem E, Poincloux R, Gauffre F et al (2010) Matrix architecture dictates three-dimensional migration modes of human macrophages: differential involvement of proteases and podosome-like structures. J Immunol 184:1049–1061
15. Ehrbar M, Sala A, Lienemann P et al (2011) Elucidating the role of matrix stiffness in 3D cell migration and remodeling. Biophys J 100:284–293
16. Bergert M, Chandradoss SD, Desai RA, Paluch E (2012) Cell mechanics control rapid transitions between blebs and lamellipodia during migration. Proc Natl Acad Sci U S A 109:14434–14439
17. Liu Y-J, Le Berre M, Lautenschlaeger F et al (2015) Confinement and low adhesion induce fast amoeboid migration of slow mesenchymal cells. Cell 160:659–672
18. Gao Y, Wang Z, Hao Q et al (2017) Loss of ERα induces amoeboid-like migration of breast cancer cells by downregulating vinculin. Nat Commun 8:14483
19. Derivery E, Fink J, Martin D et al (2008) Free Brick1 is a trimeric precursor in the assembly of a functional Wave complex. PLoS One 3:e2462
20. Beckham Y, Vasquez RJ, Stricker J et al (2014) Arp2/3 inhibition induces amoeboid-like protrusions in MCF10A epithelial cells by reduced cytoskeletal-membrane coupling and focal adhesion assembly. PLoS One 9:e100943
21. Sabeh F, Shimizu-Hirota R, Weiss SJ (2009) Protease-dependent versus -independent cancer cell invasion programs: three-dimensional amoeboid movement revisited. J Cell Biol 185:11–19
22. Parri M, Taddei ML, Bianchini F et al (2009) EphA2 reexpression prompts invasion of melanoma cells shifting from mesenchymal to amoeboid-like motility style. Cancer Res 69:2072–2081
23. Kosla J, Paňková D, Plachý J et al (2013) Metastasis of aggressive amoeboid sarcoma cells is dependent on Rho/ROCK/MLC signaling. Cell Commun Signal 11:51
24. Taddei ML, Giannoni E, Morandi A et al (2014) Mesenchymal to amoeboid transition is associated with stem-like features of melanoma cells. Cell Commun Signal 12:24
25. Koch B, Meyer AK, Helbig L et al (2015) Dimensionality of rolled-up nanomembranes controls neural stem cell migration mechanism. Nano Lett 15:5530–5538
26. Haeger A, Krause M, Wolf K, Friedl P (2014) Cell jamming: collective invasion of mesenchymal tumor cells imposed by tissue confinement. Biochim Biophys Acta 1840:2386–2395
27. Heit B, Kubes P (2003) Measuring chemotaxis and chemokinesis: the under-agarose cell migration assay. Sci STKE 2003:PL5

Chapter 4

An In Vitro System to Study the Epithelial–Mesenchymal Transition In Vitro

Natalya A. Gloushankova, Svetlana N. Rubtsova, and Irina Y. Zhitnyak

Abstract

The epithelial–mesenchymal transition (EMT) plays an important role in development and cancer progression. Upon EMT, epithelial cells lose stable cell–cell adhesions and reorganize their cytoskeleton to acquire migratory activity. Recent data demonstrated that EMT drives cancer cells from the epithelial state to a hybrid epithelial/mesenchymal phenotype with retention of some epithelial markers (in particular, E-cadherin), which is important for cancer cell dissemination. In vitro studies of the effect of growth factors (in particular, epidermal growth factor (EGF)) on cultured cells can be highly advantageous for understanding the details of the early stages of EMT. The methods described in this chapter are intended for studying intermediate phenotypes of EMT. Time-lapse DIC microscopy is used for visualization of changes in morphology and motility of the cells stimulated with EGF. The transwell migration assay allows the evaluation of the migratory activity of the cells. Studying of dynamics of a fluorescently labeled actin-binding protein F-tractin-tdTomato using confocal microscopy allows detection of EGF-induced changes in the organization of the actin cytoskeleton. Live-cell imaging of cells stably expressing GFP-E-cadherin visualizes reorganization of stable tangential E-cadherin-based adherens junctions (AJs) into unstable radial AJs during the early stages of EMT.

Key words Epithelial–mesenchymal transition, Epidermal growth factor, Migration, Adherens junctions, Actin cytoskeleton

1 Introduction

The epithelial–mesenchymal Transition (EMT), a fundamental process by which epithelial cells acquire mesenchymal phenotype, plays a key role in development. The EMT contributes to wound healing and to pathological states such as fibrosis and tumor malignancy. The EMT program is regarded as a trigger of the detachment of neoplastic cells from primary carcinomas, of their invasion, migration and metastatic dissemination [1, 2]. Recent data indicate that EMT initiation aids phenotypic and functional tumor cell plasticity [3]. The EMT program is also associated with the generation of normal and cancer cells with stem cell-like properties [4, 5].

Alexis Gautreau (ed.), *Cell Migration: Methods and Protocols*, Methods in Molecular Biology, vol. 1749, https://doi.org/10.1007/978-1-4939-7701-7_4, © Springer Science+Business Media, LLC 2018

EMT entails profound changes in morphology and behavior of epithelial cells that include loss of apical–basal polarity, disruption of cell–cell contacts and acquisition of migratory properties. Upon EMT, epithelial cells lose stable E-cadherin-based adherens junctions and reorganize their cytoskeleton to provide directional migration driven by dynamic lamellipodia [6, 7]. These changes are associated with downregulation of epithelial cell markers (e.g., E-cadherin, occludin, and cytokeratins) and upregulation of mesenchymal cell markers (e.g., N-cadherin, vimentin, and fibronectin). Microenvironmental factors such as growth factors (TGFβ, EGF, HGF, Notch, FGF, Wnt, IGF, HIF, NF-kB, and others), hypoxia, and oncogenic, metabolic, or mechanical stresses, can activate various signaling pathways involved in EMT [7, 8]. HGF, TGFβ, IGF, or FGF secreted by cancer-associated fibroblasts induce EMT of carcinoma cells [9, 10]. It was also shown that EGF and VEGFA released by cancer-associated macrophages promote migration and intravasation of cancer cells [11, 12]. Upon EMT, signaling pathways activate specific transcription factors (EMT-TFs)—SNAI1, SNAI2, TWIST1/2, ZEB1/2, and others—that play important roles in the repression of the epithelial phenotype and activation of the mesenchymal phenotype. EMT-TFs repress E-cadherin expression and upregulate genes that contribute to the mesenchymal phenotype [13, 14].

Growing evidence indicates that EMT is not a binary switch between epithelial and mesenchymal state but that epithelial cells can attain a hybrid epithelial/mesenchymal phenotype having both epithelial and mesenchymal properties (i.e., partial or intermediate EMT) [15, 16]. It is considered that this hybrid phenotype may be a metastable state [16] and is regulated by complex network of TFs, such as EMT-TFs, OVOL1/2 and GRHL2, and miRNAs that specifically inhibit translation of EMT-TFs [1, 17]. Two mutually inhibitory feedback loops—miR-34/SNAI1 and miR-200/ZEB—have been studied in detail [18, 19]. Expression of the EMT effector genes and miRNAs is also regulated by DNA methylation and histone modifications [1]. The hybrid phenotype of cancer cells allows them to adapt to environmental changes. E-cadherin-based adherens junctions may be used by cancer cells for effective collective invasion and migration [20, 21]. Upon colonization and metastatic outgrowth, cancer cells often revert to a more epithelial phenotype [3, 22, 23].

EGF plays pivotal roles in normal cell growth, development and tumor progression. Ligand binding to EGFR activates several intracellular signaling pathways (PI3K/Akt, JAK/STAT, NF-κB, PLCγ/PKC, and Ras/MAPK/ERK) and elicits a pleiotropic molecular response that affects cell proliferation, survival, and motility [8, 24]. Cells treated with EGF reorganize their cytoskeleton and acquire a front–rear polarity, which is necessary for cell migration. Mesenchymal migration is driven by the extension of a lamellipodium at the leading edge, followed by the contraction of the cell body and the retraction of the trailing edge. In vitro studies of the effect of growth factors on cultured cells are useful starting points to understand cytoskeletal mechanisms underlying various stages of EMT.

This chapter describes the methods that we use to study of EGF-mediated EMT in vitro. Time-lapse DIC microscopy is used for visualization of changes in cell phenotype and motility of the cells treated with EGF (Fig. 1). The transwell migration assay is used to evaluate the migratory activity of the cells. In this test, cells are seeded into the insert and migrate through the 8 μm pores in the membrane to the bottom of the insert. Quantitative kymographic analysis is used to assess the rate of lamellipodial protrusion. Fluorescent staining of fixed cells using primary antibodies to E-cadherin or β-catenin and phalloidin as a specific marker of filamentous actin is a first choice method for assessment of reorganization of actin cytoskeleton and adherens junctions upon EGF treatment. Study of dynamics of a fluorescently labeled actin-binding protein F-tractin-tdTomato using confocal microscopy (Fig. 2) allows detection of EGF-induced rapid changes in the organization of actin cytoskeleton and migratory behavior of cells. Detailed study of reorganization of adherens junctions in EGF-treated cells is carried out on stable cell lines expressing GFP-E-cadherin (Fig. 3).

2 Materials

2.1 Cell Culture and Treatment

1. The IAR-20 line of normal rat liver epithelial cells (*see* **Note 1**).
2. Dulbecco's Modified Eagle's Medium (DMEM) supplemented with 10% fetal bovine serum (FBS), 100 U/mL penicillin, and 100 μg/mL streptomycin.
3. Dissociation solution: 0.1% trypsin in Versene solution (0.526 mM EDTA in PBS).
4. 35-mm culture dishes or 35-mm glass bottom culture dishes.
5. Round coverslips (d = 10 mm).
6. Stock solutions of EGF. Reconstitute lyophilized EGF (Sigma-Aldrich E9644) in 10 mM acetic acid to 1 mg/mL, aliquot and freeze. To prepare working stock solution, dilute one aliquot of the 1 mg/mL solution to 20 μg/mL using 10 mM acetic acid supplemented with 0.1% BSA, aliquot and freeze (*see* **Note 2**).

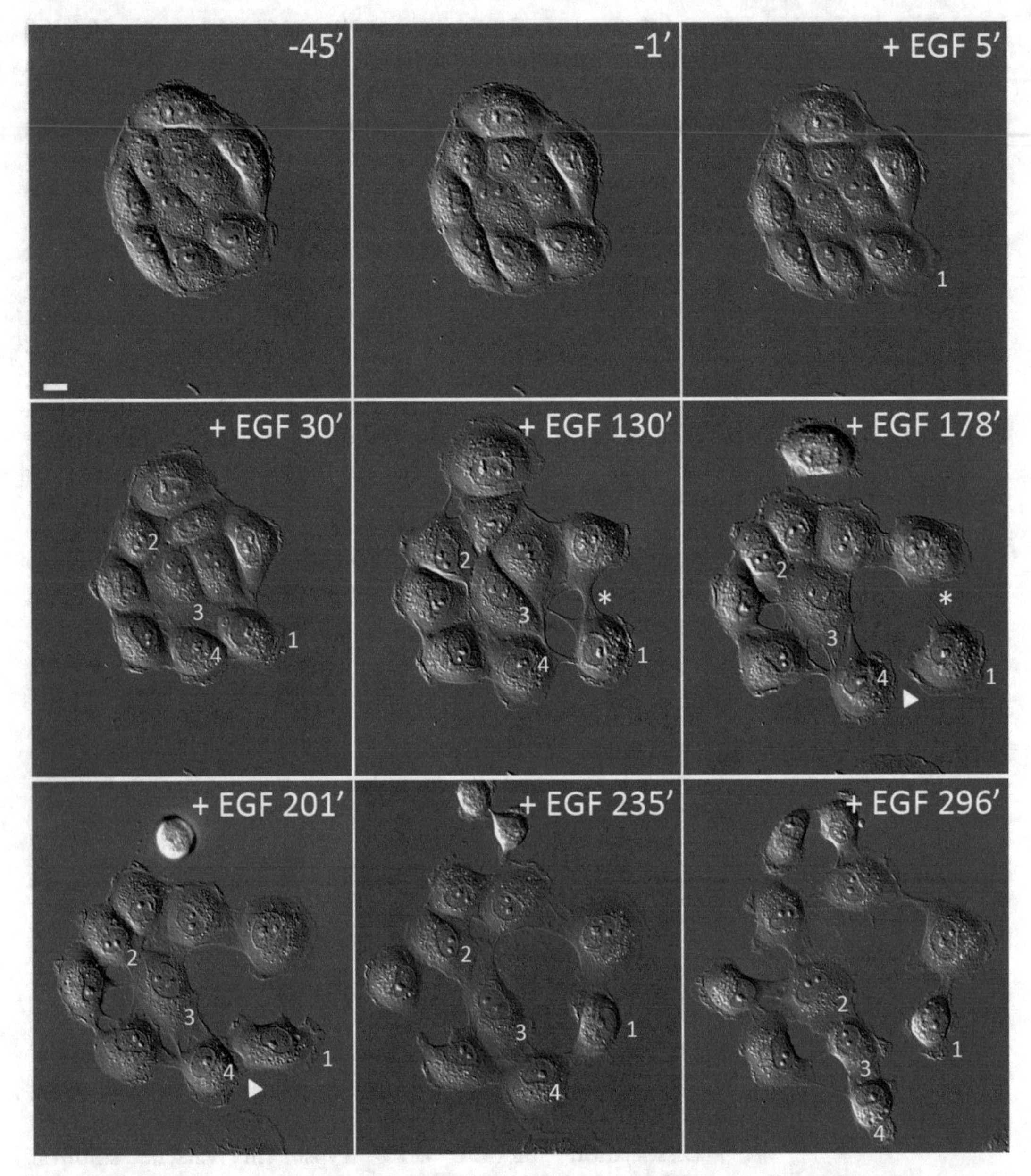

Fig. 1 Changes in cell phenotype and cell–cell interactions in IAR-20 epithelial cells treated with EGF. IAR-20 cells were seeded sparsely into glass bottom culture dishes. In the process of experiment, cells were treated with 50 ng/mL EGF. DIC microscopy (Nikon Eclipse Ti-E microscope, CFI Plan Fluor 40× objective; ORCA-ER Hamamatsu camera controlled with NIS-Elements AR2.30 software) reveals rapid changes in cell phenotype in response to EGF. Selected frames from a time-lapse video sequence are presented. In control (45′ and 1′ before treatment with EGF), cells are joined into islands with stable cell–cell contacts. After addition of EGF, cell–cell contacts are disrupted. Cells start to migrate and detach from the neighboring cells (asterisks). Later, the migratory cells can form new transient contacts with neighboring cells (arrowheads). Both individual (cell 1) and collective migration (cells 2, 3, and 4) can be observed. Scale 10 μm

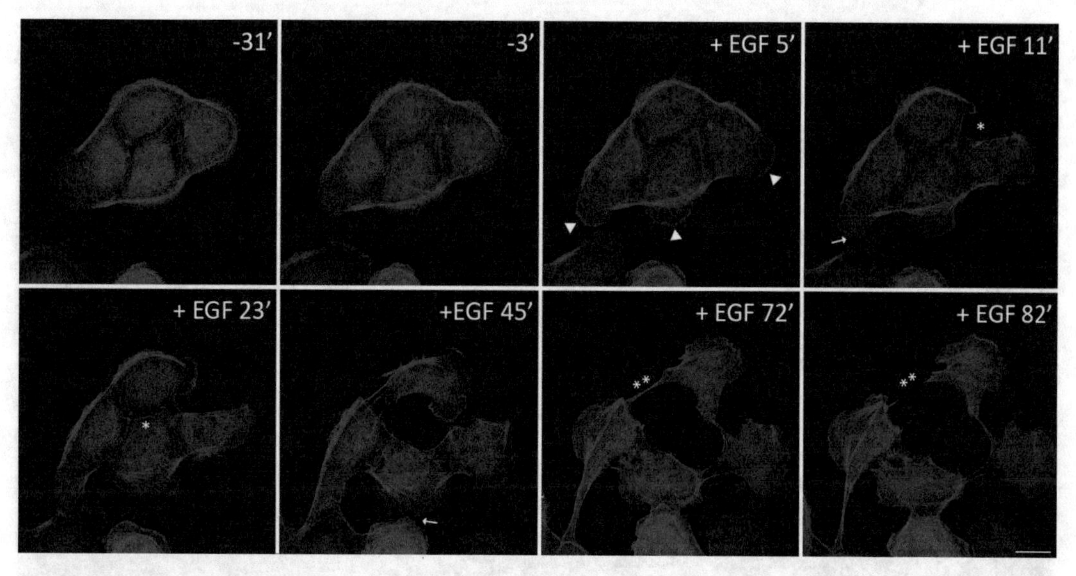

Fig. 2 Reorganization of the actin cytoskeleton in EGF-treated IAR-20 epithelial cells during EMT. IAR-20 cells were stably transfected with F-tractin-tdTomato that labels F-actin. F-tractin-tdTomato-expressing IAR-20 cells were seeded sparsely into glass bottom culture dishes and in the course of the experiment were treated with 50 ng/mL EGF. Cells were observed with a Leica TCS SP5 confocal laser scanning microscope (HDX PL APO 63×1.4 objective; Argon laser, 30% 543 nm). Selected frames from a time lapse video sequence are presented. In sparse culture, control epithelial cells form islands. Cell–cell contacts are stable. In the cells, a marginal actin bundle and junctional actin bundles are observed (31′ and 3′ before treatment with EGF). Immediately after addition of EGF (5′) multiple lamellipodia are formed (arrowheads). Cell–cell contacts are disrupted (11′ and 23′, asterisks). Cells become migratory. Migrating cells may remain connected by thin cables containing actin bundles (double asterisk, 72′), which can eventually break as cells move far apart (82′, double asterisk). Marginal actin bundles disappear. Cells may establish new cell–cell contacts with neighboring cells (11′ and 45′, arrows). Scale 20 μm

2.2 Time-Lapse Video Microscopy

1. An inverted microscope equipped with DIC-optical components, 40×, and 100×1.4 objectives, digital camera and software for image acquisition and time-lapse experiments.

   We use a Nikon Eclipse Ti-E microscope (CFI Plan Fluor 40× and CFI Plan APO 100×1.4 oil immersion objectives); ORCA-ER Hamamatsu camera controlled with NIS-Elements AR2.30 software (Nikon).

2. Temperature control system for microscopes.

   We use the Cube & Box temperature control system (Life Imaging Services, Switzerland).

3. DMEM/F12 medium without phenol red containing 15 mM HEPES. The liquid medium should be discarded after 1 month.

4. Analyze kymograph using ImageJ software with the Multiple Kymograph plugin.

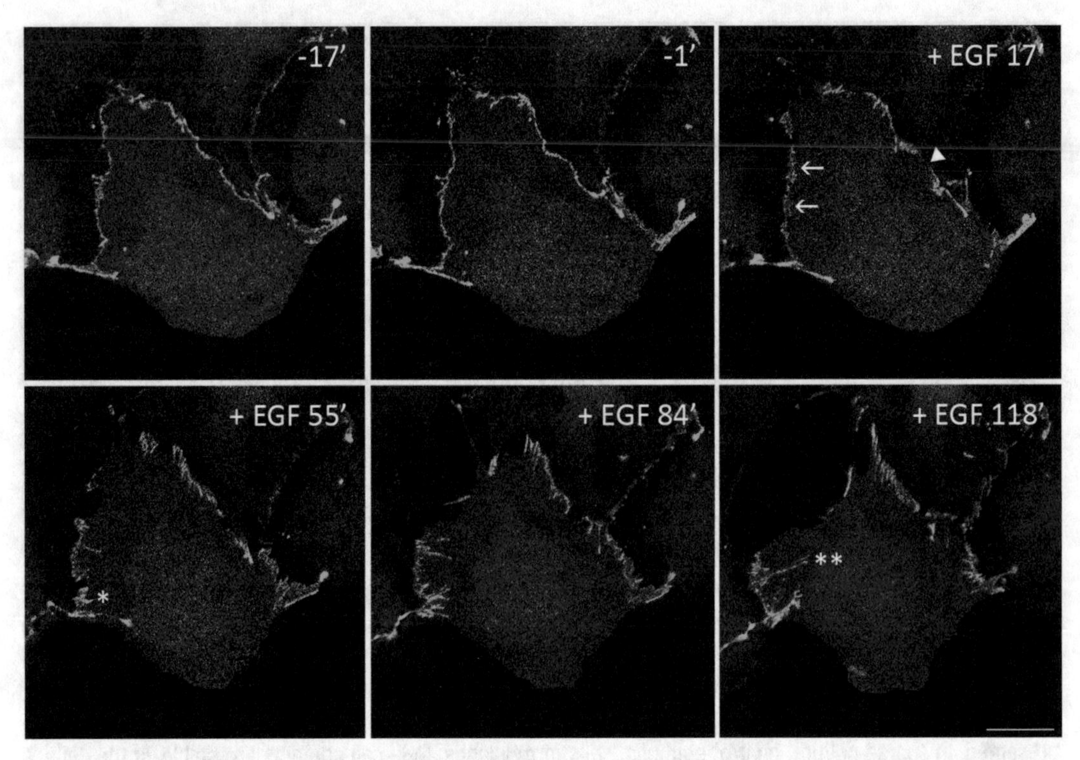

Fig. 3 Rearrangement of E-cadherin-based AJs in EGF-treated IAR-20 epithelial cells during EMT. IAR-20 cells were stably transfected with GFP-E-cadherin. GFP-E-cadherin-expressing IAR-20 cells were seeded sparsely into glass bottom culture dishes and during the course of the experiment were treated with 50 ng/mL EGF. Cells were observed with a Leica TCS SP5 confocal laser scanning microscope (HDX PL APO 63×1.4 objective; HeNe laser, 15% 488 nm). Selected frames from a time lapse video sequence are presented. In untreated IAR-20 cells stable tangential AJs (17′ and 1′ before treatment with EGF) are observed. Within several minutes after addition of EGF (shown here at 17′), AJs become irregular (arrowhead) or dissociate into E-cadherin punctas (arrows). Later (55–138′ of treatment with EGF) radial AJs form de novo (asterisk) and elongate (double asterisk). Scale 20 μm

2.3 Fluorescent Staining

1. Phosphate-buffered saline (PBS). To prepare 2 L of 20× PBS solution, mix 320 g NaCl, 8 g KCl, 44.2 g Na_2HPO_4, and 2 g KH_2PO_4, bring up to a final volume of 2 L with H_2O, warm up the mix to 50 °C with constant stirring. Store at room temperature (RT).
2. 4% paraformaldehyde (PFA) solution in PBS.

   Add 4 g of PFA to 70 mL of H_2O and warm up the mix to 70 °C with constant stirring. Add 50 μL of 4 N NaOH to dissolve PFA. Add 5 mL of 20× PBS and bring up to a final volume of 100 mL with H_2O. Adjust pH to 7.4. Aliquot PFA solution in 15 mL conical tubes and store at −20 °C. The solution may be frozen/thawed repeatedly.
3. 1% PFA in DMEM/HEPES.

   Prepare 10% PFA/PBS by adding 10 g PFA to 70 mL H_2O and warming up the mix to 70 °C with constant stirring. Add

50 μL of 4 N NaOH to dissolve PFA. Add 5 mL of 20× PBS and bring up to a final volume of 100 mL with H_2O. Adjust pH to 7.4. Aliquot 10% PFA/PBS in 15 mL conical tubes and store at −20 °C. The solution may be frozen/thawed repeatedly.

Immediately before fixation, dilute 10% PFA/PBS to 1% with DMEM supplemented with 20 mM HEPES.

4. 0.5% Triton X-100/PBS.
5. Methanol.
6. Primary antibodies for E-cadherin and β-catenin.
7. Secondary antibodies conjugated with Alexa 488 or TRITC.
8. Fluorescently labeled phalloidin (Alexa 488 or TRITC) to detect filamentous actin.

2.4 Transwell Assay

1. Migration chambers containing 8-μm pore polycarbonate membrane inserts in a 24-well tissue culture plate (Corning 353097).
2. Fluorescent dye DAPI (4′,6-diamidine-2′-phenylindole dihydrochloride).

2.5 Transfection and Establishment of Stable Lines Expressing Fluorescent Proteins

1. Lipofectamine® LTX and PLUS™ transfection reagent (Invitrogen).
2. F-tractin-tdTomato and EGFP-E-cadherin constructs.
3. DMEM medium supplemented with 10% FBS, without antibiotics.
4. Opti-MEM.
5. 100 mg/mL G418 in DMEM without FBS. Store at −20 °C.

2.6 Study of Adherens Junction Remodeling and Cytoskeleton Reorganization in Living Cells

1. An inverted microscope equipped with 63×1.4 or 100×1.4 objectives.

   We use a Leica TCS SP5 confocal laser scanning microscope (PL APO 63×1.4 objective) controlled by LAS-AF v. 2.7.3.9723 software (Leica Microsystems).
2. Temperature control system for microscopes.

   We use the Cube & the Box temperature control system (Life Imaging Services, Switzerland).
3. DMEM/F12 medium without phenol red containing 15 mM HEPES.

3 Methods

3.1 Treatment with EGF

1. Seed 1 × 10^5 cells into 35-mm culture dishes in 2 mL of DMEM supplemented with 10% FBS, 100 U/mL penicillin, and 100 μg/mL streptomycin and incubate them at 37 °C in a humidified atmosphere with 5% CO_2.

2. 24 h later replace the medium with 2 mL of fresh DMEM containing 1% FBS, penicillin, and streptomycin, and incubate cells for 16–20 h.
3. Add EGF to a final concentration of 50 ng/mL.
4. Incubate cells with or without EGF at 37 °C in a humidified atmosphere with 5% CO_2.

3.2 Time-Lapse DIC-Microscopy

1. Seed 1 × 10^5 cells into 35-mm glass bottom culture dishes and incubate them as described under Subheading 3.1 (**steps 1** and **2**). We use P35G-1.5-14-C dishes by MatTek Corporation.
2. Before starting a Time-Lapse Video experiment switch on the temperature control system and let the temperature stabilize at 36 °C overnight.
3. Replace the medium with 2 mL of DMEM/F12 without phenol red containing 15 mM HEPES supplemented with 1% FBS. Replace the plastic lid of the dish with a DIC lid with a glass insert (reusable, UV-sterilized).
4. Start microscope controlling software. In our case, NIS-Elements AR2.30 software is used.
5. Place the Analyzer filter in the beam path, open the transmitted light (DIA) shutter.
6. Place the dish on the inverted microscope stage and focus on the specimen. Set up Kohler illumination on the microscope (*see* **Note 3**) and adjust the polarizer for optimal DIC imaging.
7. Switch the light path to the camera. Select Camera Live tool. Detect the auto exposure value. Adjust light intensity so that the optimal exposure falls into the 250–300 ms range.
8. Activate background correction (*see* **Note 4**).
9. Activate PFS and adjust focus by rotating the knob on the PFS controller.
10. Open a NIS-Elements Macro, edit exposure value, folder location, and file name as necessary. Save the Macro after editing (*see* **Note 5**).
11. To record a time-lapse series Go to "Acquire" > "Capture Time lapse automatically".

Select the interval between frames. We set 1 min between frames. Select the overall duration of the time series. We set 6 h. Check "Execute command before capture", select "Run Macro", and choose the edited Macro. Check "PFS ON". Press "Run now".

3.3 Kymograph Analysis

Similar equipment is used as with Time-Lapse Video Microscopy of Cells (Subheading 3.2). A higher-magnification microscope objective (100×1.4) may be used. We use Nikon Eclipse Ti-E microscope with a 100×1.4 objective and Hamamatsu ORCA-ER camera controlled with NIS-Elements AR2.30 software to acquire good quality time-lapse sequences at 1 frame each 3 s for 25–30 min.

Kymographs are produced using the ImageJ software with Multiple Kymograph plugin. A one-pixel-wide line Region of Interest (ROI) is drawn that crosses the lamella region at the right angle to the cell edge. The software then assembles a composite image or "kymograph" by stacking sequentially the intensity along the ROI from each video frame. Thus, in a kymograph, the x axis represents time and the y axis, the distance along the ROI line. Rates of lamellipodial protrusion and retraction are determined from the angles of slopes produced by advancing leading edges. The data may be expressed as mean ± SEM where N = 15–20 cells.

3.4 Transwell Assay

1. Put transwell inserts in empty wells of a 24-well plate. Fill the bottom well of each chamber with 900 μL of DMEM with 10% FBS. Seed 6 × 10^4 cells/300 μL of DMEM with 10% FCS into the transwell insert. Place the transwell insert into the bottom well (*see* **Note 6**).
2. Incubate the plate at 37 °C in humidified atmosphere with 5% CO_2 for 10 h to allow cells to attach to the surface of the membrane.
3. Replace the medium with DMEM containing 1% FBS with or without EGF, and incubate at 37 °C in humidified air with 5% CO_2 for 16–20 h. The cells will migrate through the pores in the membrane.
4. Completely remove cells from the upper surface of membrane with a cotton swab.
5. Wash the migratory cells on the lower surface of the membrane with PBS, fix with 4% PFA in PBS for 3 min and stain with DAPI for 40 min.
6. Remove the filter with forceps. Mount the membrane onto glass slides and examine microscopically at 20× magnification by counting the numbers of stained cells on the membrane in 15 randomly selected fields. The data may be expressed as a percentage of the cells that migrated during the incubation period or as an EGF migration index (the number of cells that migrated in response to EGF relative to the number of cells that migrated in the absence of EGF).

3.5 Establishment of Stable Lines Expressing Fluorescent Proteins

1. For transfection and selection, seed 3–3.5 × 10^5 cells into a 35-mm dish and incubate them overnight as described in Subheading 3.1 (**step 1**).
2. Replace the culture medium with 2 mL of DMEM without antibiotics supplemented with 10% FBS.

3. In a sterile eppendorf tube, mix 200 μL of Opti-MEM, 0.7–1 μg of plasmid DNA and 1 μL of the PLUS™ transfection reagent. Pipette gently and incubate for 10 min at RT. Add 3 μL of the Lipofectamine® LTX to the mixture, gently vortex once and incubate for 30 min at RT. Avoid UV exposure of the transfection mixture. Add the transfection mixture dropwise, mix by gently rocking the dish and incubate for 5–6 h at 37 °C in a humidified atmosphere with 5% CO_2. At the end of the incubation, replace the medium with fresh DMEM without antibiotics supplemented with 10% FBS.
4. 24–48 h later, replace the medium with DMEM supplemented with 10% FBS, penicillin and streptomycin and add G418 to a final concentration of 1 mg/mL.
5. Culture cells for 2 weeks by replacing the medium with fresh DMEM supplemented with 10% FBS, penicillin, streptomycin, and G418 every 3 days or replating as necessary.
6. For cloning, trypsinize the cells and through serial dilutions, prepare 10 mL of cell suspension containing 10 cells/mL in DMEM supplemented with 15% FBS, penicillin and streptomycin. Add 100 μL of the suspension to each well of a 96-well plate, seal the plate with Parafilm and incubate overnight at 37 °C in a humidified atmosphere with 5% CO_2.
7. Prepare 10 mL of DMEM supplemented with 15% FBS, penicillin and streptomycin and 2 mg/mL G418. Add 100 μL of the medium to each well of the 96-well plate, seal the plate with Parafilm and incubate for 2–3 weeks, adding DMEM supplemented with 15% FBS, penicillin, streptomycin, and 1 mg/mL G418 as needed. Periodically visually assess clone growth.
8. After the clones have sufficiently grown (at least 1/3 of the well area), place a round coverslip of d = 10 mm into each well of a 24-well plate and plate the clones individually into this 24-well plate in 1.5 mL of DMEM supplemented with 10% FBS, penicillin and streptomycin. Incubate the clones for 24 h, then add G418 to 1 mg/mL.
9. When the clones have grown to subconfluent monolayers fix the cells by removing coverslips from the culture plate and dipping them individually into wells of a 12-well plate filled with 1 mL of 4% PFA. This procedure should be done in a cell culture hood. After fixing for 15 min, rinse three times with PBS for 5 min, mount the coverslips on slides and examine the clones microscopically for expression of fluorescently labeled proteins.

3.6 Study of Adherens Junction Remodeling and Cytoskeleton Reorganization in Living Cells

1. Seed 1 × 10^5 cells stably expressing GFP-E-cadherin or F-tractin-tdTomato into 35-mm glass bottom culture dishes and incubate them as described under Subheading 3.1 (**steps 1** and **2**).

2. Before starting time lapse acquisition switch on the temperature control system of the confocal microscope and let the temperature stabilize at 36 °C overnight.
3. Replace the medium with 1.5 mL of DMEM/F12 without phenol red containing 15 mM HEPES supplemented with 1% FBS. Add 1 mL of the same medium to a sterile Eppendorf tube, add 5 μL of 20 μg/mL EGF solution and place the tube at 37 °C.
4. Place the dish on the stage of the confocal microscope, remove the lid. The dish may be held in place with small pieces of Blu-tack.
5. Start LAS AF, choose Yes to Initialize DMI6000 stage.
6. Switch on the laser (argon (30%) for E-GFP or HeNe 543 for tdTomato).
7. In the "XY" tab, set up live acquisition parameters (512×512 pix, 400 Hz, pinhole Airy 1.5).
8. Set up detector range (Leica-GFP for E-GFP or Leica-TRITC for tdTomato) and laser (15% 488 nm for E-GFP or 30% 543 nm for tdTomato), set smart gain to max. Switch to pseudocolor mode and adjust laser power and offset as necessary.
9. Mark 4–7 positions using "Mark & Find" function. For acquiring stacks, in the "Mark & Find" tab uncheck "same stack for all" and define individual stacks. Review positions correcting z values if necessary.
10. In the "Acquisition mode" tab, choose "XYT" or "XYZT." In the "Time" tab define interval between frames (1–2 min) and duration of the shoot. In the "XY" tab define acquisition parameters (512×512 or 1024×1024 pix, 200 Hz, Line average 2) and press "Start." In case of focus drift, interrupt the acquisition to correct the focus.
11. To add EGF, stop the acquisition, reach into the temperature controlled chamber through the flexible flap with a 1000-μL sampler and add the medium containing EGF. Resume the acquisition (*see* **Note 7**).

3.7 Fluorescent Staining

1. Place coverslips (d = 10 mm) into 35-mm dishes and culture cells as described in Subheading 3.1 (**steps 1** and **3**).
2. Aspirate and discard the culture medium, wash three times with DMEM supplemented with 20 mM HEPES.
3. For double staining for E-cadherin and actin fix the cells by adding 2 mL of 1% PFA in DMEM with 20 mM HEPES for 15 min at RT (21–24 °C), rinse once with PBS and fix with cold methanol for 3 min. Rinse by adding PBS and gently aspirating the methanol/PBS mixture, keeping the cells under a layer of liquid. Leave for 3 h at RT to reduce nonspecific background staining.

4. For double staining for β-catenin and actin fix the cells by adding 2 mL of 4% PFA in PBS for 10 min at RT. Wash three times with PBS (5–10 min each time) and permeabilize with 0.5% Triton X-100 in PBS for 3 min. Wash three times with PBS (5–10 min each time) and leave for 3 h at RT.
5. Dilute primary antibodies in PBS.
6. Put 50 μL of diluted primary antibodies per coverslip onto a piece of Parafilm in a 10-cm dish. Place the coverslips onto the drops with cells facing the Parafilm. Incubate for 30 min at RT.
7. Remove the coverslips from the drops, place them in a 12-well plate and wash three times with PBS (5–10 min each time).
8. Repeat **steps 5**–7 for secondary antibodies (incubation should be carried out in the dark).
9. Mount the coverslips on slides and leave for 1 h at 37 °C.
10. Acquire images using a fluorescence microscope. The fluorescence is stable for approximately 2 weeks provided the slides are kept in the dark place at +4 °C.

4 Notes

1. Other cell lines can also be used for the study of EGF-stimulated EMT. EGF is a well-characterized EMT inducer in MCF10A normal human mammary epithelial cell line, MDA-MB-468 and MCF7 breast cancer cell lines, and in DU145 human prostate cancer cell line.
2. Keep a working aliquot at +4 °C, discard unused solution after 2–3 weeks. Do not repeatedly freeze/thaw either solution.
3. Focus on the specimen. Fully close field diaphragm. Move the condenser until the edges of the diaphragm are in focus. Center the condenser with the help of adjusting screws. Open the field diaphragm so that it is just outside the field of view. Close the aperture diaphragm as needed to increase the contrast.
4. In the "Live" mode, move the z-drive out of focus so that the image is featureless. Go to "Acquire"—"Background correction"–"Capture correction image". Return to "Live" mode, the image should be uniform grey. Refocus.
5. Macro text

```
Stg_SetFilterPosition(FILTER_TURRET, 0);
Stg_SetShutterState(SHUTTER_DIA, 1);
CameraSet_Exposure(1, 300.0000000);
FineGrab();
Stg_SetShutterState(SHUTTER_DIA, 0);
SaveNext_Images("E:\User\Date\Experiment", 4,
"dic", 0);
```

```
SaveNext_ImageInfo(NULL, NULL, NULL, NULL, 0, 0);
ImageSaveNext();
_CloseCurrentDocument();
```

6. Make sure that the solution touches the membrane of the insert.
7. Due to major focus drift during the first 10–15 min after addition of EGF, acquire individual images rather than series, moving between positions and refocusing before shooting every frame. After focus stabilizes, return to acquiring time-lapse series.

Acknowledgment

This work was supported by the Russian Science Foundation (Grant 16-15-10288).

References

1. Nieto MA, Huang RY, Jackson RA et al (2016) EMT: 2016. Cell 166(1):21–45
2. Diepenbruck M, Christofori G (2016) Epithelial–mesenchymal transition (EMT) and metastasis: yes, no, maybe? Curr Opin Cell Biol 43:7–13
3. Beerling E, Seinstra D, de Wit E et al (2016) Plasticity between epithelial and mesenchymal states unlinks EMT from metastasis-enhancing stem cell capacity. Cell Rep 14(10):2281–2288
4. Mani SA, Guo W, Liao MJ et al (2008) The epithelial-mesenchymal transition generates cells with properties of stem cells. Cell 133(4):704–715
5. Morel AP, Lièvre M, Thomas C et al (2008) Generation of breast cancer stem cells through epithelial-mesenchymal transition. PLoS One 3(8):e2888
6. Huang RY, Guilford P, Thiery JP (2012) Early events in cell adhesion and polarity during epithelial-mesenchymal transition. J Cell Sci 125:4417–4422
7. Lamouille S, Xu J, Derynck R (2014) Molecular mechanisms of epithelial–mesenchymal transition. Nat Rev Mol Cell Biol 15:178–196
8. Lindsey S, Langhans SA (2014) Crosstalk of oncogenic signaling pathways during epithelial–mesenchymal transition. Front Oncol 4:358
9. Vermeulen L, de Sousa EMF, van der Heijden M et al (2010) Wnt activity defines colon cancer stem cells and is regulated by the microenvironment. Nat Cell Biol 12:468–476
10. Taddei ML, Giannoni E, Comito G et al (2013) Microenvironment and tumor cell plasticity: an easy way out. Cancer Lett 341:80–96
11. Wyckoff J, Wang W, Lin EY et al (2004) A paracrine loop between tumor cells and macrophages is required for tumor cell migration in mammary tumors. Cancer Res 64:7022–7029
12. Harney AS, Arwert EN, Entenberg D et al (2015) Real-time imaging reveals local, transient vascular permeability, and tumor cell intravasation stimulated by TIE2hi macrophage-derived VEGFA. Cancer Discov 5:932–943
13. Peinado H, Olmeda D, Cano A (2007) Snail, Zeb and bHLH factors in tumour progression: an alliance against the epithelial phenotype? Nat Rev Cancer 7:415–428
14. Puisieux A, Brabletz T, Caramel J (2014) Oncogenic roles of EMT-inducing transcription factors. Nat Cell Biol 16(6):488–494
15. Yu M, Bardia A, Wittner BS et al (2013) Circulating breast tumor cells exhibit dynamic changes in epithelial and mesenchymal composition. Science 339:580–584
16. Klymkowsky MW, Savagner P (2009) Epithelial-mesenchymal transition: a cancer

researcher's conceptual Friend and Foe. Am J Pathol 174(5):1588–1593
17. Ye X, Weinberg RA (2015) Epithelial–mesenchymal plasticity: a central regulator of cancer progression. Trends Cell Biol 25:675–686
18. Burk U, Schubert J, Wellner U et al (2008) A reciprocal repression between ZEB1 and members of the miR-200 family promotes EMT and invasion in cancer cells. EMBO Rep 9(6):582–589
19. Kim NH, Kim HS, Li XY et al (2011) A p53/miRNA-34 axis regulates Snail1-dependent cancer cell epithelial-mesenchymal transition. J Cell Biol 195(3):417–433
20. Friedl P, Locker J, Sahai E et al (2012) Classifying collective cancer cell invasion. Nat Cell Biol 14(8):777–783
21. Rubtsova SN, Zhitnyak IY, Gloushankova NA (2015) A novel role of E-cadherin-based adherens junctions in neoplastic cell dissemination. PLoS One 10(7):e0133578
22. Kowalski PJ, Rubin MA, Kleer CG (2003) E-cadherin expression in primary carcinomas of the breast and its distant metastases. Breast Cancer Res 5(6):R217–R222
23. Bukholm IK, Nesland JM, Borresen-Dale AL (2000) Re-expression of E-cadherin, α-catenin and β-catenin, but not of γ-catenin, in metastatic tissue from breast cancer patients. J Pathol 190(1):15–19
24. Sibilia M, Kroismayr R, Lichtenberger BM et al (2007) The epidermal growth factor receptor: from development to tumorigenesis. Differentiation 75:770–787

Chapter 5

Detection of Migrasomes

Yang Chen, Ying Li, Liang Ma, and Li Yu

Abstract

The migrasome is a newly discovered, migration-dependent membrane-bound cellular organelle. It functions in the active release of intracellular contents into the external environment and in cell–cell communications. Migrasomes have characteristic morphological features compared with intracellular organelles and extracellular vesicles. This unit describes methods for visualizing migrasomes by fluorescence microscopy and electron microscopy.

Key words Migration, Migracytosis, Migrasome, Cell–cell communication, Fluorescence microscopy, Transmission electron microscopy, Fibronectin

1 Introduction

In 2014, Ma et al. described a new cellular phenomenon called migracytosis, a cell migration-dependent process for the active release of intracellular contents through new organelles named migrasomes [1, 2]. During cell migration, long tubular cytoplasmic extensions called retraction fibers are left behind the migrating cells. Migrasomes grow on the tips or intersections of the retraction fibers after cells migrate away. The generation of migrasomes is dependent on cell migration. The number of migrasomes increases when migration is enhanced and decreases when migration is inhibited. Once they are disassociated from the migrating cells, migrasomes are either broken to release their contents into the extracellular environment or engulfed by neighboring cells [1]. Migrasomes represent a distinct type of extracellular vesicle and are significantly different from other microparticles in size, morphology, and function [3–5]. Migrasomes may function in both spatial and temporal cell–cell communications under multiple physiological conditions and in disease models.

This chapter describes detailed methods for visualizing migrasomes in cells either by fluorescence microscopy or by electron microscopy. In order to generate migrasomes, cells are plated at

Alexis Gautreau (ed.), *Cell Migration: Methods and Protocols*, Methods in Molecular Biology, vol. 1749,
https://doi.org/10.1007/978-1-4939-7701-7_5, © Springer Science+Business Media, LLC 2018

30% confluency into culture dishes pretreated with the extracellular matrix protein fibronectin to create conditions that enhance migration. Using TSPAN4-GFP as a migrasome marker, it is possible to label migrasomes and observe their structure by confocal microscopy during migration of living cells. The corresponding protocol is described in Subheading 2.1. Transmission electron microscopy (TEM) reveals the ultrastructure of migrasomes in situ within cultured cells. The corresponding protocol is described in Subheading 2.2.

2 Materials

2.1 Observation of Migrasomes in Cultured Cells by Fluorescence Microscopy

1. Phosphate-buffered saline (PBS): 135 mM NaCl, 4.7 mM KCl, 10 mM Na_2HPO_4, 2 mM NaH_2PO_4.
2. Culture medium: DMEM/High glucose supplemented with 10% FBS, Penicillin, Streptomycin and GlutaMAX-ITM.
3. Fibronectin: purchase from Gibco (catalog number: PHE0023).
4. VigoFect: purchase from Vigorous (catalog number: T001).
5. Dynasore: purchase from Calbiochem (catalog number: D00098387).
6. Blebbistatin: purchase from Calbiochem (catalog number: D0015970).
7. 3.5 cm glass-bottom dishes from In Vitro Scientific (catalog number: D35-20-1-N).

2.2 Observation of Migrasomes in Cultured Cells by TEM

1. 0.1 M phosphate buffer (PB, pH 7.2, 100 mL): 36 mL 0.2 M $Na_2HPO_4 \cdot 12H_2O$; 14 mL 0.2 M $NaH_2PO_4 \cdot 2H_2O$; 50 mL distilled water.
2. Fixative buffer (EM grade fixative): 2.5% (v/v) glutaraldehyde in 0.1 M PB at pH 7.2.
3. Postfixative buffer: 2% (w/v) osmium tetroxide (Ted Pella) and 3% (w/v) $K_3[Fe(CN)_6]$ in water (*see* **Note 1**).
4. PON12 (SPI) resin: add the four components in the order PON 812 (50 mL), DDSA (16 mL), MNA (35.6 mL), BDMA (2 mL). Mix the resin until it is homogeneous and seal with parafilm. Incubate the mixture for 30 min, then aliquot the resin into 5 mL or 15 mL centrifuge tubes and store at −20 °C.
5. 2% (w/v) uranyl acetate (Merck) in water.
6. Reynold's lead citrate solution [6].

   Mix 1.33 g $Pb(NO_3)_2$, 1.76 g $Na_3(C_6H_5O_7) \cdot 2H_2O$, and 30 mL distilled water in a 50 mL centrifuge tube. After shaking for 30 min, add 8.0 mL 1 M NaOH and make the solution up to 50 mL with distilled water. Mix the solution by inversion. Centrifuge at 10,000 × *g* for 5 min before use. The pH of the solution is routinely 12.
7. 100 mesh copper grid coated with formvar.

3 Methods

Visualizing the structure of migrasomes is a cell biology approach, and the cell is thus the core factor in the study. Moreover, generation of migrasomes is dependent on cell migration. Ensuring that cells are in a normal migration-competent state is therefore essential for visualizing migrasomes (*see* **Note 2**).

3.1 Observation of Migrasomes in Cell Culture by Fluorescence Microscopy

1. Transfect L929 cells with plasmid TSPAN4-GFP using VigoFect transfection reagents (*see* **Note 3**).
2. Replace the medium with fresh culture medium 4 h after transfection (*see* **Note 4**).
3. 6–8 h posttransfection; replate the transfected cells into a 3.5 cm glass-bottom dish. The dish was pretreated for 1 h with 10 μg/mL fibronectin diluted in PBS. The cell density is about 30% (*see* **Notes 5** and **6**).
4. Incubate cells in a tissue culture incubator supplemented with 5% CO_2 for 15 h.
5. Use the live-cell imaging system on a confocal microscope (FV1200, Olympus) to visualize migrasomes with an excitation wavelength of 488 nm. The pinhole should be adjusted to 120 μm (Fig. 1) (*see* **Note 7**).
6. To study how cell migration influences the generation of migrasomes, cells transfected with TSPAN4-GFP were treated with the dynamin inhibitor Dynasore (working concentration 20 μM) or the myosin II inhibitor Blebbistatin (working concentration 100 μM) for 1 h. Movies of the untreated and treated cells were recorded for 4–6 h by confocal microscopy. The number of newly generated migrasomes and their migration rate were quantified from the movie recorded. Cells treated with migration inhibitors show lower migration rates and fewer migrasomes [1].

3.2 Observation of Migrasomes in Cultured Cells by TEM (Fig. 2)

1. Cell preparation: Plate NRK cells with 30% confluency in a 3.5 cm cell culture dish pretreated with 10 μg/mL fibronectin. Incubate the cells in a tissue culture incubator supplemented with 5% CO_2 for 12 h to let the cells settle down and migrate.
2. Fixation: Fix cells with a mixture of fixative buffer and cell culture medium at a ratio of 1:1 (v/v) in PB for 5 min (*see* **Note 8**). Then change to fresh fixative buffer and continue to fix for 2 h (*see* **Note 9**). Wash the samples with PB for 10 min and repeat three times. Then postfix cells using 1% osmium tetroxide and 1.5% $K_3[Fe(CN)_6]$ [7]. Wash samples with distilled water for 5 min, three times.

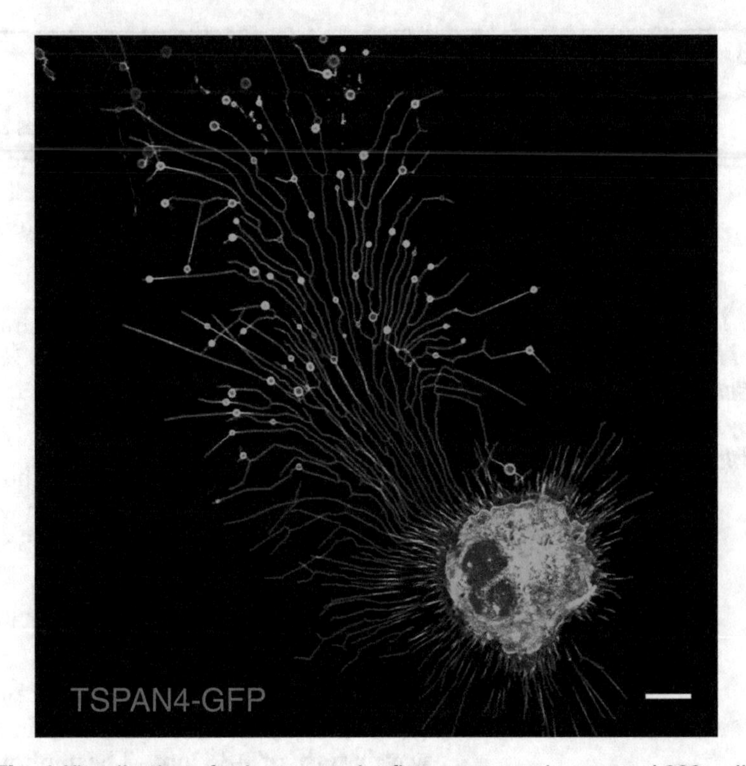

Fig. 1 Visualization of migrasomes by fluorescence microscopy. L929 cells were transfected with TSPAN4-GFP and observed by confocal microscopy. Scale bar: 10 μm

3. Staining: Use 1% uranyl acetate (UA) for in-block staining overnight and then wash the samples with distilled water for 5 min, three times.
4. Sample dehydration: Dehydrate samples in a graded ethanol series of 50%, 70%, 80%, 90%, 3×100%, 2 min each (*see* **Note 10**).
5. Resin infiltration and embedment:
   (a) Add resin to 100% ethanol at a ratio of 1:1 (v/v), then apply the mixture to the sample in the culture dish. Incubate for 30 min at room temperature and remove the resin.
   (b) Repeat the previous step with a 2:1 (v/v) ratio of resin to 100% ethanol.
   (c) Repeat the previous step with a 3:1 (v/v) ratio of resin to 100% ethanol.
   (d) Add pure resin to the sample and incubate overnight at room temperature.
   (e) Change to fresh pure resin and incubate for 1 h at room temperature.
   (f) Remove the pure resin and add 0.5 mL fresh resin to the dish.
   (g) Polymerize at 37 °C for 8 h, 45 °C for 24 h, and 60 °C for 12 h.

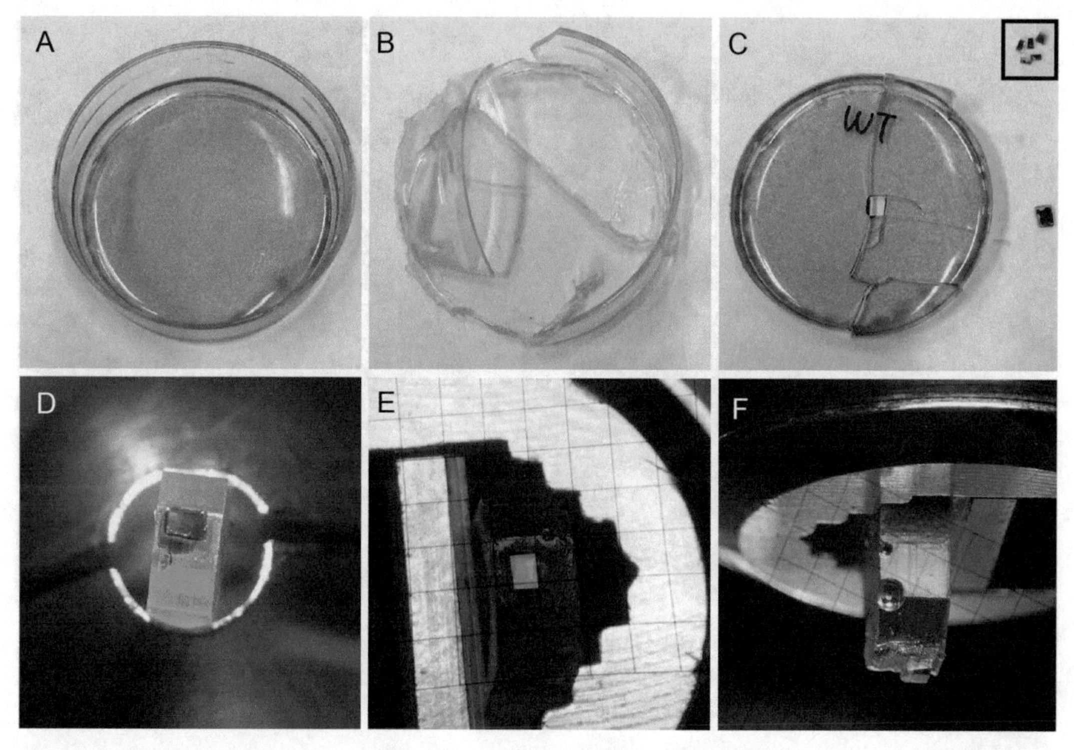

Fig. 2 Illustration of the procedures for TEM sample preparation. (**a**) Infiltrate the sample in resin and allow the resin to polymerize. (**b**) Remove the plastic dish. (**c**) Label the region of interest and cut the resin block into small pieces (see the box on the top right). (**d**) Glue the small block onto the top of a resin block made in the embedding mold. (**e**) Trim the block to about 0.5 mm^2. (**f**) Make ultrathin sections (70 nm)

6. Sectioning
   (a) Remove the plastic dish and cut the resin block into small pieces about 1 mm thick. Use adhesive to glue the small block onto the top of a resin block made in the embedding mold. Trim the block to about 0.5 mm^2 and then make ultrathin sections (70 nm). Pick the sections up using copper grids (*see* **Note 11**).
7. Staining
   (a) Stain the sections with 2% uranyl acetate for 30 min and lead citrate for 7 min.
8. Observation
   (a) Analyze the stained ultrathin sections under an H-7650 transmission electron microscope. Search the cell first at low magnification (6000×) and then focus on individual migrasomes at (80,000×). Migrasomes are found on the tips, in the middle, or at the intersections of retraction fibers (Fig. 3).

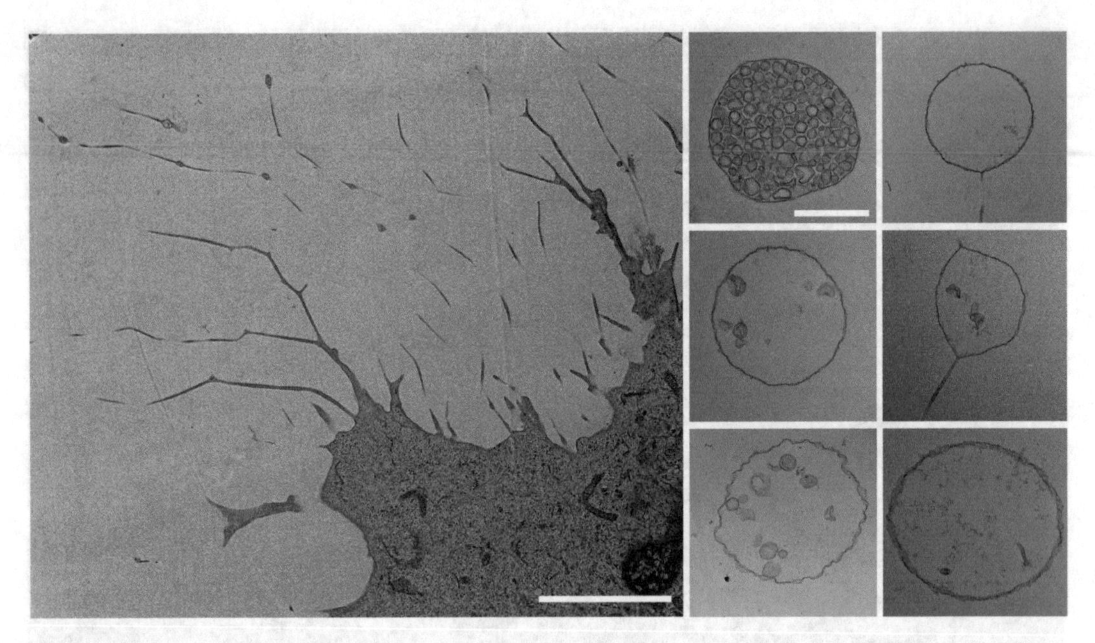

Fig. 3 Visualization of migrasomes by TEM. The left panel shows migrasomes connected with retraction fibers and the cell body. Scale bar: 5 μm, magnification 6000×. The magnified migrasomes are shown in the right panels. Scale bar: 500 nm, magnification 80,000×

4 Notes

1. Make the postfixative buffer immediately before use.
2. Not all kinds of cell show typical migrasome structures. The size and number of migrasomes varies among different cell lines. L929 is a cell line that is commonly used to visualize migrasomes. The average number of migrasomes per L929 cell cultured on fibronectin is 12. Moreover, migrasomes generated by L929 cells are usually larger and easy to visualize. NRK cells can also be used, but the number of migrasomes is much lower. Overexpression of the migrasome marker TSPAN4-GFP will increase the migrasome number, which makes NRK a suitable cell line for this study.
3. When using VigoFect to transfect cells, the density of cells should be up to 80%. Change the cell culture medium to fresh culture medium 1 h before VigoFect transfection.
4. Four hours after transfection by VigoFect, change the medium to fresh culture medium. This procedure is important to reduce the side effects of the transfection reagents and is good for migrasome generation. When exchanging the medium, gently aspirate the medium and gently pipette in the fresh cul-

ture medium prewarmed to 37 °C to avoid disruption of newly generated migrasomes.

5. The density of cells is important for visualizing migrasomes. In order to maintain a good cell status while at the same time providing the cells with enough space for migration, the density is usually 30% when plating the cells. Migrasomes are typically generated 6–9 h after cell plating and the number peaks at 12–15 h. In order to capture the generation of migrasomes by movie, the best time frame is 8–12 h (start the recording 6–9 h after plating the cells).
6. The extracellular matrix is important for migrasome generation. Different cell types use different focal adhesion molecules, and thus require different matrix proteins. For NRK and L929, pretreating the plates with fibronectin enhances cell migration, and thus boosts the number of migrasomes.
7. The currently available marker for migrasomes, TSPAN4-GFP, does not specifically label migrasomes. It also localizes to intracellular structures including the plasma membrane and intracellular organelles. If no migrasomes are observed, the localization of TSPAN4-GFP in the cell body is a good indicator of whether the marker is expressed.
8. All buffers used in the TEM protocol should be prewarmed to 37 °C unless indicated otherwise.
9. After fixation, samples can be used immediately in the next step, or kept at 4 °C until needed.
10. During sample dehydration, be careful not to leave the sample to dry out completely.
11. The sample block needs to be in parallel with the diamond knife in order to be cut in the correct direction.

References

1. Ma L et al (2015) Discovery of the migrasome, an organelle mediating release of cytoplasmic contents during cell migration. Cell Res 25:24–38
2. da Rocha-Azevedo B, Schmid SL (2015) Migrasomes: a new organelle of migrating cells. Cell Res 25:1–2
3. Trams EG et al (1981) Exfoliation of membrane ecto-enzymes in the form of micro-vesicles. Biochim Biophys Acta 645:63–70
4. Johnstone RM et al (1987) Vesicle formation during reticulocyte maturation. Association of plasma membrane activities with released vesicles (exosomes). J Biol Chem 262:9412–9420
5. Kowal J, Tkach M, Thery C (2014) Biogenesis and secretion of exosomes. Curr Opin Cell Biol 29:116–125
6. Reynolds ES (1963) The use of lead citrate at high pH as an electron-opaque stain in electron microscopy. J Cell Biol 17:208–212
7. Vermeer BJ et al (1978) Ultrastructural findings on lipoproteins in vitro and in xanthomatous tissue. Histochem J 10:299–307

Check for updates

Chapter 6

3D Endothelial Cell Migration

Kathryn A. Jacobs and Julie Gavard

Abstract

Endothelial cells have the capacity to shift between states of quiescence and angiogenesis. The early stage of angiogenesis, sprouting, occurs with the synchronized activities of tip cells, which lead the migration of the sprout, and stalk cells, which elongate this vessel sprout. Here, we describe a method to study in vitro this early and rapid stage of sprouting angiogenesis.

Key words Sprouting angiogenesis, Endothelial cell, VEGF, Tumor microenvironment, Fibrin matrix, Conditioned medium

1 Introduction

Blood vessels fuel organs and tissues throughout the body with oxygen, nutrients, hormones, and growth factors, while eliminating metabolic by-products. They also allow for circulation of immune cells that patrol the blood stream [1].

Blood vessels form a hierarchized and stereotyped network of many branches, which are lined with endothelial cells. Angiogenesis is defined as the expansion of this predefined network. This occurs, in physiological and pathological conditions, in response to changes in metabolic demands, with nutrient deprivation and a reduction in oxygen tension as the primary provocations for angiogenesis [2–4].

Endothelial cells are mainly found quiescent with a slow turnover in adult mature vessels. However, these differentiated cells remain highly plastic, with the ability to quickly switch between states of quiescence to rapid growth, i.e., vessel sprouting, when stimulated by growth factors or hypoxia. The most accepted model of vessel sprouting proposes a coordinated activity between endothelial cells in different states [4]. Schematically the leading cells—the first state—the so-called endothelial tip cells, navigate the vasculature and guide vessel elongation. The second one,

Alexis Gautreau (ed.), *Cell Migration: Methods and Protocols*, Methods in Molecular Biology, vol. 1749, https://doi.org/10.1007/978-1-4939-7701-7_6, © Springer Science+Business Media, LLC 2018

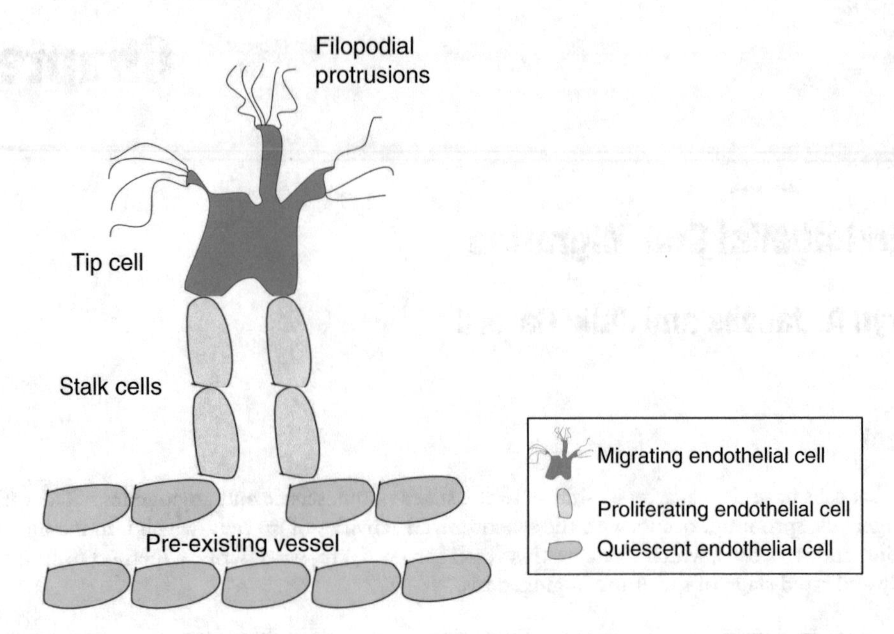

Fig. 1 Sprouting angiogenesis. A schematic model of vessel sprouting where stalk cells proliferate to expand the sprout, and the tip cells guide the vessel migration

endothelial stalk cells, elongates the branch through rapid proliferation (Fig. 1) [4].

Tip and stalk differentiation is notably regulated by VEGF (vascular endothelial growth factor) and Notch signaling. VEGF signals tip cell induction and prompts expression of Notch ligand Delta-like 4 (dll4). This activates Notch signaling in neighboring endothelial cells, and suppresses VEGF receptor 2 expression, to prevent tip cell behavior [4]. This mechanism selects therefore VEGFR2-positive dll4-positive tip cells and VEGFR2-negative Notch-positive stalk cells. However, cell fates are not permanently defined, there is a dynamic switch between tip and stalk cell phenotypes depending on the fitness of the cells [5]. From a mechanistic standpoint, this sprouting angiogenesis requires orchestrated tridimensional migration of endothelial cells together with cell invasion within a defined matrix (Fig. 1).

Angiogenesis is important for maintaining homeostasis, but it also has implications in disease. Endothelial cell dysfunction is indeed a characteristic of diabetes as a consequence of elevated oxidative stress [2, 6]. In cancer, tumor-induced angiogenesis allows tumors to grow by providing them with nutrients and oxygen [1]. How the tumor microenvironment operates on endothelial cells to drive sprouting is crucial to design antiangiogenic-based anticancer strategies.

The method presented here allows for the in vitro study of the early stages of angiogenesis, by recapitulating the endothelial behavior during sprouting angiogenesis (Fig. 2). This model can

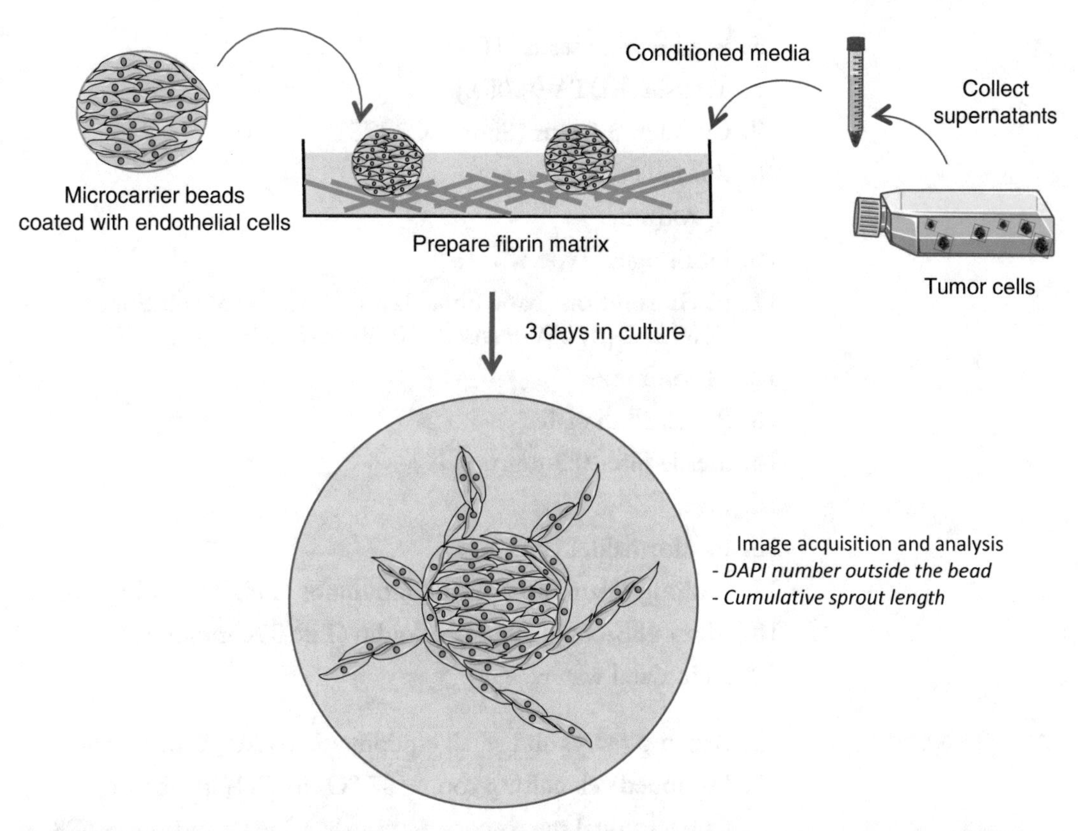

Fig. 2 In vitro sprouting assay procedure. Human umbilical vein endothelial cells (HUVEC) are coated on microcarrier beads, and allowed to sprout into a fibrin matrix, under exposure to malignant tumor cell conditioned media

have applications in a variety of disease studies, as conditions can be manipulated genetically or pharmacologically. For instance, recent studies from our lab had shown that conditioned media collected from patient-derived cancer cell cultures drive sprouting angiogenesis through secretion of growth factors [7]. In keeping with this idea, oncogenic transformation of endothelial cells also forces in vitro sprouting angiogenesis, a process that involves both the activation of intracellular aberrant signaling pathways and autocrine/paracrine cytokine action [8].

2 Materials

2.1 Reagents

1. HUVEC (human umbilical vein endothelial cells, Ea.hy926, ATCC).
2. U87-MG (human astrocytoma malignant glioma cell line, ATCC).
3. DMEM 4.5 g/L glucose.
4. GlutaMAX.

5. Fetal bovine serum (FBS).
6. Trypsin–EDTA (0.05%).
7. Cytodex-3 Beads (Sigma, C3275).
8. Rat tail collagen I.
9. Aprotinin.
10. Fibrinogen Type I.
11. bFGF solution (basic fibroblast growth factor solution, Sigma, F5392) 2 μL bFGF stock (10,000×) in 200 μL DMEM.
12. Thrombin.
13. 8-well Ibidi plate.
14. Sterile filter 0.2 μm.
15. PBS.
16. Paraformaldehyde.
17. Prolong Diamond antifade mountant (Life Technologies).
18. Alexa 488-conjugated phalloidin (Life Technologies).
19. Deionized water.

2.2 Equipment

1. Bench pipettes and small equipment (rocker, centrifuge).
2. Equipped cell culture room (37 °C/5%CO_2 incubator, hood).
3. Conventional microscope to visualize DAPI and Alexa 488.

3 Methods

3.1 Reagent Preparation

1. In order to prepare fibrinogen for use, first dissolve 2 mg/mL fibrinogen in DMEM-GlutaMAX medium. Make sure to note the clottable protein percentage and adjust accordingly. The solution should be heated in a 37 °C water bath to dissolve the fibrinogen. Filter solution through a sterile 0.2 μm filter (*see* **Note 1**).
2. To reconstitute aprotinin, dissolve lyophilized aprotinin in deionized water at 4 U/mL. Filter solution through a sterile 0.2 μm filter. Make aliquots of 1 mL each and store the solution at −20 °C (*see* **Note 2**).
3. For thrombin preparation, reconstitute thrombin in sterile water at 50 U/mL. Make aliquots of 500 μL each and store at −20 °C (*see* **Note 2**).
4. For a 10 mL fibrinogen solution, dissolve 25 mg of fibrinogen in 10 mL of DMEM, as described in **step 1**. Then add 20 μL of aprotinin (10 mg/mL) to the solution. Filter the resulting solution sterilely through a 0.2 μm filter. Finally, add 100 μL of the bFGF solution.
5. In order to prepare Cytodex beads for use, hydrate 0.5 g of dry beads in PBS (pH 7.4) for at least 3 h at room temperature.

This should be done in a 50 mL tube on a rocker, under gentle rotation. Next, let the beads settle down for approximately 15 min. Discard the supernatant and wash the beads for several minutes in 50 mL of fresh PBS. Then, discard the supernatant and replace again with fresh PBS. For 30,000 beads/mL (10 mg/mL), use 50 mL PBS.

6. Prepare all reagents for staining; paraformaldehyde (4% in PBS) for fixation, Triton X-100 (0.05% in PBS) for permeabilization. Prepare fresh solutions.

3.2 Cell Preparation

1. Grow HUVEC in DMEM-GlutaMAX +10% FBS in the days before beading. A concentration of 400 cells per bead is needed to perform the experiment. For 75 μL of bead solution, 10^6 HUVEC will be needed (*see* **Notes 3** and **4**).
2. Grow U87 in DMEM-GlutaMAX +10% FBS. To prepare U87-MG condition medium (CM), 250.000 cells are plated in 10 cm dish, grow for 2 days in DMEM-GlutaMAX, supplemented with 10% FBS. Cells are washed thrice with PBS and incubated at 37 °C overnight in DMEM-GlutaMAX serum-free media [9]. Two days later, media are decanted and cleared by centrifugation (300 × *g*, 5 min), followed by filtration through a 0.2 μm filter. CM are then used immediately or stored at −20 °C until use (*see* **Notes 2** and **5**).

3.3 Sprouting Assay

1. On day −1, coat beads with HUVEC. First, 4000 Cytodex microcarrier beads are incubated with collagen (1/50 dilution in PBS, 15 min, RT). Aspirate the supernatant and wash the beads in 1 mL of prewarmed DMEM. Trypsinize nonconfluent HUVEC and mix 75 μL of beads with 10^6 HUVEC in 1.5 mL of prewarmed DMEM in a 15 mL round tube. Make sure to place the tube vertically in the incubator (37 °C) (*see* **Note 6**).
2. Incubate the tube for 4 h at 37 °C, shaking the tube every 20 min. After 4 h, transfer the coated beads into a T75 flask, add 12 mL of DMEM and incubate overnight at 37 °C (*see* **Note 7**).
3. On day 0, coated beads should be embedded in fibrin gel. To do this, prepare the fibrinogen/aprotinin/bFGF solution (2.5 mg/mL) (*see* Subheading 3.1). Next, transfer the coated beads to a 15 mL conical tube. Let the beads settle. Wash the beads three times with 1 mL DMEM. Then, count the beads on a 10 μL coverslip and resuspend them in the fibrinogen solution at a concentration of approximately 500 beads per mL (*see* **Note 8**).
4. Add 0.625 U/mL of thrombin to each well of the IBIDI plate. Stock is at 100 U/mL, add 20 μL of the stock diluted 1–10 by adding 450 μL of DMEM to a 50 μL aliquot. Then add 400 μL

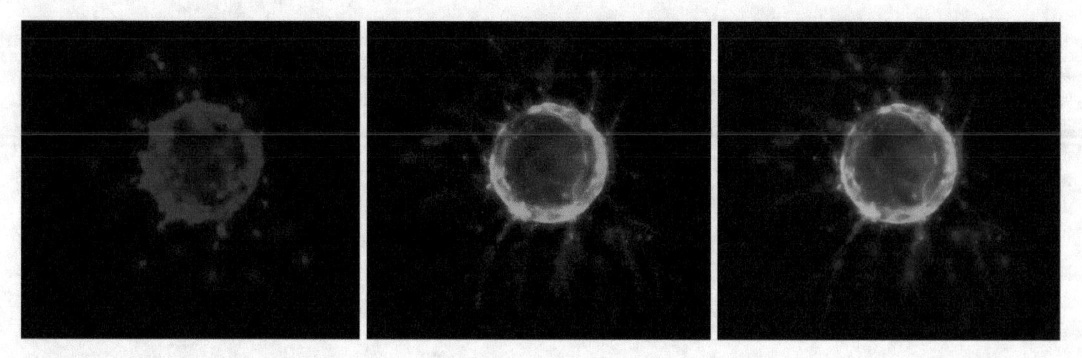

Fig. 3 Typical image of endothelial sprouting. Human endothelial cells were prepared for sprouting assays, as described and incubated for 3 days with conditioned media from malignant glioma cells. Nuclei are stained with DAPI (blue), actin cytoskeleton is visualized with Phalloidin (green)

of the fibrinogen/bead suspension to each well of the plate. Mix the thrombin and fibrinogen by pipetting up and down gently four or five times (*see* **Notes 1**, **9**, and **10**).

5. Leave the plate in the cell culture hood for 5 min. After, place plate in the incubator (37 °C) for 10 to 15 min in order to generate a clot. Once the clot is formed, add 1 mL of DMEM/bFGF dropwise to each well. Return the plate to the incubator (*see* **Note 11**).
6. On day 1, add the U87-CM on top of the fibrin gel. Return plate to incubator (37 °C) (*see* **Notes 12** and **13**).

3.4 Data Analysis

1. By day 3, sprouting should have occurred. Check under bright field microscope and harvest the experiment. Fix in paraformaldehyde 4% (15 min, RT) and permeabilized in Triton (5 min, RT). Wash once in PBS. Incubate with Alexa 488-conjugated phalloidin (1/1000 in PBS, 45 min, RT). Wash three times in PBS. Mount in DAPI-containing mounting medium (*see* **Note 14**).
2. Proceed to image acquisition, with a minimum of 5 random fields of views (Fig. 3) (*see* **Note 15**).
3. From individual bead, quantify (1) number of sprouted cells by counting DAPI-positive nuclei away from the beads, and, (2) sprout extension by measuring cumulative sprout length and mean sprout length (Fig. 2) (*see* **Note 16**).

4 Notes

1. Tubes should never be vortexed, instead mix by inverting the tube.
2. Avoid freeze–thaw cycles.

3. Antibiotics (penicillin/streptomycin) can be added to the medium.
4. Alternate endothelial cells could be used, such as from human, mouse, rat and porcine origin, as well as from any organs. Culture conditions vary and might need to be adapted.
5. Other tumor cell lines could be used. Cell density might need to be tested and adjusted. Usually 48 h is preferred to collect CM.
6. Allow the beads to settle, but do not centrifuge them.
7. Make sure to rinse the T75 flask with DMEM.
8. As a control of good coating, beads should look like golf balls.
9. Make sure to change the pipette tip each time.
10. Avoid creating large bubbles.
11. Tiny bubbles are usually formed in the fibrin gel. They should disappear by the end of the experiment.
12. Media on plate should be changed every other day. Control for evaporation, a humid chamber can be useful when CM volume is limited.
13. Do not forget negative control, such as DMEM-GlutaMAX media collected similarly to U87-CM from cell-free plates.
14. Experiment can be stopped after the fixation step and plate left in PBS, 4 °C overnight. Slowly aspirate medium by pipetting, avoid vacuum it.
15. Images can be acquired with any conventional large field microscope, equipped with DAPI and FITC filters, and automatized camera.
16. Image analysis can be performed with any image viewer software. We recommend Fiji software (Fiji is just Image J), free of use at https://fiji.sc/.

Acknowledgment

The authors are thankful to the present and past members of SOAP laboratory, in particular to Sandy Azzi and Lucas Treps who developed the sprouting assays in the team. Research in SOAP team was funded by: Fondation ARC pour la recherche contre le Cancer, Ligue Nationale contre le Cancer comité Pays-de-la-Loire, comité Maine-et-Loire and comité Vendée, and Connect Talent grant from Région Pays-de-la-Loire and Nantes Métropole. KAJ received a doctoral fellowship from Nantes Métropole.

References

1. De Bock K, Georgiadou M, Carmeliet P (2013) Role of endothelial cell metabolism in vessel sprouting. Cell Metab 18:634–647
2. Adams RH, Alitalo K (2007) Molecular regulation of angiogenesis and lymphangiogenesis. Nat Rev Mol Cell Biol 8:464–478
3. Herbert SP, Stainier DY (2011) Molecular control of endothelial cell behaviour during blood vessel morphogenesis. Nat Rev Mol Cell Biol 12:551–564
4. Potente M, Gerhardt H, Carmeliet P (2011) Basic and therapeutic aspects of angiogenesis. Cell 146:873–887
5. Jakobsson L, Franco CA, Bentley K, Collins RT, Ponsioen B, Aspalter IM, Rosewell I, Busse M, Thurston G, Medvinsky A et al (2010) Endothelial cells dynamically compete for the tip cell position during angiogenic sprouting. Nat Cell Biol 12:943–953
6. Yeh WL, Lin CJ, Fu WM (2008) Enhancement of glucose transporter expression of brain endothelial cells by vascular endothelial growth factor derived from glioma exposed to hypoxia. Mol Pharmacol 73:170–177
7. Azzi S, Treps L, Leclair HM, Ngo H, Harford-Wright E, Gavard J (2015) Desert Hedgehog/Patch2 axis contributes to vascular permeability and angiogenesis in glioblastoma. Front Pharmacol 6:281
8. Dwyer J, Azzi S, Leclair HM, Georges S, Carlotti A, Treps L, Galan-Moya A, Alexia C, Dupin N, Bidere N, Gavard J (2015) The guanine exchange factor SWAP70 mediates vGPCR-induced endothelial plasticity. Cell Commun Signal 13:12
9. Dwyer J, Hebda JK, Le Guelte A, Galan-Moya EM, Smith SS, Azzi S, Bidere N, Gavard J (2012) Glioblastoma cell-secreted interleukin-8 induces brain endothelial cell permeability via CXCR2. PLoS One 7:e45562

Chapter 7

Transmigration of Leukocytes Across Epithelial Monolayers

Penny E. Morton and Maddy Parsons

Abstract

Migration of leukocytes through epithelial monolayers represents an essential step in the generation of an inflammatory response and is often seen in inflammatory conditions such as Crohn's disease (Matthews et al., Toxicol Pathol 42:91–98, 2014) and asthma (Lambrecht and Hammad, Nat Med 18:684–692, 2012). Transepithelial migration involves adhesion to the basal surface of the epithelium before migration through the epithelial cell layer to the apical surface. Analyzing this process can present a technical challenge due to complications of using a coculture model and trying to recapitulate an intact monolayer. Here we describe two methods of assessing transepithelial migration based on a Transwell assay, the first of which measures the apical–basal migration of epithelial cells and the second "Inverted" transwell assay that measures basal–apical transmigration of leukocytes and therefore more closely mimics the in vivo process.

Key words Transepithelial migration, Transwell, Transmigration, Inflammation, Epithelial adhesion, Image analysis

1 Introduction

Following initiation of an inflammatory response, immune cells such as neutrophils and monocytes must egress from the vasculature and migrate through the epithelium to reach the site of infection in order to mount an effective immune response. This process of transepithelial migration (TEpiM) is stimulated by cytokines (e.g., IL-8 [3], TNFα [4]) and other factors derived from epithelial cells under inflammatory conditions, or after infection by pathogens [5, 6]. Epithelial cells can also directly contribute to TEpiM through regulation of surface proteins controlling the adhesion of leukocytes and passage through the epithelial cell junction. For example, exposure of epithelial cells to pathogens or proinflammatory agents leads to altered protein expression such as ICAM-1 [7] that may facilitate inflammatory leukocyte transepithelial migration and signaling events such as phosphorylation of CAR (Coxsackievirus and Adenovirus Receptor) which promotes TEpiM in bronchiolar epithelial cells [4]. A chemotactic gradient may also play a role in promoting

Alexis Gautreau (ed.), *Cell Migration: Methods and Protocols*, Methods in Molecular Biology, vol. 1749, https://doi.org/10.1007/978-1-4939-7701-7_7, © Springer Science+Business Media, LLC 2018

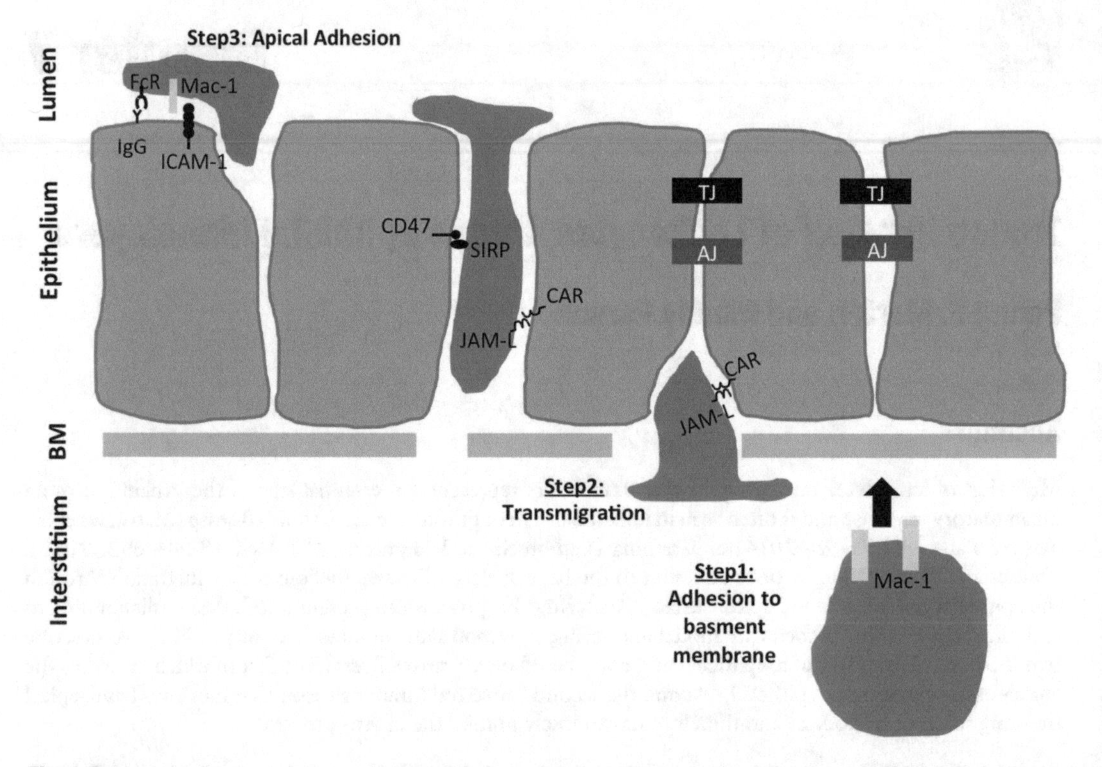

Fig. 1 Schematic diagram of steps of transepithelial migration of leukocytes. Step 1: Leukocytes adhere to the basal membrane of the epithelium, pass through the basement membrane (BM) using matrix metalloprotease mediated degradation, and adhere to epithelial proteoglycans via Mac-1. Step 2: Leukocytes migrate through intercellular spaces of cell–cell junctions. This step involves interaction of leukocytes with junction proteins expressed on the epithelial cells and dynamic changes to the behavior of the junction proteins and the cell–cell junction itself. Jam-L interaction with CAR, SIRP interaction with CD47 and homodimerization of JAM family members are all important in mediating migration of leukocytes through the intercellular space. Step 3: Postmigration events involving emergence of leukocytes from the epithelial monolayer and adhesion to the apical surface for varying periods of time. Retention of leukocytes at the apical cell surface is mediated by ICAM-1–Mac-1 and Ig–FcR (Fc receptor) interactions. *AJ* adherens junction, *BM* basement membrane, *TJ* tight junction

TEpiM but there are also known adhesion and signaling interactions that are essential for TEpiM to occur but are independent of chemotaxis.

Transepithelial migration of leukocytes consists of three sequential steps (reviewed in [8]) which are summarized in Fig. 1. Step 1 is the adhesion of leukocytes to the basal membrane of the epithelium which is general comprised of basement membrane extracellular matrix proteins (BM). Step 2 is the migration of leukocytes through intercellular spaces or cell–cell junctions. This involves interaction of leukocytes with cell–cell junction proteins expressed on the epithelial cells and dynamic changes to the behavior of the junction proteins and the epithelial cell cytoskeleton. Step 3 encompasses the postmigration events, which involves emergence of leukocytes from the epithelium and adhesion to the apical surface for varying periods of time. Although some of

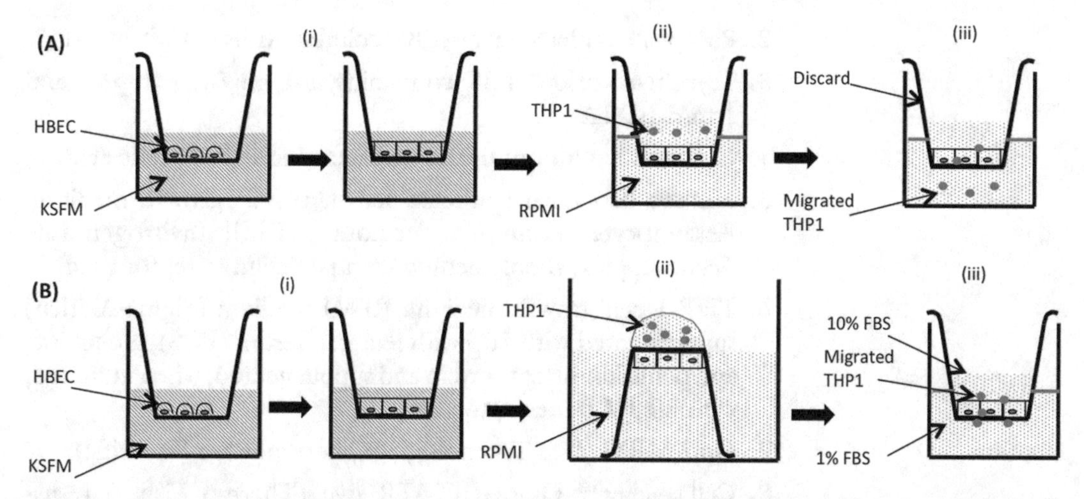

Fig. 2 Comparison of a traditional and inverted transmigration assays. (**a**) Traditional assay (i) epithelial cells are grown on top of the Transwell filter for 3 days to establish a monolayer, (ii) labeled leukocytes are added to the top well of the Transwell and allowed to adhere (iii) after 24 h the number of migrated leukocytes present in the lower chamber are counted using FACS. (**b**) Inverted assay (i) epithelial cells are grown on top of the Transwell filter for 3 days to establish a monolayer, (ii) the Transwell is inverted and leukocytes labeled with cell tracker dye are added to the underside of the Transwell in a bubble. (ii) after 1 h the leukocytes had adhered and the Transwell can be returned to the upright position. (iii) after 8 h the transwells are fixed and transmigrated cells are visualized using confocal microscopy

the events and components of transepithelial migration are now known, the cell surface adhesion molecules and molecular mechanisms by which leukocytes migrate through epithelial layers remain poorly understood.

Analysis of this process requires assays that mimic the epithelial environment and allow passage and quantification of migrated leukocytes across an intact monolayer of cells. Traditional transmigration assays involve applying leukocytes to the apical surface of epithelial or endothelial cells grown on a filter and then measuring the number of transmigrated leukocytes some time later (Fig. 2). Since leukocytes generally adhere firstly to the basal surface of epithelial cells the ideal assay involves inversion of the epithelial cells to allow adhesion of leukocytes to the basal surface and analysis of migrated leukocytes toward the apical surface/space. This "inverted" transmigration assay is technically more difficult but more closely mimics the in vivo progression of leukocytes through epithelial monolayers (Fig. 2). Here we describe the methods used to measure both apical–basal transmigration in a traditional transwell system and basal–apical migration in an inverted TEM system.

2 Materials

1. Plasticware: Transwell inserts with 8 μm pore polyester filter (*see* **Note 1** on choice of filter material and pore size).

2. Phosphate buffered saline (PBS) solution without Ca^{2+} and Mg^{2+},
3. Trypsin solution: PBS containing 20 mg/mL trypsin and 1 mM EDTA.
4. Collagen I solution: 0.01% Collagen I in 0.2 N acetic acid.
5. Human bronchial epithelial cells (HBEC) growth medium: Keratinocyte serum free medium (KFSM; Invitrogen Life Technologies) supplemented with penicillin/streptomycin.
6. THP-1 cell growth medium: RPMI medium (Sigma-Aldrich) supplemented with 10% with fetal calf serum (FCS), glutamine, and penicillin–streptomycin and supplemented, when indicated, with 50 μM β-mercaptoethanol.
7. Opti-MEM (Life Technologies/ThermoFisher Scientific).
8. CellTracker™ Orange CMTR dye (Thermo Fisher) resuspended in DMSO as per manufacturer's instructions.
9. Autoclaved glass dishes or sterile plastic ware with a clearance depth of at least 22 mm to accommodate the inverted transwells without compromising the underside of the filter once inverted.
10. Formaldehyde 37% solution.
11. PBS.
12. FACS FLOW (BD Biosciences).
13. Phalloidin-488 for the inverted assay.
14. Bovine serum albumin (BSA).

3 Methods

3.1 Traditional Transmigration Transwell Assay

The following method describes the analysis of apical to basal transepithelial migration of THP1 cells through HBEC monolayers, in the presence or absence of an inhibitor to an epithelial receptor. Since the experiment is not intended for analysis of THP1 behavior but rather the ability of the epithelial cells to facilitate transepithelial migration of the THP1 cells, no chemotactic gradient is used here. Because of the lack of specific chemotactic stimuli, not all THP1 cells will adhere and migrate, so this is compensated for in the number of THP1 cells used per Transwell insert.

3.1.1 Day 1

1. Precoat the required number of transwell inserts with Collagen I solution for 1 h, remove collagen solution and wash once with KSFM.
2. Place the coated transwell inserts into 12-well plate and add 500 μL KSFM to the lower chamber.
3. Seed 100,000 HBEC in 200 μL KSFM into the upper chamber of three transwells per condition (*see* **Note 2**). Take care to

seed the cells evenly to create an even monolayer (*see* **Note 2**). Leave one coated transwell without HBEC cells to use as a positive control.

3.1.2 Day 3

1. Carefully replace the media in the upper and lower chambers of the Transwell inserts taking care not to disturb the epithelial monolayers. As the filters in this assay are polyester rather than polycarbonate, the outlines of the epithelial cells on the filter should be visible using transmitted light on an inverted light microscope. The monolayer of epithelial cells should appear as a continuous carpet of well-adhered cells with no visible gaps. The cells are then left for a further 24 h to ensure there are no gaps in the monolayer and that intact cell–cell junctions have been formed uniformly across the entire filter.

3.1.3 Day 4

1. Replace the media in the lower chamber of the transwell inserts with 500 μL of prewarmed RPMI (without β-mercaptoethanol, *see* **Note 3**) and the upper chamber with 150 μL of RPMI (without β-mercaptoethanol). It is essential not to disrupt the epithelial monolayer and all inserts should be checked at this point for damage to the monolayer. Any inserts with incomplete monolayers should be discarded. Also add media to the upper and lower chambers of the Transwell insert reserved for the positive control. Return the transwell inserts to a 37 °C incubator for 2 h.
2. Using a hemocytometer, count the THP1 cells and aliquot sufficient cells into a falcon tube to allow for ~100,000 THP1 per condition. Spin at 2000 rpm (approx. 500 × *g*) to pellet the cells without damaging them and resuspend in 10 mL prewarmed Opti-MEM.
3. Prepare a 10 mM CellTracker Orange solution in DMSO then add CellTracker Orange solution to the THP1 cells in Opti-MEM to a final concentration of 1 μM. Mix gently by inverting 2–3 times and incubate at 37 °C for 30 min.
4. Centrifuge the cells at 2000 rpm (approx. 500 × *g*), remove supernatant and resuspend in RPMI without β-mercaptoethanol to wash and repeat v wash step three times. It is important to remove any trace of the CellTracker dye to prevent staining of epithelial cells. Then return the THP1 in RPMI to 37 °C for a further 30 min keeping a small amount in reserve to check the cell concentration again.
5. After this final incubation step, adjust the cell concentration to 50,000 cells per 100 μL media and add 100 μL to each Transwell insert directly to the top well, including the positive control. Return the plate to the 37 °C incubator for 24 h.

3.1.4 Day 5

1. Carefully remove the transwell inserts from the plate and discard. Collect the media from the wells and transfer directly into FACS tubes ensuring that no THP1 cells are left in the plate.
2. Transfer samples to the FACS machine on ice and count only the cells that are labeled with the cell tracker dye. This should ensure that no contaminating epithelial cells are included in the final analysis of transmigrated leukocytes. Alternative methods to assess the number of transmigrated cells at this stage are described in **Note 4**. As calibrations and measurement settings differ for different FACS analysis instruments it is not possible to give a detailed description of the FACS analysis here. FACS analysis should be performed according to the manufacturer's instructions. Local advice should be sought for calibration and experimental setup for the wavelength of dye used here as well as subsequent gating to retrieve transmigrated cell numbers.

3.2 Inverted Transmigration Transwell Assay

The protocol for this method is identical to the Traditional Transmigration assay protocol up to day 4.

3.2.1 Day 4

1. As described in the Traditional Transwell Assay protocol, replace the media on the transwells with RPMI without β-Mercaptoethanol and incubate for 2 h.
2. During this time label the THP1 with CellTracker Dye as described above. From this point on, the method for the inverted assay differs to that for the traditional one.
3. In order to assess basal–apical migration, the labeled THP1 need to be applied to the underside of the transwell filter. To do this, the Transwell insert needs to be inverted in sterile glass or plastic dishes with enough height that the THP1 added to the underside of the filter will not be compromised once the lid is replaced. This can be estimated by placing a new unused transwell insert inverted into a suitable 6 or 12-well tissue culture plate, adding the well plate lid and checking the distance remaining between the top of the transwell insert and the base of the lid.
4. Carefully remove each insert from the plate, invert and place into the new sterile well or plate. Add RPMI to the well such that the epithelial cells are still in contact with the media but the "underside" of the filter remains exposed.
5. Add 100 μL RPMI containing 50,000 THP1 to the "underside of the filter" (i.e., the top of the insert facing up) so that the solution containing cells is maintained on the underside of the filter as a bubble by surface tension (*see* Fig. 2). If the surface tension bubble breaks the filter can be gently washed with media and the THP1 can be readded.

   The THP1 cells will adhere at room temperature in the sterile laminar flow hood, however it is preferable to return the transwells to a 37 °C for 60 min to allow adhesion to proceed. After 30 min of incubation, the transwell inserts should be checked and extra media can be carefully added if required to prevent drying due to evaporation or fluid exchange through the filter.

6. Whilst the THP1 cells are adhering, replace the RPMI containing 10% FBS in the original Transwell plates to RPMI containing 1% FBS.
7. After 60 min, the THP1 will have adhered to the filters. Reinvert the Transwell inserts and place all except the "adhesion control" inserts back into the wells of the original plate (now containing RPMI + 1%FBS). Add 200 μL of RPMI containing 10% FBS to the top chamber of the Transwell and return the plate containing the Transwell inserts to the 37 °C incubator for 8 h.
8. Adhesion control inserts can be fixed using 4% formaldehyde for 30 min directly after the 60 min adhesion step. Then wash the inserts three times in PBS and store in PBS overnight at 4 °C.
9. After the 8 h incubation period, add formaldehyde directly to the upper and lower wells of the Transwell inserts at a final concentration of 4% and leave for 30 min. Alternatively, the inserts can be removed, washed with PBS then fix in 4% PFA, however this is more likely to result in loss of adhered THP1 or disruption of the epithelial architecture. Postfixation, the Transwell inserts can be washed three times with PBS and kept at 4 °C overnight.

3.2.2 Day 5

To visualize the epithelial cells on the inserts, fluorescently labeled phalloidin can be used to stain F-actin. This provides a useful guide as to the monolayer integrity and cell health. Note that any THP1 cells present on the filters will also be stained with phalloidin, but as they are phenotypically distinct the two cell populations are easily identified.

1. Wash the inserts three time with PBS and permeabilize the cells with 0.02% Triton-PBS before a further three washes with PBS and blocking in 3% BSA-PBS for 30 min at room temperature.
2. Remove blocking solution and incubate with 1:400 dilution of phalloidin-Alexa 488 diluted in 3% BSA-PBS for 60 min.
3. Wash filters gently and cover completely in PBS. The inserts can be kept in PBS for several days before imaging if necessary.

3.3 Confocal Microscopy Analysis of Transwell Inserts

1. Transwell inserts can be imaged directly using an inverted Confocal Microscope by placing the Transwell inserts into plastic or glass-bottomed dishes with a small amount of PBS to prevent drying (*see* **Note 5** for alternative analysis methods). Using a 40× air objective lens, acquire 5×5 tilescans using 488 nm and 561 nm excitation (or equivalent wavelengths suitable to excite

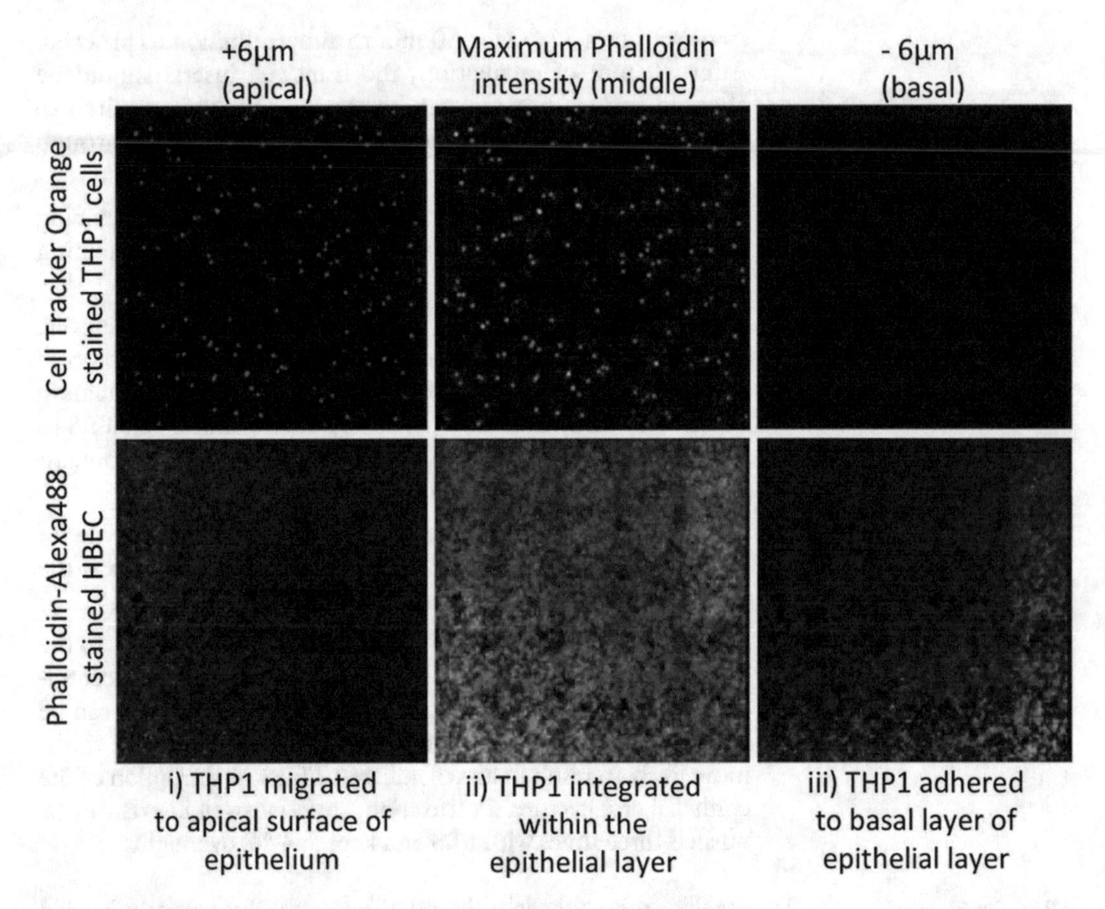

Fig. 3 Example images of inverted transwell experiments. (**a**) Example reconstructed 5×5 tilescan confocal images of CellTracker Orange stained THP1 adhered to or migrating through HBEC cells (top panels) and HBEC cells counterstained with Phalloidin (bottom panels). Images show THP1 adhered to the apical surface of HBEC (I, far left panels), integrated within the HBEC monolayer (ii, central panels) or adhered to the basal surface of the HBEC monolayer (iii, far right panels)

phalloidin-Alexa 488 and CellTracker Orange dyes) at three different focal planes (*see* Fig. 3): Middle: this can be defined by the focal plane where the phalloidin stain is maximal and is the epithelial cell monolayer. Basal and apical: this should be the focal plane where THP1 cells attached to the basal and apical membranes of the epithelial cells are visible. A distance of 6 μM above and below the maximal phalloidin signal (Middle plane) is optimal for HBEC. However for other epithelial cells lines the optimal depth for identification of adhered THP1 cells may need to be characterized.

2. The "adhesion control" transwell inserts should also be imaged in the same way however a single focal plane is required which corresponds to the location of the THP1 attached to the basal surface of the epithelial cells.

3.4 Analysis of Transmigration Images

1. In order to analyze the number of THP1 cells within each focal plane, the images from each tile scan should be recombined into one larger 5×5 image. This can then be imported into ImageJ software as a tiff file for further analysis. Figure 4 shows the resulting images and data from an example data set to illustrate the following steps.
2. Before starting go to "Analyse>Set measurement" and ensure that Area, mean gray value and integrated density are selected. This will ensure that this analysis will result in the correct information.
3. Firstly, split the channels to produce separate green and red images (phalloidin and CellTracker Orange respectively). Close the green channel images as only the red images are required for analysis steps from now on.
4. The next step is to locate the THP1. To do this, first combine the red images from all three planes using "Process>Image>colour>merge channels" (Fig. 4b). Then convert the image to 8bit using "Image>type>8bit" and adjust the threshold using "Image>adjust threshold" making sure that all THP1 cells are visible in the resulting binary image.
5. To locate the THP1 as individual objects, use "Analyse>Analyse Particles" making sure that the size is adjusted to eliminate objects smaller than THP1, and that "display results," "add to manager," and "show outlines" are selected in the "Analyse particles" dialog window (Fig. 4c). The ROI manager now contains a list of ROIs that correspond to the area occupied by THP1 cells in any Z plane (Fig. 4d).
6. These ROIs can now be applied to the red images from individual Z planes to analyze the Z location of each THP1 cell (Fig. 4e). In order to do this, select the red image from each Z plane 1 at a time and click measure within the ROI manager ensuring that "show all" and "labels" are selected so that the outlines of the ROIs will be visible on the image being analyzed (*see* Fig. 4e). The results window now contains a numbered list of the ROIs as well as Integrated Density for each ROI (Fig. 4f). This list can be saved as a .txt file and opened in Excel for data processing.

3.5 Processing of Analyzed Transmigration Data

1. In Microsoft Excel (or similar) open the .txt files for all three z planes and paste into a single data sheet with the ROI numbers in column A, and the corresponding Integrated density values from each Z plane in columns B to D.
2. If using Excel it is then possible to find the plane where each THP1 has the brightest signal but using the IF function. For example in column E type the following:
=IF((AND(B2>C2,B2>D2)), "1", "0").

This places a 1 in the column corresponding to Z plane where the THP1 signal is most intense. These columns can then be summed to provide numbers of THP1 cells in each Z position.

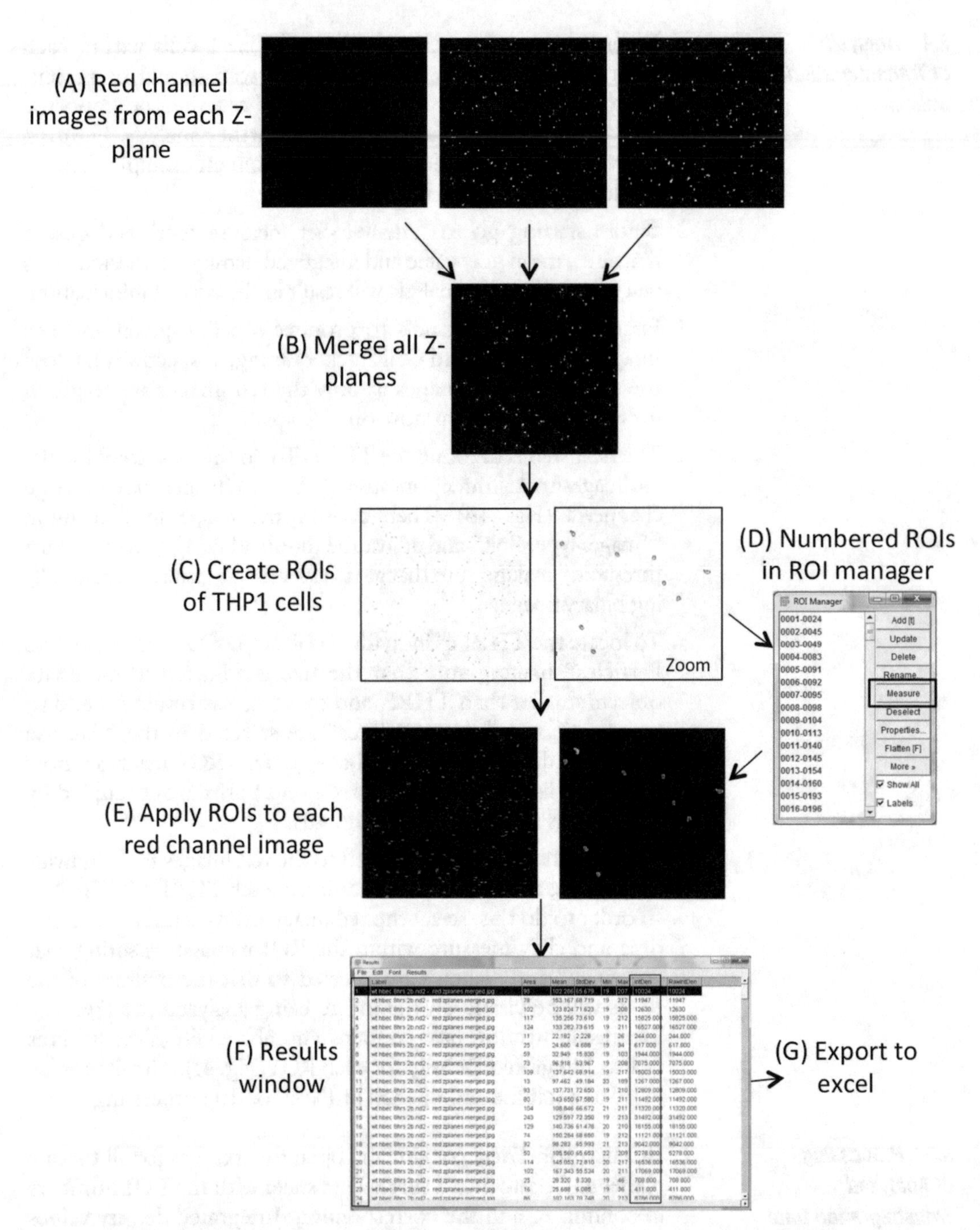

Fig. 4 Analysis of Transmigration data in ImageJ/FIJI. (**a**) Example reconstructed 5×5 tilescans of CellTracker Orange stained THP1 corresponding to above (left), within (middle) and below (right) the epithelial layer. (**b**) A merge of the three red images in **a**. (**c**) Drawing of ROIs created from "Analyse particles" in ImageJ: Full 5×5 tilescan (left) and a zoomed in area (right). (**d**) The ROI manager window showing the numbered ROIs and highlighting the measure button. (**e**) The numbered RIOs reapplied to one of the single Z planes (left) and showing the zoomed in area corresponding to the same area as in **c**. (**f**) An example results window showing the Integrated Density value corresponding to the numbered ROIs

Data can then be presented as a percentage of THP1 that have fully transmigrated, or integrated and transmigrated, as dictated by the needs of the experiment.

3.6 Analyzing Adhesion Images

Tile-scans obtained from the "Adhesion control" transwell inserts only contain a single plane and so can be analyzed more simply. Firstly open in Image J, convert the image to 8 bit and obtain the cell number in the image by first thresholding and using "Analyse Particles" function as previously described. Only the number of cells per field is needed here and if necessary can be used to normalize the Transmigration data to eliminate the effect of any differences in adhesion.

Although a total number of adhered THP1 can also be obtained by summing the THP1 within each focal place, it is also important to look at variation in initial adhesion using these "Adhesion Controls."

4 Notes

1. Transwell inserts are available in a range of pore sizes and materials. It is important to choose the best pore size for your experiment. In the experiments described here we have used 8 μM pore filters. Most primary leukocytes and leukocyte cell lines will migrate efficiently through filters with at least a 5 μM pre size and some more migratory epithelial cells will not form stable monolayers on filters as large as 8 μM. Therefore it may be preferable to choose filters with 5 μM rather than 8 μM pores, depending on the cells used. However, it should be noted that in order to visualize the epithelial cells prior to commencing the transmigration experiment, it is important to use polyester rather than polycarbonate filters and these are not available with a 5 μM pore size.
2. The seeding density of the epithelial cells must be optimized for each cell line. For HBEC, 100,000 cells grown for 3 days allows formation of an intact monolayer with no visible holes but does not lead to the formation of multiple differentiated layers that would impede migration of leukocytes through the epithelial layer. It is important to seed the epithelial cells evenly. For this reason the cells must be prediluted to the correct cell concentration and added dropwise over the filter to distribute the cells as evenly as possible. Once the cells are attached (after approximately 1 h), giving the plates a light tap will help to redistribute any non-adhered or loosely adherent cells into unoccupied areas of filter.
3. THP1 cells are routinely cultured in media containing β-mercaptoethanol to prevent differentiation. However, β-mercaptoethanol is toxic for HBEC, so coculture experiments are performed in RPMI without β-mercaptoethanol. Depending on the combination of cells to be used, the media for the trans-

migration coculture steps for each protocol may need to be optimized to ensure optimal health of both cell populations.

4. Alternative measurements to calculate numbers of transmigrated cells: it is possible to count the THP1 cells using a hemocytometer following transmigration into media. However, low numbers of cells can often lead to inaccurate counting using this method. As the THP1 cells have been labeled with a fluorescent dye, another alternative is to use a fluorescent plate reader to measure total fluorescence per well, rather than number of THP1 cells. However, this method does not discriminate between intact and dead cells, which the FACs method allows for.
5. An alternative approach of seeding epithelial cells on underside of transwells, growing to confluence and then adding leukocytes to top is possible. However a number of epithelial cell lines may not grow well under these conditions so the above described method is more likely to achieve an intact healthy epithelial monolayer.

References

1. Matthews JD, Weight CM, Parkos CA (2014) Leukocyte-epithelial interactions and mucosal homeostasis. Toxicol Pathol 42(1):91–98
2. Lambrecht BN, Hammad H (2012) The airway epithelium in asthma. Nat Med 18(5):684–692
3. Godaly G, Proudfoot AE, Offord RE, Svanborg C, Agace WW (1997) Role of epithelial interleukin-8 (IL-8) and neutrophil IL-8 receptor A in Escherichia coli-induced transuroepithelial neutrophil migration. Infect Immun 65(8): 3451–3456
4. Morton PE, Hicks A, Ortiz-Zapater E, Raghavan S, Pike R, Noble A, Woodfin A, Jenkins G, Rayner E, Santis G, Parsons M (2016) TNFalpha promotes CAR-dependent migration of leukocytes across epithelial monolayers. Sci Rep 6:26321
5. Kohler H, Sakaguchi T, Hurley BP, Kase BA, Reinecker HC, McCormick BA (2007) Salmonella enterica serovar Typhimurium regulates intercellular junction proteins and facilitates transepithelial neutrophil and bacterial passage. Am J Physiol Gastrointest Liver Physiol 293(1):G178–G187
6. Boll EJ, Struve C, Sander A, Demma Z, Krogfelt KA, McCormick BA (2012) Enteroaggregative Escherichia coli promotes transepithelial migration of neutrophils through a conserved 12-lipoxygenase pathway. Cell Microbiol 14(1):120–132
7. Langlois C, Gendron FP (2009) Promoting MPhi transepithelial migration by stimulating the epithelial cell P2Y(2) receptor. Eur J Immunol 39(10):2895–2905
8. Zemans RL, Colgan SP, Downey GP (2009) Transepithelial migration of neutrophils: mechanisms and implications for acute lung injury. Am J Respir Cell Mol Biol 40(5):519–535

Check for updates

Chapter 8

Evaluation of Tumor Cell Invasiveness In Vivo: The Chick Chorioallantoic Membrane Assay

Selma Maacha and Simon Saule

Abstract

Metastases is largely responsible for the mortality among cancer patients. Metastasis formation is a complex multistep process, which results from the propagation of cancer cells from the primary tumor to distant sites of the body. Research on cancer metastasis aims to understand the mechanisms involved in the spread of cancer cells through the development of in vivo assays that assess cell invasion. Here we describe the use of the chick chorioallantoic membrane to evaluate cancer cell invasiveness in vivo. The chick chorioallantoic membrane assay is based on the detection and quantification of disseminated human tumor cells in the chick embryo femurs by real-time PCR amplification of human *Alu* sequences.

Key words Metastasis, Migration, Invasion, Extravasation, Chorioallantoic membrane

1 Introduction

Cancer metastasis results from the dissemination of cancer cells from the primary tumors to distant sites [1]. Metastasis occurs through a series of steps, in which tumor cells first migrate from the primary tumor and locally invade the surrounding tissue, intravasate into nearby blood and lymphatic vessels, transit through the circulatory system, arrest in the capillary beds of distant sites and finally extravasate these capillaries in order to colonize distant organs [2].

Metastasis is the most deadly aspect of cancer and its inhibition is one of the most important issues in cancer research. An understanding of the mechanisms involved in tumor cell invasion may lead to limited tumor progression and, as a result, to a significant reduction in mortality for cancer patients. Thus, focusing research efforts on developing in vivo assays that assess cell invasion is crucial in cancer metastasis research.

Here we describe the use of a quantitative method to evaluate cancer cell invasiveness through the chorioallantoic membrane (CAM) of chick embryos by quantifying in vivo rates of cell inva-

Alexis Gautreau (ed.), *Cell Migration: Methods and Protocols*, Methods in Molecular Biology, vol. 1749, https://doi.org/10.1007/978-1-4939-7701-7_8,

siveness into the embryo femurs based on the detection of human *Alu* sequences by quantitative real-time PCR (Fig. 1) [3]. Several studies have established the chick embryo CAM assay as a useful model for the investigation of tumor cell invasion [4–10]. This model offers many advantages including the accessibility of the CAM, a well-vascularized extraembryonic tissue that surrounds the developing embryo. The histology as well as the physiology of the CAM allow a comprehensive recapitulation of all the steps in the metastatic cascade such as migration, invasion, angiogenesis, and colonization of distant organs. Moreover, the chick embryo is a naturally immunodeficient host that supports inoculation with a wide variety of tumor cells from various tissues making the model a powerful tool for the study of spontaneous metastasis. Finally, the chick embryo CAM assay represents a rapid and economical technique reasonably powered to readily achieve statistical significance in the evaluation of cancer cell invasiveness between groups.

Lastly, it is important to consider that the CAM assay is suitable for invasiveness assessments that do not require the observation of tumors in distant organs of the embryo, as tumor cells cannot produce macroscopic visible colonies in secondary organs due to the short postinoculation observation time (8 days).

2 Materials

2.1 Egg Preparation and Dissection

1. Egg incubator (Savimat, MG100).
2. Fertilized White Leghorn eggs.
3. Egg tray.
4. dH_2O.
5. Binocular zoom microscope with a light source.
6. Sterile 10-cm petri dishes
7. 70% ethanol (in dH_2O).
8. 18-Gauge needles.
9. 50 mL-syringe fitted with a piece of 0.125 cm Nalgene™ tubing (inner Ø).
10. Fine forceps.
11. Dissection scissors.
12. Mobcap.
13. Face mask.

2.2 Preparation of Cancer Cells

1. Tissue culture laminar flow hood.
2. Cancer cell lines of interest.
3. Complete culture medium specific to the cell lines to be used.
4. Serum free media specific to the cell lines to be used.

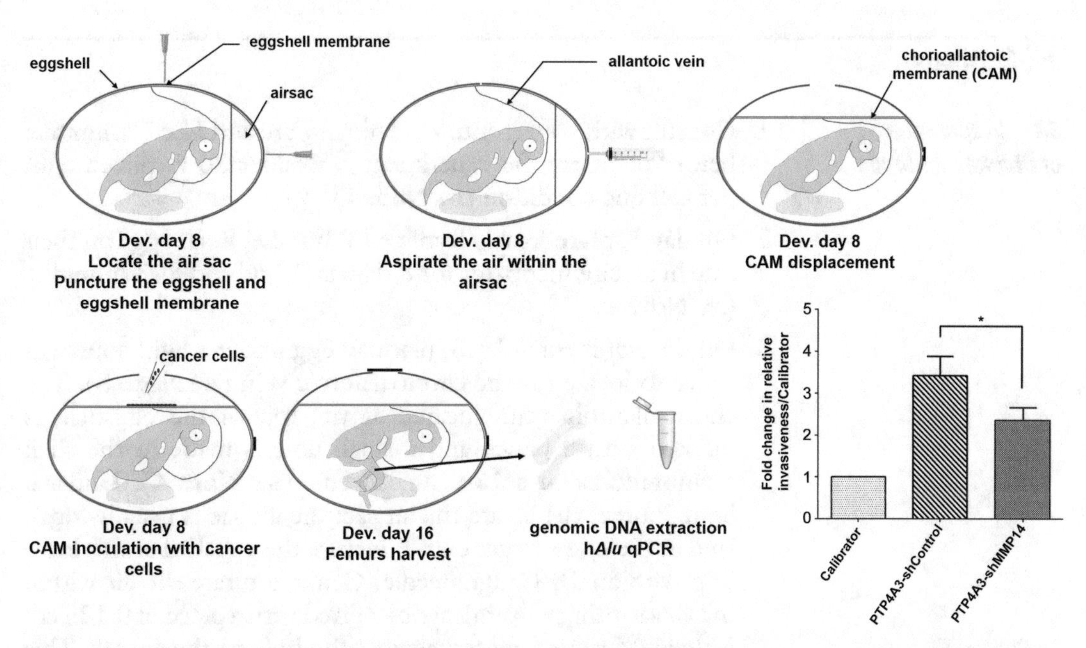

Fig. 1 Illustration of the CAM assay procedure. Fertilized eggs were incubated for 8 days before dropping the CAM underneath the eggshell membrane by aspirating the air within the air sac. Human cancer cells are then inoculated onto the CAM and allowed to disseminate for 8 additional days. On developmental day 16, the femurs of the chick embryos are harvested and genomic DNA is extracted. The presence of human tumor cells is assessed by real-time amplification of the human *Alu* sequences. Representative results obtained from 14 chick embryos per condition using the uveal melanoma cell line OCM-1 expressing PTP4A3-shControl or –shMMP14 are shown as fold change in relative invasiveness over the calibrator [3]

5. Dulbecco's PBS.
6. 0.05% trypsin–EDTA.
7. Tissue culture supplies (T-75 flasks, Falcon tubes, Pasteur pipets, serological pipets, …).
8. Tabletop centrifuge.
9. Hemocytometer or automated cell-counting machine.

2.3 Genomic DNA Extraction and Real Time PCR

1. Nucleospin® Tissue (Macherey-Nagel).
2. Eppendorf tubes.
3. UltraPure™ DNase/RNase-Free Distilled Water.
4. Spectrophotometer.
5. SYBR Green PCR Master Mix (2×).
6. PCR microplates.
7. Real time PCR thermocycler.
8. Primers: h*Alu*-sense: *5′* CACCTGTAATCCCAGCACTTT *3′*, h*Alu*-antisense: *5′* CCCAG-GCTGGAGTGCAGT *3′*, chGAPDH-sense: *5′* GAGGAAAGGTCGCCTGGTGGATCG *3′*, chGAPDH-antisense: *5′* GGTGAGGACAAGCAGTG AGGAACG *3′*.

3 Methods

3.1 Preparation of Chicken Embryos

1. Consult with your institute's Animal Care and Use Committee before planning the experiment. Consider 30 fertilized eggs per cell line condition (*see* **Note 1**).
2. On day 1, place freshly fertilized White Leghorn eggs on their side in an egg incubator for 8 days at 38 °C and 80% humidity (*see* **Note 2**).
3. On developmental day 8, place an egg under a light source in order to locate (a) the chorioallantoic vein (*see* **Note 3**). The chorioallantoic vein, located at the top of the eggshell, is marked with a pencil and a small hole is made in the shell membrane using a 18-Gauge needle (*see* **Note 4**). Under a light source, (b) locate the air sac; the air sac is usually localized at the egg's larger end. Puncture the eggshell at this location with an 18-Gauge needle. Gently aspirate the air within the air sac using a 50-mL syringe fitted with a piece of 0.125 cm Nalgene™ tubing placed against the hole in the air sac. This suction will cause the CAM to drop away from the eggshell creating an air pocket underneath (*see* **Note 5**). Seal both holes with laboratory tape and return the eggs to the incubator; immediately proceed with cancer cells preparation for the inoculation.

3.2 Preparation of Cancer Cells for Inoculation

1. Carry out cancer cell culture under sterile conditions. Perform experiments with cancer cells seeded in T-75 flasks for 24–48 h and grown to ~80–90% confluency.
2. Immediately after preparation of the embryos for inoculation, wash cancer cells with 5 mL of D-PBS. Aspirate D-PBS and detach cells using 2 mL of 0.05% trypsin–EDTA for 5–10 min at 37 °C.
3. Inactivate the trypsin solution by adding 8 mL of fresh complete culture media. Collect all cells and transfer them to a sterile falcon tube.
4. Centrifuge the cells at 300 × *g* for 5 min at room temperature. After centrifugation, discard the supernatant and gently wash cells in 5 mL of serum free media to remove any residual media, trypsin, or EDTA.
5. Centrifuge the cells at 300 × *g* for 5 min at room temperature. After centrifugation, discard the supernatant and gently resuspend cells in 1 mL of serum free media.
6. Count the cells using a Hemocytometer or automated cell-counting machine in order to determine the concentration of cells. Dilute cells to $2–4 \times 10^6$ cells/mL ($1–2 \times 10^5$ cells/50 μL) with serum-free media.

3.3 Inoculation of Cancer Cells onto the Chorio-allantoic Membrane

1. Remove the laboratory tape sealing from the hole previously made on top of the CAM. Draw a square box of 0.5 cm and carefully cut the eggshell in order to expose the underlying CAM.
2. Inoculate 50 μL of the cell suspension containing 1–2 × 10^5 cells onto the CAM vein of each embryo (*see* **Note 6**).
3. Reseal the previously cut window with laboratory tape and return the eggs to the incubator until developmental day 16.

3.4 Harvesting the Chick Femurs

The CAM assay relies on the detection of human DNA in chick femurs by PCR, it is therefore critical for the reliability of the experiment to prevent contamination with exogenous DNA. It is strongly recommended to wear a mobcap and a face mask. Dissections should be performed in a dedicated area thoroughly cleaned prior to femurs harvesting. Moreover, dissection surgical tools should be autoclaved before use and thoroughly cleaned between embryos in distilled water followed by 95% ethanol and flame sterilizing.

1. On developmental day 16, spray down the eggs with 95% ethanol to anesthetize the embryos.
2. Open the egg with surgical scissors around the CAM window big enough to retrieve the embryo and place the latter in a petri dish with 10 mL of D-PBS; sacrifice the embryo by decapitation.
3. To harvest the femurs, remove the skin from the embryo legs. Use a scalpel to cut cartilage between femur and tibia, cut away muscle from femur and around hip joint. Keep femoral head in between forceps and clean the femur by removing the muscle. Proceed in D-PBS.
4. Place the femurs in a clean Eppendorf tube, cut them in small pieces and immediately store the tubes at −80 °C.

3.5 Femoral Genomic DNA Extraction and Real Time PCR Quantification of Human Alu Sequences

Genomic DNA isolation is carried out using the Nucleospin® Tissue kit from Macherey-Nagel as we have noticed inhibition of PCR reactions when using a standard phenol-chloroform isolation procedure. In order to avoid contamination with exogenous human DNA, prepare the PCR reaction under a PCR workstation.

1. Thaw the femurs at room temperature and proceed with genomic DNA extraction using Nucleospin® Tissue kit from Macherey-Nagel according to the manufacturer's instructions (*see* **Note 7**).
2. Solubilize the DNA for 20 min at 60 °C and store the tubes at −20 °C until later use or proceed with next step.
3. Determine the concentration of the isolated genomic DNA. Store the DNA at −20 °C until later use or proceed with next step.

4. Assess the presence of human cells in the chick femurs by real time PCR of the human *Alu* sequences: Perform PCR with a SYBR Green PCR Master Mix in the presence of 100 ng of chick genomic DNA and 0.6 μM of each human *Alu* primer in a final volume of 25 μL under the following conditions: 95 °C for 3 min, followed by 30 cycles of 95 °C for 30 s, 65 °C for 30 s and 72 °C for 30 s using a real time thermocycler (*see* **Note 8**). Chick genomic DNA is used as an internal control and is amplified using chGAPDH primers under the same conditions described for the human *Alu* sequences.
5. Relative analysis of invasiveness is performed by comparing the sample of interest to a calibrator using the $2^{\Delta\Delta CT}$ method where $\Delta CT = CT_{Alu} - CT_{GAPDH}$ and $\Delta\Delta CT = \Delta CT_{Sample\ of\ interest} - \Delta CT_{Calibrator}$. The calibrator represents noninjected chick embryos and reflects the background of the experiment. Statistical significance is evaluated using the Mann–Whitney nonparametric test.

4 Notes

1. The yield is rarely 100% as embryo death may occur.
2. It is strongly recommended to immediately incubate the fertilized eggs upon receipt, however, if the experiment cannot be started on the day of receipt, place the fertilized eggs at 12 °C for up to 1 week. Do not wait more than 1 week as this will affect the survival of the embryos.
3. A viable chicken embryo should show numerous blood vessels located at the top of the eggshell when candled with a light source.
4. When making the hole in the eggshell membrane, be careful not to puncture the CAM underneath. If bleeding is noticed, the CAM is damaged and the survival of the embryo is likely compromised.
5. If the CAM has been successfully dropped away from the eggshell, the air pocket should be visible under a light source while the air sac should disappear.
6. The number of cells to be inoculated is cell-line-specific. A pilot experiment should be run in order to determine the optimal number of cells to be inoculated to see colonization of the chick embryo by cancer cells.
7. Incubate the femurs with proteinase K in T1 buffer overnight at 56 °C, vortex, transfer the supernatant to a clean Eppendorf

tube and proceed with the following steps of the manufacturer's protocol.

8. The optimal amount of template DNA to be used in order to obtain quantitative data should be optimized empirically.

References

1. Mehlen P, Puisieux A (2006) Metastasis: a question of life or death. Nat Rev Cancer 6:449–458. https://doi.org/10.1038/nrc1886
2. Hanahan D, Weinberg RA (2011) Hallmarks of cancer: the next generation. Cell 144:646–674. https://doi.org/10.1016/j.cell.2011.02.013
3. Maacha S, Anezo O, Foy M et al (2016) Protein tyrosine phosphatase 4A3 (PTP4A3) promotes human uveal melanoma aggressiveness through membrane accumulation of matrix metalloproteinase 14 (MMP14). Invest Ophthalmol Vis Sci 57:1982–1990. https://doi.org/10.1167/iovs.15-18780
4. Zijlstra A, Mellor R, Panzarella G et al (2002) A quantitative analysis of rate-limiting steps in the metastatic cascade using human-specific real-time polymerase chain reaction. Cancer Res 62:7083–7092
5. Palmer TD, Lewis J, Zijlstra A (2011) Quantitative analysis of cancer metastasis using an avian embryo model. J Vis Exp. https://doi.org/10.3791/2815
6. Kain KH, Miller JWI, Jones-Paris CR et al (2014) The chick embryo as an expanding experimental model for cancer and cardiovascular research. Dev Dyn 243:216–228. https://doi.org/10.1002/dvdy.24093
7. Kim Y, Williams KC, Gavin CT et al (2016) Quantification of cancer cell extravasation in vivo. Nat Protoc 11:937–948. https://doi.org/10.1038/nprot.2016.050
8. Lokman NA, Elder ASF, Ricciardelli C, Oehler MK (2012) Chick chorioallantoic membrane (CAM) assay as an in vivo model to study the effect of newly identified molecules on ovarian cancer invasion and metastasis. Int J Mol Sci 13:9959–9970. https://doi.org/10.3390/ijms13089959
9. Xiao X, Zhou X, Ming H et al (2015) Chick Chorioallantoic membrane assay: a 3D animal model for study of human nasopharyngeal carcinoma. PLoS One 10:e0130935. https://doi.org/10.1371/journal.pone.0130935
10. Laurent C, Valet F, Planque N et al (2011) High PTP4A3 phosphatase expression correlates with metastatic risk in uveal melanoma patients. Cancer Res 71:666–674. https://doi.org/10.1158/0008-5472.CAN-10-0605

Check for updates

Chapter 9

Analysis of Invasion Dynamics of Matrix-Embedded Cells in a Multisample Format

Marleen Van Troys, Paola Masuzzo, Lynn Huyck, Karima Bakkali, Davy Waterschoot, Lennart Martens, and Christophe Ampe

Abstract

In vitro tests of cancer cell invasion are the "first line" tools of preclinical researchers for screening the multitude of chemical compounds or cell perturbations that may aid in halting or treating cancer malignancy. In order to have predictive value or to contribute to designing personalized treatment regimes, these tests need to take into account the cancer cell environment and measure effects on invasion in sufficient detail. The in vitro invasion assays presented here are a trade-off between feasibility in a multisample format and mimicking the complexity of the tumor microenvironment. They allow testing multiple samples and conditions in parallel using 3D-matrix-embedded cells and deal with the heterogeneous behavior of an invading cell population in time. We describe the steps to take, the technical problems to tackle and useful software tools for the entire workflow: from the experimental setup to the quantification of the invasive capacity of the cells. The protocol is intended to guide researchers to standardize experimental set-ups and to annotate their invasion experiments in sufficient detail. In addition, it provides options for image processing and a solution for storage, visualization, quantitative analysis, and multisample comparison of acquired cell invasion data.

Key words Cancer, 3D migration, Spheroid, Live cell imaging, Cell tracking, CellMissy

1 Introduction

Cell migration within our bodies occurs with a purpose: neural crest cell migration is crucial during development, motile macrophages and cytotoxic T-cells are key players in immune defense, and the migration of fibroblasts underlies wound repair. Unfortunately, in the context of cancer, cells that are capable of invading, i.e., actively migrating into 3D tissue either as individual cells or collectively, become a challenging enemy [1]. Under the condition that in vitro cell invasion assays sufficiently mimic the physiological environment, they provide useful tools to understand fundamental aspects of the molecular basis of invasive behavior of tumor cells or to screen for strategies to halt invasion [2, 3].

Alexis Gautreau (ed.), *Cell Migration: Methods and Protocols*, Methods in Molecular Biology, vol. 1749,
https://doi.org/10.1007/978-1-4939-7701-7_9,

Working with cells embedded in a 3D extracellular matrix (ECM) and quantifying the heterogeneous dynamic behavior within a cancer cell population represent important steps toward physiological relevance [4, 5].

This chapter presents a detailed protocol on how these technically complex in vitro invasion assays and their subsequent analysis can be performed in a multisample format. It aims at providing practical guidelines for two frequently reported assays (Fig. 1a): a 3D cell-zone exclusion assay (3D-CEZ) and a multicellular tumor spheroid invasion assay (MCTS). These two assays can be monitored using a traditional microscope equipped for live-cell imaging or using automated Live Cell Imaging Systems, which have recently become available. In both assays, a bulk cell population is present at start [i.e., the confluent cell sheet in 3D-CEZ or the spheroid in MCTS (Fig. 1a)]. Depending on the tested cells and the experimental condition, the invasion recorded in time will be either that of this bulk population or that of individual cells escaping the bulk population (or both). The observed bulk cell invasion may rely on physical interactions between invading cells in the case of collective cell migration, or not [6]. The protocol focuses on phase-contrast time-lapse imaging since this approach demands the least cell manipulation and is also possible with primary cells (e.g., from patient tumors). It can, however, without major alteration, be applied to homogeneously fluorescent cell populations (e.g., stably expressing a fluorescently tagged protein of interest such as histone H2B [7] or stained prior to migration with live cell dye (e.g., Cell Tracker dyes)). At several places of the protocol, we explain how to exploit the possibilities offered by CellMissy (Cell Migration and Storage System), a free and open source software tool dedicated to cell migration, which we have recently developed [8, 9] (Fig. 1b).

Figure 1c shows an overview of the protocol. The experimental setup is created and stored ("CellMissy, Part I"), the multiwell plate is prepared for the assay of choice, the time-lapse imaging is run and the acquired images stored. Subsequently, the invasive process is analyzed using image processing software tools. The numerical output (raw data) is stored, visualized, and subjected to quality control ("CellMissy, Part II and III"). Finally, relevant invasion parameters (e.g., bulk cell invasion velocity, individual cell speed and directionality) are calculated, plotted and compared between the tested conditions using nonparametric statistics ("CellMissy, Part IV"). The protocol consequently also aims at promoting detailed annotations and standardization of cell invasion assays. This endeavor facilitates data reposition, sharing, and meta-analyses, which are emerging goals of the field [10] (*see* also https://cmso.science/).

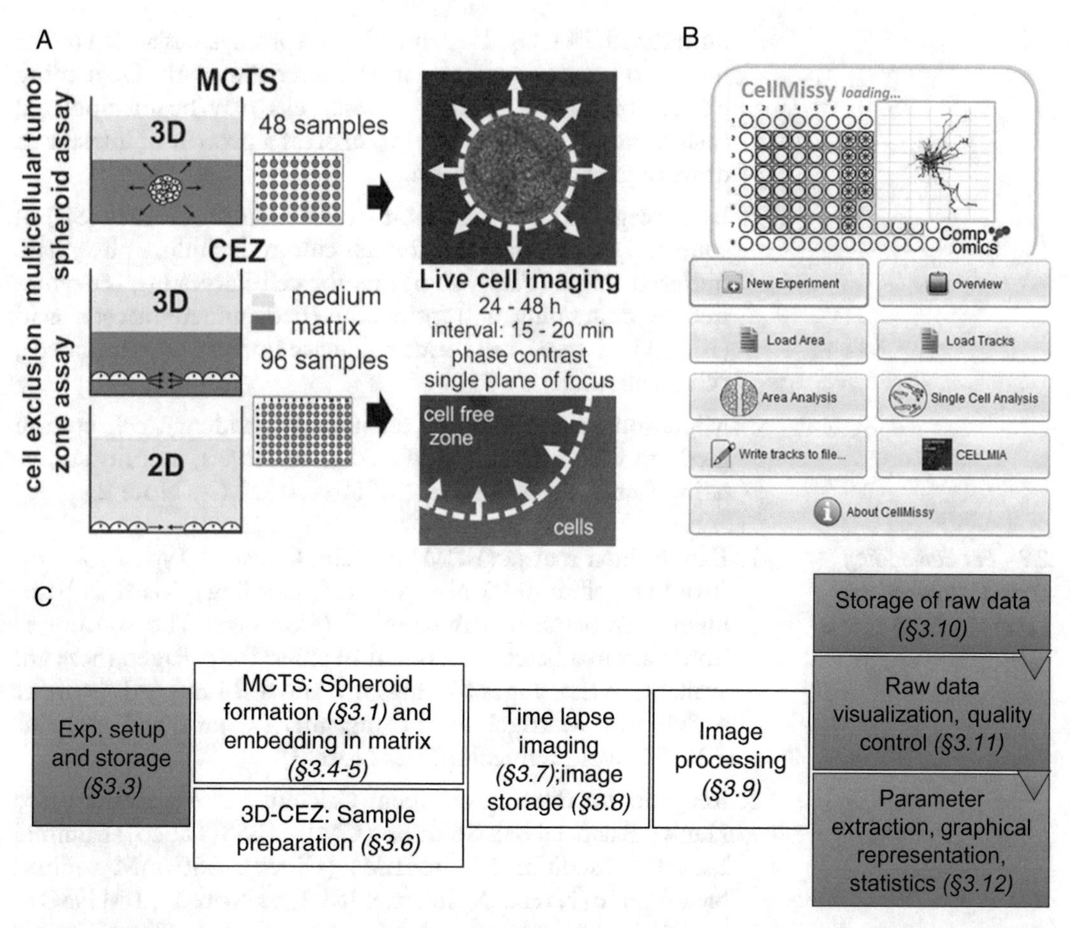

Fig. 1 Protocol overview. (**a**) Scheme of experimental setups of the multisample invasion assays. For MCTS, a tumor cell aggregate, a spheroid, is embedded in 3D–collagen gel or Matrigel. For 3D CEZ, the cells are present as a confluent cell sheet in between two ECM layers around a cell free zone (2D CEZ is shown for comparison). Top view images at time 0 for both assays are shown (for 3D–CEZ:1/4 of a well); arrows indicate the direction of invasion that will be monitored in time. (**b**) Logo and main window of the software CellMissy. (**c**) Scheme of the protocol with in gray the parts where CellMissy is used. For each step, there is a corresponding paragraph in the Methods section

2 Materials

2.1 For Cell Culture and Migration Assay Medium

1. Adherent cells of interest with corresponding serum containing cell medium: e.g., human breast adenocarcinoma fibroblast-like MDA-MB-231 (ATCC: HTB-26), human fibrosarcoma HT-1080 (ATCC: CCL-121) or human mammary adenocarcinoma epithelial-like MCF-7 (ATCC: HTB-22) cultured in Dulbecco's modified Eagle's medium supplemented with 10% fetal bovine serum (FBS), 20 mM L-GlutaMAX I. We maintain all cells in media at 37 °C in

humidified 5% CO_2. 1% penicillin–streptomycin is added to the media to prevent contaminating bacterial growth. Depending on the biological question, these cells may be manipulated (using standard procedures) to express a protein of interest or downregulate its expression.

2. Basic reagents (*see* **Note 1** on sterilization procedures (SP) of solutions) and equipment for cell culture including phosphate buffered saline (PBS), solutions for cell harvesting (enzyme-free or containing a trypsin–ethylenediaminetetraacetic acid (EDTA) mixture), cell counting device (manual or automated), cell incubator.
3. Migration assay medium (cell type dependent): cell culture medium with normal or lowered serum levels, motility stimulating factor, drug/inhibitor of interest, ... (*see* **Note 2**).

2.2 For Embedding Cells or Multicellular Spheroids in 3D Matrix

1. Extracellular matrix (ECM) protein: Collagen Type I solution (from rat tail, in 0.02 N acetic acid, Corning), Matrigel basement membrane matrix solution (Corning). The concentrations vary in a batch-dependent manner: for collagen these are available in the ranges 3–4 mg/mL and 8–11 mg/mL (store at 4 °C), for Matrigel ~8–11 mg/mL (aliquot and store at −20 °C, avoid refreezing) (*see* **Note 3**).
2. Reagents for diluting collagen: Calcium and Magnesium-free Hank's Balanced Salt Solution (CMF-HBSS, Gibco), Minimal Essential Medium 10× (MEM) (Gibco), 250 mM sodium bicarbonate ($NaHCO_3$) in water (SP2, *see* **Note 1**), 1 M NaOH in water (SP2), serum containing cell culture medium.
3. Materials for generating multicellular spheroids: 2% (w/v) agarose in water (SP1b (*see* **Note 1**), 200 mL in a 500 mL bottle (to liquefy before use: loosen cap of bottle and carefully solubilize in microwave, use short heating time, check complete melting by gentle swirling), sterile polydimethylsiloxane micromold or 3D Petri Dish® micromold (Microtissues, Sigma-Aldrich) (*see* **Note 4**), tweezers, spatula.
4. Oris™ Cell Migration Assay: includes Oris™ compatible 96-well plate, Oris™ Cell Seeding Stoppers (sets of four) and Oris stopper tool (Platypus) (*see* **Note 5**).
5. 6-, 12-, 24-, 48-, or 96-well tissue-culture treated multiwell plates (Nunc, Greiner, ...), sterile glass Pasteur micropipettes.
6. Universal heated stage at 37 °C (preferably with viewing window for use under upright microscope) and temperature control unit (e.g., from MTG – Medical Technology Vertriebs-GmbH), optional: upright microscope for use in laminar flow.

2.3 For Phase Contrast Imaging

A system suitable for live cell imaging and recording image sequences in time in phase contrast mode, equipped with temperature (set at 37 °C) and CO_2 (set at 5%) control. The environmental control is based on an incubator system (box incubator, stage top incubator, …). Alternatively, the system may operate inside a standard CO_2-incubator used for cell culture. Preferably also humidity is controlled. When working with cell populations that are (homogeneously) fluorescent, the described assays are compatible with epifluorescence imaging (*see* **Note 6**).

3 Methods

3.1 Preparing Multicellular Spheroids (See Note 7)

This step is performed 2–3 days before the invasion assay, since multicellular spheroid formation takes 48–72 h (depending on cell type, *see* below).

1. Work in a laminar flow hood. Place one or more sterile polydimethyl-siloxane micromolds in a sterile culture dish and pipet solubilized 2% (w/v) sterile agarose solution (~40–50 °C) into each micromold (~3 mL for mold with diameter 30 mm; when using commercial, smaller 3D Petri Dish® micromolds, adapt volume, *see* **Note 4**). Air bubbles trapped in the agarose in the area, where microwells will be formed, should be removed using a sterile pipet tip.
2. Allow agarose to solidify for at least 5 min at room temperature (agarose gel becomes slightly opaque). Hold the micromold with agarose upside down over the well of a multiwell tissue culture plate (6-well-plate for mold with diameter 30 mm, *see* **Note 4**) and use a pipet tip or sterile spatula to transfer the agarose gel into the well (assist this by slightly bending the mold). Submerge the agarose print with microwells entirely in growth medium (4 mL for a 6-well plate), remove air bubbles in the microwells by flushing with medium (use P1000) and equilibrate in a CO_2-cell incubator during at least 2 h (*see* **Note 8**).
3. Prepare a cell suspension of 3×10^6 cells/mL in growth medium (e.g., 10% FBS containing growth medium) (i.e., 600,000 cells/200 μL).
4. Slightly tilt the multiwell dish and carefully remove the medium (use a vacuum pump and Pasteur pipette) from the well and the agarose gel (also in the central area). Per agarose gel, add 200 μL of the prepared cell suspension (600,000 cells) on the area with microwells. Dispense the cells over the entire central area in a dropwise fashion. Let the cells settle in the microwells in the laminar flow. After ~5–10 min, check using a microscope how well the microwells are filled with cells (*see* **Note 9**).

5. Allow cells to settle into the microwells for another ~10 min before carefully adding growth medium to the well (4–5 mL in 6-well). Submerge the entire agarose gel and prevent it from floating to the surface. Incubate at 37 °C and 5% CO_2 for 48–72 h (depending on cell type) during which time the spheroids form (*see* **Note 9**).

3.2 Preparing Extracellular Matrix Solution for 3D-Invasion Assays (Time: 30 min)

1. ***Collagen type I Solution A*** (for polymerization at physiological pH 7.4): add the different components in the order indicated in Table1A (detailed for 1 mL, adjust to the total volume needed for the experiment, *see* Subheading 3.6). Table1A is for a collagen concentration of 2 mg/mL (relevant range may vary: 1–6 mg/mL). Use 1 M NaOH to adjust the pH approximately to 7.4 based on the Phenol-Red pH indicator present in the MEM solution. Prepare just before use (*see* **Note 10** and associated figure).
2. ***Collagen type I Solution B*** (for polymerization at pH >8.3): add the different components in the order indicated in Table 1B (detailed for 1 mL, adjust to the total volume needed for the experiment, *see* Table2A); use a collagen stock solution in the range of 2–4 mg/mL. Table 1B is for a collagen concentration of 1 mg/mL. 1 M NaOH is added to increase the pH to above 8.3. This diluted collagen preparation can be stored between 0–4 °C (on ice) for a couple of days (*see* **Note 10**).
3. ***Matrigel dilution***: Just before use dilute a thawed aliquot of the Matrigel stock solution (*see* Subheading 2.2, **item 1**) with ice-cold medium to the desired final concentration (e.g., 2 mg/mL or higher) and keep on ice until use (*see* **Note 11**).

Table 1A
Collagen type I Solution A

Order of addition	Component	Volume	Final conc.
1	CMF-HBSS	Use to adjust to end volume	
2	10× MEM	0.100 mL	1×
3	0.250 M $NaHCO_3$	0.033 mL	8.3 mM
4	Collagen type 1 (stock concentration X mg/mL)	(1 mL × 2 mg/mL)/X mL	2 mg/mL
5	1 M NaOH	Variable (*see* **Note 10**)	
Total volume, final pH		1 mL, pH ~7.4	

Table 1B
Collagen type I Solution B

Order of addition	Component	Volume	Final conc.
1	CMF-HBSS	Use to adjust to end volume	
2	10× MEM	0.072 mL	0.72×
3	0.250 M $NaHCO_3$	0.072 mL	18 mM
4	Serum containing growth medium	0.200 mL	1/5 volume
5	Collagen type 1 (stock concentration X mg/mL, use stock with conc. of 2–4 mg/mL)	(1 mL × 1 mg/mL)/X mL	1 mg/mL
6	1 M NaOH	0.013 mL	
Total volume, final pH		1 mL, pH > 8.3	

3.3 Using CellMissy (Part I): Experimental Setup of Multisample Invasion Assays

1. The 3D CEZ and MCTS invasion assays are multisample assays: they allow the evaluation and comparison of different conditions (e.g., different cell types, cells expressing different siRNA, control versus drug treated, and different drug concentrations). Because of variability in sample preparation and inherent variable cell behavior, multiple technical replicates per condition are included: we typically use 4–10 for the 3D CEZ assay and, since multiple spheroids can be tested per well (*see* Subheading 3.4), sufficient replicate wells to monitor 10–12 spheroids in MTCS assay. Figure 2 shows a typical experimental setup for a 3D CEZ invasion assay in which 60 samples are analyzed in parallel (*see* **Note 12**). The experimental setup shown is output of the free CellMissy tool (https://github.com/compomics/cellmissy, *see* **Note 13** for new users). The CellMissy module used to set up/annotate a new cell migration/invasion experiment is the "Experiment Manager" [8, 9]. CellMissy is a user-friendly tool that allows easy sharing and reusing of experimental setups. Furthermore, it stores the setups together with the experimental data and metadata (*see* below).
2. In CellMissy, create a new experiment, indicate to which existing or new project it belongs and provide a number and a short description for the experiment. You then get to choose a format for the plate [e.g., 96 (8 × 12)] and can start annotating the tested conditions. Click on condition 1, select the replicate wells in the plate view that will belong to this condition (by

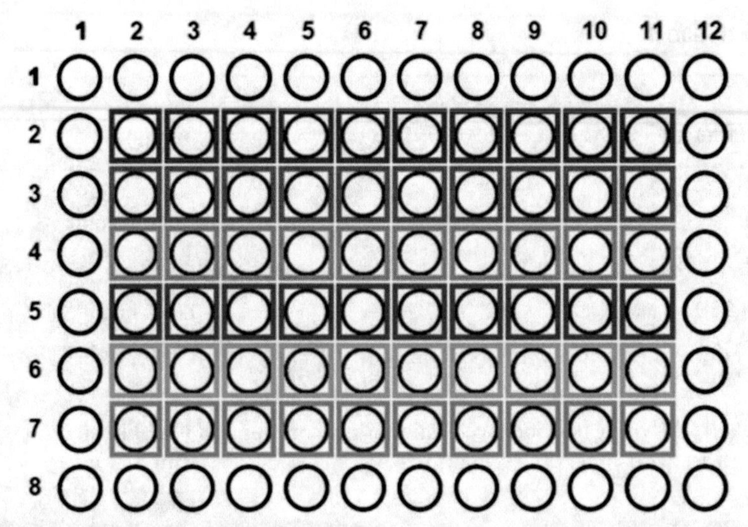

Condition	Cell line	ECM	Treatment
	MDA-MB-231	Collagen I 2.0 mg/ml	control
	MDA-MB-231	Collagen I 2.0 mg/ml	10 μM ROCK inhibitor
	MDA-MB-231	Collagen I 2.0 mg/ml	10 μM Rac inhibitor
	MCF7	Collagen I 2.0 mg/ml	control
	MCF7	Collagen I 2.0 mg/ml	10 μM ROCK inhibitor
	MCF7	Collagen I 2.0 mg/ml	10 μM Rac inhibitor

Fig. 2 Example setup for a 3D CEZ experiment. The experiment consists of 60 samples (6 conditions, each of 10 replicates). The image is adapted from the CellMissy setup report. The table details cell line, extracellular matrix (type, concentration), cell treatment (drug, concentration). This is only a small selection of the sample annotation that is stored in the CellMissy database

drafting a box around them) and annotate details on cell line, assay type, ECM, and treatment. Click to add a second condition to the list on top, indicate the associated wells and annotate the condition. Repeat these steps for each tested condition. Store the annotation and the experimental template in the CellMissy database and generate a pdf setup report with plate view useful for preparing the plate in the cell culture room (Subheadings 3.4–3.6) (*see* **Note 14**).

3.4 Spheroid Invasion (MCTS) Assay in Collagen 3D Matrix: Multisample Preparation

This procedure is adapted from [11] and takes 2.5–3 h (*see* **Note 15**).

1. Place a 12, 24, 48 or 96 well plate (that is prewarmed at 37 °C for at least 1 h) on the preheated stage (37 °C) with microscopy viewing port (*see* Subheading 2.2, **item 6**). The aim is to perform all subsequent steps at 37 °C so that both nucleation and growth phase of the collagen polymeric network formation occur at this temperature. This will ensure generating a reproducible hydrogel with specific pore size (*see* **Note 10**).
2. Prepare collagen solution B (pH >8.3) (*see* Subheading 3.2, **step 2**, *see* **Note 10**). Table 2A shows the volume of this solution required per well of a 12, 24, 48 or 96 well plate. Calculate the total volume required depending on the number of wells used (add at least 10–25% extra volume). Keep on ice until use in **steps 3** and **5**.
3. **Create the bottom collagen matrix layer**: Add the volume of collagen solution B indicated in column "Bottom" of Table 2A to the wells of the warmed plate while keeping it at 37 °C (for example add 300 μL per well of a 12-well plate). Leave for exactly 10 min on the 37 °C plate to allow initiation of matrix polymerization (*see* **Note 16**).
4. **Collect the spheroids**: Take the plate with formed multicellular spheroids (present in the microwells of the agarose gel, Subheading 3.1). Put ready a number of test tubes (1.5 mL); each tube will serve to collect the spheres for one well of the plate prepared in the previous steps. Pick a number of uniformly sized spheroids together with some medium using a Pasteur pipette with rubber bulb and transfer the spheroids to a tube. Picking the spheres is facilitated if you can view the microwells using a microscope in the horizontal laminar flow and by holding the pipette under an angle of 45° while gently touching the edge of the microwells with the pipette tip (*see* also **Note 17**). Add to each tube the desired number of spheroids, e.g., 12 when working with a 12-well plate (*see* Table 2B). If too little medium remains in or around the microwells, carefully add more before continuing to collect the spheres.

   Allow the spheroids to sink to the bottom of the tubes by gravitation (usually less than 5 min, this is visible by eye). Remove medium (and possibly air bubbles) from the tubes without disturbing the spheres (leave a minimal amount of medium, ~25–40 μL depending on the number of spheres) (for 96-well plate, *see* **Note 17**).
5. **Create the top collagen matrix layer containing the spheroids**: Following **step 3**, keep the multiwell plate in which the first layer of collagen gel is present, at 37 °C. Add a volume of collagen solution B, as indicated in the column "top" of

Table 2
Sample Preparation for the MCTS Assay

	2A			2B	2C
Plate type	Total matrix volume/well (mL)	Bottom matrix layer[a] (μL)	Top matrix layer (μL)	n° of spheres/well	Medium on top of matrix (mL)
6-well	2.25	1000	1250	15–20	1.5
12-well	0.66	300	360	8–15	0.75
24-well	0.33	150	180	6–12	0.5
48-well	0.22	100	120	1–6	0.3
96-well	*0.060*	*20*	*40*[b]	*1*	*0.15*

[a]*see* **Note 15**
[b]*see* **Note 16**

Table 2A, to the spheroids in a first tube using a P1000 tip or a shortened p200 tip (remove 1 cm of the tip end with a sterile scalpel to increase the tip opening). Bring spheroids in suspension by gently pipetting up and down just once and immediately pipet the volume dropwise into the well on top of the bottom collagen layer. This will create the top matrix layer in which the spheroids will be embedded. Try to distribute the spheroids optimally over the well area (with sufficient inter-spheroid distance, avoid clusters or touching spheres). To evaluate spheroid positions in the well, either use a microscope (with heated stage) in the lamellar flow or monitor by eye. If needed, use a pipet tip or carefully repipet some of the "top" solution to redistribute spheres that are too close together. Work fast as polymerization initiates rapidly. Repeat this procedure for the next well, etc.

6. Keep the plate with samples on the 37 °C stage for a constant time (e.g., 10–15 min) before carefully transferring it to a 5% CO_2-incubator at 37 °C. At this point, the spheres are already immobilized by the polymerizing network and should stay in position. Leave the multiwell plate with samples for 1 h at 37 °C and 5% CO_2 to allow full polymerization of the collagen matrix with embedded spheroids.
7. In a laminar flow hood, carefully add to each well the indicated volume of assay medium (Table 2C) (possibly containing compounds of interest, *see* Subheading 2.1, **item 3**, *see* **Notes 2** and **15**). The plate is now ready for imaging (*see* **Note 21**).

3.5 Spheroid Invasion Assay in 3D Matrigel: Multisample Preparation

The steps described in Subheading 3.4 can also be performed using Matrigel as a 3D ECM. We work for example with 2 and 5 mg/mL as final concentration in 48–96-well plates. The hydrogels composed of Matrigel at low concentrations (≤2 mg/mL) are not very practical in larger area wells (in 6–24 wells) as, in our experience, they frequently detach from the plate or from the bottom Matrigel layer during setup or during the invasion assay.

3.6 Cell Exclusion Zone Invasion (3D-CEZ) Assay in Collagen Matrix or Matrigel: Multisample Preparation

This procedure is adapted from [12] and protocols present at Platypus Technologies (*see* **Note 18**). **Steps 1–3** require 1.5–2 h; **steps 4–6** require approximately 2–2.5 h (*see* **Note 19**).

1. Prepare 1 mL of collagen solution A (*see* Subheading 3.2, **step 1**) or Matrigel solution (Subheading 3.2, **step 3**) at the desired concentration (*see* **Note 19**).
2. Pipette 100 μL of the collagen or Matrigel solution into the first four sample wells of the 96-well plate (following the experimental set-up Subheading 3.3). Avoid air bubbles. Subsequently remove the collagen (or Matrigel) solution from these wells (leaving only a thin coating) and return it to the source solution. *Immediately* populate these four wells with sterile Oris™ Cell Seeding Stoppers (Platypus) (*see* **Note 19**). Use tweezers or the Oris Stopper Tool (*see* **Note 5**) to push the stoppers firmly down. Proceed with the next four wells. Repeat this procedure for the remainder of the plate. Incubate the plate in a humidified chamber (37 °C, 5% CO_2) for 30 min to allow polymerization of the thin layer of collagen gel/Matrigel (*see* **Note 19**).
3. Collect cells using either enzyme-free or trypsin-containing solutions (depending on the cell line) and prepare, in cell culture growth medium containing normal serum levels, a cell suspension of the optimal seeding concentration (e.g., 400,000 cells/mL) (*see* **Note 20**). Leaving the Oris™ Cell Seeding Stoppers in position, pipette 100 μL (containing for example 40,000 cells) of the cell suspension into each well. The Oris™ Cell Seeding Stoppers are designed such that they allow pipetting solutions in the well via openings between well wall and either side of the stopper (*see* **Note 5**). Incubate the seeded plate containing the Oris™ Cell Seeding Stoppers in a humidified chamber (37 °C, 5% CO_2) overnight (or if performing sample preparation in 1 day at least 2 h) to permit cell attachment in the well area around the stopper. An option is to put weight on top of the plate during the adhesion phase (*see* **Note 20**).
4. Prepare 60 μL (+5–10% extra) of collagen solution A per well (*see* Subheading 3.2, **step 1**) or Matrigel solution (Subheading 3.2, **step 3**) at the desired concentration (*see* **Note 19**).

5. Use the Oris™ Stopper Tool to remove all stoppers. Do not use the stopper tool as a lever, but carefully lift it vertically. Remove media with a pipette; the central cell free zone is now visible by eye. Add 40 μL of collagen solution A or Matrigel to each well. Pay attention to prevent air bubbles from forming during pipeting. Work one column at a time, so that cells are not without medium for a significant amount of time. Incubate the plate at 37 °C and 5% CO_2 for 45–60 min to permit polymerization of the matrix.
6. Add 200 μL of migration assay medium on top of the collagen gel or Matrigel containing the confluent cell layer. Optionally, the medium can be supplemented with invasion inhibitors or promoters or any drug of interest (*see* **Notes 2** and **15**).
7. Add 270 μL PBS to all unused wells to avoid evaporation of the medium in the sample wells. The plate is now ready for imaging (*see* **Note 21**).

3.7 Imaging

1. Make sure the conditions in the microscope incubator or imaging system are optimal, i.e., 37 °C, 5%CO_2.
2. Select the x-y positions you want to image across the entire plate. Imaging is performed using phase contrast in one focal plane (*see* **Note 22**) at a low magnification (5, 10 or 20×) to image a sufficiently large population of invading cells (*see* **Note 6**). In the 3D CEZ assay, the imaging focal plane is the focal plane of the confluent cell layer. Either the entire cell-free zone with part of the confluent cell sheet or one quadrant of the well can form the starting image (Fig. 1a, bottom right). In the MCTS assay, the optimal focal plane is the largest diameter of the sphere (Fig. 3, *see* **Note 23** and associated figure). Depending on the microscope or imaging system used, you may rely on autofocus for setting the z-position or set the optimal z for each x-y position manually. This results in a list of xyz positions (e.g., a 60 position list associated with the colored wells in Fig. 2) that will be reused in each time step for the total duration of the time-lapse recording. Since in the MTCZ

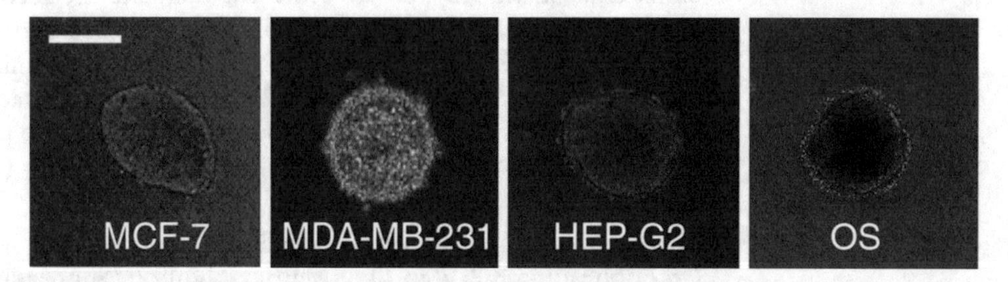

Fig. 3 Collagen embedded spheres. Spheres of cell lines of different origin at the focus (z) chosen to start imaging; "OS": low passage cell line from osteosarcoma patient tumor. Scale bar: 100 μm

assay more than one spheroid may be present in one well (*see* Table 2B) multiple xyz positions may be imaged within one well. Make sure the order of imaging across the plate is also stored to be able connecting an image sequence to a condition in a later stage.

3. Set the parameters for image acquisition and start the time-lapse imaging. Exposure time for phase contrast imaging using the 10× objective is about 10 ms (*see* **Note 6**). The time interval of imaging needs to be chosen in correlation with the expected cell invasion speed (*see* **Note 24**). A typical time interval for cancer cells is 15 min to 1 h (short interval if the aim is single cell tracking); immune cells that move faster should be imaged with an even smaller time interval, between 10 s to 2 min, to allow cell tracking. The total imaging time is cell-type dependent and usually varies from 6 to 48 h (*see* **Note 25**).

3.8 Image Storage

Store image data (TIFF file, MultiTIFF files, or proprietary file format) and metadata in a structured way: create your own local system or explore the possibilities offered by the imaging software used or those offered by central repository systems such as OMERO (https://www.openmicroscopy.org/site/products/omero). In any case, make sure that image acquisition metadata files generated by the imaging system software are also stored and remain associated with the imaging data or are traceable (*see* **Note 26**).

3.9 Image Processing: Extraction of Quantitative Data

Both commercial and open source software tools that can process the image sequences obtained from the MCTS- or 3D-CEZ assay exist. Full image processing should aim to generate quantitative data, e.g., area versus time or xyt data (*see* **Note 27** on possible software tools).

1. ***Area measurements (pix^2 or μm^2) at the different time points of the sequence (area vs time data)***. Area can either indicate the area covered by the cell sheet (3D–CEZ) or by the multicellular tumor spheroid (MCTS). As the cells invade the matrix, this area will increase. Conversely, in the 3D CEZ assay, the area of the cell free zone decreases. In general terms, this step includes image preprocessing tasks followed by segmentation of the cell sheet, cell spheroid or cell-free zone in each image of the sequence and quantification of the size of the segmented area.

2. ***Cell trajectories (xyt data)***. In addition to image preprocessing and individual cell segmentation, image processing here includes cell tracking. This means finding for each segmented cell i, its position $x_i y_i$ at time t_i, then its new position at the next time point t_{i+1}, etc.

3. These numerical data extracted from the processed images (either 'area vs time' or 'xyt' data) form the input for the subsequent quantitative analysis (parameter extraction and comparison over different test conditions, detailed in Subheadings 3.11 and 3.12).

3.10 Using CellMissy (Part II): Storage of Quantitative Data from Image Processing [8, 9]

1. In Subheading 3.3, we detailed how CellMissy is used to setup an experiment using a multiwell plate configuration and how this results in the storage of the experimental metadata coupled to each sample (or well) in the relational database of CellMissy. In Subheading 3.9 the image sequences obtained from each well are translated by the image processing software in numerical data. These can be stored in the same database, as detailed here. Data import requires that one connects the text files containing the numerical output from the image processing for a specific time sequence to the corresponding well in the experimental set-up in CellMissy and, by doing so, couples it to the specific biological condition tested. This process is implemented in CellMissy through a simple "drag and drop" procedure.
2. Table 3 details the steps required for data import, describes the format of the input text files and the input of image metadata. Alternatively, data loading by CellMissy can be fully automated, once tailored to a customized combination of imaging system and imaging processing software (*see* **Note 28**).
3. After batch import, all data are stored in the CellMissy database and ready for further analysis.

3.11 Raw Data Visualization and Quality Control: Using CellMissy (Part III) [8, 9]

1. Open CellMissy. In the main menu (Fig. 1b, or under "File") choose "area analysis" or "single cell analysis." Select project and experiment number. Within the experiment, select the dataset of interest for analysis and an imaging type: phase contrast or fluorescence (*see* **Note 30**).
2. ***Area versus time data***

   Indicate the units in which the area is expressed (*see* Table 3). One can select whether area is based on quantifying the cell-covered or cell-free area (*see* **Note 32**). Retrieve the data from the database by clicking "start."

   Plot the area over time data. Table 4 and Fig. 4a, b detail the different ways this can be done. Prior to plotting, the area is converted into normalized area (*see* **Note 32**).

Table 3
Importing Invasion Data in CellMissy

	Invasion of bulk cell population (area vs time data)	Invasion of individual cells (xyt-data)	*see* Note
Loading option	Load area	Load tracks	28
Choose project and experiment (annotated)			
Input of image metadata; Press "start"	Add number of time frames, time interval (and unit), duration (hours)		29
Enter "data loader"	Select well and add data or datasets to it (one or more, drag and drop procedure), repeat for all wells		30
Dataset format	Tab or comma separated text file (.txt, .csv) or .xls	Tab or comma separated text file (.txt, .csv) or .xls	
Structure of the input text file/sample	2-column: Time; area	4-column: Trajectory ID; time; x coordinate; y coordinate	31
Units of input	Minutes or time step; μm^2, pixels or %area	None; minutes or time step; μm or pixel; μm or pixel	
Import data to database by clicking "finish"			

Remove outliers. CellMissy identifies (1) outliers in area increase between consecutive time steps for a specific replicate (these may be artifacts due to technical issues or segmentation errors) and (2) it identifies replicates that are outliers within a condition (*see* **Note 33**).

3. **Cell trajectories (xyt data)**

Indicate the units (pixel or μm) in which the xy positions are expressed (*see* Table 3). Retrieve the data from the database by clicking "start" and show specific data of a condition by clicking on the condition in the plate view or the list.

Plot cell tracks. Table 5 and Fig. 4c, d detail the different possibilities that can be accessed via the tabs "overview," "global view condition," "explore tracks," "global view experiment." For raw data and derived data, you can use a plate heat map (*see* Subheading 3.12).

Discard artifacts in the xyt data sets by using the flexible CellMissy data filtering options (*see* **Note 36** and associated figure).

Table 4
Visualization options for area vs time data in CellMissy

Plot area over time for...	How? *See* [8], CellMissy manual and information lines in software for more detail	
All replicates of one condition	Use "Data preprocessing/normalized area"	Figure 4a *see* **Note 32**
	Use "Data preprocessing/corrected area"	*See* **Note 33**
All conditions of the experiment	Use "Global view" for a user-selected time interval U (use raw or corrected data; Each condition is shown as the median of replicates)	Figure 4b *See* **Note 34**

3.12 Cell Invasion Parameter Extraction, Graphical Representation and Multisample Statistics: Using CellMissy (Part IV) [8, 9]

1. **From "Area versus time data" to median invasion velocity of bulk cell population**
   In CellMissy, pass from "Global view" (*see* Fig. 4b, Table 4) to "Linear regression model" tab to perform data fitting (with R^2 value) and obtain the slope of the area over time plots for all replicate wells per condition. The slope is a measure for the velocity of the bulk cell population in μm^2/min. Derived from these slopes, the median velocity per condition is calculated and listed and a bar plot of these values for all or a user-selected set of conditions in the experiment is produced (Fig. 5a).
2. **From "cell trajectories (xyt data)" to distribution of displacements, speed and directionality**. Table 6 lists all extractable invasion parameters. These can be considered cell-centric (C) or step (S)-centric (*see* **Note 37**). Plot the parameters at the level of the well, the condition or the experiment. Table 6 indicates whether mean and median values are displayed in the software and it lists the options for displaying the distributions of parameter values. Figure 5b–e illustrates these different options for graphical representation (*see* **Note 38**).
3. **Statistical comparison**. Compare parameter distributions pairwise between conditions of the multisample experiment (e.g., compare velocity distribution of bulk cell population or compare the velocity or end-point directionality for individual cell invasion). Select the conditions to compare, give groups names, add groups to the analysis and perform statistical comparison using the nonparametric Mann–Whitney test. Inspect in the new window the pairwise *p*-value table and adapt the significance threshold and/or type of correction for multiple testing (*see* **Note 39**) (Fig. 5f). Saving the analysis ensures it will be integrated in the PDF report.
4. **Produce a PDF report of the quantitative analysis** with user defined options (*see* **Note 40**).

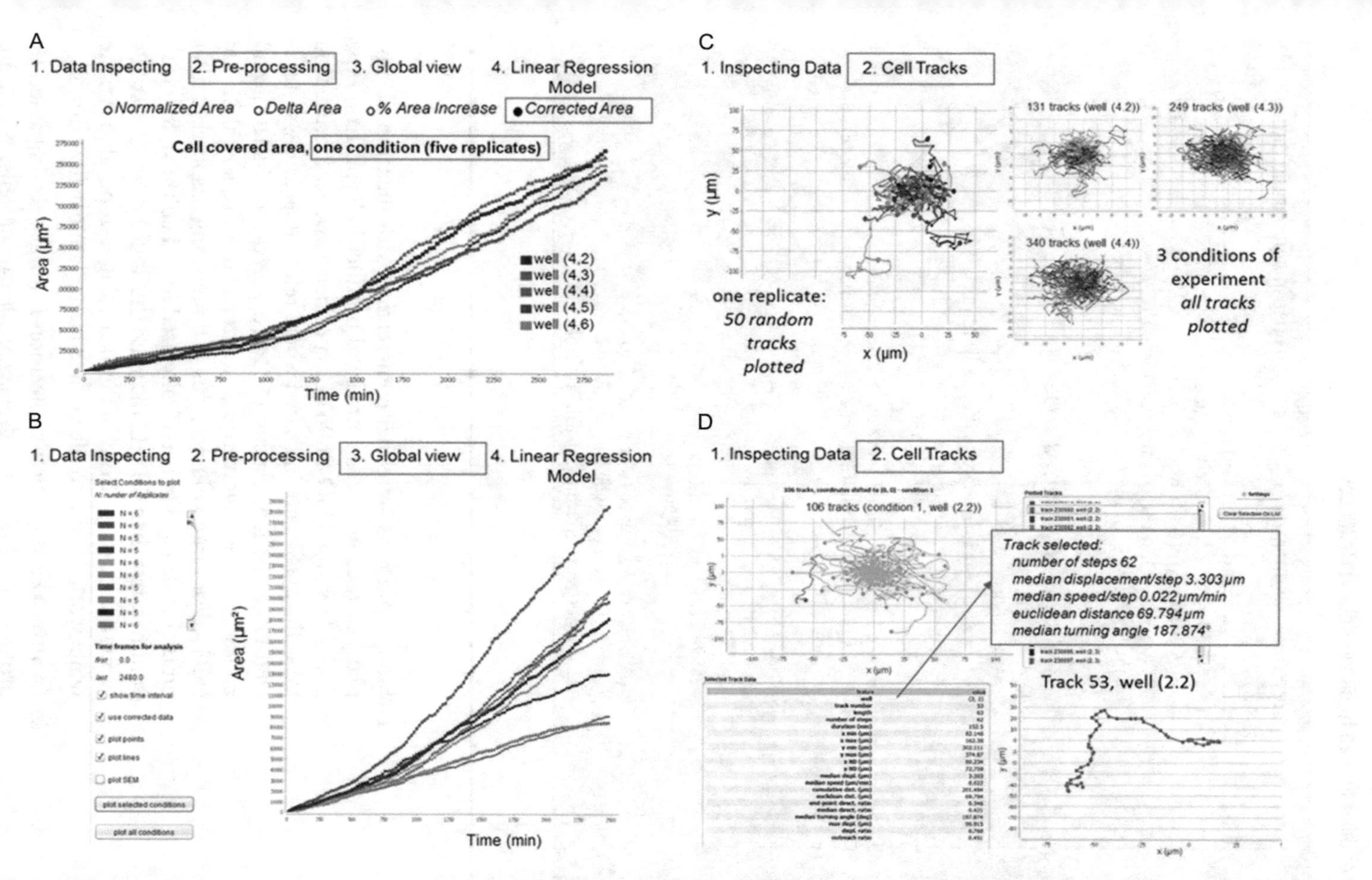

Fig. 4 Data visualization and quality control. All panels are adapted from data analysis performed using CellMissy (https://github.com/compomics/cellmissy). (**a**, **b**) Bulk cell invasion: (**a**) area versus time plots for the replicates of one condition (**b**) plot of median area versus time for all conditions of the multisample experiment (median calculated over technical replicates); (**c**, **d**) Single cell invasion xyt data: (**c**) Cell track visualization for one replicate of a condition (right) and for three individual conditions of an experiment (global view experiment, left) (**d**) Explore track window

Table 5
Visualization options for cell track xyt data in CellMissy

Plot of:	How? *See* [9], manual and information lines in software for more detail	
Cell tracks of one well	**Use "Cell tracks/overview" tab**; choose to shift track origins to (0,0) for display, choose the well of interest within the condition, choose to display a number of randomly selected tracks or plot all tracks	Figure 4c *see* **Note 35**
Cell tracks of one condition	Idem as above, use **"cell tracks/global view condition" tab** to see replicates sides by side (or choose **condition** (instead of well) under overview tab for one plot containing all replicates of the condition).	
Cell tracks of all conditions of the experiment side by side	**Use "Cell tracks/global view experiment" tab**; Displays plots of all conditions in an array format of choice (e.g., 3 × 2 for 6 conditions).	Figure 4c *see* **Note 35**
Individual track	**Use "Cell tracks/explore" tab**. Choose a track by clicking on its end point or in track list. Explore a.o. detailed track data, replay how a selected cell moved in time, display track hull area, ...	Figure 4d
Plate heat maps	**Use "Cell tracks/plate heat map"**: Shows a color indication per well to quickly evaluate similarities and differences (*see* also Table 6)	

4 Notes

1. Sterile solutions are either commercial (with mention of supplier) or user-prepared from solid product in Type 1 or Type 2 Grade water (ISO 3696) or specified solvent. Sterilization procedures (SP) are indicated where relevant as follows: sterilized either by autoclaving (SP1a (dry cycle) or SP1b (liquid cycle) or by passage through a filter with pore size 0.22 μm (SP2). For large volumes, this is done using disposable filter units (bottle top filters typically used in cell culture); for small volumes, less than 50 mL, dispense the liquid through a sterile syringe filter (e.g., from Millipore) using a disposable syringe while working in a sterile environment (laminar flow hood).
2. In vitro, cell migration or invasion is always accompanied by proliferation [13, 14]. Especially for cancer cells, which are less contact-inhibited, proliferation is usually significant during the time-lapse measurement in part because ongoing

Table 6
Extracted invasion parameters: Options for calculation and visualization

Parameter (*see* Note 38)	Cell-(C) and step (S)-centric (*see* Note 37); mean (M), median (Md)	Graphical presentation: Kernel density distribution (K) /box plot (BP)/plate heat map (pHM)	
		Well/condition level	Experiment level
Displacement	C, S, M, Md	K (Fig. 5b), BP	BP
Speed	C, S, M, Md	K, BP, pHM (Fig. 5c)	BP
Turning angle	C, S, M, Md	(Angular) histogram (Fig. 5d)	
End point directionality	C, M, Md	K, BP (Fig. 5e), pHM	BP

migration also creates space for proliferation. We observed this in both assays presented here. In this perspective, what is usually expressed as "invasion" rather reflects "invasive growth." For some cell lines, proliferation can, however, be reduced and migration measured more exclusively by lowering the serum concentration. Invasion of MDA-MB-231 cells can for example be measured in medium with only 1% serum (FBS) supplemented with 1 nM epidermal growth factor (EGF) (Sigma). Under these conditions, the cells still invade with a significant speed but have lower proliferation rates than in the same medium supplemented with 10% serum. Other cell lines, however, migrate poorly when serum levels are reduced, or conversely migrate sufficiently fast under normal growth conditions so that assay time (and thus cell proliferation) can be reduced. Upon assaying a new cell line, we usually compare invasion levels in cell medium containing 10% serum and 1% serum in the presence and absence of a relevant motility inducing growth factor (e.g., EGF at 1–5 nM or hepatocyte growth factor (HGF) at 25–50 ng/mL) to ultimately decide on the migration assay medium.

With relation to the exclusion zone assay, we refer to Glenn et al., 2016 [15] for a method based on staining cells with the lineage-tracing vital fluorescent CellTrace Dyes (ThermoFisher Scientific). This method proposes to obtain a measurement of migration efficiency not affected by proliferation. Note that this method requires fluorescent imaging and relies only on cell counts as read-out of migration (not speed measurements) and is only validated in a 2D setting.

3. Collagen stock solutions should be stored at 4 °C and kept on ice when used. The commercial collagen stock solutions used

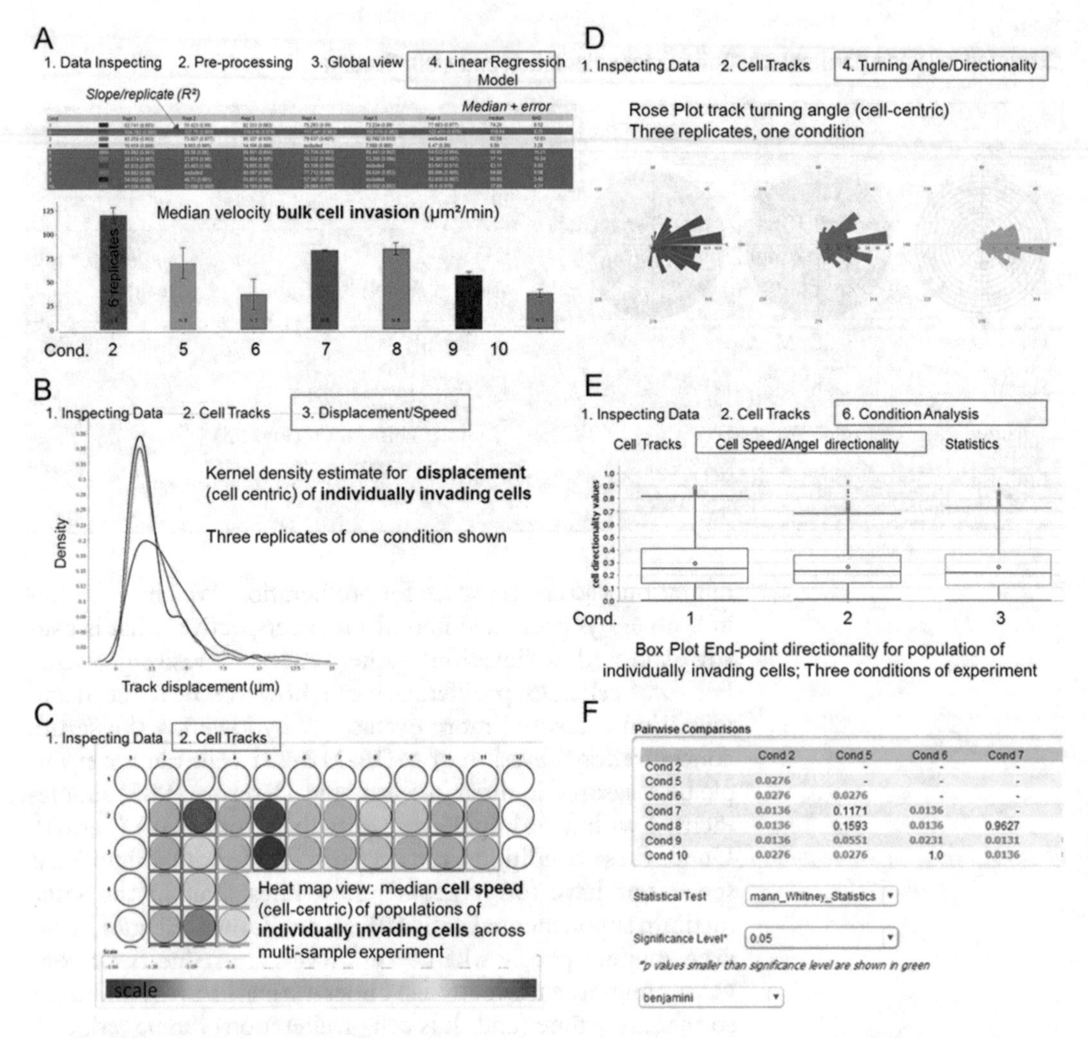

Fig. 5 Selection of extracted cell invasion parameters, graphical representation of parameters and multisample statistics. All panels are adapted from data analysis performed using CellMissy (https://github.com/compomics/cellmissy). (**a**) Median velocity of invading bulk cell population, (**b**–**e**) different parameters of individually invading cells (**f**) part of statistics window

here are acid extracted from tendon, but not pepsin-treated. Thus this solution contains collagen monomers with short peptides at the ends (telopeptides, 11–26 residues), which are critical for fibril cross-linking and for ordered assembly into a structured network [16]. Pepsin-treated collagens have lost C-terminal telopeptide tails. They are also commercially available (e.g., Nutragen, Advanced Biomatrix). The gels they form are different from their non-pepsin-treated counterparts especially in relation to pore size [17, 18]. These differences can influence invasion speed and modes.

Matrigel is mainly composed of laminin, collagen Type IV, and enactin, and contains traces of other ECM components and growth factors (*see* https://www.corning.com/worldwide/en/products/life-sciences/products/surfaces/matrigel-matrix.html). It is stored at −20 °C in aliquots (0.5–1.5 mL) to limit repeated freezing. Always thaw Matrigel on ice. Especially for Matrigel, it is recommended to acquire a sufficiently large stock of an identical batch number to perform a set of related experiments.

4. As described in Subheading 3.1, the micromold is used to prepare an agarose print with an array of microwells that is used as a self-generated dish (now termed by MicroTissues: 3D Petri Dish®) that allows preparing multicellular spheres. Currently the commercially available micromolds (MicroTissues, Sigma-Aldrich) allow to prepare multicellular spheres of different diameters by preparing agarose dishes that fit in 12- and 24-wells. We typically use the first generation of molds that generate agarose dishes with 330 microwells, each with a diameter of 800 μm, which fit in a 6-well. The volumes mentioned in Subheading 3.1 are for these large micromolds. The MicroTissues micromold with an array of 265 microwells that fit in a 12-well plate generates spheres of similar sizes as described here. For MicroTissues 256 well mold, the following volumes are needed: 500 μL agarose solution, a cell suspension of 1.3–2.6×10^6 cells/mL, 190 μL for cell seeding (containing 250,000–500,000 cell/190 μL). We refer the reader to https://www.microtissues.com/3dcellculture_protocols or https://www.microtissues.com/3dcellculture_protocols/Microtissues-Sigma-Aldrich-3D-Petri-Dish-Facts.pdf for volumes to be used with the various commercial molds.

   The micromolds are sterilized before use (SP1b) and handled in the laminar flow using clean tweezers. Hold the mold using ethanol rinsed gloved hands only when needed, e.g., during spatula-aided removal of agarose from the mold (*see* Subheading 3.1, **step 2**).

5. Oris™ Cell Seeding Stoppers (Platypus) are small silicone tools that - when inserted in a 96-well – allow to shield an area of 3.14 cm²/well in the center of four neighboring wells. For details on inserting the stoppers in a ORIS-compatible 96-well plate, see Platypus Technologies Oris Cell Seeding Stopper Demo (https://vimeo.com/158378983). In the ORIS 96-well compatible plates (Platypus), the stoppers fit tightly. As reported by others [15], we sterilize the stoppers (ST1b) and reuse them either in ORIS compatible plates or in other 96-well plates. Note that this is only possible a limited number of times and in both cases (and especially with other plates), this may result in a less tight fit of the stoppers in the well, so that the cell free zone is not always cell-free at start or not perfectly circular. In this case more replicates per condition

(4–10) need to be implemented to allow eliminating "bad" wells or to reduce variation between wells.

The Oris™ Stopper tool is a handy tool provided in a.o. the Oris™ Cell Migration Assay that can be used to remove the stoppers from the wells of 96-well plate.

6. The imaging system can be built around a traditional microscope. We use a CellM Imaging Station (Olympus) with an upright phase contrast and epifluorescence microscope (frame IX81 which is now discontinued and replaced by newer IX83 configuration) equipped with a motorized stage (xyz control) and a box-type incubator (built around microscope) with temperature- and CO_2-control (ACU, Evotec). We work with a long working distance 10× objective CPLFLN10XPH (Olympus, Numerical Aperture (NA) 0.3, working distance W.D. (mm) 9.5, Field Number 22). The longer working distance is optimal for 3D assays. Alternative phase contrast (high throughput) live cell imaging systems are those that operate inside a CO_2-incubator or are stand-alone imaging systems and these are now available from multiple companies (for a useful selection *see* http://www.biocompare.com/BioImaging-Microscopy/Microscope-and-Cell-Imaging-Systems/?said=0&soids=2254335,2254290,2254288,2254347,2254344 on BioCompare (search with: "Cell Imaging System, Digital, Live Cell Imaging, Phase Contrast, Time Lapse Microscopes").
7. Multicellular spheres are considered relevant in vitro models of tumor biology [19]. To form spheres, conditions must be created in which cells prefer to adhere to each other rather than to surfaces or substrates in their environment (refs in [20, 21]). We here detail a method in which multicellular spheroids are generated in microwells of a discrete diameter in agarose molds. This method was initially described by Napolitano and coworkers [22] and relies on the nonadhesive nature of the microwells. Alternative methods rely on continuous stirring (described in detail in [11]) or using the hanging drop method; *see* ref. 20 for a comparison of methods for spheroid generation. In our experience, both the hanging drop and nonadherent surface dependent method are user-friendly and work well for most cells. In general, different cell lines generate spheroids that vary in compactness and size (Fig. 3). Epithelial cells most efficiently form compact large spheroids (e.g., MCF7 breast cancer cells, A431 epidermoid carcinoma, HepG2 hepatocellular carcinoma, HT116 colorectal carcinoma) but some cell lines that in 2D–culture appear mesenchymal also form solid spheres (e.g., HT1080 fibrosarcoma, BT-549 ductal carcinoma). For cells with inherent low self-adherence, such as E-cadherin negative cells like MDA-MB-231, the gentler methods, nonadherent surface or hanging drop, are abso-

lutely required and even then the formed spheroids frequently display low compactness and easily fall apart upon handling. Addition of 0.12% w/v methylcellulose (Sigma, M 6385; GJ Bakker, Radboud University Medical Center, NL, personal communication), collagen (~30 μg/mL, [23]), or Matrigel (2.5% in medium, [20]) has been described to improve spheroid compactness. As alternative to the agarose microwells, other nonadherent surfaces have been reported mainly under the form of ULA-plates (U-bottom low attachment plate (96 or 384 wells), e.g., from Corning).

Using the micromold method, we have been able to generate spheres from primary cells such as low passage cells cultured from patient tumors (*see* Fig. 3) or from mouse xenographs, as also described by others [24].

8. Well-equilibrated agarose gels are pink when using phenol-red containing medium. Agarose gels with microwells can be prepared in advance and kept for several days in culture medium in the CO_2-incubator before cell seeding.
9. The microwells should appear completely filled with a dense mass of cells after ~5 min. Residual cells in suspension will still be present at the edges of the central area of the mold. These do not need to be removed, since they are scattered upon subsequent medium addition to the well. If the microwells are not sufficiently filled, an additional 100 or 200 μL of the cell suspension can be added. Note that in the larger diameter micromolds that we use, the best spheroids of homogenous size (150–250 μm diameter) only form in the outer wells of the array (2–3 last rows).

   The spheroids formed in this manner have a diameter around 150–250 μm. Larger spheres (400–500 μm diameter) develop a necrotic core [19]. This may be useful in some applications but note that drug/compound infiltration is also limited in depth. Too small spheres (<100 μm) are not optimal for subsequent manipulations or for monitoring invasion.

   The number of days that cells are allowed to form spheroids depends on cell line or type, with 3 days as a good starting point. It is our experience that for some cell types, incubation in the mold for longer than 72 h yields poorly invasive spheroids.
10. The final density of the collagen 3D matrix is a strong determinant for the cell invasion levels that will be measured. The matrix density (related to network structure, pore size, fiber diameter) is dependent on collagen concentration but also on the conditions (ionic strength, pH, and temperature) during the self-assembly process as further discussed below [16, 25, 26]. It is, thus, of major importance to keep these conditions

constant across experiments and also to report these conditions in detail together with the invasion data.

Effect of pH: the pH of the collagen solution influences the polymerization process and consequently the density of the formed gel. In particular, the mean fiber diameter, the average pore size and the pore area fraction of the resulting matrix all decrease, when the pH increases (*see* ref. 16 for detail).

Effect of temperature: a less dense collagen matrix (with smaller fiber diameter and larger pores) will also be formed if the gelation process occurs at lower temperature. Yang et al. describe that especially the temperature during the nucleation process is crucial in this respect [27]. These authors report that for a 1 mg/mL collagen solution an increasingly less dense matrix is formed by increasing the incubation time at 22 °C (~room temperature). Subsequent incubation at 37 °C does not affect this (*see* ref. 27 for details). Although this creates possibilities for making collagen gels that pose very different challenges for invading cells, in more practical terms it implies that both nucleation temperature and time need to be tightly controlled and standardized in order to reproducibly prepare the samples on multiwell plates.

Collagen stock solutions are very viscous and pipetting errors consequently also introduce variability in matrix production. Give special attention to correct transfer of the desired volumes; for example, allow sufficient time for liquid inflow and outflow and rinse pipet tip with the diluted solution after pipetting in the stock solution. It can help to use pipet tips of which the extreme end is cut off using a sterile scalpel blade.

We use two collagen solutions A and B. These differ in ionic strength but the main differences are in the final collagen concentration and in the pH (*see* Tables 1A and 1B). Based on the invasive behavior of cells in gels established from collagen solutions A or B (Subheading 3.2), the matrix density is very similar (despite different collagen concentrations) but the gel from the high pH solution B appears more rigid and thus more suitable for larger gels (e.g., in 6- or 12-well plate).

Solution A (2 mg/mL, pH ~7.4) has as advantage that cells are always at physiological pH, but setting the pH is difficult. To reach a pH of 7.4, NaOH is carefully added to the point where the phenol-red indicator reaches a salmon-pink color (*see* Fig. 6 as guide). The amount needed will depend on the volume of the collagen stock solution used. As a guideline: for a 1 mL of 2 mg/mL collagen solution A and starting from a 8.25 mg/mL collagen stock, 5–10 μL of 1 M NaOH is needed. When the correct indicator color appears to be reached, keep the solution on ice for at least 5 min and check

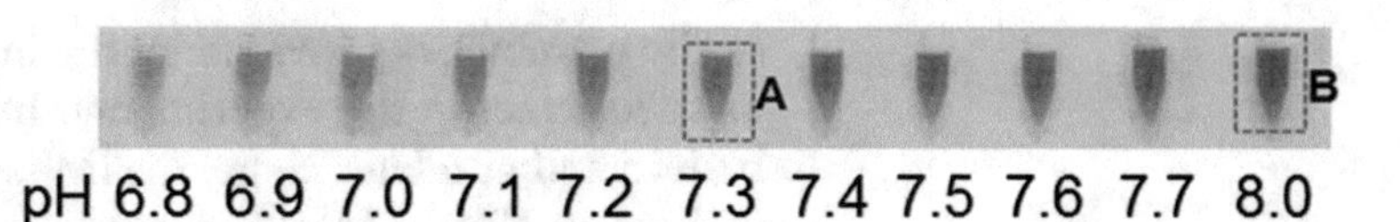

Fig. 6 Color guide for pH setting of collagen solutions A and B. Figure adapted from images contributed by Max Schwalbe, under the Creative Commons Attribution-Share Alike 4.0 International license

whether the correct pH has indeed been reached; if not, add more NaOH.

Solution B (1 mg/mL, pH > 8.3), initially described by [11] has the advantage that it is easier to prepare in a reproducible manner (a constant amount of NaOH is added, the solution should be fuchsia, *see* Fig. 6) but this implies that the cells are temporarily at higher pH during the procedure.

11. Matrigel rapidly polymerizes between 22 and 37 °C. Therefore, work on ice during all manipulations and always use ice-cold solutions and precooled tubes to prepare Matrigel dilutions. For first time users or when working with relatively high final concentrations (5 mg/mL and higher), it is recommended to also work with precooled pipet tips (put the tip box in −20 °C). It is our experience that Matrigel dilutions below 2 mg/mL take a long time to form a gel at 37 °C and that the resulting gel still remains very jelly-like and easily detaches from plastic surfaces.
12. In setting up a multisample time-lapse experiment, the number of testable conditions will be limited by the number of positions that can be imaged in the chosen cycle which is the time interval between consecutive images of each sample. Imaging speed varies with the imaging system used and one technical determinant is the speed of x y z positioning and autofocus, if used. Prior to setting up the experiment, make sure that all samples can be imaged within the cycle time (*see* Subheading 3.7, **step 3** and *see* **Note 24** for cell speed as main biological determinant of cycle time).
13. The free software tool CellMissy (download from: https://github.com/compomics/cellmissy) relieves the user of the task of keeping experimental metadata together with the subsequent quantitative data (e.g., "this well in which the cells were under these conditions rendered these 'numbers' after analyzing the images") as it employs a user friendly plate view to annotate the experimental setups and will finally store not only this metadata but also the extracted quantitative data itself in a relational database. As an alternative, spreadsheet software

or the possibilities provided by the imaging software itself can be used to setup the experiments. In this case, CellMissy can be initiated at a later stage. CellMissy will run on any system with a Java Virtual Machine version 1.8.0 or above (Windows, Linux, and Mac OS-X). First time users need to perform a number of steps: download software, connect to MySQL and run the SQL script "cellmissy_schema.sql" (you find it on the software homepage). Download the CellMissy application and run it by double-clicking the executable .jar file present in the "CellMissy" folder (unzip the compressed CellMissy folder before you execute the .jar file). You will also need to configure the database (DB) connection for CellMissy: edit the properties that establish this connection as detailed in the manual. Also edit the main directory (default C:/M) to the site where you store your image data, if you opt for an automated coupling between output of image processing and CellMissy. In a single-user setup, CellMissy can run on minimal hardware (laptop or desktop PC); if the software is to be used as a shared system between many users, it will be more practical to set up a central DB (on any modern desktop machine with sufficient storage space) that all users can access simultaneously. The CellMissy manual is present on the homepage (https://github.com/compomics/cellmissy#usage) and describes all steps for a first user (§1 1. HOW TO RUN CELLMISSY FOR THE FIRST TIME) including basic user managements. Note that CellMissy is a constantly evolving software tool (current version 1.1.0), so check for updates and new possibilities regularly.

14. Annotate each experimental condition by filling in the details under the tabs: "Cell Line" (cell line name, growth condition, seeding density, …), "Assay_ECM" (type of assay, e.g., 3D_ORIS for cell exclusion zone assay or spheroid, note that polymerization time and temperature as well as polymerizing pH can be added as meta-data (*see* **Note 10**)), "Treatments" (expressed proteins in cells, silencing targets, perturbations (drugs, growth factor, …), drug solvent, …). How to perform the annotation is documented in detail in the CellMissy manual (https://github.com/compomics/cellmissy#usage) (§2.1 EXPERIMENT MANAGER - CREATE A NEW CELL MIGRATION EXPERIMENT). The manual also contains information on how to reuse previously designed experimental templates or how to share them with other users.
15. In the spheroid invasion assay, the spheroids are embedded in a collagen matrix that is polymerized on top of a thinner bottom collagen matrix [11]. The latter prevents that the matrix-embedded spheroids settle too close to the rigid surface of the well plate as this is reported to influence the invasive behavior

[28]. Once polymerized, medium (with possible drugs, inhibitors, ...) is added on top of the spheroid containing matrix. Multiple spheroids can be embedded per well of a 6–48-well plate (*see* Table 2B). We aim at monitoring at least 10–12 spheroids/experimental condition in 1 or 2 wells/condition. In this way, a similar experimental throughput can be obtained compared to working in a 96-well plate.

If any of the additions (of drugs, ...) to the migration assay medium implies that an organic solvent is present in the assay medium, keep the final solvent concentration as low as possible and perform a control sample with the identical final solvent concentration. For the frequently used solvent dimethylsulfoxide (DMSO), we work at 0.1% or lower, because higher DMSO concentration affects cell invasion efficiency.

16. It may be opportune to initiate the laborious, time consuming **step 4** before **step 3** as the polymerization time of 10 min in **step 3** needs to be respected as it is crucial for reproducibility (*see* **Note 10**). In addition, the possible loosening of the top layer during the course of the assay is prevented by allowing this partial, 10 min, polymerization of the bottom matrix. This is likely due to the fact that inter-layer crosslinking still occurs after the top layer is added. Make sure that the volume for the bottom layer is spread over the entire well surface (tap sides of plate to attain this).

    The thickness of the bottom layer needs to be limited to allow focusing on the spheres contained in the top layer. Note that the volumes in Table 2A are optimal for the objective and imaging system described in **Note 6**.

17. Alternative method for spheroid picking: flush spheroids from the microwells using medium; remove the agarose mold and collect the spheres from the medium. Note for the 96-well format (Table 2): for the top layer (40 μL per well) in Table 2A, it is better to prepare a collagen stock solution B of 1.5× the final desired concentration. Pick one sphere in 10 μL (best via alternative method, use cut P20 tip) and add to 30 μL of 1.5× collagen solution, mix (no air bubbles) and transfer this to well of 96-well plate containing bottom layer.

18. In the 3D-CEZ protocol described here, the cell-free zone is created using silicone stoppers and the cells are present as a cell layer in between two matrix layers (bottom and top) around the central zone. Tools, kits and various protocols are available via http://platypustech.com/. In the Oris™ Pro 96-well Invasion Assay the stopper is replaced by a drop of biocompatible BioGel positioned in the well center. This BioGel is dissolved before/as the top collagen layer is applied. The BioGel has the advantage that automation is easier. More recently, the Oris™ 3D Embedded Invasion Assay is proposed.

Here individual cells are embedded in a thicker collagen gel around the central BioGel. After dissolving the BioGel, the central opening in filled with 10 μL of collagen gel.

19. The procedure describes a multisample preparation over 2 days: day 1 for creating the bottom collagen gel (**step 2**), cell seeding and overnight cell adherence (**step 3**); day 2 for adding the top collagen matrix (**step 5**). The collagen solution A is best prepared separately for each day. Alternatively, the multisample preparation can be completed in 1 day allowing 2–4 h for cell adherence (make sure cells are well spread; remember spreading time is cell type dependent): In this case prepare 100 μL collagen/well (+5–10% extra) collagen solution A.

    The immediate addition of the Oris™ Cell Seeding Stoppers (Platypus) implies this happens before complete polymerization of the collagen in the well with the aim of creating an indentation in the bottom gel coating (**step 2**) that will be filled by the upper layer added in **step 5**. Evidence in Fig. 7a supports that the cells are confronted with the upper layer upon invading the central region and do not use the interlayer zone.

20. After cell adhesion, the cell layer in the well periphery should be fully confluent up to the edge of the central area (the edge of the stopper). For most frequently used cancer cell lines 40,000–50,000 cells per well is suitable; for myoblasts, we used 20,000 cells/well. For a new cell line or cell type, the optimal cell number per well can be tested using a dilution series (20,000–65,000 cells/well).

    To seed the cells, hold the pipet tip vertical and add the cell suspension carefully via the port at the side of the Cell Seeding Stopper without disturbing the stopper itself. After seeding, lightly tap the plate at all sides while keeping it on the work bench to evenly distribute cells over the nonshielded well area. When only part of the plate is used, add spare stoppers at the plate peripheries to ensure that the plate lid (once replaced) does not disturb the stopper positions in the sample wells. Adding a weight on the plate lid (e.g., extra small plastic plate +50 mL small glass bottle) during cell adherence in the incubator helps to prevent that cells populate the area under the stopper during this step. This is especially useful when not using the ORIS-compatible 96-well plates.

21. Imaging in phase contrast (or epi-fluorescence mode) is done in the plastic multiwell plate with lid on top. When the incubator used allows controlling humidity, all wells of a 96- or 48-well multiwell plate can be used for the invasion assays without risking volume loss by evaporation and thus position-dependent artifacts. If humidity control is not possible or

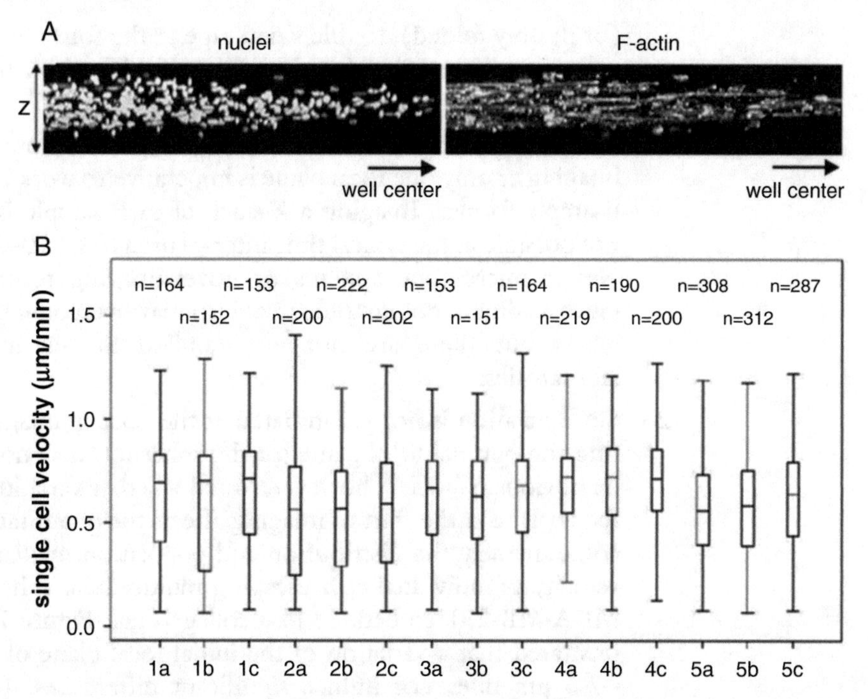

Fig. 7 (**a**) Side view of migration pattern in 3D CEZ assay. Confocal microscopy images (z-stack, cross-section) of MDA-MB-231 cells, stained with DAPI (left) or Alexa-694-phalloidin (right), after 36 h of invading the collagen matrix (2 mg/mL) in 3D–CEZ assay; total z is 90 μm. The original cell layer is broadened due to invasion in the z-direction, indicating that in their movement toward the center of the well, the cells display xyz invasion and do not merely follow the surface of the bottom matrix layer. (**b**) Flexibility in choice of focal plane (z-position) in MCTS. Five spheroids (1–5) were imaged (phase-contrast, 24 h, 20 min interval). For each spheroid, this was imaged in triple (a–c) using the same xy position but with varied z-positions: (*a*) at the optimal focus plane z, (*b*) at z −7.5 μm and (*c*) at z + 7.5 μm. The cells invading out of the spheroid were tracked. The distribution of mean cell velocity of the tracked cells at the three focus planes is shown (*a*–*c*). The output of the triple imaging (*a*–*c*) is not significantly different (95% confidence level)

insufficient, we advise not using the outer wells (upper and lower row and right and left column, *see* for example Fig. 2) of the multiwell plate but instead fill these wells with PBS (270 μL per well of 96-well plate). In addition, evaporation can in part be prevented by sealing the closed plate on all sides with Micropore Rayon Synthetic Surgical Tape (3 M, allows exchange of O_2 and CO_2). When working in a box-type incubator, temperature gradients may exist between plate and lid: this results in condensation and drop formation on the lid of the plate. During time lapse, this will disturb imaging. One way out is replacing the plate lid with a glass plate that is pretreated with a commercial antifog/anticondensation spray on the side oriented toward the sample (we use Cressi Antifog Spray 0% alcohol). Evenly spray the glass and dry by wiping with paper towel (repeat three times). To allow gas exchange, we attach the glass plate on the multiwell plate using thick (or doubly folded) double sided tape at the four corners and subsequently seal the closed plate at all sides with Micropore Tape.

22. With a classical phase contrast microscope, the limitation of imaging at only one focal plane is imperative to work in a multisample format. Imaging a *Z*-stack of each sample is indeed not possible in the typical time interval used (e.g., 10–20 min). Newer microscope systems or novel imaging technologies (such as digital holographic imaging) may overcome this limitation but these are not yet standard in cell migration laboratories.
23. Since invasion is not yet initiated at the start of imaging, setting the optimal focal plane for the spheroids may not always be obvious. We have, however, tested whether variation in the focus plane at the start of imaging affects the quantitative outcome, namely the distribution and population median of the velocity of individual cells escaping multicellular spheroids of MDA-MB-231 embedded in a collagen gel. Figure 7b demonstrated that a variation of the initial focal plane of +7.5 or −7.5 μm does not induce significant differences. It is not expected (at least in this experiment with MDA-MB-231) that users would choose a focus plane that deviates more than the tested range since focus quality declines rapidly beyond these limits.
24. By adapting the time interval to cell speed, one first of all ensures that the changes in xy coordinates of the moving cells between consecutive images are sufficiently small so that the tracking software (*see* Subheading 3.9) performs well, i.e., that a single cell trajectory remains associated with the same cell. It also ensures that the derived cell trajectories are as close as possible to the actual cell paths and consequently allows to extract invasion parameters with high accuracy.

    Note that the time needed for imaging of all set xy positions needs to be smaller than the time interval. This limits the throughput.
25. When longer imaging is needed (e.g., 5–6 days), one may opt to image at 5–10 discrete time points (e.g., every 16–24 h) and place the multisample plate in CO_2 incubator in between image acquisitions.
26. Each commercial imaging software stores image acquisition settings in a proprietary file. Check where this metadata file is stored and check whether your imaging software allows storing your images as OME-TIFF, the image format specification of the Open Microscopy Environment (OME), to facilitate

cross-platform image exchange including the associated, valuable metadata.

27. We developed the customized software tool, **Cell Migration and Invasion Analysis CELLMIA (or MIA)**, which works well on phase contrast images generated from cells embedded in 3D matrix as for example in the 3D-CEZ and MCTS assay. It is a proprietary software generated in collaboration with DciLabs (Belgium) and has been applied in two recently reported studies [9, 29]. CELLMIA simultaneously measures area changes over time and cell trajectories (thus both image features mentioned in Subheading 3.9), which is unique to our knowledge. In addition, it handles images of sequences generated in either of the two experimental setups described here (Fig. 8). CELLMIA performs equally well when collective invasion occurs or when migration of sparsely seeded individual cells is imaged. The user can batch-load all

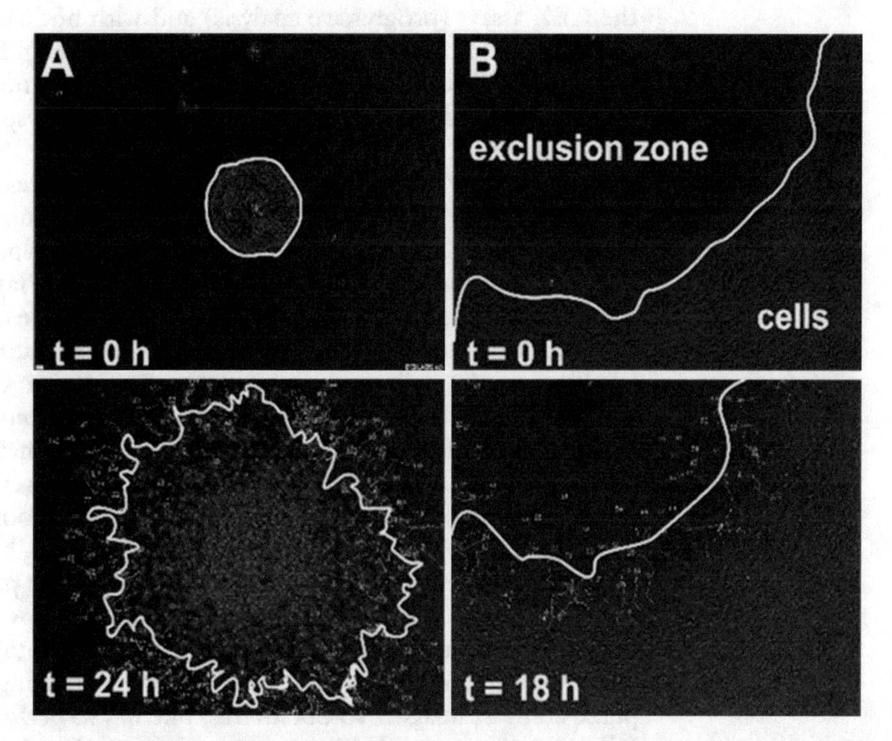

Fig. 8 Dynamic monitoring of cell invasion/migration using the double segmentation approach in CELLMIA. Only two time points are shown, but all time points (possibly up to several hundreds) of a loaded time sequence are analyzed automatically by the software. The panels show the image analysis overlaid on the raw image for invasion of MDA-MB-231 cells in collagen matrix in (**a**) MCTS and (**b**) 3D CEZ. The area covered by the bulk cell population, i.e., the expanding spheroid (**a**) or the peripheral confluent cell layer invading the central exclusion zone (**b**), is delineated progressively in each time frame (thin red line drawn by CELLMIA, here for better visualization accentuated by thicker white line). In the lower panels, multicolored numbered tracks outside of the bulk cell area indicate single cell trajectories

time sequences generated in one experiment (e.g., 60 for the experiment in Fig. 2); thereafter the analysis for all time points of all these samples is fully automated. It generates text files that can be imported in CellMissy (Subheading 3.10 CellMissy Part II).

We list here a selection of alternative analysis methods. For the MCTS assay, a number of commercial imaging systems contain software that allows analysis of phase contrast and fluorescent images, e.g., IncuCyte (Essen Bioscience), Gen5™ Data Analysis Software on Cytation3 (Biotek), and Celligo Image Cytometer (http://www.nexcelom.com/Celigo/migration-invasion.php). Several authors have used the "concentric circle method" (PlugIn within ImageJ/FIJI) for analysis of spheroids in which the nuclei are made fluorescent for example using Hoechst staining [30, 31]. Kumar et al. [32] developed an open source software for both CEZ assays and for the MCTS assay (*see* web link in publication) to use on fluorescent images of cells with cytosolic staining for the CEZ assays (progressive analysis) and with nuclear staining for the MTCS assay (discrete time point analysis). For the latter, the cells escaping from the spheroid are counted and the distance of the escaping cells from the estimated spheroid center is measured [32]. For the CEZ assay, the Platypus website (http://platypustech.com/application-notes) provides a number of quantification methods via commercials systems and one based on Image J (http://platypustech.com/wp-content/uploads/2016/01/ApNote_Platypus_ImageJ.pdf) that quantifies the number of cells in the initially cell-free zone. In all cases, these analyses rely on fluorescent labelling of the cells (e.g., with green fluorescent Calcein AM). These analyses remain quite labor intensive from a user perspective, at least for a multisample setup. Note that when the analysis is only performed at a number of discrete time points, differences between conditions may be overlooked.

FIJI (FIJI is just Image J, batteries included, [33], http://fiji.sc/) is an open source software platform for image processing. Using only a number of image processing steps within FIJI, the area values of the spheroid core for the entire time sequence of the MCTS assay can be derived starting from phase contrast images. To obtain this, one has to perform the following steps: load the time sequence by dragging the image stack in the FIJI window, set the "Image Type" to 8 bit, "Process/Binary/Make Binary" (choose black background, check the box to apply binarization to all time points), use 2–3 consecutive steps of "Process/Binary/Erode" to close remaining gaps in the core of the sphere, in "Analyze/Set measurement" select area and perform "Analyze/Analyze Particles": the results table immediately shows the area values for each time point. This approach is most suitable for cell types only displaying bulk/collective cell invasion. Alternatively, if individual cells are invading the matrix, one can quantify both core and escaping cells in one measurement of relative area in FIJI: perform the same preprocessing steps as above, in "Analyze/ Set measurement" select "% area" and perform "Analyze/ Measure". Record a macro (Under PlugIns/Macros) to easily repeat these measurements for all time points. Both FIJI methods deliver area versus time values for the MTCS and are expected to also work on the image sequences from the 3D CEZ assay.

FIJI also contains different plugins for cell tracking (TrackMate, Toast, …) but we have not explored their capacities of tracking the cells in the presence of an expanding bulk populations, since this is possible with CELLMIA.

28. Table 3 describes the generic data loading based on user selection of the txt or excel files containing the data. CellMissy also has capabilities to automatically loading the data generated by the image processing software by retrieving metadata of the raw and processed images and when provided with information on the imaging order, i.e., the position list generated in Subheading 3.7, **step 2** as a text file. This makes the software amenable to high-throughput data processing. In the current version, this automated loading is established for the specific workflow of our group: i.e., imaging on the CellM system (Olympus) which generates multitiff images and an *.obsep* metafile/experiment; image processing by the custom imaging software CELLMIA described in **Note 27**, and storage of all data (raw, processed and position list) on a local server in a hierarchical structure. Automated import from any other (high-throughput) workflow is possible but will require customization via programming of an adapter. The developers of the CellMissy software are committed to supporting such work and to assist new users in establishing such interfaces.

    During automated import, CellMissy retrieves and stores all imaging metadata: number of images, image resolution in pixel and/or μm as provided, time interval used during time sequence, magnification of imaging, imaging types (phase contrast, fluorescence), number of imaging types, name of position list, and so on. Based on this, it automatically assigns the data txt files (area in time data, xyt data) to the wells and conditions on the plate. All the user has to do is click on the well/sample that was first imaged. Once the import is finished, the cell migration data are saved to the database by clicking the "*Finish*" button.

29. For bulk cell migration: analysis is also possible if the data consists of only a limited number of time points because of

the manner of image acquisition or method used for image analysis.

30. Different area over time data may have been generated per well of the multiwell plate. In case of the MCTS assay this may be the imaging of several spheroids at different xy positions per well (*see* Table 2B, these are considered as technical replicates). Alternatively, one xy position may have been imaged by different imaging types (e.g., phase-contrast and fluorescence) or the image sequence of one xy position may have been analyzed by different imaging processing software or different settings of a specific software. Note that this combination of datasets can differ from well to well within the experiment. Within CellMissy it is possible to load all these datasets (and subsequently analyze them in parallel, *see* Subheadings 3.11 and 3.12).

31. Example input files for area in time data are available on the CellMissy website https://github.com/compomics/cellmissy#downloads and described in the manual. For example track (xyt) data *see* https://figshare.com/articles/cell_trajectories_text_files/3413848.

32. In case "cell-covered area increase" forms the input, the start area is set to 0 and all values are normalized to this (area at time 0 is subtracted). If the input is "cell free area reduction," this can also be processed by the software: the start area (time 0) is in this case set to 100, all values are normalized to this value at time 0 and expressed in % of the original cell free area that becomes covered by cells.

33. **Quality control and data filtering in area over time data**. Bulk cell invasion is expected to occur at an approximately constant speed and thus with a fairly constant % increase relative to the area in the previous time frame. As documented in detail in the Supplement to [8] and in the manual, CellMissy identifies and corrects for each replicate the outliers in the distribution of the % area increase caused by technical errors during imaging, errors in area segmentation, and so on. The difference in the distribution plots, displayed as Kernel probability density functions, between raw and corrected data for all replicates of a condition can be visually compared under "pre-processing/% area increase." Outliers are also highlighted in the data table and the raw and corrected area vs. time plots are both shown under preprocessing/corrected area.

    In Subheading 3.3, it is advised to include per condition 4–10 replicates for 3D CEZ and 10–12 spheroids in MTCS assay. The software inspects in an unbiased manner whether some replicates deviate (using Euclidean distance as similarity metric, Fig. 9a) and suggests their removal. The user

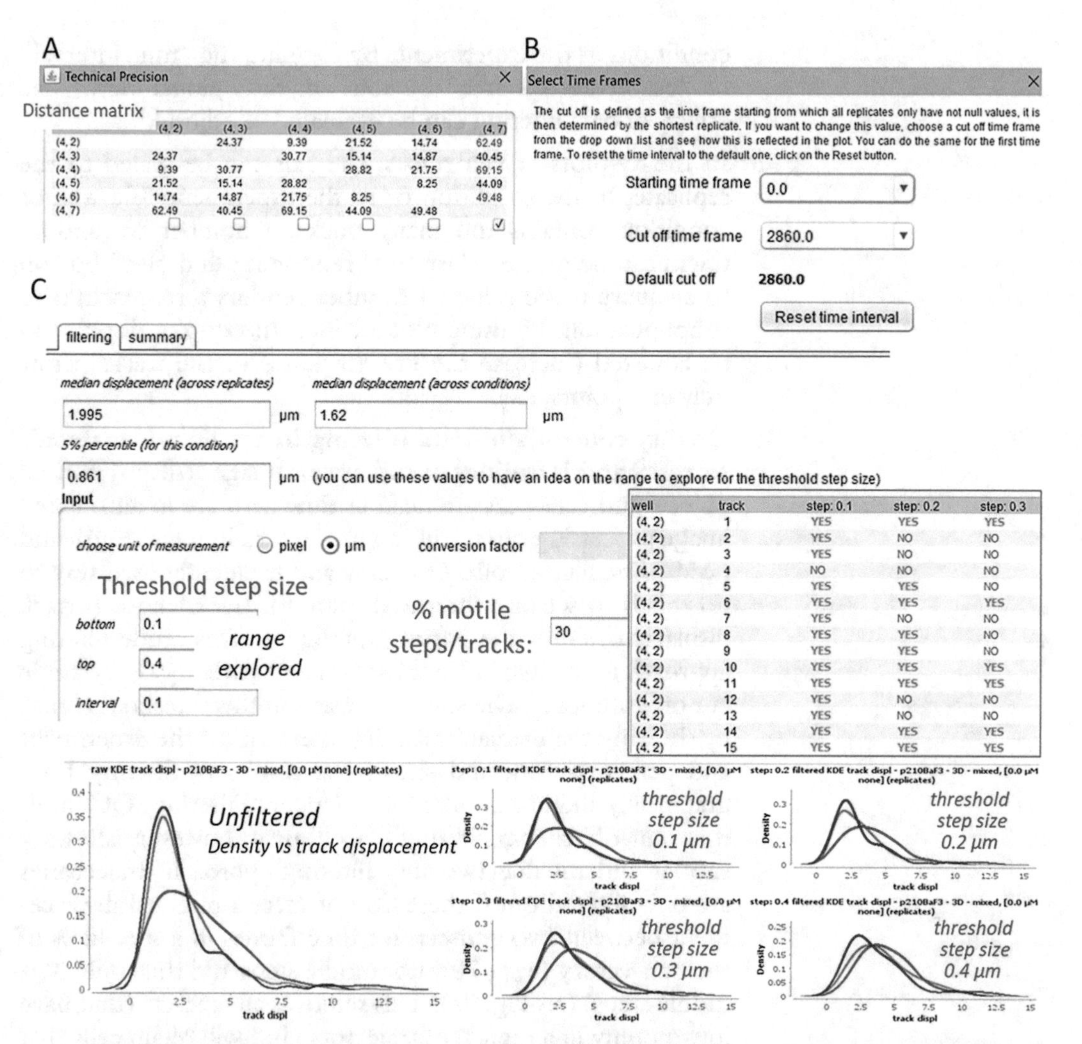

Fig. 9 Quality control and data filtering (**a**, **b**) Bulk cell invasion: window illustrating the identification of replicate outliers using a similarity matrix (**a**) or possibility to select the time interval, during which area vs. time data is linear (**b**). (**c**) Filtering of cell track data using multiple cutoff check. Median displacement values across replicates or conditions are shown. A range of thresholds for minimal step size in combination with a % of motile steps per track is set and tested. Results are visualized in a table (upper right) or using graphs of density distribution (bottom)

can either accept or decline the suggestions made by the software [8].

34. Ideally select a time interval in which the area increase is linear for subsequent data analysis (Subheading 3.11) since velocity is derived from the slope of the linear fit on area versus time data (Fig. 9b). Start and end time point can be set (e.g., 20–1000 min). Note that if the limitation in time range is set in one biological condition, this automatically implies that this time range will be used in the data analysis of all biological

conditions in the experiment. By toggling the "time interval" or "use corrected area" buttons on the "global view," the effects of data filtering can be evaluated or adjusted.

35. In the xy-plots of cell tracks, the axis can be scaled to the replicate, to the condition or to the experiment. If a well or condition contains too many tracks, a number of random tracks can be plotted. Use the "randomize and plot" button to evaluate if the selected number renders a representative subpopulation. By using plot settings, the xy-plot display can be adjusted (increase the line thickness of the tracks, show only end points, …).
36. **Quality control and data filtering in xyt data (for details *see* ref.** 9 **and info-lines in software)**. Image artifacts, noncell objects and cells that are dead at time zero are in automated image processing often still assigned a certain track length and need to be filtered out. One easy and frequently used way to do this is to set one threshold value for track length (=track displacement) for the entire experiment. This crude filtering approach is possible in CellMissy under "Filtering/QC/single cut off criterion." The software assists in this threshold choice by showing the median track displacement for the experiment and per condition, and shows the results of filtering both graphically and in a tabular form. Under "Filtering/QC/multiple cut-off check/filtering", CellMissy, however, allows a subtler and tunable two-step filtering approach: trajectories are only filtered out if these do not meet a minimal displacement between two consecutive time frames in a specific % of their trajectory (e.g., 70% nonmotile steps and thus only 30% motile steps) (*see* Fig. 9c). This removes all artifacts (that have low motility in a high % of trajectory) but will retain cells that pause or display the so-called "juggling" phenotype (stop, pause, stop, …) during their trajectory. The choice for the minimal translocation in pixels or μm is guided by the reported median displacement per condition and across conditions, and can be tested in CellMissy for a range of thresholds (the result can be visually compared graphically and in tabular form). A cutoff can be applied to just one condition or to the experiment. The summary tab shows the effect of filtering: % of retained tracks and the distribution for two essential invasion parameters [cell displacement and track speed, *see* Subheading 3.11)] after filtering. In [9], the two methods of data filtering have been compared for a specific experiment.
37. CellMissy uses two different ways to compute the cell migration parameters [9]. **"Cell centric parameter" (= "trajectory-centric parameter")**: for each tracked cell a specific parameter value is calculated (e.g., the cell velocity based on the total track length divided by the time the cell is tracked) is considered. The mean or median of the distribution of these cell-based

parameter values for all cells in the measured population (of a replicate or condition) is reported. "**Step-centric parameters**": the values for a specific parameter (e.g., displacement) for all steps of all trajectories in a measured cell population (of a replicate or condition) are assembled in one distribution for which a mean or median can be determined. *See* Fig. 5 and detailed discussion on advantages and limitations of two approaches (and thus benefit of having both types of parameters in the output) in ref. 9.

38. From one trajectory three basic parameters are extracted on the level of each step: Displacement (the distance the cell travelled between two consecutive time frames, also known as instantaneous displacement), instantaneous speed (instantaneous displacement divided by the time interval) and the turning angle α between two consecutive steps using:

$$(\alpha = \tan^{-1}(\Delta y/\Delta x)).$$

These parameters are directly used in obtaining the distributions of step-centric values (*see* **Note 37**) and are also used to calculate similar properties for each trajectory (total and net distance travelled by the cell, mean/median speed and turning angle). Median turning angle is a measure of cell directionality; alternatively, directionality is expressed as end point directionality (net distance/total distance). The later values are used for describing the distributions of cell-centric parameters.

39. Non parametric tests are available in CellMissy for pairwise comparison between conditions within the multisample experiment and two types of corrections for multiple testing can be used (Benjamini or the more stringent Bonferroni correction). The *p*-value table indicates in green which pairwise comparisons are significantly different, using the user-set cutoff value 0.1, 0.05, or 0.01 (Fig. 5f).

40. All tables and plots can be exported using simple copy-paste. In addition, a PDF report can be generated. In multiple windows of the software, the option exists to add a certain plot to the PDF report. When generating the PDF report, the user can in addition choose to include specific contents.

Acknowledgments

The authors acknowledge funding from the European Union's Horizon 2020 Programme under Grant Agreement 634107 (PHC32–2014). L.H. was funded by Bijzonder Onderzoeksfonds UGent 01J04806. M.V.T. acknowledges VIB TechWatch for funding of software acquisition.

References

1. Clark AG, Vignjevic DM (2015) Modes of cancer cell invasion and the role of the microenvironment. Curr Opin Cell Biol 36:13–22. https://doi.org/10.1016/j.ceb.2015.06.004
2. Kramer N, Walzl A, Unger C, Rosner M, Krupitza G, Hengstschläger M, Dolznig H (2013) In vitro cell migration and invasion assays. Mutat Res 752:10–24. https://doi.org/10.1016/j.mrrev.2012.08.001
3. Masuzzo P, Van Troys M, Ampe C, Martens L (2016) Taking aim at moving targets in computational cell migration. Trends Cell Biol 26:88–110. https://doi.org/10.1016/j.tcb.2015.09.003
4. Nath S, Devi GR (2016) Three-dimensional culture systems in cancer research: focus on tumor spheroid model. Pharmacol Ther 163:94–108. https://doi.org/10.1016/j.pharmthera.2016.03.013
5. Fayzullin A, Tuvnes FA, Skjellegrind HK, Behnan J, Mughal AA, Langmoen IA, Vik-Mo EO (2016) Time-lapse phenotyping of invasive glioma cells ex vivo reveals subtype-specific movement patterns guided by tumor core signaling. Exp Cell Res 349:199–213. https://doi.org/10.1016/j.yexcr.2016.08.001
6. Friedl P, Locker J, Sahai E, Segall JE (2012) Classifying collective cancer cell invasion. Nat Cell Biol 14:777–783. https://doi.org/10.1038/ncb2548
7. Stewart MD, Jang CW, Hong NW, Austin AP, Behringer RR (2009) Dual fluorescent protein reporters for studying cell behaviors in vivo. Genesis 47:708–717. https://doi.org/10.1002/dvg.20565
8. Masuzzo P, Hulstaert N, Huyck L, Ampe C, Van Troys M, Martens L (2013) CellMissy: a tool for management, storage and analysis of cell migration data produced in wound healing-like assays. Bioinformatics 29:2661–2663
9. Masuzzo P, Lynn H, Simiczyjew A, Ampe C, Martens L, Van Troys M (2017) An end-to-end software solution for the analysis of high-throughput single-cell migration data. Sci Rep 13:42383. https://doi.org/10.1038/srep42383
10. Masuzzo P, Martens L (2015) An open data ecosystem for cell migration research. Trends Cell Biol 25:55–58. https://doi.org/10.1016/j.tcb.2014.11.005
11. De Wever O, Hendrix A, De Boeck A, Eertmans F, Westbroek W, Braems G, Bracke ME (2014) Single cell and spheroid collagen type i invasion assay. Methods Mol Biol 1070:13–35. https://doi.org/10.1007/978-1-4614-8244-4_2
12. Hulkower KI, Herber RL (2011) Cell migration and invasion assays as tools for drug discovery. Pharmaceutics 3:107–124. https://doi.org/10.3390/pharmaceutics3010107
13. Simpson MJ, Treloar KK, Binder BJ, Haridas P, Manton KJ, Leavesley DI, McElwain DLS, Baker RE (2013) Quantifying the roles of cell motility and cell proliferation in a circular barrier assay. J R Soc Interface 10:20130007–20130007. https://doi.org/10.1098/rsif.2013.0007
14. Johnston ST, Simpson MJ, McElwain DLS, (2014) J R Soc Interface 11:20140325. https://doi.org/10.1098/rsif.2014.0325
15. Glenn HL, Messner J, Meldrum DR (2016) A simple non-perturbing cell migration assay insensitive to proliferation effects. Sci Rep 6:31694. https://doi.org/10.1038/srep31694
16. Raub CB, Unruh J, Suresh V, Krasieva T, Lindmo T, Gratton E, Tromberg BJ, George SC (2008) Image correlation spectroscopy of multiphoton images correlates with collagen mechanical properties. Biophys J 94:2361–2373. https://doi.org/10.1529/biophysj.107.120006
17. Sabeh F, Shimizu-Hirota R, Weiss SJ (2009) Protease-dependent versus-independent cancer cell invasion programs: three-dimensional amoeboid movement revisited. J Cell Biol 185:11–19. https://doi.org/10.1083/jcb.200807195
18. Wolf K, Alexander S, Schacht V, Coussens LM, von Andrian UH, van Rheenen J, Deryugina E, Friedl P (2009) Collagen-based cell migration models in vitro and in vivo. Semin Cell Dev Biol 20:931–941. https://doi.org/10.1016/j.semcdb.2009.08.005
19. Kunz-Schughart LA, Kreutz M, Knuechel R (1998) Multicellular spheroids: a three-dimensional in vitro culture system to study tumour biology. Int J Exp Pathol 79:1–23. https://doi.org/10.1046/j.1365-2613.1998.00051.x
20. Nagelkerke A, Bussink J, Sweep FCGJ, Span PN (2013) Generation of multicellular tumor spheroids of breast cancer cells: how to go three-dimensional. Anal Biochem 437:17–19. https://doi.org/10.1016/j.ab.2013.02.004
21. Fennema E, Rivron N, Rouwkema J, van Blitterswijk C, De Boer J (2013) Spheroid culture as a tool for creating 3D complex tissues. Trends Biotechnol 31:108–115. https://doi.org/10.1016/j.tibtech.2012.12.003
22. Napolitano AP, Dean DM, Man AJ, Youssef J, Ho DN, Rago AP, Lech MP, Morgan JR (2007) Scaffold-free three-dimensional cell

culture utilizing micromolded nonadhesive hydrogels. BioTechniques 43:494–500. https://doi.org/10.2144/000112591

23. Leung BM, Lesher-Perez SC, Matsuoka T, Moraes C, Takayama S (2015) Media additives to promote spheroid circularity and compactness in hanging drop platform. Biomater Sci 3:336–344. https://doi.org/10.1039/C4BM00319E
24. Surolia R, Li FJ, Wang Z, Li H, Liu G, Zhou Y, Luckhardt T, Bae S, Liu R, Rangarajan S, de Andrade J, Thannickal VJ, Antony VB (2017) 3D pulmospheres serve as a personalized and predictive multicellular model for assessment of antifibrotic drugs. JCI Insight 2:e91377. https://doi.org/10.1172/jci.insight.91377
25. Raub CB, Suresh V, Krasieva T, Lyubovitsky J, Mih JD, Putnam AJ, Tromberg BJ, George SC (2007) Noninvasive assessment of collagen gel microstructure and mechanics using multiphoton microscopy. Biophys J 92:2212–2222. https://doi.org/10.1529/biophysj.106.097998
26. Gobeaux F, Mosser G, Anglo A, Panine P, Davidson P, Giraud-Guille MM, Belamie E (2008) Fibrillogenesis in dense collagen solutions: a physicochemical study. J Mol Biol 376:1509–1522. https://doi.org/10.1016/j.jmb.2007.12.047
27. Yang YL, Motte S, Kaufman LJ (2010) Pore size variable type I collagen gels and their interaction with glioma cells. Biomaterials 31:5678–5688. https://doi.org/10.1016/j.biomaterials.2010.03.039
28. Fraley SI, Feng Y, Krishnamurthy R, Kim D, Celedon A, Longmore GD, Wirtz D, Louis S (2010) A distinctive role for focal adhesion proteins in three-dimensional cell motility. Nat Cell Biol 12:598–604. https://doi.org/10.1038/ncb2062.A
29. Bertier L, Boucherie C, Zwaenepoel O, Vanloo B, Van Troys M, Van Audenhove I, Gettemans J (2017) Inhibitory cortactin nanobodies delineate the role of NTA- and SH3-domain-specific functions during invadopodium formation and cancer cell invasion. FASEB J 31:2460–2476. https://doi.org/10.1096/fj.201600810RR
30. Mitchell CB, O'Neill G (2016) Cooperative cell invasion: matrix metalloproteinase–mediated incorporation between cells. Mol Biol Cell 27:3284–3292
31. Blacher S, Erpicum C, Lenoir B, Paupert J, Moraes G, Ormenese S, Bullinger E, Noel A (2014) Cell invasion in the spheroid sprouting assay: a spatial organisation analysis adaptable to cell behaviour. PLoS One 9:e97019. https://doi.org/10.1371/journal.pone.0097019
32. Kumar KS, Pillong M, Kunze J, Burghardt I, Weller M, Grotzer MA, Schneider G, Baumgartner M (2015) Computer-assisted quantification of motile and invasive capabilities of cancer cells. Sci Rep 5:15338. https://doi.org/10.1038/srep15338
33. Schindelin J, Arganda-Carreras I, Frise E, Kaynig V, Longair M, Pietzsch T, Preibisch S, Rueden C, Saalfeld S, Schmid B, Tinevez J-Y, White DJ, Hartenstein V, Eliceiri K, Tomancak P, Cardona A (2012) Fiji: an open-source platform for biological-image analysis. Nat Methods 9:676–682. https://doi.org/10.1038/nmeth.2019

Chapter 10

Using Systems Microscopy to Understand the Emergence of Cell Migration from Cell Organization

Staffan Strömblad and John G. Lock

Abstract

Cell migration is a dynamic process that emerges from fine-tuned networks coordinated in three-dimensional space, spanning molecular, subcellular, and cellular scales, and over multiple temporal scales, from milliseconds to days. Understanding how cell migration arises from this complexity requires data collection and analyses that quantitatively integrate these spatial and temporal scales. To meet this need, we have combined quantitative live and fixed cell fluorescence microscopy, customized image analysis tools, multivariate statistical methods, and mathematical modeling. Collectively, this constitutes the systems microscopy strategy that we have applied to dissect how cells organize themselves to migrate. In this overview, we highlight key principles, concepts, and components of our systems microscopy methodology, and exemplify what we have learnt so far and where this approach may lead.

Key words Cell migration, Systems microscopy, Quantitative, Statistics, Modeling, Systems biology, Cancer cell

1 Introduction

Understanding how complex and dynamical states of cellular organization produce cell migration remains a major research challenge. This reflects several interrelated characteristics of the cell migration system, including: its composition by a vast array of molecular components; its emergence over multiple scales in space and time; and its mixture of heterogeneity and plasticity.

First, the challenge of establishing a systems-level understanding of cell migration is made exceedingly complicated simply due to the huge number of molecular components that comprise and modulate the process. For instance, cell–matrix adhesions represent just one important macromolecular assemblage in the broader machinery of the cell migration system, yet mass spectrometry analyses reveal that at least several hundred distinct protein species are associated with these complexes [1, 2]. Not only is the number of components enormous, but their parallel deployment in

Alexis Gautreau (ed.), *Cell Migration: Methods and Protocols*, Methods in Molecular Biology, vol. 1749,
https://doi.org/10.1007/978-1-4939-7701-7_10, © Springer Science+Business Media, LLC 2018

numerous functional networks, which themselves experience intricate and extensive cross talk, adds copious layers of complexity.

Second, cell migration is a dynamical process that arises over extended spatial and temporal scales. In three-dimensional (3D) space, functional units exist: at the molecular level; at the subcellular level in the form of macromolecular machineries; and collectively at the cellular level. Similarly, the temporal dynamics that compose migration arise on time scales ranging from milliseconds to days. Therefore, understanding cell migration requires the integration of multiscale data spanning a highly extended, four-dimensional (4D) feature space.

Third, cell migration displays profound heterogeneity and plasticity. Even given basic visual inspection, cells under equivalent conditions typically display extensive natural heterogeneity in their rates of motion, directional stability, and morphology. Given more systematic quantification, it is revealed that even clonally identical cells can spontaneously produce and switch between alternate migration modalities, which reflect distinct underlying regulatory strategies and patterns of molecular organization [3]. Furthermore, to meet extrinsic (e.g., changes in ECM microenvironment) or intrinsic (e.g., changes in Rho GTPase signaling) challenges, cells may adaptively exploit even more extensive plasticity in the cell migration system; either by switching the balance between such preexisting migration modes [3], or by transitioning to profoundly different forms of migration [4].

Given these challenges, understanding how cells organize themselves for motility demands research strategies that integrate enormous complexity spanning the four dimensions of space and time in a manner that not only captures heterogeneity, but leverages it as a critical source of information. To this end, we have developed a multidisciplinary approach that we refer to as systems microscopy, built on a combination of fluorescence microscopy, quantitative image analysis, multivariate statistical interrogation and mathematical modeling. Using this strategy, we and others have begun to build a systematic and extensible understanding of cell migration. With continued elaboration and the quantitative integration of new research, Systems Microscopy has the potential to illuminate a comprehensive picture of how cell migration emerges from underlying cell organization.

We have focused on analyzing single cell migration upon two-dimensional (2D) substrates using human cancer cells as a model system. This model is most relevant to cancer cell dissemination in settings were migration occurs along 2D surfaces such as myotubes, blood vessels, and in premade pores of the ECM [5]. This relatively simple and optically accessible experimental model makes it possible to simultaneously record cell behavior and organization across the spatial and temporal scales that constitute the cell migration system, enabling integrated analysis of dependencies over

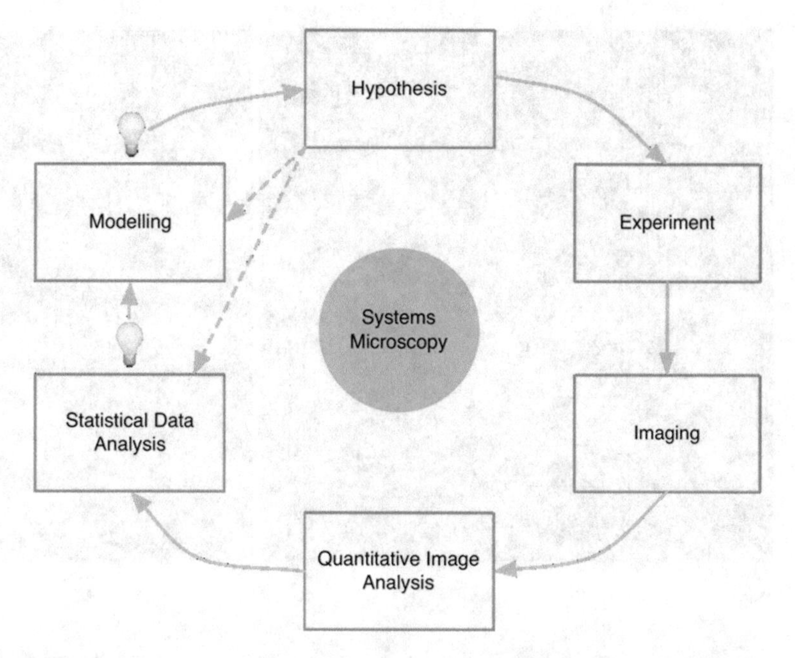

Fig. 1 Overview of systems microscopy. Main components of the systems microscopy research strategy are depicted. Note that new insights (lightbulbs) are principally derived through analytical phases involving statistical interrogation and data modeling, resulting in new hypotheses. These may be tested either through new experiments, or through modified analysis of existing data. This "synthetic" experimental cycle is both rapid and efficient, and is facilitated by the extremely rich, high-dimensional nature of systems microscopy data. The figure is evolved from Lock & Strömblad 2010

these scales. Using this model, we recorded and analyzed hundreds of quantitative features that relate to cell migration, with a special focus on cell–matrix adhesions, mediators of the attachment necessary for mesenchymal cell migration (typically demarcated by EGFP-Paxillin) and, to a lesser extent, the actin cytoskeletal system (demarcated by RubyRed-LifeAct) that provides the forces driving migration [6, 7]. An overview of our systems microscopy strategy is illustrated in Fig. 1.

The recorded variables include morphological features (e.g., area, shape), pixel intensity and variance features, dynamic features (e.g., turnover, movement, rates of change) and localization features (e.g., distance to cell border or centroid). Within these categories, data describes cells as a whole, individual objects within each cell (e.g., adhesion complexes), and aggregate descriptors of intracellular object populations. Collectively, this produces approximately 200 features that broadly sample information spanning molecular, subcellular and cellular scales, simultaneously capturing data on both the behavior (migration dynamics) and organization of the cell migration system (Fig. 2). This simultaneity allows the

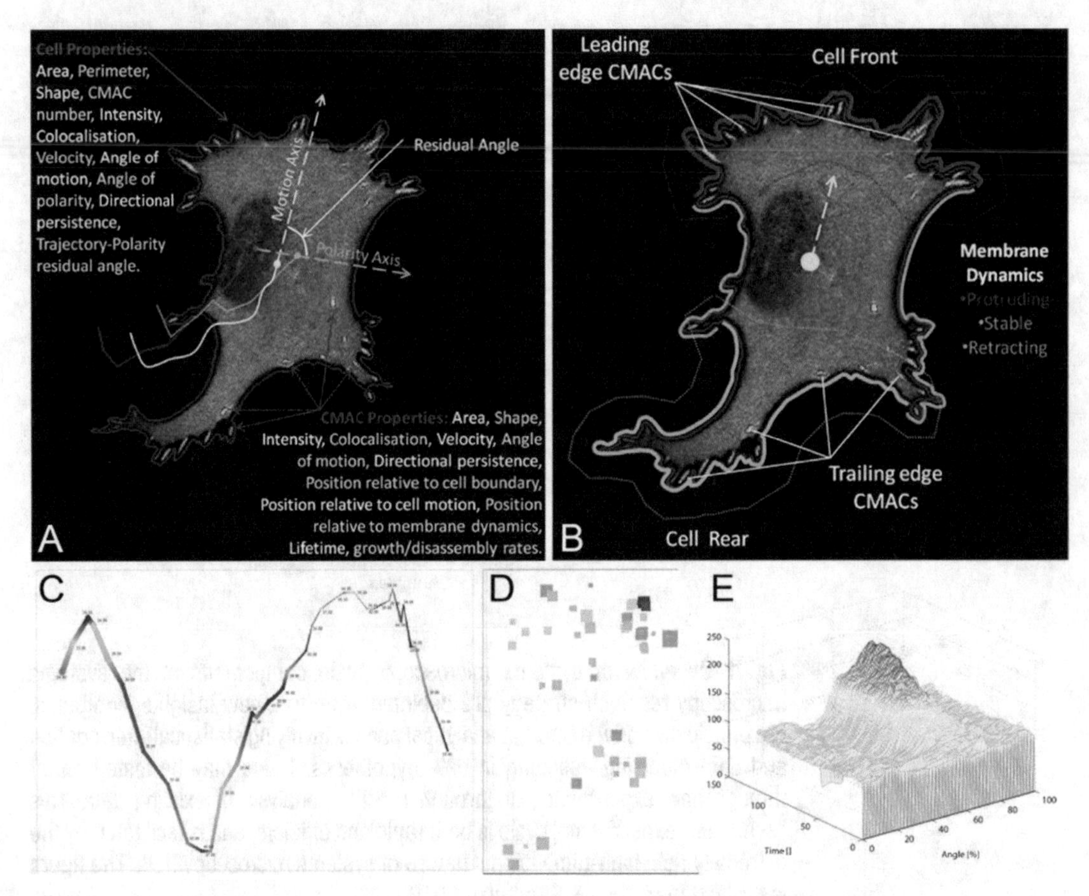

Fig. 2 Feature extraction and visualization. (**a**) The PAD software enables extraction of ~200 quantitative features per cell, which span spatial scales from the cell-level down to molecular complexes such as cell–matrix adhesion complexes (CMACs), and temporal scales from minutes (e.g., CMAC assembly) to hours (e.g., cell migration). The displayed cell is an H1299 lung cancer cells stably expressing EGFP-Paxillin as a marker for CMACs. (**b**) Postprocessing allows extraction of membrane dynamics [10] and orientation. (**c**–**e**) Diverse Systems Microscopy data necessitate a wide range of visualization approaches: (**c**) to concurrently assess cell migration trajectories (line), cell area (color), average CMAC area (thickness), and CMAC number (text); (**d**) CMAC position (coordinate), CMAC area (size), and protein density (color); or (**e**) peripheral membrane displacement (color and *Z* axis) over time (*X* axis) relative to cell motion angle (*Y* axis)

assessment of correlative and causal relationships between features of cell organization and behavior, using statistical techniques that leverage covariation in feature values [3, 8–12]. This may include specific, pairwise relationships (e.g., between adhesion complex size and cell migration speed), or more extensive inference of influence networks. Moreover, spontaneous (or induced) variations in feature values reveal the structural patterns of heterogeneity within and between cells. This can indicate, for example, whether variations in cell migration behavior denote distinct forms of migration, or simply variation within a continuum [3]. The abilities to recognize distinct subpopulations based on a broad sampling of cellular features, and to discern the dependencies between such features, are powerful benefits of Systems Microscopy.

2 Methodology

2.1 Preparation of Cells

In our experience, seemingly subtle differences in cell culture handling can give rise to remarkable variability in the quantitative features extracted from migrating cells. It is therefore vitally important to standardize cell handling before imaging. This includes cell culturing conditions during passages before replating. While use of the same medium and serum batch are obvious standardization measures, it is also important to standardize the range of passages to be used, cell density at all stages of culturing and the timing of population splitting. These measures greatly reduce variability between experiments, ultimately enhancing biological signal-to-noise ratios in experimentation.

In some cases, cells have been perturbed by pretreatments such as by transfections and/or use of chemical inhibitors. Needless to say, the perturbation protocols must be standardized to obtain reproducible results. As many others, we have perturbed cells via siRNA transfection prior to imaging, revealing the effects of reducing or eliminating specific proteins. However, while typical siRNA experiments address aggregate population responses to near complete depletion of a target protein, systems microscopy enables deeper and more nuanced analyses of depletion effects. In particular, extraction of data from single cells allows for the concurrent analysis of heterogeneity in both (residual) protein levels and response features of interest (migration speed, adhesion area, etc.). Moreover, by coupling this capability with use of low siRNA concentrations, giving high variation in target knockdown efficiency between cells, we were able construct "titration response" curves for each feature of interest given varied residual levels of target protein. This converts RNAi-based knock down approaches from a binary analysis promoting linear inferences to an analogue analysis permitting nonlinear inferences, which is potentially critical in understanding complex cellular mechanisms. For instance, this approach permits a distinction to be drawn between proteins whose natural range of expression level variation is or is not functionally significant. In our hands, this analogue analysis strategy revealed the effects of natural variations within the physiological range of talin protein expression levels, and moreover, illuminated both nonlinear (e.g., sigmoid) and nonmonotonic responses to talin expression level variation [12]. Such insights would not have been accessible using a typical siRNA knockdown approach with aggregate analyses.

To enable live cell imaging, we have also transfected cells with various genetically tagged fluorescent markers. To obtain the most uniform expression levels possible, we followed transfection and antibiotic selection with multiple rounds of FACS sorting. In general, low to moderate expression levels are selected. When observing adhesions based on a cytoplasmic marker protein (e.g., paxillin), low expression levels typically give the best signal-to-noise ratio at the imaging stage.

2.2 The Imaging Environment

For imaging, we typically use 96-well glass bottom dishes optimized for high resolution microscopy spanning multiple wells (conditions) per experiment. We precoat wells with purified ECM proteins to provide a well-defined and reproducible substratum for cell adhesion and migration. We most commonly employ fibronectin purified from human plasma, but we have also utilized purified collagen type I, vitronectin, and laminin. This choice typically reflects the integrin heterodimer(s) of interest, or a desire to stimulate alternative mechanisms of migration. Before performing experiments at any large scale, we analyzed ECM coating evenness in wells by using a mixture of fluorescently labeled and unlabeled fibronectin. We found unexpectedly large variations between coating conditions and between plate manufacturers; we thereafter selected conditions and plates that resulted in the most even and reproducible ECM coating, thus limiting unintended sources of local variation within wells. This was a crucial step to increase experimental reproducibility.

Similarly, at the replating step, cell seeding density is critical. In biological terms, cell density has a profound effect on the organization, behavior, and health of individual cells and cell populations. From a technical perspective, high cell density can be problematic, since it increases the cell collision frequency, thus distorting cell migration behaviors and patterns. Conversely, low cell density obviously limits the number of cells per microscope imaging field, thus demanding a larger number of fields, potentially limiting imaging frequency or simply reducing data production. The optimal balance between these trade-offs varies across cell types and conditions, and is another vital parameter to be optimized prior to full-scale data acquisition. Given these factors, replating conditions must be optimized for each cell model, well size and coating condition, among other variables.

As others, we routinely avoid using wells positioned at the outer rim of the plate, since these wells are subject to more rapid evaporation during imaging. This is despite use of either small (stage-mounted) or large (stage and objective-encompassing) environmental regulation chambers that control humidity (as well as temperature and CO_2). Typically, when using high resolution oil objectives (e.g., 60× oil; N/A 1.4, see below), we limit imaging to two neighboring wells for each imaging session, enabling acquisition of one control and one perturbed condition. While this results in low throughput, repositioning of the objective between only two wells per session provides the best chance to retain a stable autofocus function during imaging sessions ranging from 6 to 24 h. Even given this restriction, slow stage/objective movement is nonetheless key to retaining effective contact with viscous immersion oils over such periods.

To enable consistent tracking and segmentation (including watershed splitting) of cells, membrane dyes (e.g., Cell Mask Deep Red; Thermofisher) and nuclear markers (e.g., Draq5; Thermofisher, Hoechst 33342; Thermofisher) are typically added approximately 30 min prior to imaging. Minimal concentrations (typically 0.1–0.01 times the manufacturer's instructions) are used to prevent unwanted biological effects.

2.3 Live Cell Microscopy

Microscope automation is vital to enable the recording of image data spanning multiple positions within multiple wells across multiple fluorescence channels over time. We typically image a mosaic of adjacent objective fields in each well (e.g., 4 × 4 adjacent fields), allowing software-aided stitching of fields during or postacquisition. This not only increases the number of cells observed, but allows tracking of cells migrating between imaging fields, thereby providing spatially and temporally extended trajectories that would otherwise be truncated. We restrict imaging to the central area of each well to avoid edge effects including cell crowding. Another critical component of acquisition performance is reliable real-time autofocus. While autofocus based on image analysis is also possible, this is typically slow and requires repeated illuminations of the sample, potentially causing phototoxicity. Accordingly, the so-called "hardware" autofocus systems are preferable, since they are extremely rapid and use long-wavelength lasers to reliably optimize focus without phototoxic effects. While we found such hardware autofocus to be relatively robust, loss of autofocus nonetheless occurs intermittently and appears virtually unavoidable at a low frequency. We mitigate this problem through careful curation of data prior to quantitative analysis (see below).

When setting up microscopy experiments, there is nearly always a compromise between speed and resolution. We have chosen to use high resolution confocal microscopy (60× oil; NA = 1.4) to be able to resolve and segment subcellular features such as cell–matrix adhesion complexes that form at or below the diffraction limit and are located in close proximity to each other. We typically employ four- or eight-fold scan averaging by line to enhance signal-to-noise ratios, improving intensity quantification. It is critical that minimal laser power be used so as to limit light exposure and thereby avoid phototoxicity and photobleaching effects.

Another critical experimental setting corresponds to the frequency and duration of imaging. This must be optimized so that dynamic features of interest can be captured with sufficient detail as well as with a robust safety margin, e.g., sampling frequently enough to accurately quantify dynamics, while minimizing the influence of occasional data gaps and limiting light exposure to nonperturbative levels. This problem can be formally assessed using analysis of dynamical event frequencies and the Nyquist criterion, although it is impractical to perfectly optimize acquisition of multiple dynamical processes with different frequencies (e.g., cell migration and adhesion turnover). Thus, in practice, we have typically found a compromise between the requirements of multiple overlapping biological processes-of-interest, as well as other competing parameters including image resolution, light-exposure, and intensity signal-to-noise ratios. Thus, to study migration in H1299 lung cancer cells, we typically image cells at 5 min intervals over 8–12 h, thereby producing time series of 96–144 frames per image field.

2.4 Immunofluorescence Labeling

In specific cases, we have followed live cell imaging with rapid on-microscope fixation/permeabilization, followed by use of liquid handling robotics to perform automated immunofluorescence labeling. By subsequently returning the same sample to the microscope, it is then possible to assess the disposition of labeled molecular markers in the same individual cells that were previously imaged live. Thus, we can link the organizational state of each cell—augmented by information from additional molecular markers—to the known behavioral history of the same cell(s). This provides a basis to leverage natural heterogeneity and covariance in cell behavior and organization as a source of information. Notably, liquid handling robotics proved important in this workflow, as they provided dramatically improved consistency (and hence quantitative robustness given subsequent image analysis) and greatly reduced damage to cell samples (resulting from fluid shear stress in 96-well plates when manually labeled/washed).

2.5 Quantitative Data Extraction

A range of software solutions now directly support the image analysis requirements of Systems Microscopy as detailed herein; simultaneous multiscale segmentation and tracking of parent objects (cells) and intracellular objects (e.g., adhesions), maintaining relational hierarchies. When we initiated this approach, however, it was necessary to develop such a solution. To this end, we engaged in codevelopment of the PAD software with a commercial entity, Digital Cell Imaging Laboratories (Belgium). This software is effective for segmentation and tracking of adhesion complexes and their "parent" cells, thereafter extracting approximately 200 quantitative features spanning morphological, intensity, positional, and dynamical properties at the cellular and adhesion scales. Examples of cell features include cell area, perimeter, shape variables, number of adhesion complexes, object intensity, pairwise channel colocalization, cell migration velocity, angle of motion, and directional persistence; similar features are simultaneously extracted for each segmented adhesion complex (Fig. 2; described in detail by Lock et al. [9]). In addition, we performed postprocessing to identify protruding and retracting membrane domains as well as membrane domain (or "pixel") age, as described by Kowalewski et al. [10]. The performance of the segmentation in terms of the smallest segmentable size and the resolution of adjacent adhesion complexes was estimated by Hernandez-Varas et al. [11].

2.6 Parsing of Data

To analyze image derived quantitative data, we established an analysis platform through which data are processed in a stepwise manner, starting with data parsing and continuing with a range of complementary statistical analyses (Fig. 3). The step of data parsing is central to the accuracy of the final interpretations. Parsing steps include error corrections and elimination of obvious measurement noise, along with quantitative standardization and normalization steps. Notably, the PAD software allows manual inspection and curation of segmentation and tracking errors. This is setup so that the correction of an erroneous segmentation or tracking event will be propagated to automatically correct the segmentation for the same cell or object in the following time series, avoiding the need to manually correct every single frame.

Gaps in time sequence data also require curation and/or correction. Such gaps may be caused by, for example, temporary loss of autofocus or erroneous object segmentation. For correction, it should first be considered what maximal time gap should be considered for joining. We limit manual data curation to gaps of 2–3 frames and fill these small gaps by interpolating between the existing time points surrounding the gap. However, for larger gaps, we consider data independently in order to limit spurious sources of dynamical data. Moreover, we only consider objects that occur at a minimum of four consecutive time points, ensuring that neither shot noise nor transient cytoplasmic signals contribute to quantitative analyses.

For cell migration quantification, we have been interested to relate the instantaneous cell migration behavior, such as instantaneous cell migration speed or changes in motion angle, to features defining cell organization. However, given that we use changes in cell centroid (geometric center of area) to determine cell migration speed, this measure could be influenced by transient protrusions and retractions that do not constitute genuine cell migration (i.e., whole cell displacement). To mitigate such an effect, we optimized (for each project) spline fitting to smooth cell trajectories, thus reducing the influence of small, short-lived alterations in centroid position. This smoothing has the added advantage of converting velocity data from a discrete (integers of pixel dimension) to a continuous distribution (fractions of pixel

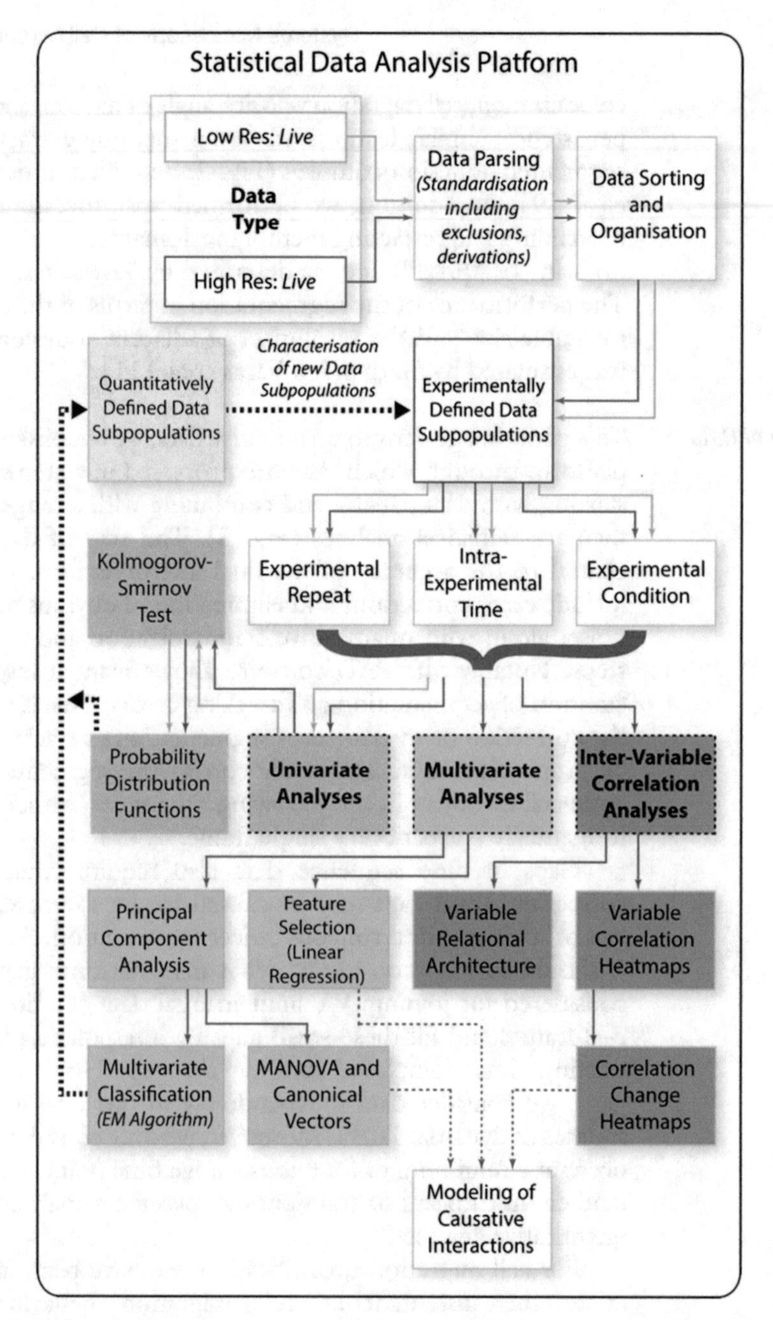

Fig. 3 Statistical Data Analysis Platform. Pathways for analysis of data derived from Low resolution (blue path; captures cellular-scale features only) and High resolution (red path; also captures features of each intracellular object of interest, e.g., adhesions) imaging. Data parsing, sorting, and organization precede a range of complementary statistical analysis methods. Established data subpopulations (e.g., different experimental repeats or conditions) drive comparison of data using univariate (e.g., distribution analyses and non-parametric Kolmogorov–Smirnov testing) and multivariate techniques (e.g., multivariate feature selection via regression modeling; or dimension reduction via principal components analysis). Dimension reduction methods facilitate unsupervised/supervised multivariate classification of new, quantitatively defined data subpopulations, promoting new analysis cycles. MANOVA and canonical vectors assess the statistical validity of subpopulation definitions and their arrangement in high-dimensional space. Correlations reveal pairwise interfeature relationships (heat maps), relationship plasticity (e.g., between experimental conditions), and correlative networks defining variable relational architectures. Features highlighted via regression-based feature selection, canonical vectors analysis or as sensitive features in correlation change heat maps are then candidates for modeling of causal interactions, e.g., using Granger causality modeling or Bayesian methods

dimension). This substantially improves the performance of several downstream statistical methods, especially those based on regression.

Other data parsing steps addressed measurement of object lifetimes, e.g., adhesion complex lifetimes indicative of stability. To accurately interpret this measure, we removed all adhesions detected at the start or end of image sequences, since the true lifetimes of these objects are inevitably underestimated and ultimately unknown. Furthermore, though microscope laser settings are standardized, detection intensity nonetheless varies between imaging sessions. Therefore, to allow comparison of intensity measures, we normalized adhesion complex fluorescent intensities relative to the median of all individual adhesion intensities for adhesions between 0.15 and 0.2 μm^2, for each recorded channel, in each experimental repeat of the relevant control condition. This size range incorporates numerous adhesions and is above both the optical and digital resolution limits of our acquisition systems, thus ensuring that intensities will not vary simply as a function of undetected differences in object size.

2.7 Statistical Analyses

We have integrated and applied a complementary array of statistical methods to analyze the rich, high-dimensional data derived through imaging and quantitative image analysis (Fig. 3). Despite this emphasis on quantitative data and techniques, the value of actually looking at the images should never be underestimated. This is an invaluable step toward revealing robust and easily recognizable patterns of interest, which may serve to focus subsequent quantitative analyses.

Statistical analyses are performed in a stepwise manner, and follow a logical structure such that each step informs us as to the need (or no need) for subsequent steps, as well as how such steps should be specified. Analyses begin by recognizing the identities of different experimentally defined data subpopulations, reflecting different experimental repeats (or replicates), different time-periods within an experiment, and/or alternative experimental conditions. These definitions provide the first rational basis for data comparisons, which address either individual features (univariate analyses), correlative relationships between pairs of features (correlation analyses), or arrays of features (multivariate analyses).

For elucidation of univariate questions (e.g., how specific features differ between conditions), plotting of the data distributions for each of the measured features, such as in probability density plots, allows quick manual comparison of responses. Such visualisation often reveals interesting differences between conditions (e.g., perturbation effects) that can then be the focus of deeper analysis. Plotting and analysis of univariate data distributions can also help to guide the choice of subsequent statistical methods (e.g., parametric versus non-parametric), and even provide insights into classes of potential mechanisms that may underlie emergence of the distribution. For instance, power law distributions (similar to those observed for adhesion size) are occasionally produced by "rich get richer" mechanisms or so-called "preferential attachment." Such information may illuminate new lines of analysis. Within our data, we found that most features displayed non-parametric distributions, dictating a reliance on non-parametric statistical methods, such as the Kolmogorov–Smirnov test for statistical difference between distributions.

Beyond univariate analyses, we assess intervariable correlations in a systematic manner through global pairwise measurement of Spearman's correlation coefficient. Spearman's correlation coefficient is less sensitive to linearity in relationships (as compared to Pearson's correlation coefficient), since it is rank ordered. Again, this makes it a preferred match for non-parametric data containing nonlinear dependencies. Considering the challenge posed by the large number of measured features, we visualize these pairwise correlations as matrix heat maps, facilitating the rapid identification of interesting correlations [9]. To quickly identify relationships that change between cell subpopulations (e.g., under alternate conditions), we also generate heat maps visualizing differences in these correlation matrices.

In addition, we apply multivariate methods to assess both systemic and specific characteristics of cells and adhesions. In the first instance, this is typically achieved through use of dimension reduction methods, such as principal components analysis (PCA; applied to rescaled data only). PCA is a powerful tool for exploratory data analysis that, by maximizing the proportion of total variance translated from the original high-dimensional data to the two or three-dimensional visualization, allows us to rapidly identify novel subpopulations or outliers in data sets containing hundreds or thousands of observations. PCA can therefore be used as the substrate for supervised subpopulation classification. PCA also provides a basis to identify correspondences and divergence between features, and indeed, which features have relatively high or low values corresponding with particular subpopulations of the observed data. This provides a rapid mechanism to interpret the "structure" of multivariate heterogeneity across observations, as well as which features particularly contribute to defining this structure.

As well as being used to compare cell populations between experimental conditions, we have also used PCA to control, at a multivariate level, for potential effects of phototoxicity and photobleaching. This was achieved by binning different time periods throughout an imaging session, and then projecting this data into a PCA space. This revealed that there was no progressive drift of data populations, which would have implied time (or light)-dependent changes in cells *en masse*, over time [9].

Another dimensionality reduction method that we utilized is canonical vector analysis (CVA). As opposed to PCA, CVA is sensitive to the identities of predefined groups of observations, e.g., comparing different conditions, and attempts to maximize the separation of these groups by optimally weighting features within each canonical vector (axis). This supervised approach allows us to ask whether substantial differences are quantitatively detectable between defined data populations, and can also be used to rapidly identify which features are most altered (in value) between populations. This provides a useful basis for automated selection of features that distinguish defined subpopulations. To select features that strongly contribute to the prediction of another feature value (e.g., cell speed), we utilized Elastic Net regression [13]. This progressively minimizes the complexity of a model, while attempting to maximize estimation precision. By measuring an adjusted metric for precision (that trades off model complexity versus error size), it was possible to identify a small set of features that provide the maximum information about cell migration speed, while still maintaining a level of interpretability. Interestingly, when configured to assess changes in cell speed, it was noteworthy that both CVA and Elastic Net regression-based feature selection methods displayed a strong overlap [9].

Different migratory behaviors or "modes" reflect different underlying cellular organizations. Moreover, in these modes, it is likely that distinct regulatory mechanisms exert control on cell migration features such as speed or directionality. The identification of cell subpopulations that employ different migration modes is therefore vitally important, since disaggregation of fundamentally different cellular (or cellular component) behaviors is a prerequisite for their understanding. Therefore, extending on the simple observation of subpopulations (data clusters) displayed via dimension reduction techniques (PCA, CVA), formal classification techniques provide unbiased approaches to determine the composition of subpopulations. To this end, we have employed the expectation maximization (EM) algorithm to automate definition of cell and adhesion subpopulations across a range of datasets [3, 9, 11].

2.8 Modeling

A key advantage of time series data sets, such as those derived from time lapse microscopy of migrating cells, is that temporal information can reveal directionality in relationships between features. Accordingly, we have used cross-correlation analysis to reveal time correlated patterns in stochastic fluctuations of force and adhesion growth dynamics [11]. Moreover, we have utilized the Granger causality concept realized through vector autoregression modeling to reveal directional influences between features of the cell migration system.

Utilizing our time-resolved quantitative data, we applied Granger causality modeling, a statistical method originally developed for the Econometrics field [14]. Specifically, we used Granger causality modeling to detect directed patterns of influence between pairs of cell migration system features [9]. Granger causality modeling reveals the degree to which one variable can improve prediction of the future state of a second variable, when compared to the ability of the second variable to predict itself. This essentially tests whether the first variable contains external, preemptive information about the future state of another variable, which is defined as evidence for Granger causality. This approach also allows assessment of the temporal structure of causality, i.e., if effects are immediate or time-delayed. Thus, using Granger causality modeling, we can assess whether one feature acts upstream or down-stream of another, if it has a positive or inhibitory influence, and whether the effect is immediate or time-delayed. This provides a major advance upon canonical correlation analysis methods.

We have also utilized probabilistic/stochastic modeling approaches to explore the experimentally quantified relationship between cell adhesion size and mechanical force application across the adhesion component vinculin [11]. This modeling reconstituted empirical feature distributions observed across the lifetime of adhesion complexes, confirming its biological validity. Furthermore, the model explained links between the tension-size relationship and inherent limits of adhesion complex size and lifetime. Even more remarkably, this modeling predicted that focal adhesions could become locked in an attractor state wherein normal size and force fluctuations diminished, implying also that of such "locked" adhesions would ultimately disassemble via a specific, force-independent mechanism. Using this prediction, we experimentally tested whether catastrophic microtubule-mediated adhesion disassembly could account for this force-independent disassembly mechanism, and indeed found compelling evidence to this effect [11].

3 Future Perspectives

The utilization of quantitative imaging is rapidly becoming standard in cell biology research. Combining this quantitative imaging with various multivariate statistical methods and modeling—as described herein—provides a powerful mechanism for performing systems biology analyses resolved in time and space. Nonetheless, significant scope remains for development of these methodologies, to enhance their efficacy and efficiency.

Firstly, we may be able to utilize our hard-earned data in more efficient ways. Given the enormous complexity of cell migration and its underlying cellular organization, including different migration strategies (migration modes) and context dependencies, one strategy to resolve this complexity more efficiently would be to integrate data and analyses from different laboratories. Developing such integrated meta-analyses would add significant value to each contributing data set, and allow both accelerated and extended progress beyond the scope of individual laboratories. However, a number of prerequisites are required to facilitate this capacity. Specifically, cell migration meta-analysis would benefit from establishment of designated data repositories for quantitative cell migration data, entailing both migration specific variables and migration related cellular and subcellular features. However, for such repositories to efficiently facilitate meta-analysis, agreement is first needed to standardize data to reflect a controlled vocabulary, such as for descriptors of experimental conditions and quantified features. Standards are also needed for minimal reporting requirements and file formats, including a standard for the hierarchical organization of quantitative multiscale time series data. There is presently an international effort to achieve such standards for the cell migration research field, embodied in the Cell Migration Standardization Organization (CMSO) [15, 16]. If successful, this effort could significantly accelerate cell migration research. In addition, development of standardized measures of the variability within and between multivariate datasets would be a useful quality indicator that can be used to determine which datasets that may be suitable for inclusion in a specific meta-analysis.

We could also learn a lot more about the cellular organization underlying cell migration by adapting and combining already existing tools. For example, recent developments in deep learning may provide powerful tools for unraveling inherent relationships within migrating cells at different scales and over time [17]. In addition, combining the capture of multivariate migration features with analyses of exerted forces, such as by traction force microscopy [18], could help us unravel fundamental relationships between cellular organization, force application, and cell migration behavior. Furthermore, the development and ongoing improvement of biosensors, e.g., FRET sensors for specific signaling components [19, 20], will allow capturing and analysis of how specific signaling event resolved in space and time contribute to specific cell migration events. There is also rapid development in optogenetic tools allowing the activation or inhibition of specific signaling components at particular cellular localizations [20]. The combination of biosensors to detect signal dynamics at specific locations, and optogenetics, to perturb the same signals in a highly controlled spatial and temporal manner, will allow the production of conclusive evidence as to how specific signaling events control cell migration. A complementary strategy may be to use recently developed image multiplexing technologies for in situ expression profiling, such as MERFISH [21], as well as the rapidly developing capacity of single cell sequencing. These technologies may provide important tools in situations where cell populations display heterogeneity with regards to their cell migration strategy. For example, this could reveal the differences in gene expression that are linked to particular cell migration behaviors, and could also reveal the heterogeneity in gene expression patterns that persists within each cell migration modality.

We are convinced that utilization of these and other developments, in combination with the systems microscopy strategies outlined herein, will greatly advance our understanding of cell migration by unraveling in detail how cells organize themselves to migrate in different contexts.

References

1. Horton ER, Byron A, Askari JA et al (2015) Definition of a consensus integrin adhesome and its dynamics during adhesion complex assembly and disassembly. Nat Cell Biol 17:1577–1587
2. Schiller HB, Friedel CC, Boulegue C, Fässler R (2011) Quantitative proteomics of the integrin adhesome show a myosin II-dependent recruitment of LIM domain proteins. EMBO Rep 12:259–266
3. Shafqat-Abbasi H, Kowalewski JM, Kiss A et al (2016) An analysis toolbox to explore mesenchymal migration heterogeneity reveals adaptive switching between distinct modes. Elife 5:e11384
4. Odenthal J, Takes R, Friedl P (2016) Plasticity of tumor cell invasion: governance by growth factors and cytokines. Carcinogenesis 37:1117–1128
5. Alexander S, Weigelin B, Winkler F, Friedl P (2013) Preclinical intravital microscopy of the tumour-stroma interface: invasion, metastasis, and therapy response. Curr Opin Cell Biol 25:659–671
6. Lock JG, Wehrle-Haller B, Strömblad S (2008) Cell-matrix adhesion complexes: master control machinery of cell migration. Semin Cancer Biol 18:65–76
7. Callan-Jones AC, Voituriez R (2016) Actin flows in cell migration: from locomotion and polarity to trajectories. Curr Opin Cell Biol 38:12–17
8. Lock JG, Strömblad S (2010) Systems microscopy: an emerging strategy for the life sciences. Exp Cell Res 316:1438–1444
9. Lock JG, Mamaghani MJ, Shafqat-Abbasi H et al (2014) Plasticity in the macromolecular-scale causal networks of cell migration. PLoS One 9:e90593
10. Kowalewski JM, Shafqat-Abbasi H, Jafari-Mamaghani M et al (2015) Disentangling membrane dynamics and cell migration; differential influences of F-actin and cell-matrix adhesions. PLoS One 10:e0135204
11. Hernández-Varas P, Berge U, Lock JG, Strömblad S (2015) A plastic relationship between vinculin-mediated tension and adhesion complex area defines adhesion size and lifetime. Nat Commun 25:7524
12. Kiss A, Gong X, Kowalewski JM et al (2015) Non-monotonic cellular responses to heterogeneity in talin protein expression-level. Integr Biol 7:1171–1185
13. Zou H, Hastie T (2005) Regularization and variable selection via the elastic net. J R Stat Soc Series B Stat Methodol 67:301–320
14. Granger CWJ (1969) Investigating causal relations by econometric models and cross-spectral methods. Econometrica 37:424–438
15. Masuzzo P, Martens L, Cell Migration Workshop Participants (2014) An open data ecosystem for cell migration research. Trends Cell Biol 25:55–58
16. Masuzzo P, Van Troys M, Ampe C, Martens L (2016) Taking aim at moving targets in computational cell migration. Trends Cell Biol 26:88–110
17. Kraus OZ, Grys BT, Ba J et al (2017) Automated analysis of high-content microscopy data with deep learning. Mol Syst Biol 13:924
18. Style RW, Boltyanskiy R, German GK et al (2014) Traction force microscopy in physics and biology. Soft Matter 10:4047–4055
19. Aoki K, Kamioka Y, Matsuda M (2013) Fluorescence resonance energy transfer imaging of cell signaling from in vitro to in vivo: basis of biosensor construction, live imaging, and image processing. Develop Growth Differ 55:515–522
20. Sample V, Mehta S, Zhang J (2014) Genetically encoded molecular probes to visualize and perturb signaling dynamics in living biological systems. J Cell Sci 127:1151–1160
21. Chen KH, Boettiger AN, Moffitt JR, Wang S, Zhuang X (2015) RNA imaging. Spatially resolved, highly multiplexed RNA profiling in single cells. Science 348:aaa6090

Chapter 11

Neuronal Precursor Migration in Ex Vivo Brain Slice Culture

Fu-Sheng Chou and Pei-Shan Wang

Abstract

Neuronal migration during fetal brain development is a well-coordinated process between the migrating neurons and their substrates, the basal processes of the radial glial cells (RGCs). The progeny–progenitor relationship between the migrating neurons and the RGCs in the developing fetal brain may make interpretations of the results difficult, because the variable in question may affect both the RGCs and the migrating neurons in different ways. A transplantation assay combining migrating cells and the scaffolding tissue from two different sources may circumvent this issue. We developed an ex vivo brain slice transplantation assay that allows recording of migrating neurons in real time.

Key words Neuronal migration, Neurosphere culturing, Neurosphere labeling, Brain slice culture, Cerebral cortex development

1 Introduction

Telencephalic development begins at neural tube closure and the transitioning of proliferative neuroepithelial cells into neurogenic radial glial cells (RGCs) [1]. The RGCs produce postmitotic neurons either directly or indirectly through intermediate progenitor cells. Additionally, RGCs serve as scaffolds for migrating neurons during development. To this end, the basal processes of the RGCs extend their membrane protrusions from the apically positioned cell body to the pial (basal) surface across the entire span of the cortex. Newly born postmitotic neurons are initially located on the apical surface in the ventricular or the subventricular zone. They then migrate basally toward the pial surface. Neuronal migration expands the telencephalon layer by layer: late-born neurons migrate a greater distance toward the basal surface than the early-born neurons do. The addition of neuronal layers also induces gyrification. Neuronal migration disorders are a group of birth defects that affect proper positioning of neurons during fetal brain development, resulting in a spectrum of structural abnormalities including lissencephaly, schizencephaly, polymicrogyria,

Alexis Gautreau (ed.), *Cell Migration: Methods and Protocols*, Methods in Molecular Biology, vol. 1749,
https://doi.org/10.1007/978-1-4939-7701-7_11,

and heterotopia. These patients suffer from a wide range of symptoms including motor dysfunctions, seizure activities, and cognitive impairment [2].

Migration defects and ensuing disorders can be caused by various factors: defective migration of the migrating neurons, abnormal development of RGC basal processes, defective communication between migrating neurons and corresponding RGCs, or a combination of the above. The progeny–progenitor relationship between the migrating neurons and the RGCs renders difficult to study neuronal migration in a cell type-specific manner by taking a traditional genetic approach alone. Alternatively, an ex vivo neurosphere transplantation assay allows the scaffolding RGCs and the migrating cells to be derived from different genetic backgrounds or to subject them to different pretreatments. Therefore it is an ideal approach to distinguish among various possibilities. An additional benefit of an ex vivo transplantation assay is the ability to record progress of cell migration in real time using time-lapse imaging techniques. Here, we will detail the technique we recently developed [3].

2 Materials

2.1 Mouse Embryonic Neural Stem Cell Harvesting

1. Timed pregnant mice with embryonic dates (E) between 13.5 and 15.5.
2. Dissecting tools: Spring scissors, standard scissors, microdissection forceps with ultrafine points (*see* **Note 1**), weighing spatula.
3. Dissecting microscope.
4. 70% ethanol.
5. Hanks' balanced salt solution (HBSS).
6. 10 cm petri dish.
7. 35 mm tissue culture-treated dish.
8. Hemocytometer.
9. Light microscope.
10. Tissue culture equipment.
11. Neural growth medium (NGM): DMEM/F12 supplemented with 20 μg/mL insulin, 0.5% N2, 1% B27, 20 ng/mL epithelial growth factor (EGF), and 20 ng/mL basic fibroblast growth factor (bFGF) (*see* **Note 2**).

2.2 Neurosphere Culturing and Single-Cell Dissociation

1. NGM (as described in Subheading 2.1, **item 11**).
2. Tissue culture-treated 6-well plates.

3. Dulbecco's PBS (DPBS).
4. Accutase® cell detachment solution.

2.3 Lentiviral Production

1. Constructs for lentiviral production are available from Addgene (*see* **Notes 3–5**).
2. 293T cells.
3. 293T cell culture medium: DMEM with 10% fetal bovine serum (FBS) supplemented with Penicillin/Streptomycin.
4. Calcium-phosphate DNA precipitation solutions: 2× Hepes buffered saline (HBS) at pH 7.1, 2 M $CaCl_2$, nuclease-free H_2O (*see* **Note 6**).
5. 0.22 μm syringe filters.
6. 10 mL syringes.

2.4 Labeling Dye

1. Vybrant® DiI Cell-Labeling Solution (ThermoFisher Scientific).
2. HBSS.
3. NGM (as described in Subheading 2.1, **item 11**).

2.5 Embryonic Brain Slicing and Culturing

1. Mouse fetal brains of desired genetic background.
2. Low melting-point (LMP) agarose.
3. Sterile-filtered (with 0.22 μm syringe filter) 1× Krebs buffer (1× Krebs, 1 M HEPES, 50 μg/mL gentamicin).
4. Sterile-filtered (with 0.22 μm syringe filter) 50% sucrose in distilled H_2O.
5. Plastic embedding molds (*see* **Note 7**).
6. Leica VT1200 Vibratome.
7. Sterile razor blades.
8. 0.5 in. flat end watercolor paintbrush.
9. Sterile weighing spatula (bended to about 30° for ease of use).
10. Shredded ice.
11. Brain slice collection medium: MEM, 10% FBS, 0.5% glucose, 50 μg/mL gentamicin.
12. Brain slice culturing medium: Neurobasal medium, 2% B27, 0.5% sucrose, 50 μg/mL gentamicin.
13. Millicell® Cell Culture Inserts for 6-well plates.
14. Small water bath with temperature set to 43 °C.

2.6 Neurosphere Transplantation

1. P10 pipetman.
2. Inverted fluorescence microscope with a live imaging chamber connected to compressed air and CO_2, long-distance 10× objective and autofocus function.

3 Methods

3.1 Mouse Embryonic Neural Stem Cell Harvesting

1. Euthanize an E13.5-E15.5 timed pregnant mouse with CO_2 following institutional guidelines.
2. Place the euthanized pregnant mouse in a supine position, and spray the abdomen with 70% ethanol to wet the fur and to clean the area.
3. Make a 1-in. midline incision in the abdomen to expose the abdominal contents.
4. Use a dissecting probe to locate and bring the entire uterus out of the abdominal cavity.
5. Cut the connecting tissues and vessels with standard scissors.
6. Use spring scissors and microdissection forceps to tear apart the uterine wall and the amniotic sacs in order to harvest the embryos. Following dissociation from the placenta, transfer the embryos to a 10 cm petri dish filled with cold HBSS immediately.
7. Use microdissection forceps to decapitate the embryos, followed by transferring the heads to a new 10 cm petri dish filled with clean cold HBSS.
8. Remove skull bones and dissect the fetal brains under a dissecting microscope using microdissection forceps with extra-fine points (*see* **Note 8**).
9. Use microdissection forceps to isolate mouse cortices by removing meninges, followed by tearing brain tissues into small pieces.
10. Transfer dissected brain tissues in HBSS into a 15 mL conical tube, centrifuge at 100 × *g* for 3 min.
11. Remove supernatant and add ice-cold NGM (0.5 mL/brain).
12. Gently triturate brain tissues with P1000 pipetman until big tissue chunks are completely broken down and no resistance is felt during aspiration.
13. Leave cell suspension on ice for 2 min, followed by passing through a 40 μm cell strainer.
14. Take an aliquot for cell counting.
15. Dilute cell suspension with NGM to 5×10^4 cells/mL.
16. Plate 2×10^5 cells (4 mL) in each well in a 6-well plate.
17. Culture cells for 3–4 days for neurosphere formation (*see* **Note 9**).
18. Unused cells may be cryopreserved in NGM with 10% DMSO.

3.2 Neurosphere Subculturing

1. Collect neurospheres with medium into 15 mL conical tubes and spin down at 100 × *g* for 1 min, remove supernatant.
2. Add 5 mL DPBS to resuspend neurospheres.

3. Spin down again at 100 × *g* for 1 min, remove supernatant.
4. Add 1 mL Accutase® cell detachment solution to the cell pellet, gently triturate to break up the cell pellet.
5. Place the mixture in a 37 °C tissue culture incubator for 3 min.
6. Triturate neurospheres gently using P1000 pipetman until large clumps no longer exist (approximately 20–30 times).
7. Add 2 mL NGM to neutralize cell detachment solution.
8. Take an aliquot for cell counting (*see* **Note 10**).
9. Spin down single cells at 160 × *g* for 4 min, remove supernatant.
10. Resuspend cells in NGM at a density of 2×10^4–3×10^4/mL. Plate 4 mL cell suspension into each well in a 6-well plate.
11. Replace half medium volume with fresh NGM every other day.
12. Repeat neurosphere subculturing when the sphere size reaches 150 μm (*see* **Note 11**).

3.3 Cell Labeling for Transplantation: Lentiviral Transduction

1. On Day 1, seed 2×10^6 293T single cells evenly in a 10 cm tissue culture-treated dish. Culture cells overnight to allow attachment and spreading (*see* **Note 12**).
2. On Day 2, mix 5 μg of lentiviral construct, 3 μg of gag/pol construct, and 2 μg of envelope construct using a calcium-phosphate DNA precipitation method.
3. Drop calcium phosphate–DNA precipitates on 293T cells evenly. Swirl the dish gently before replacing the dish in the tissue culture incubator (*see* **Note 13**).
4. On Day 3, remove calcium phosphate–DNA precipitate-containing medium and replace with 5 mL fresh 293T culturing medium. Incubate for 24 h (*see* **Notes 14** and **15**).
5. On Day 4, collect medium in a 10 mL syringe and pass through a 22 μm syringe filter to remove cell debris.
6. Collect filtered medium (viral supernatant) in a 15 mL conical tube, place on ice while preparing neurospheres for viral transduction.
7. Prepare single-cell suspension from neurospheres as described in Subheading 3.2, **steps 1**–**7**.
8. Remove supernatant and resuspend single cells with NGM at 5×10^6/mL density.
9. Transfer 100 μL cells (5×10^5 cells) into a 96-well plate. Add 100 μL viral supernatant to the same well, and place the plate in the 37 °C tissue culture incubator for 6 h (*see* **Note 16**).
10. Resuspend transduced cells in neurosphere culturing medium at 2×10^4/mL for 24 h.
11. Enrich transduced cells using the appropriate marker selection method (cell surface marker antigen–antibody-based magnet

selection, antibiotics resistance, or fluorescent protein-based cell sorting).

12. Culture transduced cells in NGM. Subculture using procedures described in Subheading 3.2.

3.4 Cell Labeling for Transplantation: Dye Labeling (See Note 17)

1. Collect neurospheres into 15 mL conical tubes and spin down at 100G for 1 min, followed by removing supernatant.
2. Resuspend neurospheres at a density between 1 mL in HBSS (approximately 1×10^5–1×10^7 cells) (*see* **Note 18**).
3. Add 5 μL of the cell-labeling solution supplied per mL of cell suspension. Mix well by gentle pipetting.
4. Incubate for 5 min at 37 °C followed by 15 min at 4 °C (*see* **Note 19**).
5. Spin down at 100G for 1 min, remove dye-containing HBSS, and wash once with 5 mL NGM.
6. Resuspend in NGM in the same volume.

3.5 Embryonic Brain Slicing and Culturing

1. Dissolve low melting-point agarose in sterile PBS to a final concentration of 4%. Boil until melted and leave the melted agarose at 43 °C water bath for later use.
2. Harvest fetal brains as described in Subheading 3.1, **steps 1–5**.
3. Place melted agarose in a disposable embedding mold at a depth of 1 cm, followed immediately by placing the harvested fetal brain in the agarose while still in the liquid form. Place the mold on ice to allow solidification to occur. Make sure that the agarose hardens fully.
4. Trim the solidified agarose with a razor blade to the margin of the fetal brain.
5. Place the embedded brain with the cortex orienting toward the razor on a Leica VT1200 Vibratome to section the brain coronally at a thickness of 300 μm.
6. Add ice-cold Krebs buffer to the Vibratome well. Surround the well with shredded ice (*see* **Note 20**).
7. Using the watercolor paintbrush and the weighing spatula, collect slices that contain visible ventricles and the neocortex into a 6-well plate containing sterile-filtered Krebs buffer on ice (*see* **Note 21**).
8. Transfer the slices in the laminar flow tissue culture hood to Millicell® Cell Culture Inserts placed in a 6-well plate containing 1 mL brain slice collection medium, followed by placing the plates in a CO_2 incubator (37 °C, 5% CO_2) for 1 h (*see* **Note 22**).
9. Replace the brain slice collection medium with 1 mL 37 °C brain slice culturing medium and place it back to the incubator.

3.6 Neurosphere Transplantation and Time-Lapse Imaging

1. Take the brain slices out of the incubator and remove any medium surrounding the tissue on the membrane (*see* **Note 23**).
2. Pick labeled neurospheres manually (>150 μm in diameter with bright fluorescence signal) using a P10 pipetman (set at 1 μL) under an inverted fluorescence microscope (*see* **Notes 24** and **25**).
3. Place the neurosphere inside the ventricle of the brain slice (*see* **Note 26**).
4. Gently push the neurosphere to allow attachment to the ventricular surface dorsally.
5. Place the brain slices with transplanted neurospheres back to the tissue culture incubator for 2 h.
6. Observe the transplanted neurosphere under an inverted fluorescence microscope and use the brain slice with integrated neurospheres (cells attached and migrated into the brain slice) (*see* **Note 27**).
7. Transfer the insert with the transplant to a glass bottom dish containing 1 mL serum-free medium for time-lapse imaging (*see* **Notes 28** and **29**).

4 Notes

1. We use Dumont #5 Forceps with biology tip.
2. After adding EGF and bFGF, the medium has a refrigerator life of approximately 1 week.
3. We use pSicoR-GFP lentiviral construct, along with the murine leukemia virus Gag/Pol plasmid and the pCMV-VSV-G plasmid for envelope protein.
4. Choice of selection markers (common types of selection markers include surface markers, fluorescent proteins, or antibiotic resistance genes) depends on experimental needs. A lentiviral construct that contains a fluorescent protein-expressing gene may facilitate time lapse imaging.
5. The choice of the fluorescent protein marker depends on the availability of the excitation lasers and the wavelength filters on the cell sorting machine as well as on the microscope for time-lapse imaging.
6. Calcium-phosphate transfection kits are commercially available from various vendors.
7. We use Peel-A-Way T-12, 12 × 12 × 20 mm, available from VWR 15160-157.
8. Keep brain tissues from each embryo separate if genotyping is needed.

9. Replace half medium volume with fresh NGM every other day.
10. Under light microscopy, assess whether cells are in single-cell suspension. If numerous neurospheres are still present, consider repeating Subheading 3.2, **steps 3**–7.
11. We typically subculture every 4 days.
12. Even distribution of 293T single cells across the dish improves plasmid transfection efficiency and viral load.
13. Black dots mimicking bacteria or cell debris should be visible under light microscopy few minutes after adding calcium phosphate–DNA precipitates to the dish. The dot sizes generally vary significantly, as opposed to contaminated bacteria which are uniform in size.
14. If a lentiviral construct containing a fluorescent protein marker is used, >50% 293T cells should turn into the corresponding color under fluorescence for a transfection reaction to be considered an efficient one. Transfection efficiency correlates with viral load in our experience.
15. 293T cells may begin to round up and detach from the dish bottom. This finding usually correlates with good viral load in the viral supernatant.
16. Avoid prolonged transduction period to avoid differentiation induction of neural cells.
17. The benefit of dye labeling is convenience. However, the dye labeling technique cannot guarantee labeling all cell types. Therefore there is a chance the result may be biased. Various dyes for short-term or long-term labeling are available from ThermoFisher Scientific.
18. The neurospheres can also be resuspended in NGM for dye labeling.
19. Low temperature slows down endocytosis and allows the dye to label the plasma membrane without appearing on the membranes of the cytoplasmic vesicles.
20. Replace existing Krebs buffer with fresh cold Krebs solution frequently and add shredded ice to surround the well periodically in order to avoid heat build-up.
21. If separation of brain tissues from the agarose mold is noticed, replace the Vibratome blade or adjust the Vibratome frequency and speed accordingly.
22. Slices should be placed in a CO_2 incubator within 2 h of sacrificing the pregnant dam.
23. Pay special attention to any residual medium inside the ventricles, as any remaining medium will prevent transplanted neurospheres from staying with the brain slice tissue and will make it difficult to adjust their positions after transplantation, remove if present.

24. Avoid subculturing neurospheres within 3 days of brain slicing, so that the neurospheres will have sufficient time to grow to a size of 150 μm on the day of transplantation.
25. Make sure the neurospheres are properly labeled using an inverted fluorescence microscope.
26. Only place one neurosphere per ventricle of the brain slice.
27. We always perform at least six transplantations at a time to increase yield. This will ensure that at least one transplant is suitable for time-lapse imaging. Unused transplants may be cultured for various durations and fixed for additional analysis such as immunofluorescence staining.
28. We take images every 5–10 min for 8–16 h.
29. Before imaging, place the glass bottom dish in the chamber and find the point of interest and leave it for at least 30 min to allow the temperature of the dish to equilibrate with the imaging chamber. This will prevent image shift due to thermal expansion and/or contraction.

Acknowledgment

This work is funded by Children's Mercy-Kansas City Children's Research Institute.

References

1. Paridaen JT, Huttner WB (2014) Neurogenesis during development of the vertebrate central nervous system. EMBO Rep 15:351–364
2. Neuronal Migration Disorders Information Page. National Institute of Neurological Disorders and Stroke. https://www.ninds.nih.gov/Disorders/All-Disorders/Neuronal-Migration-Disorders-Information-Page. Accessed 24 May 2017
3. Wang PS, Chou FS, Ramachandran S et al (2016) Crucial roles of the Arp2/3 complex during mammalian corticogenesis. Development 143:2741–2752

Chapter 12

In Vitro Models to Analyze the Migration of MGE-Derived Interneurons

Claire Leclech and Christine Métin

Abstract

In the developing brain, MGE-derived interneuron precursors migrate tangentially long distances to reach the cortex in which they later establish connections with the principal cortical cells to control the activity of adult cortical circuits. Interneuron precursors exhibit complex morphologies and migratory properties, which are difficult to study in the heterogeneous and uncontrolled in vivo environment. Here, we describe two in vitro models in which the migration environment of interneuron precursors is significantly simplified and where their migration can be observed for one to 3 days. In one model, MGE-derived interneuron precursors are cultured and migrate on a flat synthetic substrate. In the other model, fluorescent MGE-derived interneuron precursors migrate on a monolayer of dissociated cortical cells. In both models, cell movements can be recorded by time-lapse microscopy for dynamic analyses.

Key words Embryonic neurons, Interneurons, Cerebral cortex, Medial ganglionic eminence, Tangential migration, Migration assay, Culture, Cocultures, Culture dishes for videomicroscopy

1 Introduction

In the developing brain, postmitotic cells actively migrate away from the proliferative area surrounding the future ventricles to reach their target region where they will differentiate and establish functional connections. Two modes of neuronal migration coexist, either radial or tangential, according to the direction of movement. After leaving the proliferative region from its basal side, most neural cells migrate perpendicularly to the central lumen, e.g., radially. Alternatively, some populations migrate tangentially, e.g., parallel to the surface of the brain. They form several streams distributed at the surface or in deeper parts of the developing brain [1].

The tangential mode of migration was first discovered by lineage tracing experiments in the late 1980s [2, 3], and confirmed by pioneering time-lapse experiments in organotypic slices [4]. Genetic studies in the late 1990s established that GABAergic neurons use this tangential mode of migration and can travel very long

Alexis Gautreau (ed.), *Cell Migration: Methods and Protocols*, Methods in Molecular Biology, vol. 1749,
https://doi.org/10.1007/978-1-4939-7701-7_12, © Springer Science+Business Media, LLC 2018

distances to colonize the whole brain from restricted proliferative zones located in ventral regions [5, 6]. In particular, the inhibitory GABAergic cortical interneurons, which control the activity of cortical circuits in the adult, are born in the ganglionic eminences (GEs) of the basal telencephalon and migrate a long distance to reach the developing cortex. The correct positioning of these neurons in their cortical target, which critically influences their function, largely depends on their migratory properties. Therefore, impaired migration of cortical interneurons can lead to neurodevelopmental pathologies such as neuropsychiatric disorders [7].

We use interneuron precursors born in the medial ganglionic eminence (MGE) as a model system to study the cellular and molecular mechanisms controlling the tangential migration of embryonic neurons. MGE-derived interneurons exhibit complex morphologies with branched leading processes [8, 9] and the precise characterization of their migratory properties is often difficult in the complex 3D physiological environment. For this reason, we have designed in vitro models where MGE-derived cells migrate on flat substrates. Two types of in vitro migration assays have been developed: culture of MGE-derived interneuron precursors on a synthetic substrate [10, 11] or coculture of MGE-derived interneuron precursors on a monolayer of dissociated cortical cells [12].

Cultures of MGE explants on a synthetic substrate have been designed to test the role of adhesive contacts on the migratory properties of interneurons, in particular the role of integrin and N-cadherin (Ncad) mediated adhesive interactions. We describe here a mixed synthetic substrate of laminin and Ncad onto which MGE exhibit an active migration [10] (*see* Subheading 3.2 below). These cultures allow the visualization of isolated neurons migrating in a controlled 2D adhesive environment, and of their intracellular organization (cytoskeleton, adhesive contacts or signaling molecules).

We have designed another in vitro model in which MGE cells are allowed to migrate on a flat monolayer of dissociated cortical cells themselves plated on a laminin substrate. To identify and image MGE cells in these cocultures, we are using donor embryos expressing a fluorescent marker, either GFP or RFP [12] (*see* Subheading 3.3 below). MGE cells migrate here in a simplified environment that no longer shows any complex 3D structure, but in which functional interactions with cells from the cortical target are preserved. These cocultures can be easily treated with pharmacological compounds or used to analyze how mutations in either cortical or MGE cells can influence the migratory behavior of interneuron precursors.

In both types of cultures, MGE-derived interneurons migration can be imaged by time-lapse videomicroscopy and subsequently analyzed.

2 Materials

2.1 Preparation of Culture Dishes for Videomicroscopy

1. Round glass coverslips diameter 20 mm, thickness No. 1.5.
2. Stainless steel racks for glass coverslips with removable handle.
3. Glass coloration boxes with glass covers.
4. Glass petri dishes diameter 60 mm.
5. Plastic culture dishes (40 mm TPP or any 35 mm dish with thick bottom).
6. Hole punch set (hand tool in metal for making holes).
7. Deburring tool (to deburr inner hole edges).
8. Addition-curing duplicating silicone for dentistry (Twinsil speed 22 blue/yellow, Rotec).
9. Nitric acid 67%.
10. Ethanol 95°.
11. MilliQ water.
12. Drying oven.

2.2 Coating

Stock solutions of coating proteins

1. Poly-L-lysine (MW 538000): Stock solution at 2 mg/mL in water. Aliquots of 100 μL stored at −20 °C.
2. Laminin (from mouse sarcoma): Stock solution at 1 mg/mL in DMEM/F12, aliquots of 10 or 20 μL stored at −20 °C.
3. Recombinant Mouse or Human N-cadherin Fc Chimera (R&D systems, sterile solution). Aliquots of 2 or 4 μg stored at −80 °C.
4. Anti-human IgG Fc produced in goat (Sigma). Buffered stock solution at 2 mg/mL. Aliquots of 10 μL stored at −20 °C.
5. Bovine Serum Albumin (BSA): sterile, cultured tested, lyophilized powder.

   Buffers and solutions

6. Sterile milliQ water.
7. Dulbecco's Modified Eagle Medium: Nutrient Mixture F-12 (DMEM/F-12). Store at 4 °C.
8. Borate buffer: Prepare Borax solution 0.1 M by dissolving 3.8 g of Borax ($Na_2B_4O_7$, 10 H_2O, MW: 381) in 100 mL of milliQ water. Prepare HCl solution 0.1 M (1 mL of HCl 36% in 100 mL of milliQ water). Place a pH-meter into the Borax solution and pour HCl solution in borax solution until pH 8.2. Filter at 0.2 μm under a hood, and store in a sterile bottle at 4 °C for months.

9. Saturation solution for N-Cadherin coating: prepare stock solution of BSA at 1.5% in sterile borate buffer. Store aliquots of 1 mL at −20 °C.

Material

10. Laminar flow hood.
11. CO_2 incubator at 37 °C with water Jacket.
12. Sterile filtering bottle equipped with a 0.2 μm filter.
13. pH meter.

2.3 Dissection and Culture

Buffers and solutions

1. Dulbecco's Modified Eagle Medium: Nutrient Mixture F-12 (DMEM/F-12). Store at 4 °C.
2. Dulbecco's Modified Eagle Medium: Nutrient Mixture F-12 (DMEM/F-12) without phenol red. Store at 4 °C.
3. Leibovitz's L-15 Medium (L-15). Store at 4 °C.
4. GlutaMAX supplement, 100× solution. Store at 4 °C.
5. HEPES buffer solution 1 M. Store at 4 °C.
6. Glucose 30% in water, filtered at 0.2 μm under a laminar flow hood. Aliquots of 665 μL stored at −20 °C.
7. N2 supplement 100×. Aliquots of 100 μL stored at −20 °C.
8. B27 supplement 50×. Aliquots of 200 μL stored at −20 °C.
9. Penicillin–streptomycin solution (10,000 U/mL). Aliquots of 5 mL stored at −20 °C.
10. Sodium pyruvate solution 100 mM stored at 4 °C.
11. Fetal Calf Serum (FCS) decomplemented (30 min at 56 °C). Aliquots of 1 mL stored at −20 °C.

Culture and dissection medium (prepared just before the dissection or a day before, in sterile conditions under a laminar flow hood). Culture medium were adapted from [13].

12. Dissection medium: L-15 with 1% penicillin–streptomycin (5 mL for 500 mL L-15).
13. Dissociation medium: 9 mL DMEM/F-12; 1 mL CFS. Can be used for 1 week.
14. Culture Medium 1×: 50 mL DMEM/F-12, 500 μL GlutaMAX 100×, 500 μL HEPES 1 M, 500 μL of penicillin–streptomycin (10,000 U/mL), 665 μL of glucose 30%. Can be used for 2 weeks if the color remains salmon.
15. Complete Culture Medium: 10 mL Culture Medium 1×, 100 μL of penicillin–streptomycin, 100 μL of N2 Supplement 100×, 200 μL B27 Supplement 50×. Can be used for 1 week.

16. Video Culture Medium 1×: 50 mL DMEM/F12 without Phenol Red, 500 μL GlutaMAX 100×, 250 μL HEPES 1 M, 500 μL of penicillin–streptomycin, 665 μL of glucose 30%, 500 μL of sodium pyruvate 100 mM. Can be used for 1 week.
17. Complete Video Culture Medium: 10 mL Video Culture Medium 1×, 100 μL of N2 Supplement 100×, 200 μL B27 Supplement 50×. Can be used for 1 week.

Dissection

18. Dissection tools: 1 pair of iris scissors, 1 pair of spring scissors with 3–4 mm blades, 2 pairs of Dumont #5 forceps (tip Biology), 1 pair of Dumont #3 forceps, 1 pair of curved forceps, 2 tungsten needles with the tip bent at 90° (we use tungsten needles for extracellular recording). *After each dissection, tools are carefully cleaned with a toothbrush and detergent (TDF4), rinsed, dried and stored in a clean box.*
19. Sterile plastic pipettes (3 mL).
20. Beaker of 150 mL filled with ethanol 95°.
21. Beaker of 150 mL filled with sterile PBS 1×.
22. Isothermal box filled with ice and covered with a clean aluminum foil.
23. Large (2× diameter 100 mm), medium (5× diameter 60 mm) and small (2× diameter 35 mm) sterile plastic petri dishes filled with dissection medium and stored on ice.
24. Dissecting binocular microscope.
25. Fluorescent binocular microscope or microscope with a low magnification objective.
26. Counting chamber.
27. CO_2 incubator at 37 °C with water Jacket.

3 Methods

3.1 Preparation of Culture Dishes for Videomicroscopy

The culture dishes for videomicroscopy have to be prepared in advance. Our protocol is adapted from the protocol of Carol Mason's lab [14].

3.1.1 Cleaning of the Glass Coverslips

1. Distribute the coverslips in stainless steel racks. Using the removable handle, put the racks in a clean and dry glass coloration box and, under a chemical hood, cover the coverslips with nitric acid 67%. Close the box with a glass cover and put a label "DANGER" on it. Leave the coverslips overnight under the hood.
2. The next day, under the hood, transfer the racks using the removable handle in another glass coloration box filled with MilliQ water. Discard nitric acid with a glass funnel in a trash

bottle. Use several sets of glass coloration boxes to rinse the coverslips at least ten times, until the smell of acid has disappeared.

3. Put the racks in a clean and dry coloration box and cover them with ethanol 95° for 2–3 h.
4. Rinse the coverslips with distilled water five times. Then transfer the racks in a clean and dry glass coloration box kept slightly open. Leave it under a laminar flow hood until the coverslips are completely dry (this can take more than 1 day).
5. Distribute the dry coverslips in clean and sterile glass petri dish and sterilize them in a dry oven at 210 °C.

3.1.2 Culture Dishes for Videomicroscopy (See Fig. 1)

1. With the hole punch tool, drill a hole of 14 mm diameter in the center of the bottom of 40 mm TPP plastic dishes (*see* **Note 1**). Then, smooth the inner edge of the hole with a deburring tool.
2. Turn the dishes bottom up and place on the outside of the dish a 20 mm cleaned glass coverslip centered on the hole to close it (*see* Fig. 1a).
3. Prepare a small amount of twinsil silicone by mixing equal volumes of blue and yellow components (*see* **Note 2**). While the mix is still liquid (only a couple of minutes after mixing), create a thin ring of silicone around the coverslip to seal it on the external side of the plastic dish (*see* Fig. 1b). To avoid leakage of silicone below the coverslip, gently press it on the dish. After the mix hardens, the coverslip is efficiently sealed at the bottom of the box creating a small well in the dish (*see* Fig. 1c).

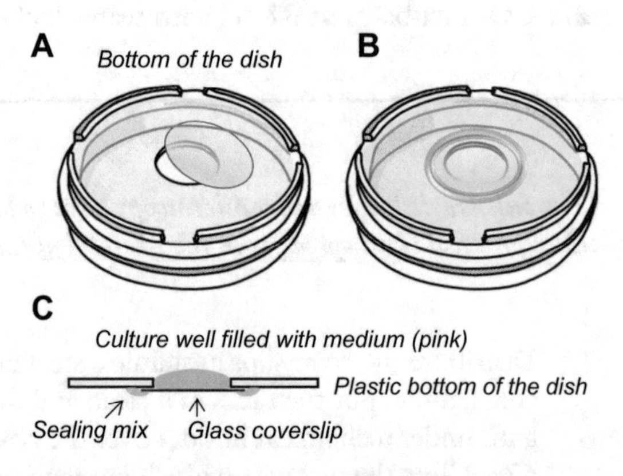

Fig. 1 Culture dishes for videomicroscopy. (**a**) An acid nitric treated glass coverslip closes the central hole performed with a hand punch tool in the bottom of a 40 mm plastic culture dish. (**b**) The coverslip is sealed with a ring of silicone (blue) on the external side of the dish. (**c**) Cross-sectional view shows the small culture well thus created in the dish, equipped with a glass bottom

4. When the twinsil silicone is dry, immerse the dishes and their covers in a beaker full of ethanol 95° for 30 min minimum. Rinse them three times with distilled water and let them dry under the laminar flow hood. Close the boxes and store them in a hermetic box (*see* **Note 3**).

3.2 Protocol 1: Culture of MGE-Derived Interneurons Migrating on a Synthetic Substrate

3.2.1 Coating: N-Cadherin and Laminin (See Note 4)

All steps should be done in sterile conditions under a laminar flow hood. Let aliquots of frozen coating proteins thaw in the fridge before use. The protocol to coat coverslips with Ncad is adapted from [15].

1. *The day before the dissection, coat with poly-L-lysine (PLL)*
   (a) Dilute the stock PLL in sterile water at a final concentration of 200 μg/mL. Put approximately 150 μL of the PLL solution on each coverslip in the culture dishes. The solution should cover the whole coverslip and fill the small well above (*see* Fig. 1c). Close the box and incubate at least 3 h in a CO_2 incubator at 37 °C (*see* **Note 5**).
   (b) Aspirate the PLL solution, and rinse the coverslip three times with sterile water. Let the coverslip dry under the laminar flow hood for at least 30 min, boxes open (*see* **Note 6**). Dry boxes can be closed and stored at 4 °C in a sterile airtight box for 1 or 2 weeks.
2. *The night before the dissection, incubate with laminin (LN) and anti-human Fc (hFc)*
   (a) Dilute the stock LN at a final concentration of 8 μg/mL and the stock hFc at a final concentration of 4 μg/mL in borate buffer. Put approximately 150 μL of the LN + hFc solution on each coverslip to cover it. Close the box and incubate overnight at 4 °C (*see* **Note 7**).
3. *The morning of the dissection, coat with N-Cadherin (NCad)*
   (a) Dilute the stock NCad in borate buffer to have a final concentration of 0.25 μg per cm^2 of coverslip (the coverslip surface of a 14 mm hole is rounded up to 1 cm^2).
   (b) Aspirate the LN + hFc and rinse it once with borate buffer. At this point the coverslip should never dry. Put approximately 150 μL of the NCad solution on each coverslip to cover it. Close the box and incubate 2 h at 37 °C in the CO_2 incubator. Start the dissection (*see* Subheading 3.2.2) during this incubation time.
   (c) Aspirate the Ncad solution and rinse once with borate buffer. To saturate hFc antibody sites unbound with recombinant Ncad, incubate the coverslips with the saturation buffer (1.5% BSA in borate buffer) for 10 min. Aspirate and replace directly with 200 μL of Complete Culture Medium. Close the boxes and let them wait at 37 °C in the CO_2 incubator, if necessary.

3.2.2 Dissection and Culture (See Fig. 2)

Use a pregnant mouse with gestation dated at embryonic day E12.5 or E13.5.

1. *Embryonic brain dissection*
   (a) Put the petri dishes filled with the dissection medium on ice (*see* Subheading 2.3). Put the clean tools in the beaker filled with ethanol 95°, and use the beaker filled with sterile PBS to rinse ethanol on tools before using them.
   (b) On a clean bench outside the culture room, prepare a sheet of absorbent paper, a 100 mm diameter petri dish filled with cold dissection medium, a pair of iris scissors and the Dumont #3 forceps previously stored in the beaker of ethanol. Kill the pregnant mouse by cervical elongation (according to validated ethical procedures). Place the mouse on the paper, abdomen up, and wash the skin of the abdomen with ethanol 95°. Open the abdomen with a long sagittal cut (first skin, then muscles). Collect the uterine horns containing the embryos in the petri dish filled with cold dissection medium. Transport the petri dish in the culture room and put it on ice (*see* **Note 8**). Discard the dead mouse in a freezer, and quickly clean the tools.
   (c) Transfer the uterine horns in a large petri dish filled with clean dissection medium. With the clean iris scissors, cut the uterine muscle all along the horns to expose embryos in their amniotic envelope. With the tips of Dumont #5 forceps, break the amniotic bag, extract the embryos and cut their head. Transfer heads using the curved forceps in a 60 mm petri dish with clean cold dissection medium.
   (d) To extract the brain, remove the skin and the connective tissue around it. To do so, gently insert the tip of a Dumont #5 forceps at the basis of each telencephalic vesicle and peel off the skin and the connective tissue. When the brain is uncovered, section it at the back between the cerebellum and the spinal cord with the fine forceps. Gently detach the brain from the brain cavity with the Dumont #5 forceps, progressing from caudal to rostral. Transfer the dissected brains in another 60 mm petri dish of clean cold dissection medium.
2. *MGE dissection* (*see* Fig. 2)
   (a) The MGE protrudes in the lateral ventricle of the telencephalic vesicle. To access the MGE, open the telencephalic vesicle dorsally with the spring scissors, and section the vesicle along the medial midline. Pursue the cut both rostrally and caudally in order to completely detach the lateral wall of the telencephalic vesicle, which comprises the cortex and the MGE (*see* Fig. 2b).

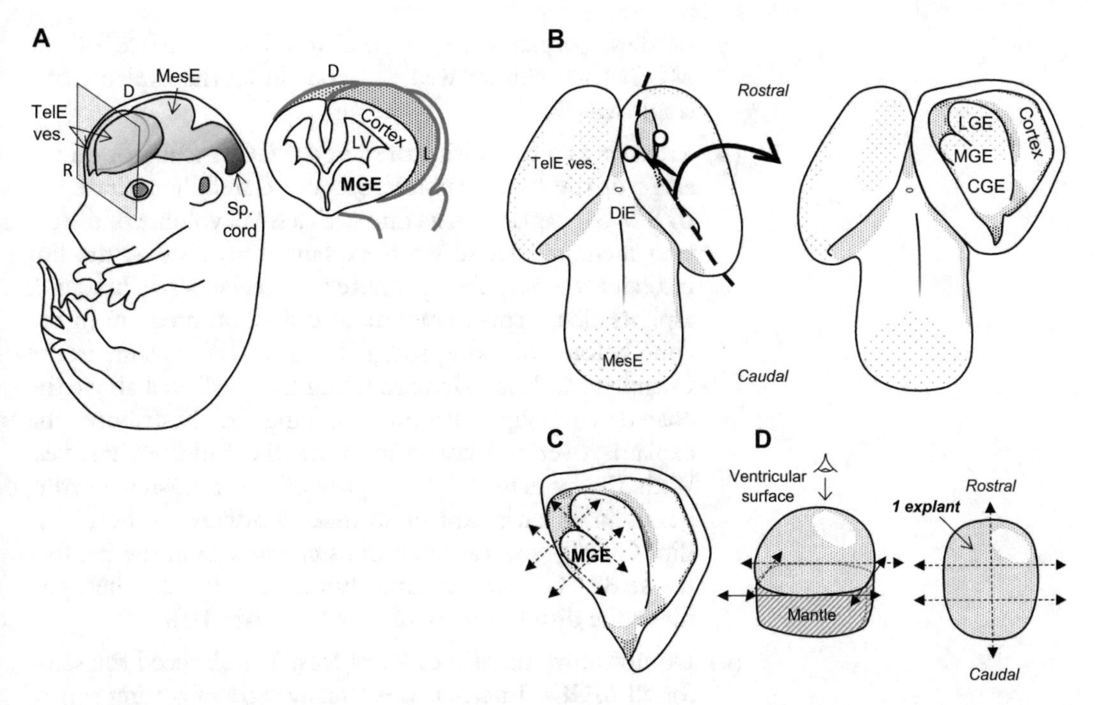

Fig. 2 Dissection of the Medial Ganglionic Eminence (MGE). (**a**) Scheme showing the brain of an E12.5/E13.5 mouse embryo. On the right, a coronal section through both telencephalic vesicles (TelE Ves.) shows the embryonic cortex and the MGE protruding in the lateral ventricle (LV). (**b**–**d**) illustrate the main steps of the MGE dissection. (**b**) In a telencephalic vesicle opened by its dorsal side, the cortical and MGE regions are clearly visible on the ventricular side. (**c**, **d**) The MGE is removed with 4 cuts (**c**), detached from the mantle zone (**d**) and cut in 6–8 pieces (explants) perpendicular to the ventricular surface. *MesE* mesencephalon, *DiE* diencephalon, *Sp. Cord* spinal cord, *LGE* lateral ganglionic eminence, *D* dorsal, *L* lateral, *R* rostral

(b) When the lateral wall of the telencephalic vesicle is turned ventricular side up, the MGE appears as a small "ball." To isolate the MGE, perform 4 cuts around it with a thin tungsten needle, perpendicular to the ventricular surface (*see* Fig. 2c). Then, detach the region of the MGE protruding in the ventricle with a cut parallel to the ventricular surface: the mantle zone (postmitotic structure) should be removed, because it will contaminate the culture with postmitotic neurons (*see* Fig. 2d). Transfer the dissected MGEs in a 35 mm diameter petri dish filled with cold dissection medium using a sterile plastic pipette. Store MGEs on ice.

3. *Culture*

(a) Right before transferring a MGE in a culture dish, cut it in 6/8 pieces (called explants). Each explant should contain a small part of the ventricular and subventricular zone (*see* Fig. 2d and **Note 9**). Sections are made with the tungsten needles, perpendicular to the ventricular surface of the MGE. Each culture well will contain all the explants of a single MGE.

(b) Transfer a culture dish stored in the CO_2 incubator on the stage of the binocular microscope. Carefully aspirate the 6/8 MGE explants in a cone with a small volume of dissection medium. These small explants dissociate at the liquid/air interface. To make sure that explants stay in liquid, aspirate first a small amount of dissection medium in the cone before collecting them. Then deliver explants in the Complete Culture Medium filling the small well above the coated coverslip without touching it. Distribute the explants over the coverslip using the tungsten needles. With the noncutting bent part of the tungsten needle, gently press each explant to make it adhere to the coverslip. Quickly and carefully transfer the dish in the incubator at 37 °C. The explants should not detach when you move the dish to the incubator (*see* **Note 10**).

(c) Do not move the dishes for at least 1 h. Proceed the same for all MGEs. Interneurons usually start migrating out of the explant after 4–6 h in culture (*see* **Notes 11** and Fig. 3a).

3.3 Protocol 2: Culture of MGE-Derived Interneurons Migrating on Dissociated Cortical Cells

As in Subheading 3.2.1, all the steps should be done in sterile conditions under a laminar flow hood. Let coating proteins thaw in the fridge before use.

3.3.1 Coating: Laminin

1. *The night before the dissection, coat with poly-L-lysine (PLL)*

   Same protocol as in Subheading 3.2.1, **step 1a** except the duration of the incubation: here, incubate PLL overnight in a CO_2 incubator at 37 °C.

2. *The morning of the dissection, coat with laminin (LN)*

   (a) Aspirate the PLL solution, and rinse the coverslip three times with sterile water. Let the coverslip dry under the laminar flow hood at least 30 min, boxes open.

   (b) Dilute the stock LN at a final concentration of 20 μg/mL in F12/DMEM. Put approximately 150 μL of LN solution on each coverslip to cover it. Close the box and incubate 3 h (no longer than 4 h) in the CO_2 incubator at 37 °C. Start the dissection (*see* Subheading 3.3.2) during this incubation time.

   (c) Aspirate the LN solution and replace it with 200 μL of Complete Culture Medium. At this point the coverslip should never dry. Let the boxes wait at 37 °C until dissociated cortical cells are ready to be plated.

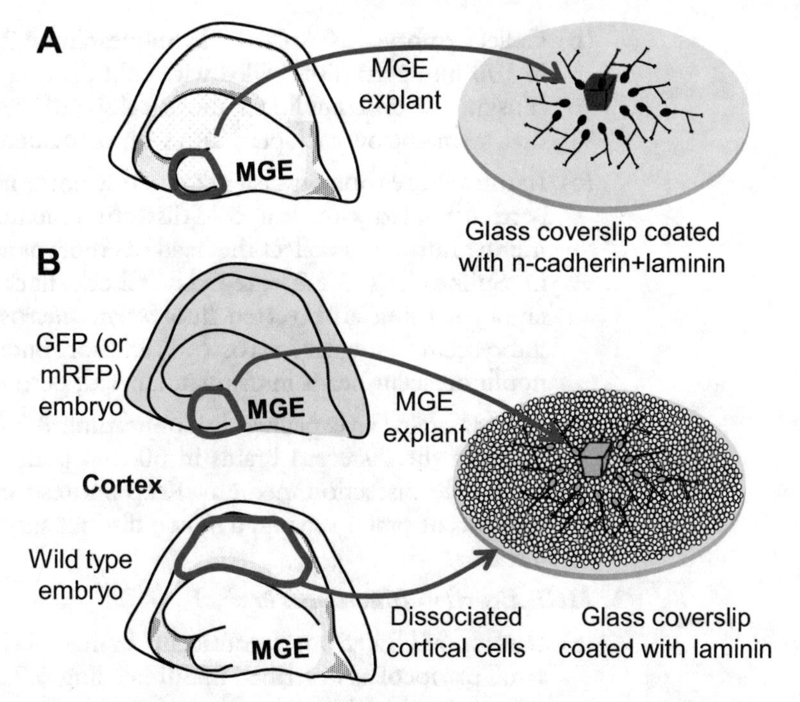

Fig. 3 In vitro models to analyze the migration of MGE-derived interneurons. Schemes on the left show the ventricular side of telencephalic vesicles dissected from E12.5/E13.5 mouse embryos. Schemes on the right illustrate the distribution of MGE cells that migrated away from a MGE explant after several hours in culture. In **a** (Subheading 3.2), the MGE explant was cultured on a glass coverslip coated with adhesive proteins. In **b**, (Subheading 3.3), the MGE explant was cultured on a glass coverslip coated with laminin and covered by a monolayer of dissociated cortical cells

3.3.2 Dissection and Culture (See Fig. 2)

These cocultures should be performed with fluorescent MGEs and nonfluorescent cortical cells, or with MGEs and cortex expressing two different fluorophores.

To minimize the use of animals, we take embryos at stage E13.5 from Swiss female mated with a heterozygote transgenic male expressing either the GFP or the mRFP under the control of the ubiquitous pCAG promoter. The MGEs are collected in fluorescent brains and the substrate of dissociated cortical cells is prepared with nonfluorescent brains (see Fig. 3b*).*

Alternatively, we can also use fluorescent embryos expressing two different fluorophores for cortical cells and MGE donors, or wild type and fluorescent mutant embryos for cortical cells and MGE donors. These experiments can also be done with cortical flaps collected at E13.5 and MGEs at E12.5. In all cases, the embryos of the two distinct litters are dissected successively the same day (See **Note 12**).

1. *Embryonic brain dissection.*
   (a) Prepare petri dishes and tools as explained in Subheading 3.2.2, **step 1a**.

(b) Collect embryos as explained in Subheading 3.2.2, **step 1b** in 100 mm petri dishes filled with cold dissection medium. If using two different litters, use one dish per litter and write their name below each petri dish with permanent ink.

(c) In the culture room, transfer the uterine horns in a 100 mm petri dish filled with clean cold dissection medium. Dissect uterine horns and collect the head of embryos as explained in Subheading 3.2.2, **step 1c**. Check fluorescence in embryos using an inverted fluorescent microscope (or a fluorescent binocular microscope) and sort fluorescent and nonfluorescent heads in distinct, labeled, petri dishes.

(d) Extract brains as explained in Subheading 3.2.2, **step 1d**. Transfer the dissected brains in 60 mm petri dishes with clean cold dissection medium. Keep fluorescent and non-fluorescent brains separated in two distinct annotated petri dishes.

2. *MGE dissection (fluorescent brains)*

(a) Dissect MGEs from fluorescent brains, following the same protocol as described in Subheading 3.2.2, **steps 2a** and **2b** (*see* Fig. 2).

3. *Cortex dissection and dissociation of cortical cells (nonfluorescent brains)*

Cortices dissected from 2 or 3 vesicles provide enough dissociated cells to prepare the substrate of one coverslip. Overall, 12 brains at E13.5 are enough to prepare 8 culture dishes for videomicroscopy.

(a) Dissect the nonfluorescent brains as explained in Subheading 3.2.2, **steps 2a** and **2b** in order to detach from each telencephalic vesicle the lateral wall that contains the dorsal cortical anlage and the GEs (*see* Fig. 2b). Turn each lateral wall with the ventricular surface up, and make a cut with the bent end of the tungsten needle, at the frontier between the cortical anlage and the lateral GE. Stop before sectioning the meninges, and peel off the cortex from the meninges by moving the needle away from the lateral GE. Transfer the flaps of cortex using a sterile plastic pipette in a 35 mm petri dish filled with clean cold dissection medium.

(b) Just before mechanical dissociation, re-section the large flaps of cortex into smaller pieces (0.5–1 mm in size) with the tungsten needles. Under the laminar flow hood, transfer these small pieces with dissection medium in a 2 mL eppendorf tube using a sterile plastic pipette. Let cortical pieces sediment. Aspirate the dissection medium and add 1 mL of dissociation medium in the eppendorf. Pipette up and down at least ten times with a 1 mL cone to dissociate cortical cells. Then centrifuge 5 min at 800 rpm at 4 °C.

Remove the dissociation medium above the pellet. Add 500–1500 μL of Complete Culture Medium, depending on the size of the pellet, and resuspend the cells.

(c) Count the cells with a counting chamber and adjust the volume to have a concentration of 4000 cells/μL. Quickly remove the solution above the coverslip in the culture well and replace it with 200 μL of cortical cell suspension (800,000 cells are deposited per coverslip) (*see* **Note 13**). Incubate at least 10 min at 37 °C in the CO_2 incubator to let cells sediment and attach to the coverslip. If necessary complete with 50–100 μL of Complete Culture Medium and culture cortical cells for at least 2 h before placing the MGE explants.

4. *Culture*

(a) Each culture well will contain all the explants from a single MGE. Just before putting the MGE in culture, section it in 6/8 pieces as explain in Subheading 3.2.2, **step 3a** (*see* Fig. 2d) and transfer the explants in the culture medium above the cellular substrate as explained in Subheading 3.2.2, **step 3b**. Distribute evenly the explants by picking them up with the sharp tip of the tungsten needle, and moving them above the cortical cells, without touching them. Then gently press each explant with the bent part of the tungsten needle, to make them adhere to the substrate, and carefully move the culture dish in the CO_2 incubator at 37 °C. Usually, MGE explants adhere less efficiently on dissociated cortical cells than on synthetic substrates.

(b) Do not move the dishes for at least 1 h. Proceed the same for all MGEs. Interneurons usually start migrating out of the explant after 12–18 h in culture (*see* **Note 10** and Fig. 3b).

3.4 Live Cell Imaging by Time-Lapse Videomicroscopy

The videomicroscope should be warmed several hours in advance (10–12 h) to work on a thermally stable system the day of the experiment.

When the interneurons start migrating out of the explant:

1. Carefully replace the Complete Culture Medium by pre-warmed Complete Video Culture Medium under a laminar flow hood: aspirate half of the volume of culture medium and replace it by the same amount of Complete Video Culture Medium (the culture should never dry). Repeat the process five times until the red color of the Complete Culture Medium is no longer visible.
2. Distribute 1 mL of Video Culture Medium along the lateral wall in the culture dish, in order to saturate the atmosphere and prevent the culture from drying. This medium should not mix with the culture medium in the culture well.

3. Close the dish and seal it with a thin band of Parafilm. Carefully clean the outside of the coverslip with ethanol.
4. Transfer carefully the culture dish on the stage of an inverted videomicroscope equipped with a spinning disk and a stage motorized in X,Y,Z. Wait at least 1 h before starting acquisitions. MGE explants cultured on a synthetic substrate are imaged each 1 or 2 min using phase contrast microscopy. MGE explants cultured on dissociated cortical cells are imaged each 3–5 min using confocal fluorescent microscopy. Cells can be imaged with 20×, 40× or 63× objectives. We usually images 10–15 fields per culture. Z-stacks comprise 7–11 planes separated by 0.5 or 1 μm, depending on magnification.

4 Notes

1. To prepare smaller culture wells, or 2 wells in the same dish, it is possible to drill two holes of 11 mm diameter in the bottom of the plastic dish, 5 mm apart from each other, and to seal 14 mm coverslips over each hole. The volume of coating solutions, the number of dissociated cortical cells, and the volume of culture medium have then to be adapted.
2. The coverslips can also be sealed with a mix of Vaseline and paraffin (ratio 1:3 w/w, *see* ref. 14). Etch the external surface of the plastic dish around the hole with sandpaper. Warm up the Vaseline–paraffin on a hot plate to melt it. Put four little drops of the Vaseline–paraffin mix around the hole, on the external side of the dish, with a warmed glass Pasteur pipette. Center a clean glass coverslip on the hole and press it gently on the drops to adhere. Then turn the bottom of the box on the hot plate. The Vaseline–paraffin drops will melt and spread between the glass and the plastic to seal the coverslip on outside of the dish. Adapt the volume of the drops to avoid overflow of Vaseline–paraffin inside the hole on the glass coverslip.
3. The treated coverslips and culture video boxes can be prepared in advance and stored for weeks. Just clean the boxes again in ethanol the day before the experiment.
4. The nature and the quality of the coating are crucial in these experiments.

   We can use three different types of coating: a mixed substrate of laminin/Ncad for the cultures of MGE explants (described in Subheading 3.2), a pure substrate of laminin for the cocultures of MGE explants on dissociated cortical cells (described in Subheading 3.3), or a pure substrate of Ncad for the cultures of MGE explants. The pure substrate of Ncad (1 μg/cm^2) is prepared according to the protocol described in Subheading 3.2 except that the laminin is removed and the

PLL at 4 μg/mL is incubated with the anti hFc at 4 μg/mL overnight at 4 °C.

MGE cells cultured on a pure substrate of laminin are well polarized but do not migrate long distances, and tend to form chains. On the substrate of pure recombinant Ncad, the motility of MGE cells is activated but cells are no longer directional as detailed in ref. 10. The mixed substrate of laminin and Ncad is used to promote a directed and active migration. Varying concentrations of laminin and Ncad have been tested and we have found that a substrate made of laminin 8 μg/mL and Ncad 0.25 μg/cm^2 is a good compromise to observe an active and directional migration of MGE cells.

5. It is possible to use less coating proteins by putting a drop of approximately 50 μL in the center of the coverslip. However, in this case, be careful that the drop does not touch the edges of the wall when moving the box, otherwise a meniscus will form, leaving the center of the coverslip dry and preventing a correct coating. We use an excess of PLL whose incubation time can vary. However, it should not be shorter than 3 h, and should not exceed overnight.
6. The drying step of PLL is mandatory.
7. The recombinant N-cadherin is a fusion between the extracellular domain of N-cadherin and the human Fc receptor [15]. Therefore, the N-cadherin coating comprises two successive steps: (1) incubation with IgG anti human Fc, (2) incubation with N-cadherin.
8. After killing the pregnant mice, the dissection of embryos has to be done as quickly as possible. The petri dishes containing the heads, brains and brain pieces should be stored on ice during waiting periods. The harvesting and dissection of the brains is performed under a binocular dissection microscope equipped with a very good optic, and installed in a clean and quiet part of the culture room. Dissecting brains and explant under a hood did not improve the quality of our short-term cultures.
9. This type of cutting ensures that all explants contain the same structures (primary and secondary proliferative zones) in order to reduce variability as much as possible.
10. This step is tricky and may require some training. Press the explants enough to avoid detachment when moving the box, but do not destroy them. Putting a small volume of Complete Culture Medium in the well can help to minimize flow movements during the transfer to the incubator and to keep explants attached.
11. After a few hours in culture, the explants will reorganize with the proliferating cells located at the periphery. Postmitotic cells exit from the bottom of the explant. Exit time can vary accord-

ing to the coating conditions, dissection time, strain/genotype/age of the mouse used.

12. The two experimental models described in this paper were designed to analyze the migration of MGE-derived interneuron precursors (*see* Fig. 3). Each model presents specific advantages and drawbacks. The cultures of MGE explants on synthetic substrates are much faster to prepare than the cocultures, and can be performed with any kind of mouse embryos, whereas the MGEs cultured on the dissociated cortical cells have to express a fluorescent marker. The two models moreover differ by the lifetime of the preparation. On the synthetic substrate, the migration of MGE cells stops after 1 day of culture, whereas the migration of MGE cells on the cortical substrate can last more than 3 days. In addition, the migratory behavior of MGE cells on the cortical cells strongly resembles the migratory behavior of MGE cells in organotypic brain slices. Finally, the cultures performed on a synthetic substrate are compatible with phase-contrast videomicroscopy and immunocytochemistry analyses are easier to perform.
13. The dissociated cortical cells quickly sediment in the eppendorf. Homogenize the suspension before pipetting the 200 μL to deposit in the culture well.

References

1. Marín O, Valiente M, Ge X, Tsai LH (2010) Guiding neuronal cell migrations. Cold Spring Harb Perspect Biol 2(2):a001834
2. Price J, Thurlow L (1988) Cell lineage in the rat cerebral cortex: a study using retroviral-mediated gene transfer. Development 104(3):473–482
3. Walsh C, Cepko CL (1988) Clonally related cortical cells show several migration patterns. Science 241(4871):1342–1345
4. O'Rourke NA, Dailey ME, Smith SJ, McConnell SK (1992) Diverse migratory pathways in the developing cerebral cortex. Science 258(5080):299–302
5. Anderson SA, Eisenstat DD, Shi L, Rubenstein JL (1997) Interneuron migration from basal forebrain to neocortex: dependence on Dlx genes. Science 278(5337):474–476
6. Fode C, Ma Q, Casarosa S, Ang SL, Anderson D, Guillemot F (2000) A role for neural determination genes in specifying the dorsoventral identity of telencephalic neurons. Genes Dev 14(1):67–80
7. Marin O (2012) Interneuron dysfunction in psychiatric disorders. Nat Rev Neurosci 13(2):107–120
8. Martini FJ, Valiente M, López Bendito G, Szabó G, Moya F, Valdeolmillos M, Marín O (2009) Biased selection of leading process branches mediates chemotaxis during tangential neuronal migration. Development 136(1):41–50
9. Britto JM, Johnston LA, Tan SS (2009) The stochastic search dynamics of interneuron migration. Biophys J 97(3):699–709
10. Luccardini C, Hennekinne L, Viou L, Yanagida M, Murakami F, Kessaris N, Ma X, Adelstein RS, Mège RM, Métin C (2013) N-cadherin sustains motility and polarity of future cortical interneurons during tangential migration. J Neurosci 33(46):18149–18160
11. Luccardini C, Leclech C, Viou L, Rio JP, Métin C (2015) Cortical interneurons migrating on a pure substrate of N-cadherin exhibit fast synchronous centrosomal and nuclear movements and reduced ciliogenesis. Front Cell Neurosci 9:286
12. Bellion A, Baudoin JP, Alvarez C, Bornens M, Métin C (2005) Nucleokinesis in tangentially migrating neurons comprises two alternating phases: forward migration of the Golgi/centrosome associated with centrosome splitting and myosin contraction at the rear. J Neurosci 25(24):5691–5699
13. Fishell G, Blazeski R, Godement P, Rivas R, Wang LC, Mason CA (1995) Tracking fluores-

cently labeled neurons in developing brain. FASEB J 9(5):324–334

14. Gregory WA, Edmondson JC, Hatten ME, Mason CA (1988) Cytology and neuron-glial apposition of migrating cerebellar granule cells in vitro. J Neurosci 8(5):1728–1738

15. Lambert M, Padilla F, Mège RM (2000) Immobilized dimers of N-cadherin-Fc chimera mimic cadherin-mediated cell contact formation: contribution of both outside-in and inside-out signals. J Cell Sci 113(Pt 12): 2207–2219

Chapter 13

Cell Migration in Tissues: Explant Culture and Live Imaging

Ralitza Staneva, Jorge Barbazan, Anthony Simon, Danijela Matic Vignjevic, and Denis Krndija

Abstract

Cell migration is a process that ensures correct cell localization and function in development and homeostasis. In disease such as cancer, cells acquire an upregulated migratory capacity that leads to their dissemination throughout the body. Live imaging of cell migration allows for better understanding of cell behaviors in development, adult tissue homeostasis and disease. We have optimized live imaging procedures to track cell migration in adult murine tissue explants derived from: (1) healthy gut; (2) primary intestinal carcinoma; and (3) the liver, a common metastatic site. To track epithelial cell migration in the gut, we generated an inducible fluorescent reporter mouse, enabling us to visualize and track individual cells in unperturbed gut epithelium. To image intratumoral cancer cells, we use a spontaneous intestinal cancer model based on the activation of Notch1 and deletion of p53 in the mouse intestinal epithelium, which gives rise to aggressive carcinoma. Interaction of cancer cells with a metastatic niche, the mouse liver, is addressed using a liver colonization model. In summary, we describe a method for long-term 3D imaging of tissue explants by two-photon excitation microscopy. Explant culturing and imaging can help understand dynamic behavior of cells in homeostasis and disease, and would be applicable to various tissues.

Key words Tissue explant, Cell migration, Two-photon imaging, Ex vivo culture, Gut, Tumor, Liver

1 Introduction

Cell migration is a hallmark of homeostasis and development of multicellular organisms. In homeostasis, cell migration ensures the correct positioning of cells and allows for fulfilment of their function, such as immune surveillance. Migration is also indispensable for the homeostatic epithelial renewal in the healthy, adult gut tissue: epithelial cells migrate from the intestinal crypts, where they proliferate, to the top of villi, where they are shed into the gut lumen. Deregulated migration accompanies pathological conditions such as cancer [1]. Cell migration is required throughout the metastatic cascade, from intratumoral migration and invasion of the surrounding stromal compartment to metastatic seeding at distant locations. Indeed, during the initial steps of carcinoma

Alexis Gautreau (ed.), *Cell Migration: Methods and Protocols*, Methods in Molecular Biology, vol. 1749, https://doi.org/10.1007/978-1-4939-7701-7_13,

metastasis, cancer cells traverse the basement membrane that encapsulates the tumor. Cancer cells then invade into the surrounding stromal compartment, which is composed of stromal cells and extracellular matrix [2]. After intravasation into blood vessels, malignant cells flow through the vasculature before arresting in distant tissues [3]. Then cancer cells extravasate and migrate into the stroma of the newly colonized tissue, where they need to survive and proliferate to give rise to metastases [3].

Intravital imaging has allowed description of a diversity of cancer cell migration modes in living animals, ranging from single cell to collective migration [4, 5]. In cancer biology, intravital imaging is preferentially used for organs that are easily accessible, such as breast or skin [6, 7]. Tumors located in deep abdominal organs, such as intestine or liver, are difficult to image using intravital microscopy due to their limited accessibility. For such tissues, we propose an alternative method that allows imaging of tissue explants over long periods of time.

We applied this approach to study cell migration dynamics in the small intestine, in primary intestinal tumors, and in the liver—a common metastatic site.

To track epithelial cells in the gut, we use a compound reporter mouse, derived by crossing the mT/mG reporter mouse [8] with the Villin-CreER^{T2} mouse [9]. Gut epithelium-specific expression of membrane-targeted GFP can be induced in a mosaic fashion, allowing for tracking of individual cells over time.

In order to address migration in intestinal tumors, we use an inducible and tissue-specific Cre/lox mouse carcinoma model [10]. In this model, tumorigenesis is driven by expression of activated Notch1 receptor and deletion of the p53 tumor suppressor in the intestinal epithelium (NICD/p53$^{-/-}$ mice), after tamoxifen induction. Intestinal carcinoma spontaneously form several months after induction, and metastasize to the lymph nodes, peritoneum, and liver [10]. Tamoxifen induction also triggers expression of GFP with nuclear localization, which allows tracking cancer cells over time.

To analyze cancer cell migration during metastasis to the liver, we perform splenic injections of cancer cells [11]. This gives tumor cells direct access to the portal circulation, allowing fast and directed colonization of the liver, which is the main target organ during metastasis in colorectal cancer [12]. For splenic injections, we use the mT/mG reporter mouse [8], which enables effective visualization of all cells in the tissue by membrane-targeted tdTomato. To visualize tumor cell migration, we injected a murine colon carcinoma cell line (CT26) stably expressing LifeAct-GFP reporter.

In summary, this technique allows precise characterization of cell migration behavior in different tissues and represents an attractive alternative to intravital microscopy in organs of difficult access.

2 Materials

1. Mice of the desired genotype.
2. Dissection tools: scissors, tweezers, scalpel, spatula, perforated spoon (Moria).
3. Cannula (e.g., feeding tube for mice, Ecimed).
4. Razor blades (Gillette Blue Extra).
5. Petri dish (100 mm, Corning).
6. Organotypic Cell Culture Insert (30 mm, hydrophilic PTFE, 0.4 μm; Millicell).
7. Glass-bottom Cell Culture Dish (35 mm, Fluorodish; World Precision Instruments).
8. Tissue anchors (SHD-26GH/10; Harvard Apparatus, UK).
9. Tissue cassette (at least 5 mm deep; Electron Microscopy Sciences).
10. Instant glue (Super Glue3, Loctite).
11. 50 mL syringe (for gut flushing).
12. 1 mL syringe (for agarose injection).
13. Insulin syringe (29 G).
14. Suture (Ethicon Vicryl 3-0).
15. DMEM-GlutaMAX (Gibco).
16. Williams E medium (Gibco).
17. Leibovitz's L-15 medium (Sigma-Aldrich).
18. Trypsin (Gibco).
19. Fetal bovine serum (FBS, Life Technologies).
20. Insulin (Sigma-Aldrich).
21. Human holo-Transferrin (Holo, Human plasma, Calbiochem).
22. Recombinant murine epidermal growth factor, EGF (Peprotech).
23. Anti-anti: 100× antibiotic–antimycotic solution, containing amphotericin B, penicillin, and streptomycin (Life Technologies).
24. Imaging medium: DMEM-GlutaMAX supplemented with 2.5% FBS, 0.25 U/mL insulin, 100 μg/mL transferrin, 10 ng/mL murine EGF and 1% anti-anti.
25. Working solution: Leibovitz's L-15 medium supplemented with 1% anti-anti. Keep at 4 °C.
26. Phosphate-buffered saline solution (PBS).
27. Low melting point agarose (UltraPure, Life Technologies).

28. Anesthetic and analgesic (e.g., mix of ketamine 8 mg/mL; xylazine 1.5 mg/mL; flunitrazepam 0.1 mg/mL; dose: 100 μL/10 g of body mass).
29. Tamoxifen (MP Biomedicals).
30. Acetone.
31. Ethanol, 70%.
32. Vibratome (Leica VT1000S).
33. Two-photon inverted microscope (e.g., Leica SP8 microscope), equipped with a stage top incubator (Okolab).
34. Long working distance objective (e.g., Leica 25×/0.95NA water).
35. Image processing tools (ImageJ, Imaris).

3 Methods

3.1 Imaging Epithelial Cell Migration in the Gut

1. One day before the experiment, inject a low dose of tamoxifen (0.5 mg/kg) i.p. into Villin-CreER^{T2} [9] × mT/mG mouse [8], to induce mosaic expression of membrane-targeted GFP.
2. Prepare 4% (w/v) low melting point agarose in the working solution. Let it cool down at 37 °C in a water bath (*see* **Note 1**).
3. Sacrifice the mouse and extract the desired gut region (*see* **Note 2**).
4. Thoroughly flush the intestine with the cold working solution using the 50 mL syringe fitted with a cannula.
5. Fill the tissue cassette with liquid, low melting point agarose.
6. Cut the intestine into small fragments (about 5 mm), using a scalpel.
7. Fill the lumen of each fragment with liquid agarose using the 1-mL syringe fitted with a cannula, and immediately immerse the tissue into the cassette prefilled with liquid agarose. Adjust the orientation of immersed gut fragment to facilitate sectioning (Fig. 1, left panel).
8. Place cassettes at 4 °C for 5 min to allow gelation of the agarose.

3.2 Imaging Cancer Cell Migration in Intestinal Tumors

1. Prepare 4% (w/v) low melting point agarose in the working solution. Let it cool down at 37 °C in a water bath (*see* **Note 1**).
2. Sacrifice the tumor-bearing mouse.
3. Open the abdomen and keep it humidified throughout the procedure with cold PBS supplemented with 1% anti-anti.

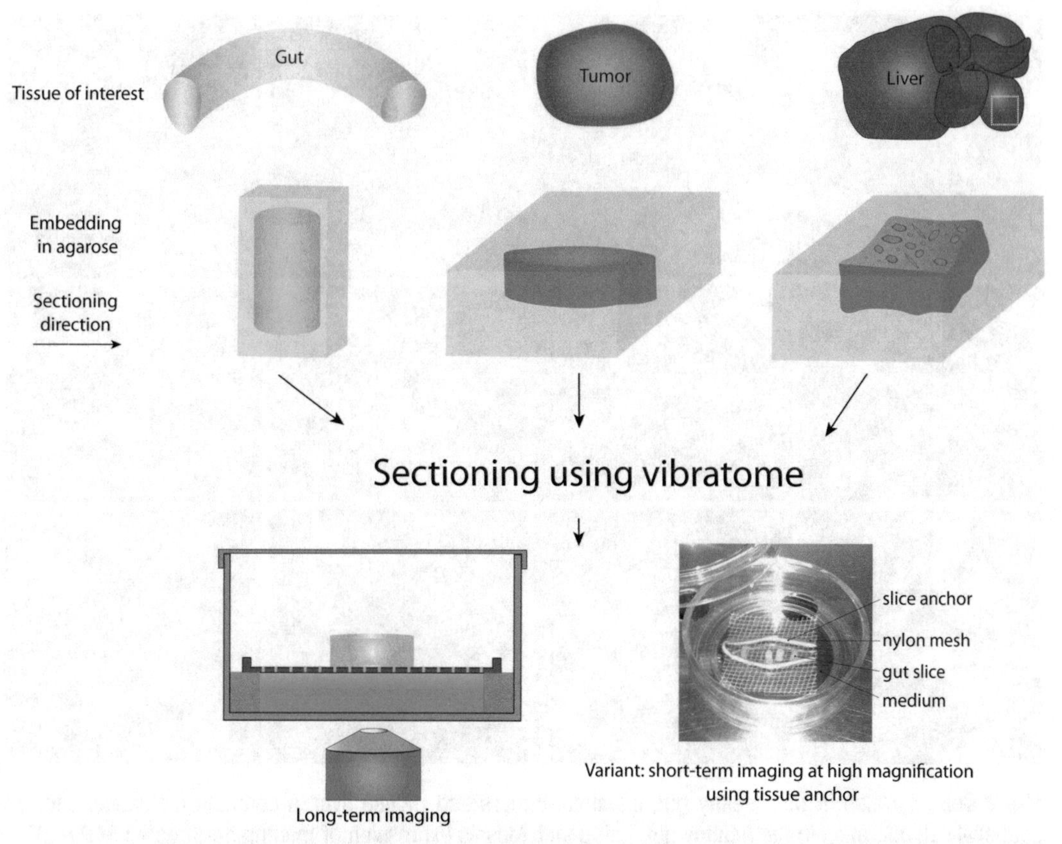

Fig. 1 Sample preparation for explant imaging. Procedure for explant preparation and imaging is depicted. The tissue slice (orange) is deposited on a cell culture insert (violet) and is in contact with the culture medium (pink)

4. Cut out the intestinal tumor with 1 cm of healthy intestine on both sides and put it in a petri dish filled with cold PBS/1% anti-anti (*see* **Note 2**).
5. Flush the intestinal tumor with cold PBS/1% anti-anti by inserting a syringe fitted with a cannula in the intestine.
6. Cut the intestine and tumor longitudinally using scissors in order to have the luminal side of intestine and tumor facing you.
7. Remove the healthy intestine from the tumor using a sterile scalpel.
8. Cut 3–6 mm tumor pieces and make sure to have at least one flat side to allow for an easier cutting later on.
9. Embed the tumor pieces in the melted agarose solution using a tissue cassette. Make sure that a side with a flat cut surface is facing the bottom of the cassette (Fig. 2, middle panel).
10. Place cassettes at 4 °C for 5 min to allow gelation of the agarose.

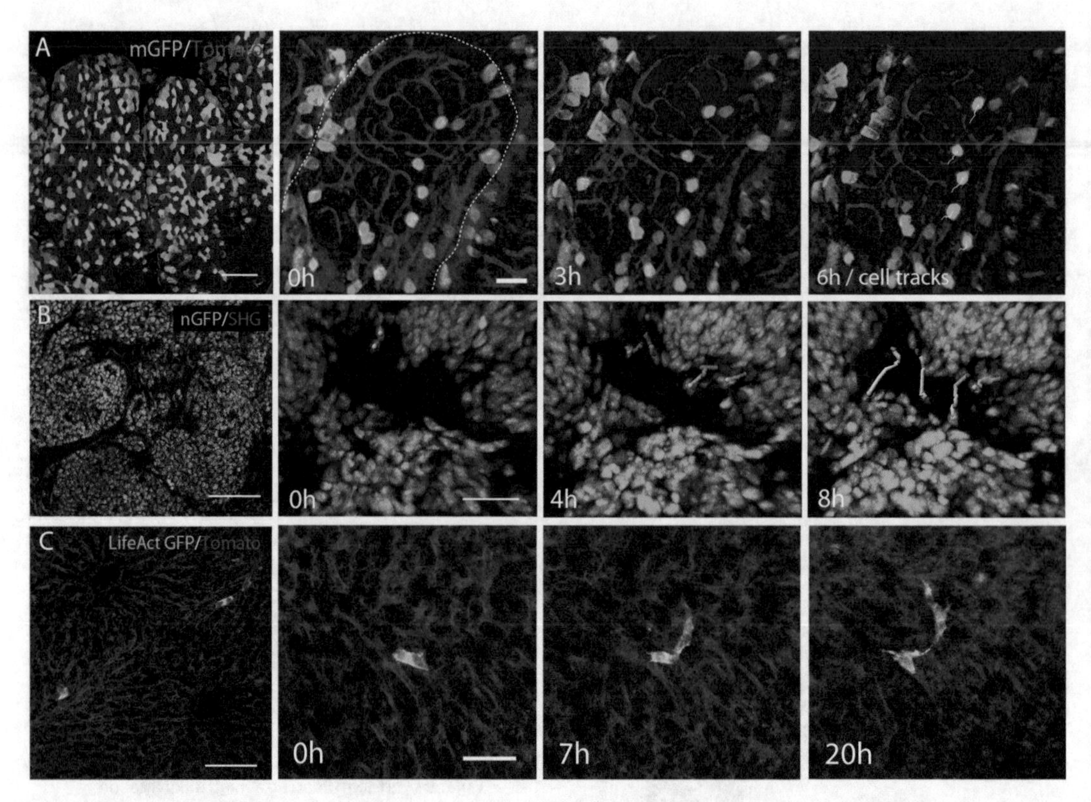

Fig. 2 Cell migration in the healthy gut, intestinal tumors and mouse liver, a common metastatic site. (**a**) Epithelial cell migration in the healthy gut. Left panel: Mosaic expression of membrane-targeted GFP (mGFP) in the gut epithelium (Villin-CreER^{T2} × mT/mG mouse). 3D rendering of the confocal z-stack was done in Imaris (Surpass view, Blend mode). Scale bar: 50 μm. Right panels (montage): Montage of a time-lapse movie showing migration of mGFP-labeled enterocytes along the villus (dashed line) in a Villin-CreER^{T2} × mT/mG gut explant. The 6-h time-point is also shown superimposed with tracked cells (cyan) and the corresponding cell tracks. 3D rendering, cell segmentation and tracking were done in Imaris (Surpass view, MIP (max) mode). Scale bar: 30 μm. (**b**) Cancer cell migration in primary intestinal tumors. Left panel: NICD/p53$^{-/-}$ tumor organization. Cancer cell nuclei (nuclear GFP, green) and fibrillar collagen (SHG, magenta). Scale bar: 100 μm. Right panels: Montage of a time-lapse movie showing cancer cell migration in intestinal tumor. Four cells of interest are manually tracked in 3D using Imaris. Mean speeds of cells are respectively, for cells from left to right: 0.15 μm/min; 0.22 μm/min; 0.21 μm/min; 0.15 μm/min. Scale bar: 50 μm. (**c**) Cancer cell migration during liver colonization. Left panel: Liver from mT/mG mouse after splenic injection of CT26 LifeAct-GFP cells (green). All cell membranes from mT/mG mouse are red. Scale bar: 100 μm. Right panels: Montage of a time-lapse movie showing CT26 LifeAct-GFP cells (green) migrating through the liver vasculature after intrasplenic injection in mT/mG reporter mice (cell membranes in red). We can appreciate that cancer cells are able to divide inside liver sinusoids (left panel), generating two cell daughters that subsequently migrate through two different vessels, away from the initial position (center and right panels). Scale bar: 50 μm

3.3 Imaging Intrahepatic Migration

1. Cell culture: CT26-LifeAct-GFP cells are cultured using DMEM/10% FBS. Immediately before cell injections, trypsinize and count cells. Resuspend cells at 10^6 cells/100 μL in PBS supplemented with 1% FBS. Load 100 μL of the suspension into an insulin syringe.
2. For injection, weigh the mouse and inject an appropriate dose of anesthetic. After verifying the loss of response to reflex stimulation (toe or tail pinch with firm pressure), perform a small incision in the left abdominal side, inferior to the rib cage, and exteriorize the spleen.
3. For intrasplenic injection, make a node at the tip of the spleen using suture thread and insert the needle into the nod, on the splenic tip. Pull on the treads to close the node around the needle and then inject the cell suspension. This is to prevent cell suspension leakage. As a sign of effective injection, the spleen will swell and go back to its initial shape once the cell solution is flowed toward the portal circulation (*see* **Note 3**).
4. After injection, place the spleen back into the peritoneal cavity. Humidify the peritoneal cavity with PBS/1% anti-anti during this time.
5. After 30 min, euthanize the mouse (*see* **Note 4**).
6. Open the abdomen and keep it humidified throughout the procedure with cold PBS/1% anti-anti.
7. Cut out the liver and place it on a petri dish with Williams medium containing 1% anti-anti, on ice (*see* **Note 2**).
8. Cut the liver in small cubes, of approximately 3 × 3 × 3 mm.
9. Prepare 4% low melting point agarose in the working solution. Let it cool down at 37 °C in a water bath (*see* **Note 1**).
10. Embed the liver pieces in the warm agarose solution using a tissue cassette. Make sure that a side with a flat cut surface is facing the bottom of the cassette (Fig. 1, right panel).
11. Place the cassettes at 4 °C for 5 min to allow gelation of the agarose.

3.4 Tissue Sectioning

1. Clean a single-use razor blade using papers soaked with the following solutions in this order: acetone, water, ethanol 70%.
2. Insert the blade into the knife holder on the vibratome.
3. Put ice into the cooling bath of the vibratome.
4. Take the agarose block with embedded explant out of the cassette. Trim the agarose block if necessary, but leave at least 2 mm agarose around the tissue. Be careful not to remove the tissue from the agarose block.
5. Fix the agarose block onto the cutting surface using instant glue. Make sure the flat side of the explant that was previously

facing the bottom of the cassette is now facing you. For gut slices: in order to get transversal gut sections, position the agarose block so that the gut fragment is sitting upright on the cutting surface (Fig. 1, left panel). For liver slices: positioning the liver tissue as depicted in Fig. 1 will facilitate vibratome sectioning.

6. Fill the buffer tray with cold working solution for gut and tumor slices, or with Williams medium 1% anti-anti for liver slices. Make sure that the agarose block is entirely covered.
7. We recommend the following vibratome settings for gut and tumor slices: speed 3 (0.125 mm/s), frequency 7 (70 Hz). Vibratome speed can be adjusted according to the stiffness of your tissue (*see* **Note 5**). For liver tissue, use speed 3 and frequency 8–9. Speed and frequency for the liver varies depending on the mouse strain (*see* **Note 6**).
8. Section at 250–360 μm thickness.
9. Collect the agarose-embedded sections from the working solution gently, using a perforated spoon or a spatula (*see* **Note 7**). Transfer the sections to a petri dish with working solution, on ice.
10. Gently remove the agarose around the slices using tweezers.

3.5 Mounting the Explants, Live-Cell Imaging and Image Analysis

1. Put 1 mL of culture and imaging medium into the glass-bottom cell culture dish.
2. Place the cell culture insert carefully into the Fluorodish, to avoid bubble formation. The 30 mm cell culture insert should sit on the inner ring of the Fluorodish and be in contact with the medium (Fig. 1, bottom left).
3. Position up to five tissue slices on the cell culture insert (*see* **Note 7**). Slices are at the air-liquid interface, exposed to air, but in contact with cell culture medium through the perforated membrane insert (Fig. 1, bottom left, and *see* **Notes 8** and **9**).
4. For high-resolution imaging of gut slices, short working distance objectives are required (e.g. 40×/1.3NA or 63x/1.4NA) for addressing membrane dynamics; thus it is necessary that the specimen is placed as close as possible to the objective lens. For this approach, we recommend placing gut slices directly into the dish and using low medium condition (200–300 μL medium per 35-mm dish) to promote gas exchange. A tissue anchor can be placed on top of the tissue to stabilize it and to prevent the drift (Fig. 1, bottom right). However, this approach is used only for short-term imaging (1–2 h), as tissue tends to deteriorate more rapidly.

5. For cell migration analysis, time-lapse movies are acquired with a two-photon confocal microscope (*see* **Note 10**). We use an inverted Leica SP8 microscope coupled to a femtosecond Chameleon Vision II laser (680–1350 nm; Coherent, Inc.), using a Leica 25×/0.95NA water immersion objective (*see* **Note 11**). Imaging of the explant mounted as described requires use of a long working distance objective. For live imaging, the microscope should be equipped with a stage top incubator (e.g., Okolab) to allow gas, temperature and humidity control. Tissue is maintained at 37 °C, in 95% O_2/5% CO_2 atmosphere, with 95% humidification. Image acquisition and microscope control are performed using the Leica LAS X software.
6. Detectors with suitable filters for the fluorophores in use are required. We use three non-descanned Leica Hybrid detectors, that have the following filters: 525/25 nm (GFP), 575/25 nm (Tomato) and <492 nm (Second harmonic Generation, SHG). The excitation wavelength used is 960 nm for GFP and Tomato, 910 nm for SHG.
7. Stacks are recorded at intervals suitable for the dynamics of the cell behavior you wish to observe. For cell migration analysis, we record stacks every 15–30 min, up to 48 h. The step size can be adapted according to the desired resolution in the *z* plane. We obtained 3D stacks at a step size of 1–2 μm intervals, using 25× objective.
8. Time-lapse processing and analysis are done using LAS X (Leica), ImageJ (NIH) or Imaris (Bitplane). All images are corrected for brightness and contrast using the ImageJ software. 3D registration can be done using the "Correct 3D drift" plugin [13]. Cells can be tracked either manually (Fig. 2), or automatically, provided that image segmentation can be implemented. In primary tumors, we manually track cell nuclei in 2D using the MTrackJ plugin [14] and in 3D using Imaris (Bitplane). 3D time-lapse movies of migrating GFP-expressing cells in the gut were segmented and automatically tracked in 3D using Imaris (Fig. 2).

4 Notes

1. Work fast with liquid agarose, as it will rapidly polymerize at room temperature.
2. For better viability, make sure to keep the tissue on ice at all steps prior to incubation and imaging.
3. Injection in the spleen is quite demanding and another person's help may be required.

4. In order to label blood vessels in the liver metastasis assay, it is possible to inject labeled lectin (100 μL, 0.5 mg/mL of Lectin DyLight 649, Vector laboratories) in the spleen, right before mouse euthanasia. As lectin gets internalized by phagocytic cells during long-term imaging, we recommend its use for short-term imaging.
5. In case of tissue with higher stiffness, vibratome speed can be increased and frequency can be decreased. You will need to find optimal settings for your tissue of interest.
6. For nude mice livers, you can use higher speeds and lower frequencies.
7. Always manipulate the tissue slices gently, preferably with blunt-edge tweezers and/or a perforated spoon.
8. Tissue viability and dynamics vary in different tissues of interest. For example, intestinal tumor tissue has a viability of maximum 48 h, liver 30 h and normal gut mucosa <10 h.
9. For purposes other than live imaging, gut and tumor slices can be cultured in a tissue incubator at 37 °C in 95% O_2/5% CO_2 atmosphere. This allows a variety of applications, such as labeling (e.g., BrdU/EdU for cell proliferation), transfection, or drug treatments.
10. Two-photon microscopy is used for long-term deep tissue imaging with reduced phototoxicity. Confocal imaging with visible lasers is also possible, but will yield more photobleaching and shorter imaging depth.
11. Imaging of the explant mounted as described requires the use of an inverted microscope.

Acknowledgments

We would like to thank Basile Gurchenkov and Fatima El Marjou for assistance with microscopes and mice. The authors greatly acknowledge the Cell and Tissue Imaging (PICT-IBiSA), Institut Curie, member of the French National Research Infrastructure France-BioImaging (ANR10-INBS-04). This work was supported by the Fondation pour la Recherche Médicale (FRM N° DGE20111123020), the Canceropole-IdF (nis2012-2-EML-04-IC-1), INCa (Cancer National Institute, n° 2011-1-LABEL-IC-4). The authors would like to acknowledge the Cell and Tissue Imaging Platform - PICT-IBiSA (member of France–Bioimaging, ANR-10-INBS-04) of the Genetics and Developmental Biology Department (UMR3215/U934) of Institut Curie for help with image analysis.

This work is funded by Institut Thématique Multi-organismes Cancer–Plan Cancer 2014–2019 and Ecole Doctorale Frontières du Vivant (FdV)–Programme Bettencourt (RS), Marie Curie Individual Fellowship (FiBRO) (JB), and ERC starting grant (DMV).

References

1. Lambert AW, Pattabiraman DR, Weinberg RA (2017) Emerging biological principles of metastasis. Cell 168:670–691. https://doi.org/10.1016/j.cell.2016.11.037
2. Chambers AF, Groom AC, MacDonald IC (2002) Dissemination and growth of cancer cells in metastatic sites. Nat Rev Cancer 2:563–572. https://doi.org/10.1038/nrc865
3. Valastyan S, Weinberg RA (2011) Tumor metastasis: molecular insights and evolving paradigms. Cell 147:275–292. https://doi.org/10.1016/j.cell.2011.09.024
4. Friedl P, Wolf K (2010) Plasticity of cell migration: a multiscale tuning model. J Cell Biol 188:11–19. https://doi.org/10.1083/jcb.200909003
5. Clark AG, Vignjevic DM (2015) Modes of cancer cell invasion and the role of the microenvironment. Curr Opin Cell Biol 36:13–22. https://doi.org/10.1016/j.ceb.2015.06.004
6. Kedrin D, Gligorijevic B, Wyckoff J, Verkhusha VV, Condeelis J, Segall JE, van Rheenen J (2008) Intravital imaging of metastatic behavior through a mammary imaging window. Nat Methods 5:1019–1021. https://doi.org/10.1038/nmeth.1269
7. Alexander S, Koehl GE, Hirschberg M, Geissler EK, Friedl P (2008) Dynamic imaging of cancer growth and invasion: a modified skin-fold chamber model. Histochem Cell Biol 130:1147–1154. https://doi.org/10.1007/s00418-008-0529-1
8. Muzumdar MD, Tasic B, Miyamichi K, Li L, Luo L (2007) A global double-fluorescent Cre reporter mouse. Genesis 45:593–605. https://doi.org/10.1002/dvg.20335
9. El Marjou F, Janssen KP, Chang BHJ, Li M, Hindie V, Chan L, Louvard D, Chambon P, Metzger D, Robine S (2004) Tissue-specific and inducible Cre-mediated recombination in the gut epithelium. Genesis 39:186–193. https://doi.org/10.1002/gene.20042
10. Chanrion M, Kuperstein I, Barrière C, El Marjou F, Cohen D, Vignjevic D, Stimmer L, Paul-Gilloteaux P, Bièche I, Tavares SDR, Boccia G-F, Cacheux W, Meseure D, Fre S, Martignetti L, Legoix-Né P, Girard E, Fetler L, Barillot E, Louvard D, Zinovyev A, Robine S (2014) Concomitant Notch activation and p53 deletion trigger epithelial-to-mesenchymal transition and metastasis in mouse gut. Nat Commun 5:5005. https://doi.org/10.1038/ncomms6005
11. Barbazán J, Alonso-Alconada L, Elkhatib N, Geraldo S, Gurchenkov V, Glentis A, Van Niel G, Palmulli R, Fernández B, Viaño P, Garcia-Caballero T, López-López R, Abal M, Vignjevic DM, Barbazan J (2017) Liver metastasis is facilitated by the adherence of circulating tumor cells to vascular fibronectin deposits. Cancer Res 77(13):3431–3441. https://doi.org/10.1158/0008-5472.CAN-16-1917
12. Van Cutsem E, Nordlinger B, Adam R, Köhne C-H, Pozzo C, Poston G, Ychou M, Rougier P (2006) Towards a pan-European consensus on the treatment of patients with colorectal liver metastases. Eur J Cancer 42:2212–2221. https://doi.org/10.1016/j.ejca.2006.04.012
13. Parslow A, Cardona A, Bryson-Richardson RJ (2014) Sample drift correction following 4D confocal time-lapse imaging. J Vis Exp. https://doi.org/10.3791/51086
14. Meijering E, Dzyubachyk O, Smal I (2012) Methods for cell and particle tracking. Methods Enzymol 504:183–200. https://doi.org/10.1016/B978-0-12-391857-4.00009-4

Check for updates

Chapter 14

Intravital Imaging of Tumor Cell Motility in the Tumor Microenvironment Context

Battuya Bayarmagnai, Louisiane Perrin, Kamyar Esmaeili Pourfarhangi, and Bojana Gligorijevic

Abstract

Cancer cell motility and invasion are key features of metastatic tumors. Both are highly linked to tumor microenvironmental parameters, such as collagen architecture or macrophage density. However, due to the genetic, epigenetic and microenvironmental heterogeneities, only a small portion of tumor cells in the primary tumor are motile and furthermore, only a small portion of those will metastasize. This creates a challenge in predicting metastatic fate of single cells based on the phenotype they exhibit in the primary tumor. To overcome this challenge, tumor cell subpopulations need to be monitored at several timescales, mapping their phenotype in primary tumor as well as their potential homing to the secondary tumor site. Additionally, to address the spatial heterogeneity of the tumor microenvironment and how it relates to tumor cell phenotypes, large numbers of images need to be obtained from the same tumor. Finally, as the microenvironment complexity results in nonlinear relationships between tumor cell phenotype and its surroundings, advanced statistical models are required to interpret the imaging data. Toward improving our understanding of the relationship between cancer cell motility, the tumor microenvironment context and successful metastasis, we have developed several intravital approaches for continuous and longitudinal imaging, as well as data classification via support vector machine (SVM) algorithm. We also describe methods that extend the capabilities of intravital imaging by postsacrificial microscopy of the lung as well as correlative immunofluorescence in the primary tumor.

Key words Tumor microenvironment, Motility, Intravital imaging, Correlative immunofluorescence, Invadopodia, Invasion, 4D multiphoton fluorescent microscopy, Second harmonic generation, Photoconvertible proteins, Support vector machine classification

1 Introduction

Metastasis remains the leading cause of breast cancer-related deaths. Unlike the primary tumor, which can be managed in the clinic with chemotherapy and surgery, there is currently no effective treatment for metastasis [1]. Metastasis is a complex, multistep

Electronic Supplementary Material: The online version of this chapter (https://doi.org/10.1007/978-1-4939-7701-7_14) contains supplementary material, which is available to authorized users.

Alexis Gautreau (ed.), *Cell Migration: Methods and Protocols*, Methods in Molecular Biology, vol. 1749,
https://doi.org/10.1007/978-1-4939-7701-7_14, © Springer Science+Business Media, LLC 2018

process, during which cancer cells dissociate from the rest of the primary tumor, move through the tissue and disseminate to secondary organs [2]. Chemical and physical cues from the tumor microenvironment, i.e. host cells [3] and the extracellular matrix (ECM) [4–6], play a significant role in shaping cancer cell behaviors related to metastasis, namely epithelial-to-mesenchymal transition (EMT), motility, invasion, homing, etc. Complex and dynamic interactions between tumor cells and the tumor microenvironment result in shifting the balance between cell behaviors accessible from the current genetic and transcriptomic landscape as well as the emergence of new cancer cell phenotypes [7, 8]. To understand such interactions at the single cell level in a quantitative fashion, it is essential to image cancer cells in real time in the context of their native niche.

With the emergence of multiphoton microscopy [9], it has become possible to study in vivo dynamic events in real time in mouse models, a notable example being intravital imaging of cancer cell motility [10]. Compared to confocal microscopy, multiphoton imaging allows deeper penetration and decreased light scattering as a result of excitation cross-sections being in the near-infrared range of wavelengths. Furthermore, the use of femtosecond pulse lasers allows acquisition of second harmonic generated (SHG) signals, produced by noncentrosymmetric structures, including collagen I fibers abundant in most solid tumors. Unlike fluorescent signals, which need to be exogenously supplied to the sample, SHG signal is endogenous and originates from the dipole orientation of collagen fibers [11, 12].

Continuous intravital imaging in mouse models is generally limited to several hours, which allows high quality measurements of motility parameters in tumor cells as well as host immune cells [13, 14] or fibroblasts [15]. In addition, blood vessel flow and collagen architecture can be measured simultaneously. However, to fate-map specific cells or niches within the tumor, observations need to include metastatic events such as intravasation and lung colonization, which occur on a scale of days. As common intravital microscopy in tumors does not allow for freely moving animals, animals are anesthetized [16, 17] and imaging is limited to 5–24 h. Hence, longitudinal imaging [18, 19] is necessary to map the journey from primary to secondary tumor site. To ensure monitoring of the same tumor regions over multiple (static or continuous) imaging sessions, the utilization of photoconvertible fluorophores, such as Dendra2, is optimal [20, 21]. Dendra2 irreversibly switches green-to-red following 405 nm exposure, enabling us to track tumor cells longitudinally.

In this chapter we provide a guide to performing intravital imaging of cell motility in photoconvertible Dendra2-labeled breast carcinoma tissues and image analysis. We introduce the surgical procedures and lay out the steps for real-time, continuous imaging and analysis of 4-color, 4D stacks demonstrating two different motility phenotypes of tumor cells: fast- and slow- (invadopodia-driven) locomoting cells. Microenvironmental features are extracted from the same 4D stacks and finally, SVM classification identifies the microenvironmental conditions amenable to presence of fast- or slow-locomoting cells. We describe how photoconversion can be used in vivo for longitudinal imaging of the same regions of interest. Lastly, we discuss postsacrificial approaches that can extend the capabilities of intravital imaging of lung metastases and correlative immunofluorescence of the primary tumor sections.

2 Materials

2.1 Intravital Imaging and Photoconversion

1. Mouse strains:
   (a) *MMTV-PyMT* × *MMTV-iCre/CAG-CAC-Dendra2* (cytoplasm of all tumor cells labeled with photoconvertible Dendra2).
   (b) *MMTV-PyMT* × *MMTV-iCre/CAG-CAC-Dendra2* × *c-fms-CFP* (cytoplasm of all tumor cells and macrophages labeled).
   (c) MDA-MB-231-Dendra2 cells orthotopically injected into SCID mice.
2. Olympus FV1200MPE multiphoton laser scanning microscope.
3. UPLSAPO 30× objective with silicone oil immersion, NA 1.05.
4. ThorLabs PM200 Handheld optical power and energy meter.
5. Temperature control environmental chamber.
6. Infrared heating pad.
7. Isoflurane.
8. Anesthesia mask.
9. Surgical drape.
10. Tissue forceps.
11. Micro scissors.
12. Trimmer.
13. Ocular lubricant ointment.
14. 70% ethanol.
15. Sterile 1× PBS Dulbecco's phosphate-buffered saline (1× PBS).
16. Dextran, Texas Red, 70 kDa (Molecular Probes).
17. MMPSense 680 Fluorescent Imaging Agent (PerkinElmer).

18. Transfer pipettes.
19. Cotton-tipped applicators.
20. Insulin syringe.
21. Super glue liquid, longneck bottle.
22. Labeling tape.

2.2 Immunofluorescence

1. 1× Dulbecco's phosphate-buffered saline (1× PBS).
2. Fixative: 4% paraformaldehyde, 1× PBS.
3. O.C.T. (optimal cutting temperature) Compound.
4. 30% w/v sucrose solution.
5. Isopentane (2-methylbutane).
6. Dry ice.
7. Disposable cryomolds.
8. Positively charged microscope glass slides.
9. Blocking Solution: 1% bovine serum albumin (BSA), 1% fetal bovine serum (FBS), 1× PBS.
10. Liquid Blocker Super Pap Pen.
11. Permeabilization Solution: 0.1% Triton X-100, 1× PBS.
12. Acetone.
13. Antibodies and fluorescent dyes: Anti-Ki67 (Abcam, cat. abcam15580, 1:200), Anti-fibronectin (Abcam, cat. ab6328, 1:100), Phalloidin conjugated to Alexa Fluor 633.
14. Fluoromount-G mounting medium.
15. Cover glass.
16. Nail polish.

3 Methods

3.1 Surgical Preparation

The animals must be surgically prepared for imaging by removing skin and exposing the cells to be imaged. Here, we briefly describe the mammary skin flap procedure, which is suitable for continuous imaging. For the skin flap preparation, the mammary tumor tissue of the fourth inguinal mammary gland is separated from the peritoneum on a skin flap (Fig. 1a). The fourth inguinal mammary gland is sufficiently distant from the chest area, which is most heavily affected by breathing. Separation from the body with the skin flap further reduces breathing disturbance on imaging. This approach is technically simple and is suitable for short, one-time imaging sessions only. The imaging time is limited to generally 6–8 h due to inflammation and blood vessel damage as a result of prolonged exposure of the tissue to the outside environment. This duration can be increased to 24 h with careful monitoring of vital

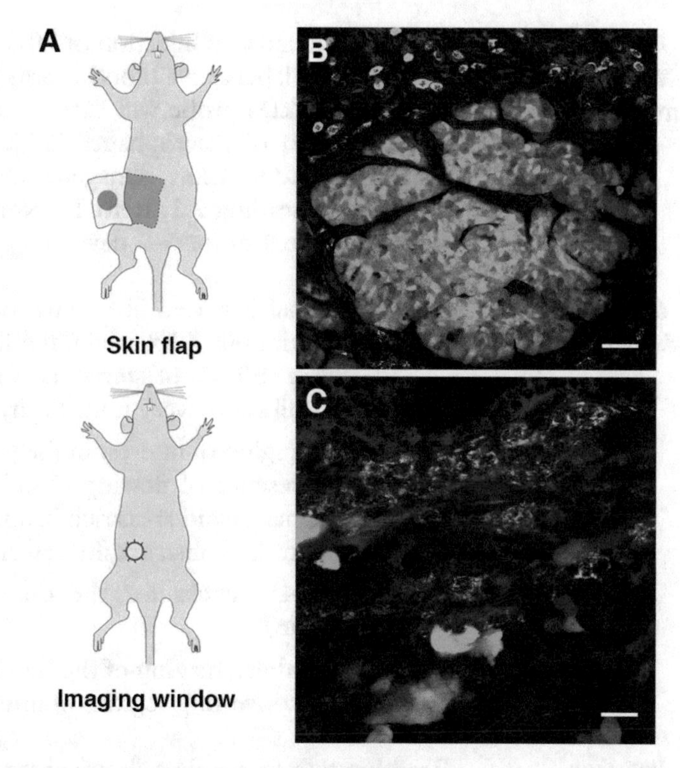

Fig. 1 Surgical preparation and intravital images of the tumor microenvironment in transgenic and orthotopic xenograft mouse models of breast carcinoma. (**a**) Surgical preparations of mice for intravital imaging: skin flap (top) and mammary imaging window (bottom). (**b**) Intravital image of tumor cells and the surrounding tumor microenvironment in carcinoma stage of the *MMTV-PyMT x MMTV-iCre/CAGCAC-Dendra2* transgenic mouse at 13 weeks. Tumor cells (green), blood vessels (red), collagen fibers (magenta), macrophages (cyan). Scale bar 50 μm. The image is reprinted with permission from [22]. (**c**) Intravital image of tumor cells and the surrounding microenvironment in the orthotopic xenografts of MDA-MB-231-Dendra2 cells. Tumor cells (green), blood vessels (red), collagen fibers (magenta), macrophages (cyan). Scale bar 25 μm

signs [18, 23]. Repeated imaging is not advised and the animal is commonly sacrificed after imaging.

In the event that repeated longitudinal imaging is desired, mammary imaging window preparation is more appropriate (Fig. 1a). The imaging window, which consists of a glass coverslip on top of a plastic or a metal ring [19, 24], is sutured into the skin on top of the tumor tissue. The animal is allowed to heal for 3 days, after which it is available for continuous, noninvasive imaging and daily, longitudinal monitoring. The advantage of this approach is the prolonged monitoring (up to 21 days) of a developing tumor. With additional points of reference (photoconvertible fluorophores, fluorescent microbeads, photo-tattoo [25]), which can be used to monitor the same region of interest. However, this method requires additional surgery training for personnel and a three-day post surgery recovery time for the animal.

3.2 Labeling Blood Vessels and/or Macrophages

Intravenous, tail vein injection of 70/155 kDa fluorescent dextran (100 μg) will label active blood vessels [26]. 2–6 h following injection, the 70 kDa probe will extravasate and label the phagocytic subpopulation of macrophages [27]. For imaging total macrophage population, using transgenic mouse models is advised [23, 28] (*see* Subheading 2.1, **item 1**). Note that macrophage labeling will remain present for 5–7 days.

3.3 Image Acquisition

1. The animal is placed in an environmental chamber at 37 °C with continuous 1.5% isoflurane flow for the maintenance of anesthesia. 50 μL of sterile 1× PBS is administered intravenously (tail vein) every hour for hydration.
2. Locate a region of interest in the primary tumor, characterized by the presence of flowing blood vessels and bright fluorescence signal. Avoid areas rich in adipocytes as they scatter excitation light and obstruct fluorescence (Fig. 1b, c).
3. Acquire 4D stacks for the duration of interest (generally 30–60 min).
4. If longitudinal imaging of the acquired area is needed, proceed to photoconverting regions of interest (Subheading 3.8).

3.4 Detection and Quantification of Fast-Locomoting Cells

Translocating or motile cells are characterized by the extension of the cell front, movement of the center and the contraction of the rear.

1. Open a 4D stack in ImageJ.
2. Correct for movement by running HyperStackReg plugin to align sections in XY plane [29].
3. Visually score each z-section of the 4D stack movies for the morphological determinants of tumor cell fast locomotion. For the ease of processing, field of view (FOV) can be divided into smaller (1/4) sections (Fig. 2a, left panels). Select one z-section for further analysis.
4. Subtract frame taken at time 0 from the frame taken at time 60 min (*Process > Image Calculator > Subtract images*). Acquire an image with all pixels translocated during the time period. This step can be performed for differences between any two time points (Fig. 2a, middle panels).
5. Threshold the image to remove background fluorescence (*Image > Adjust > Threshold*).
6. Using Particle Analysis plugin in ImageJ, count the number of locomoting cells and obtain the mask of cell translocation. By adjusting the values for particle size and circularity, false positives, such as cell edges due to XY shift attributed to breathing artifacts, can be removed (Fig. 2a, right panels) (*Analyze > Analyze Particles > Size (15 μm²-infinity), Circularity (0.1–1)*).

7. Compare the mask with the 4D stack and ensure that all locomoting cells have been captured.
8. If not all locomoting cells are captured, readjust values for size and circularity in the Particle Analysis tool.
9. Manually remove remaining false positives, such as cell edges due to motion artifacts (Draw ROI > *Edit* > *Fill*).
10. Run Particle Analysis tool on the final processed image to obtain the count of locomoting cells.
11. Merge the obtained mask of locomoting cells with the frame at time 0 to visualize initial tumor cell position and track of movement (*Image* > *Color* > *Merge Channels*).

3.5 Detection and Quantification of Invadopodia in Slow-Locomoting Cells

Invadopodia are highly dynamic, small protrusions on the surface of tumor cells, characterized by finger-like morphology, <3 μm wide and <7 μm in length, and extension/retraction cycles. In vivo, invadopodia go through extension and retraction cycles of 3–20 min [30, 31], as well as change the position of the growing tip, making the selective identification via motility analysis possible. Importantly, only the cells exhibiting invadopodia-driven slow locomotion phenotype will successfully metastasize to the lung [31, 32].

1. Open a 4D stack in ImageJ.
2. Correct for movement by running HyperStackReg plugin to align sections in XY plane [29].
3. Visually score each z-section of the 4D stack movies at 2–4× zoom for the morphological determinants of tumor cells with invadopodia. For the ease of processing, cells can be analyzed individually (Fig. 2b).
4. Subtract frame taken at time 0 from the frame taken at 3 min to 15 min (*Process* > *Image Calculator* > *Subtract images*) (*see* **Note 1**). Acquire an image with all pixels translocated during the time period.
5. Threshold the image to remove background fluorescence (*Image* > *Adjust* > *Threshold*).
6. Using Particle Analysis, count the number of invadopodia per cell and obtain the mask of the dynamics of invadopodia over the time period. By adjusting the values for particle size and circularity, one can remove false positives such as cell edges due to XY shift (*Analyze* > *Analyze Particles* > *Size (0.5* μm^2*-infinity), Circularity (0–1)*).
7. Compare the mask with the 4D stack and ensure that all invadopodia have been captured.
8. If not all invadopodia are captured, readjust values for size and circularity in the Particle Analysis tool.

9. Manually remove remaining false positives (Draw ROI > *Edit* > *Fill*).
10. Run Particle Analysis tool on the final processed image to obtain the count of invadopodia.
11. Overlay the obtained mask of translocated protrusions with the frame at time 0 to visualize the extension/retraction of invadopodia (*Image* > *Color* > *Merge Channels*) (Fig. 2b).
12. Migration of the entire tumor cell body of invadopodia-driven slow-locomoting cells can be captured by extending the time-lapse imaging to 5–8 h.

3.6 Analysis of the In Vivo Microenvironmental Parameters

Features we routinely monitor in the tumor microenvironment include the density of collagen fibers, tumor cells and phagocytic macrophages, as well as the number and diameter of flowing blood vessels present in the FOV. Directionality and alignment of collagen fibers [33] and speed of blood vessel flow [34] can also be incorporated into the analysis without additional labeling. In animals where total macrophage population is labeled via *c-fms* promoter (*see* Subheading 2.1, **item 1a**), the number and speed of (non)phagocytic macrophages can also be quantified. Finally, the amount of degraded ECM can be measured via the injection of the protease-activatable imaging probe MMPSense 680, whose fluorescence increases >10-fold following proteolysis by key matrix metalloproteinases (MMP-2, -3, -9, and -13) [31] .

1. Open a multicolor 4D stack in ImageJ.
2. Z project all frames using the maximal intensity mode (*Image* > *Stacks* > *Z Project*).
3. Visually score the entire 4D hyperstack and duplicate the 3D, 4-color stack at first time point (*see* **Note 2**) (Fig. 1c).
4. Separate channels and apply a smoothing filter (*Process* > *Smooth*). To measure collagen fiber alignment, save the image corre-

Fig. 2 (continued) individual z-section from the 4D stack (at 0 and 60 min; 0′ and 60′). Middle panels, 0′ was subtracted from 60′, resulting in Δ60 (top); image was then thresholded/binarized (bottom). Right panels, results of motility analysis including quantification of fast locomoting cells (top) and the overlay with the 0′ image (bottom). Scale bar 50 μm. (**b**) Identifying and quantifying invadopodia in slow locomoting cells. Raw images of a cell at time 0, with fully extended invadopodium at 3 min and partially retracted invadopodium at 15 min. Overlays Δ3 and Δ15 show invadopodia extension highlighted in magenta. Scale bar 10 μm. (**c**) Binarized images used for extraction of microenvironmental parameters. Image in Fig. 1c was separated and thresholded, resulting in binary images of collagen (magenta), tumor cells (green), macrophages (cyan) and blood vessels (red). (**c′**) Collagen fiber map (left) and dimensionless straightness histogram (right) from ctFIRE software. (**d**) 3D projection of SVM classification results. Red spheres denote slow locomotion, blue-fast locomotion, green-misclassifications. The size of the spheres indicates the number of locomoting cells in the FOV. D_{max} (μm) stands for the diameter of the largest blood vessel in the FOV, macrophages (%) and collagen (%) stand for the thresholded area in respective channels

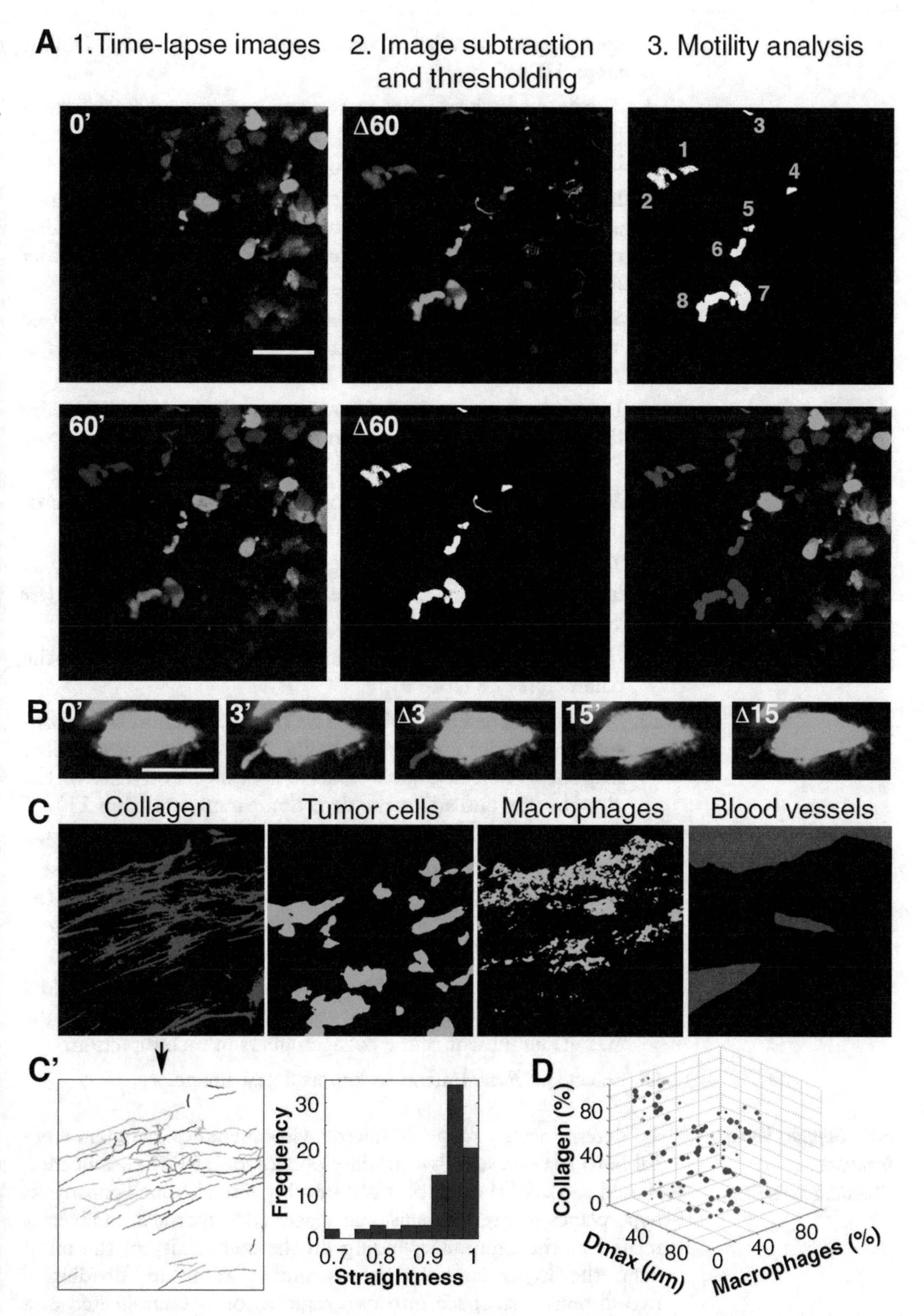

Fig. 2 Intravital Systems Microscopy: Multiparametric SVM classification of tumor cell locomotion and micro-environmental parameters (**a**) Identifying and quantifying fast-locomoting cells. Left panels, raw images of an

sponding to the collagen channel using ".tiff" format and follow **steps 10–14**.

5. Threshold each channel to remove background fluorescence (*Image* > *Adjust* > *Threshold*) (*see* **Note 3**) and obtain binary images for each parameter (Fig. 2c).
6. Measure the density of collagen fibers and tumor cells, defined as the percent area above threshold in each corresponding channel (enable the function "Limit to threshold" under *Analyze* > *Set Measurements*).
7. Separate macrophages from blood vessels on the basis of size and morphology in the dextran channel (*Analyze* > *Analyze Particles*).
8. Determine the number of flowing blood vessels and measure the diameter of the largest flowing blood vessel visible in the FOV using the "Straight Line" tool.
9. Manually count the number or measure the density of macrophages labeled with 70 kDa dextran in the FOV.
10. To measure collagen fiber alignment, import the image saved in **step 4** to ctFIRE software [33] (*Import image/data*) (*see* **Note 4**).
11. Optimize the accuracy of fiber detection by adjusting the parameters (*see* **Note 5**).
12. Run fiber detection analysis. The nonoverlaid and the reconstructed images are displayed (Fig. 2c′). If the overlaid fiber map shows non-negligible errors in fiber detection, click on the *Reset* button and adjust the detection parameters (**step 11**).
13. When fiber detection is satisfactory, check "Straightness histogram & values" and "Angle histrogram & values" under section "Select Output," and run the postprocessing analysis (*see* **Note 6**), which will return the histograms for Straightness (Fig. 2c′) and Angle (not shown).
14. Use the corresponding data set created in the output folder "...\ctFIREout" under the working directory to compute the average alignment of the collagen fibers in each direction.
15. Select the *Reset* button to import a new image.

3.7 Support Vector Machine Classifications

To determine the range of microenvironmental parameters amenable to fast- versus slow-motility phenotypes, SVM classification is used [35]. SVM is a supervised learning model, which constructs hyperplanes in high-dimensional spaces [36] that are nonlinearly related to the input. Depending on the complexity of the input data, the hyperplane can be as simple as a line dividing a two-dimensional space into two regions, or as complicated as a set of multidimensional nonlinear planes dividing the space into multiple zones. The discrete regions created by hyperplane(s) represent distinct classes that SVM recognizes among the input data. SVM algorithms are available for many mathematical packages, including Rstudio which is freely available. In this work, a nonlinear SVM from R-package "e-1072" [36] is used for data classification.

The source code (MotilityClassifier.R) for data classification is provided for download as a Supplementary File. The code receives the input data as a tab delimited text file and creates a SVM model with a "radial" kernel vector. Within the code, the details of each step are provided as comments. Briefly, the code randomly selects 90% of the input data for the learning process, based on which it generates a SVM model. The model is further tuned by iterative changing of the "cost" and "gamma" values until the "bestmodel" with the least error is achieved. Using this model, the code predicts the motility phenotype of the remaining 10% of the data and prints a table reporting the precision of the classification by indicating the percentage of misclassified data (green spheres, Fig. 2d). A low misclassification percentage (<5% in Fig. 2d) means that the input variables of two classes are statistically different and can be used for predicting the type of cell motility.

3.8 Photoconversion for Longitudinal Intravital Imaging

3.8.1 Determining Optimal Photoconversion Settings

1. Using the optical power meter, measure the power level of the 405 nm laser that enters the sample.
2. Define a range to test between 0 and 600 μW (Fig. 3a). In this case, the power settings are 20% (1 μW), 30% (4 μW), 40% (64 μW), and 50% (520 μW).
3. Balance the intensities of green to the background of red in a nonsequential scan before photoconverting.
4. Open "Live Plot" window to track the fluorescence intensities of red and green channels.
5. In Time Scan window, set Interval to "Free Run" and Number to "20."
6. Set the 405 nm laser to the power level being tested and start scanning. Monitor the Live Plot window and stop scanning as soon as red fluorescence intensity hits a plateau or green fluorescence intensity drops (photobleaching).
7. Out of four settings tested, 20% (1 μW) and 30% (4 μW) laser powers are too weak to induce photoconversion (Fig. 3b′, b″). 40% (64 μW) laser power is optimal with efficient photoconversion and with minimal photobleaching of the green fluorescence (Fig. 3b, b″). 50% (520 μW) laser power is too strong and results in complete photobleaching of the fluorescent signal (Fig. 3b, b″).

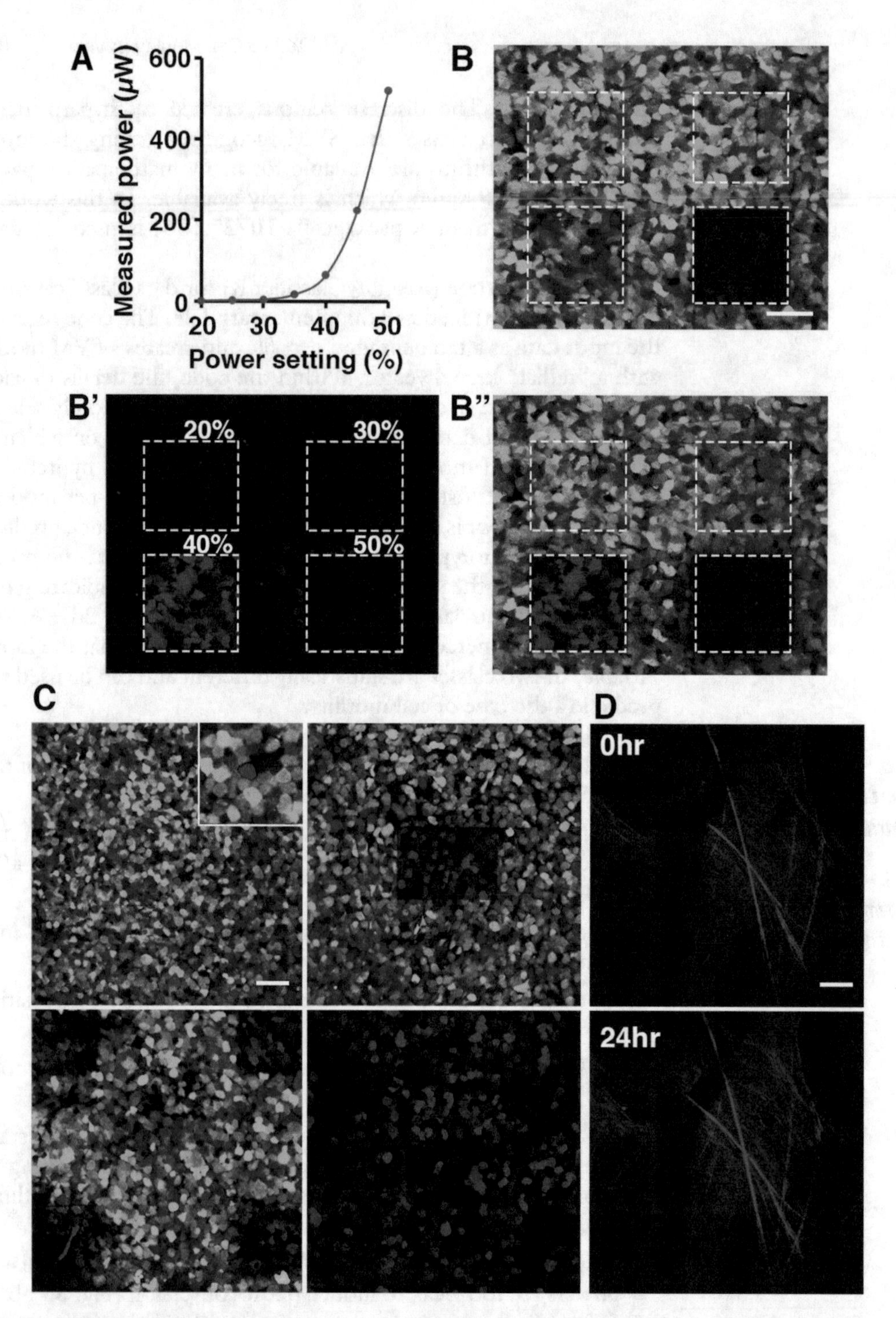

Fig. 3 Intravital photoconversion of Dendra2-labeled tumor cells (**a**) Relationship of the 405 nm laser power setting to laser power entering the tumor tissue. Measurement was done at the focal plane of UPLSAPO 30× objective. (**b**) Photoconversion efficiency at power settings 20–50% in a MDA-MB-231-Dendra2 xenograft section. (**b**) Green channel shows the effect of photobleaching at each power setting. (**b**′) Red channel shows the emission of photoconverted red Dendra2 at each power setting. (**b**″) Merge. Scale bar 50 μm. (**c**) Photoconversion of a single cell (upper left panel, zoom in the insert), 150 × 100 μm region containing ~100 cells (upper right panel), four 100 × 100 μm regions (lower left) and the entire FOV (lower right). Scale bar 50 μm. (**d**) Collagen fibers imaged at 0 h and relocated at 24 h using photoconverted region as a reference point. Scale bar 50 μm

3.8.2 Photoconversion of Tumor Cells In Vivo

After continuous imaging of the region of interest, decide on the area to be photoconverted

1. For a single cell, select the Point tool and place it at the center of the cell.
2. Using the optimal settings determined in Subheading 3.8.1 (64 μW laser power), scan until red fluorescence signal plateaus.
3. The photoconverted proteins diffuse and fill the entire cell (Fig. 3c, top left panel).
4. For a single area, use the selection tool to draw a rectangular region of interest (*see* **Note 7**) and scan (Fig. 3c, top right panel).
5. For photoconverting several regions (Fig. 3c, bottom left panel) or all cells in that field of view (FOV) (Fig. 3c, bottom right panel, *see* **Note 8**) the same tool as in **step 4** can be used (*see* **Note 9**).
6. Using photoconverted regions as tissue landmarks, the same FOV can be located at multiple imaging sessions, allowing longitudinal imaging of tumor cells and/or microenvironmental parameters (Fig. 3d).

3.9 Ex Vivo Imaging of Lung Metastases

Breast cancers commonly metastasize to the lung, liver, brain, and bone. Likewise, in our mouse models, we observe single cells and micrometastases in the lung. Counting the number of single cells and colonies is suitable for assessing total number of metastases following euthanasia of the animal at the end of imaging. Additionally, the number of red metastases can be measured at a set time point (2–7 days) following photoconversion, enabling the measurement of dynamics of tumor cell homing, proliferation, or dormancy [37] in the lung. Dendra2 red variant is highly stable and protein turnover in vivo is negligible ([21, 24]; therefore, with each cell division red fluorescence is diluted by 50%. As the non-photoconverted green variant of Dendra2 is continuously being synthesized under the CMV promoter, cells will become orange and yellow following cell division(s) (Fig. 4a′). The origin of green cells cannot be delineated, as they can be both descending from yellow cells, or arriving from the nonphotoconverted green regions of the primary tumor.

1. Immediately after sacrificing, secure the mouse in supine position by taping the legs to the surgical surface.
2. Make a midline incision on the chest, open the skin and peritoneum, exposing the ribcage.
3. Cut through the bones and cartilage to open up the ribcage. Absorb blood with tissue paper.

4. Harvest the lung tissue and transfer it to a dish with 1× PBS.
5. Separate two lobes of the lung and quickly rinse with fresh 1× PBS.
6. In a drop of 1× PBS, transfer the lung to a coverslip or a glass-bottom microscope dish for imaging.
7. Image metastatic lesions in 20 random FOV (Fig. 4a).
8. Using image processing, quantify the number of metastases:
   (a) Open image in ImageJ.
   (b) Remove background fluorescence and set threshold (*Image > Adjust > Threshold*).
   (c) Count the number of cells (*Analyze > Analyze Particles > Size (15 μm²-infinity), Circularity (0.1–1)*).

3.10 Correlative Immunofluorescence

1. After imaging and photoconversion, harvest the tumor tissue and immediately fix in 4% paraformaldehyde solution for 1 h at room temperature and then overnight at 4 °C.
2. Wash the tissue in cold 1× PBS for 1 h at 4 °C.
3. Incubate tissue in 30% sucrose solution overnight at 4 °C.
4. In a cryomold, immerse the tumor tissue in O.C.T compound, orient it and freeze it in an isopentane bath (*see* **Note 10**).
5. The frozen tissue can be stored at −80 °C until ready to be sectioned.
6. Using a cryostat, section the tissue sample into 6–10 μm thick slices and mount onto positively charged microscope glass slides.
7. The slides can be stored at −80 °C until ready to be stained.
8. Wash off O.C.T. compound from tissue sections by repeatedly dipping slides in water.
9. To permeabilize tissue sections, immerse slides in permeabilization solution for 10 min. Alternatively, sections can be permeabilized for 10 min in ice-cold acetone (procedure we use for labeling microenvironmental parameters, such as endomucin, laminin, collagen IV, collagen I, or fibronectin).
10. Briefly air-dry slides.
11. Draw a circle around each tissue section with a liquid blocker pen to provide a barrier for solutions.
12. Block tissues with 50–100 μL drops of the blocking solution for 2 h at room temperature. Alternatively, block tissues overnight at 4 °C. To prevent evaporation, place slides into a humidified chamber.
13. Incubate with primary antibodies diluted to appropriate concentrations in 50–100 μL of blocking solution for 3 h at room temperature, in a humidified chamber.

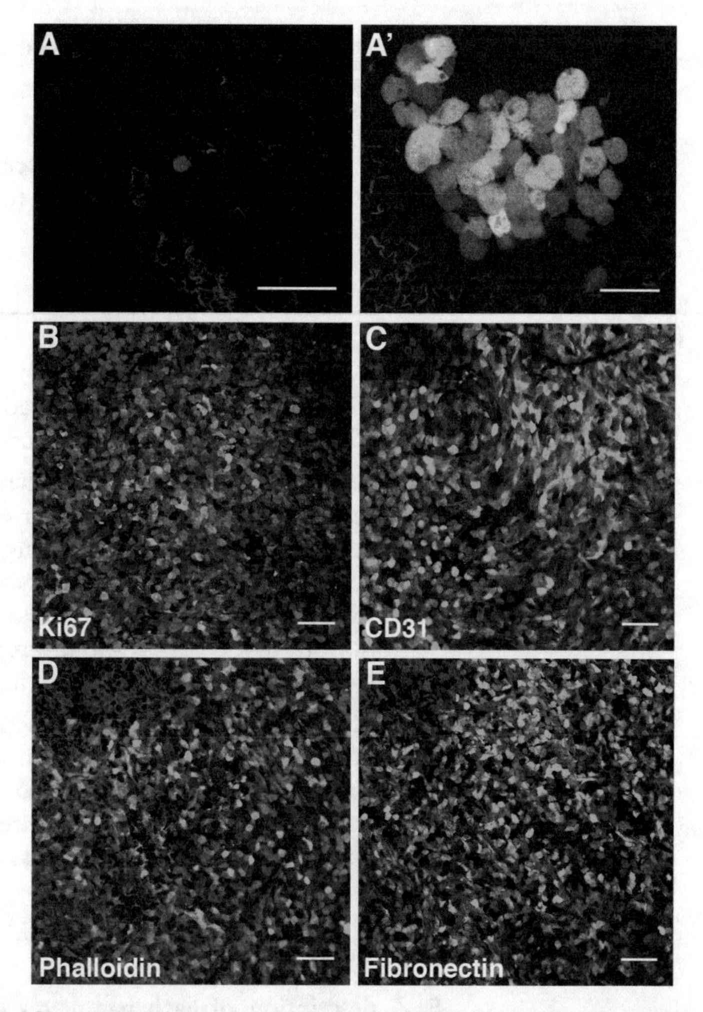

Fig. 4 Photoconversion as a multiplexing tool: Intravital imaging and photoconversion followed by imaging of lung explants and immunofluorescence of primary tumor (**a**) Ex vivo imaging of lungs of animals imaged and photoconverted in vivo. A single photoconverted red cell in the lung (**a**) and a metastatic colony (**a′**) observed at 90 h postphotoconversion. Collagen fibers are shown in magenta. Scale bar 50 μm. The image in **a′** is reprinted with permission from [31]. (**b–e**) Immunofluorescence in cryosections of primary tumors imaged and photoconverted in vivo. Additional labels are shown in purple: proliferation, Ki67 (**b**), blood vessels, CD31 (**c**), actin cortex, phalloidin (**d**), ECM, fibronectin (**e**). Scale bar 50 μm

14. Wash by dipping slides in 1× PBS.
15. Incubate with secondary antibodies and DAPI diluted in 50–100 μL of blocking solution for 2 h at room temperature in a humidified chamber and protected from light.
16. Wash by dipping slides in 1× PBS.
17. Briefly air-dry slides.

18. Place a drop of Fluoromount-G on each tissue section.
19. Avoiding air bubbles, place a glass coverslip and seal with nail polish.
20. The glass slides with tissue sections can be stored at 4 °C and protected from light, until ready to be imaged.
21. Image cryosections (Fig. 4b–e).

4 Notes

1. The choice of time points for invadopodia analysis depends on the frequency of extension–retraction cycles.
2. Identify a time-point when all microenvironmental parameters of interest are clearly visible. For example, to visualize blood vessels, it is advised to start imaging immediately after injection of fluorescent dextran. For phagocytic macrophages, inject the animal with fluorescent dextran >3 h prior to the imaging session. Both blood vessels and macrophages will be visible in the time window of 1–3 h following injection.
3. For the first set of images in the series, suitable threshold value is set manually. At that time, users can decide on a satisfactory automated thresholding method (*Image* > *Adjust* > *Auto Threshold* drop down list). The threshold value will need to be adjusted for each imaging session if the acquisition parameters differ.
4. The ctFIRE software requires MATLAB compiler runtime (MCR 7.17 2012a) installation.
5. A description of each parameter can be found in the User Manual (available at *http://loci.wisc.edu/software/ctfire*). We recommend starting with default values. The parameters and the options under the sections "Output Figure Control" and "Select Output" can be modified after fiber detection, during the postprocessing step (*see* **step 13**).
6. Additional outputs can be selected: "Overlaid fibers," "Non-overlaid fibers," "Angle histogram & values," "Length histogram & values," and/or "Width histogram & values."
7. Areas containing >20 cells are easier to locate later in cryosections and determine the orientation of the tissue.
8. Photoconverting all four corners of the FOV is a useful method if one plans to perform correlative immunofluorescence. This way, the FOV can be located by the photoconverted corners, while leaving the rest of the FOV available for labeling in the red channel. Photoconversion of the entire FOV is useful if the number of photoconverted cells needs to be maximized, such as postsacrificial imaging of lung metastases.

9. Alternatively, all cells in the FOV can be photoconverted using the mercury or LED lamp excitation through the DAPI filter [38]. Locate the area of interest and expose it to the lamp on full power for 5–10 min. This method yields photoconverted FOV with diffuse edges. If an area larger than a single FOV is needed, custom-built LED array can be used to expose the entire surface under the mammary imaging window to 405 nm light [24, 31].
10. To prepare the isopentane bath, fill a stainless steel container with isopentane and place it in dry ice. Drop a few pellets of dry ice into isopentane and wait for the boiling to stop. At this time isopentane is chilled to about –90 °C and is ready for freezing the sample. Slowly lower the sample and drop it in the isopentane bath. Leave it until the entire block is frozen through and the sample sinks to the bottom of the isopentane bath.

Acknowledgments

We thank Dr. Aviv Bergman for his contribution in establishing the SVM algorithm for classification. This work was supported by grants from the NIH 5K99CA172360 and Concern Foundation Award to B.G.

References

1. Steeg PS (2016) Targeting metastasis. Nat Rev Cancer 16:201–218. https://doi.org/10.1038/nrc.2016.25
2. Vanharanta S, Massagué J (2013) Origins of metastatic traits. Cancer Cell 24:410–421. https://doi.org/10.1016/j.ccr.2013.09.007
3. Quail DF, Joyce JA (2013) Microenvironmental regulation of tumor progression and metastasis. Nat Med 19:1423–1437. https://doi.org/10.1038/nm.3394
4. Conklin MW, Eickhoff JC, Riching KM et al (2011) Aligned collagen is a prognostic signature for survival in human breast carcinoma. Am J Pathol 178:1221–1232. https://doi.org/10.1016/j.ajpath.2010.11.076
5. Gehler S, Ponik SM, Riching KM, Keely PJ (2013) Bi-directional signaling: extracellular matrix and integrin regulation of breast tumor progression. Crit Rev Eukaryot Gene Expr 23:139–157
6. Levental KR, Yu H, Kass L et al (2009) Matrix crosslinking forces tumor progression by enhancing integrin signaling. Cell 139:891–906. https://doi.org/10.1016/j.cell.2009.10.027
7. Marusyk A, Almendro V, Polyak K (2012) Intra-tumour heterogeneity: a looking glass for cancer? Nat Rev Cancer 12:323–334. https://doi.org/10.1038/nrc3261
8. Bergman A, Condeelis JS, Gligorijevic B (2014) Invadopodia in context. Cell Adhes Migr 8:273–279. https://doi.org/10.4161/cam.28349
9. Denk W, Strickler J, Webb W (1990) Two-photon laser scanning fluorescence microscopy. Science 248:73–76. https://doi.org/10.1126/science.2321027
10. Condeelis J, Segall JE (2003) Intravital imaging of cell movement in tumours. Nat Rev Cancer 3:921–930. https://doi.org/10.1038/nrc1231
11. Mohler W, Millard AC, Campagnola PJ (2003) Second harmonic generation imaging of endogenous structural proteins. Methods 29:97–109
12. Zipfel WR, Williams RM, Christie R et al (2003) Live tissue intrinsic emission microscopy using multiphoton-excited native fluorescence and second harmonic generation. Proc

Natl Acad Sci 100:7075–7080. https://doi.org/10.1073/pnas.0832308100

13. Li JL, Goh CC, Keeble JL et al (2012) Intravital multiphoton imaging of immune responses in the mouse ear skin. Nat Protoc 7:221–234. https://doi.org/10.1038/nprot.2011.438
14. Lelkes E, Headley MB, Thornton EE et al (2014) The spatiotemporal cellular dynamics of lung immunity. Trends Immunol 35:379–386. https://doi.org/10.1016/j.it.2014.05.005
15. Hirata E, Girotti MR, Viros A et al (2015) Intravital imaging reveals how BRAF inhibition generates drug-tolerant microenvironments with high integrin β1/FAK signaling. Cancer Cell 27:574–588. https://doi.org/10.1016/j.ccell.2015.03.008
16. Helmchen F, Fee MS, Tank DW, Denk W (2001) A miniature head-mounted two-photon microscope. Neuron 31:903–912. https://doi.org/10.1016/s0896-6273(01)00421-4
17. Ghosh KK, Burns LD, Cocker ED et al (2011) Miniaturized integration of a fluorescence microscope. Nat Methods 8:871–878. https://doi.org/10.1038/nmeth.1694
18. Ewald AJ, Werb Z, Egeblad M (2011) Dynamic, long-term in vivo imaging of tumor-stroma interactions in mouse models of breast cancer using spinning-disk confocal microscopy. Cold Spring Harb Protoc 2011:pdb.top97. https://doi.org/10.1101/pdb.top97
19. Alieva M, Ritsma L, Giedt RJ et al (2014) Imaging windows for long-term intravital imaging. Intravital 3:e29917–e29916. https://doi.org/10.4161/intv.29917
20. Gurskaya NG, Verkhusha VV, Shcheglov AS et al (2006) Engineering of a monomeric green-to-red photoactivatable fluorescent protein induced by blue light. Nat Biotechnol 24:461–465. https://doi.org/10.1038/nbt1191
21. Chudakov DM, Lukyanov S, Lukyanov KA (2007) Tracking intracellular protein movements using photoswitchable fluorescent proteins PS-CFP2 and Dendra2. Nat Protoc 2: 2024–2032. https://doi.org/10.1038/nprot.2007.291
22. Génot E, Gligorijevic B (2014) Invadosomes in their natural habitat. Eur J Cell Biol 93:367–379. https://doi.org/10.1016/j.ejcb.2014.10.002
23. Egeblad M, Ewald AJ, Askautrud HA et al (2008) Visualizing stromal cell dynamics in different tumor microenvironments by spinning disk confocal microscopy. Dis Model Mech 1:155–167. https://doi.org/10.1242/dmm.000596
24. Kedrin D, Gligorijevic B, Wyckoff J et al (2008) Intravital imaging of metastatic behavior through a mammary imaging window. Nat Methods 5:1019–1021. https://doi.org/10.1038/nmeth.1269
25. Ritsma L, Vrisekoop N, van Rheenen J (2013) In vivo imaging and histochemistry are combined in the cryosection labelling and intravital microscopy technique. Nat Commun 4:2366. https://doi.org/10.1038/ncomms3366
26. Harney AS, Arwert EN, Entenberg D et al (2015) Real-time imaging reveals local, transient vascular permeability, and tumor cell Intravasation stimulated by TIE2hi macrophage-derived VEGFA. Cancer Discov 5:932–943. https://doi.org/10.1158/2159-8290.CD-15-0012
27. Wyckoff J, Wang W, Lin EY et al (2004) A paracrine loop between tumor cells and macrophages is required for tumor cell migration in mammary tumors. Cancer Res 64:7022–7029. https://doi.org/10.1158/0008-5472.CAN-04-1449
28. Wyckoff J, Gligorijevic B, Entenberg D et al (2011) High-resolution multiphoton imaging of tumors in vivo. Cold Spring Harb Protoc 2011:pdb.top065904. https://doi.org/10.1101/pdb.top065904
29. Thevenaz P, Ruttimann UE, Unser M (1998) A pyramid approach to subpixel registration based on intensity. IEEE Trans Image Process 7:27–41. https://doi.org/10.1109/83.650848
30. Oser M, Mader CC, Gil-Henn H et al (2010) Specific tyrosine phosphorylation sites on cortactin regulate Nck1-dependent actin polymerization in invadopodia. J Cell Sci 123:3662–3673. https://doi.org/10.1242/jcs.068163
31. Gligorijevic B, Bergman A, Condeelis J (2014) Multiparametric classification links tumor microenvironments with tumor cell phenotype. PLoS Biol 12:e1001995. https://doi.org/10.1371/journal.pbio.1001995
32. Gligorijevic B, Wyckoff J, Yamaguchi H et al (2012) N-WASP-mediated invadopodium formation is involved in intravasation and lung metastasis of mammary tumors. J Cell Sci 125:724–734. https://doi.org/10.1242/jcs.092726
33. Bredfeldt JS, Liu Y, Conklin MW et al (2014) Automated quantification of aligned collagen for human breast carcinoma prognosis. J Pathol Inform 5:28. https://doi.org/10.4103/2153-3539.139707
34. Dasari S, Weber P, Makhloufi C et al (2015) Intravital microscopy imaging of the liver following Leishmania infection: an assessment of hepatic hemodynamics. J Vis Esp e52303. doi: https://doi.org/10.3791/52303

35. Vapnik VN (1999) An overview of statistical learning theory. IEEE Trans Neural Netw 10:988–999. https://doi.org/10.1109/72.788640
36. Karatzoglou A, Meyer D, Hornik K (2006) Support vector machines in ddR. J Stat Softw 15. https://doi.org/10.18637/jss.v015.i09
37. Fluegen G, Avivar-Valderas A, Wang Y et al (2017) Phenotypic heterogeneity of disseminated tumour cells is preset by primary tumour hypoxic microenvironments. Nat Cell Biol 19:120–132. https://doi.org/10.1038/ncb3465
38. Baker SM, Buckheit RW, Falk MM (2010) Green-to-red photoconvertible fluorescent proteins: tracking cell and protein dynamics on standard wide-field mercury arc-based microscopes. BMC Cell Biol 11:15. https://doi.org/10.1186/1471-2121-11-15

Chapter 15

Using the Zebrafish Embryo to Dissect the Early Steps of the Metastasis Cascade

Gautier Follain, Naël Osmani, Cédric Fuchs, Guillaume Allio, Sébastien Harlepp, and Jacky G. Goetz

Abstract

Most cancers end up with the death of patients caused by the formation of secondary tumors, called metastases. However, how these secondary tumors appear and develop is only poorly understood. A fine understanding of the multiple steps of the metastasis cascade requires in vivo models allowing high spatiotemporal analysis of the behavior of metastatic cells. Zebrafish embryos combine several advantages such as transparency, small size, stereotyped anatomy, and easy handling, making it a very powerful model for cell and cancer biology, and in vivo imaging analysis. In the following chapter, we describe a complete procedure allowing in vivo imaging methods, at high throughput and spatiotemporal resolution, to assess the behavior of circulating tumor cells (CTCs) in an experimental metastasis assay. This protocol provides access, for the first time, to the earliest steps of tumor cell seeding during metastasis formation.

Key words Zebrafish, Circulating tumor cells (CTCs), Metastasis, Injection, Live imaging

1 Introduction

Metastasis progression is a complex multistep process leading to the colonization of distant organs by tumor cells, which will further develop into life-threatening secondary tumors [1]. Cell migration and invasion drive most of the steps involved in the formation of metastases. Once cancer cells acquire an invasive potential, they will (1) colonize the surrounding stroma, (2) reach neighboring vessels and enter blood circulation (intravasation), (3) be transported to distant organs, (4) stop and exit (extravasation) and (5) invade the organs and settle to secondary metastatic sites.

Gautier Follain and Naël Osmani contributed equally to this work.

Electronic Supplementary Material: The online version of this chapter (https://doi.org/10.1007/978-1-4939-7701-7_15) contains supplementary material, which is available to authorized users.

Alexis Gautreau (ed.), *Cell Migration: Methods and Protocols*, Methods in Molecular Biology, vol. 1749, https://doi.org/10.1007/978-1-4939-7701-7_15, © Springer Science+Business Media, LLC 2018

Several in vitro approaches modeling these steps have shed light on the cellular dynamics as well as the molecular pathways involved [2]. However, they often fail to reproduce the complexity of the in vivo microenvironment and thus lack physiological relevance. On the other hand, the mouse model has been extensively used for the past decades to elucidate cancer progression in a more physiopathologically relevant context. These approaches include (1) xenograft of tumor cells in mice to study stromal invasion, and intravasation [3, 4], (2) intracardiac or tail vein injection of tumor cells to model circulating tumor cells (CTC) blood transport, extravasation, and metastatic niche colonization [5, 6] or (3) genetically engineered mouse models, which resume all the steps from the initial tumor formation to metastasis formation in specific organs [7]. However, these approaches often lack the time resolution required to study the complex dynamics of metastasis as it is extremely difficult, money and time-consuming to track and record these rare events in mice. Most importantly, none of these methods allow tracking of the earliest steps of tumor cell seeding in vascular structures. Moreover, these methods are unable to tune the microenvironmental parameters (such as blood flow forces) while offering the possibility to compile observations over a high number of individuals.

Therefore, the use of small transparent organisms, which are more adapted to intravital imaging, is now emerging as new models in cancer biology. The xenograft of tumor cells into the chick embryo chorioallantoic membrane (CAM), which has no immune system during the early developmental stage, has been used to monitor specifically tumor angiogenesis as well as the metastatic cascade [8, 9]. However, the lack of transgenic tools makes the CAM model less suitable for specific intravital imaging. Moreover, the nonstereotyped vasculature hinders high-throughput visualization of single cell behaviors. The zebrafish, and its embryo, have recently emerged as a promising model for cancer research [10–18]. Its embryo is transparent with a high-breeding ability and has an immature immune system making it compatible with tumor xenograft similarly to the CAM. It displays a highly stereotyped vasculature allowing high throughput analysis of tumor cell behavior over several embryos. In addition, there is also a large panel of molecular tools. These include the use of morpholinos [19] to transiently knock down gene expression or the use of knockout/knockin (TALEN and CRISPR/Cas9) to specifically alter gene expression or create cell-type specific fluorescent transgenic lines [20–22].

In this chapter, we propose an original method using the zebrafish embryo as an experimental metastasis model (Fig. 1), which allows the dissection of (1) the early steps of metastasis formation (cell arrest and stable adhesion to the endothelium), (2) CTC exit, through extravasation, from the blood stream and (3) the seeding of the nearby stroma and the formation of micrometastases with high spatiotemporal resolution. This method enables, for the first time, the study of the behavior of single CTC

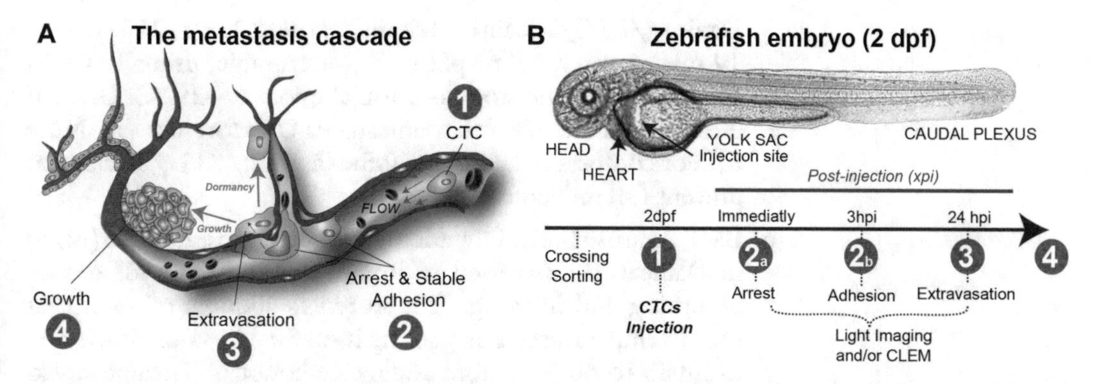

Fig. 1 Zebrafish embryos as a new experimental metastasis model. (**a**) Schematic representation of the last steps of the metastatic cascade: circulating tumor cells, carried by the blood flow, arresting and stably adhering to the endothelial cells before extravasating and forming metastasis. (**b**) Zebrafish embryo, 2 days post-fertilization (dpf) (transmitted light imaging) and global experimental timeline from breeding to CTC injection and imaging of the embryos

or small clusters of CTC in perfused vessels, and thereby offers an unprecedented window to the understanding of the earliest mechanisms that tumor cells have evolved to stop and escape from shear forces, preceding metastatic outgrowths.

2 Materials

2.1 Zebrafish Transgenic and Cell Lines

Ideally, one should use ZF transgenic lines with fluorescent reporters stably expressed in the endothelium to label the vasculature. We mostly use the *Tg(fli1a:EGFP)* transgenic line [23] (*see* **Note 1**). This protocol is designed for studying the metastatic behavior of tumor cells (cell lines or primaries; from zebrafish, mouse or human origin). Each line will differ in its ability to adhere and extravasate and one could assess this potential by comparing cells with knock-down or overexpression of any gene of interest (using siRNA or shRNA approaches). In addition, this protocol is perfectly suited for assessing the behavior of any protein/organelle of interest, provided that a corresponding stably fluorescent cell line can be established.

2.2 Zebrafish Handling and Injection Solutions

1. Danieau stock (30×): 1.74 mol/L NaCl, 21 mmol/L KCl, 12 mmol/L $MgSO_4$, 18 mmol/L $Ca(NO_3)_2$, HEPES 150 mmol/L in water.
2. Danieau (0.3×): The stock solution is diluted 1:100 in water before use (0.3×).
3. PTU stock (50×): 15 g of phenylthiourea (Sigma) in 100 mL of Danieau 0.3×, dissolve by heating at 60 °C under agitation. Stock is stored as 10 mL aliquots at −20 °C.
4. Danieau/PTU: 1 aliquot of PTU is added to 490 mL Danieau 0.3× before use (*see* **Note 2**).

5. Danieau/PTU/tricaine: Tricaine stock (25×): 210 mg in 49 mL of water, adjust pH to 7 (for example, using Tris 1 M pH 9.4). Store the stock as 1 mL aliquots at −20 °C. One aliquot is added to 24 mL Danieau/PTU before use (*see* **Note 3**). For confocal imaging, use 0.5× Danieau/PTU/Tricaine to prevent fish movement.
6. LMP Agarose 0.8%: Low Melting Point agarose at 0.8% (m/v) in Danieau 0.3× is used to immobilize the embryos during mounting and imaging. It is stored as aliquots of 24 mL at room temperature. For use, melt at 80 °C, cool down and maintain to 40 °C before adding 1 aliquot of Tricaine stock. Further proceed with mounting.
7. Mineral oil (Sigma) is needed to fill the injection glass capillary before injection (*see* **Note 4**).

2.3 Cell Preparation

Use the appropriate medium for your cell line of choice, supplemented with usual additives. Cells can be prepared for injection in their cell culture medium or in PBS. PBS 1× or PBS 1× supplemented with 0.526 mM EDTA. EDTA is used to wash the cell before trypsinization. Use the trypsin concentration usually chosen to detach cells (for us: trypsin–EDTA 0.05% to 0.25% in PBS).

2.4 Equipment (Fig. 2)

1. Classical culture room equipment: To prepare the cells, classical culture room equipment is needed: hood, incubator, centrifuge, cell counter and microscope.
2. Capillary puller: To pull the microneedle from glass capillaries, a micropipette puller is needed (Shutter instrument). We recommend using glass capillaries with 0.53 mm inner diameter (ID) and 1.14 mm outer diameter (OD) (Drummond). After pulling, microneedles are microforged to reach a 20–25 μm ID tip (*see* **Note 5**).
3. Embryos mounting (Fig. 2a, b): Heating block, transfer pipets, tweezers, petri Dishes (10 to 15 cm in diameter) and 35 mm glass-bottom dishes, uncoated (MatTek).
4. Nanoinjection (Fig. 2c): We used the Nanoject II (Drummond) that is perfectly suited for injecting ZF embryos. We combine the microinjector with the following stereomicroscope.
5. Microscopy (Fig. 2d): The stereomicroscope Leica M205FA equipped with the LAS AF 4.0 software is used for injection and live imaging. The detailed setup composition is the following: fluorescent excitation lamp (Mercury short-arc reflector lamp, EL6000), GFP filter (Ex. 450–490/Em. 500–550) and ET-C filter (Ex. 533–557/Em. 570–640), Plan APO objective 20x (10450028), camera (DFC3000 G), heating plate system (MATS).
6. Confocal microscopy: Any confocal microscope of choice would be appropriate. We are familiar with SP2/5/8 confocal

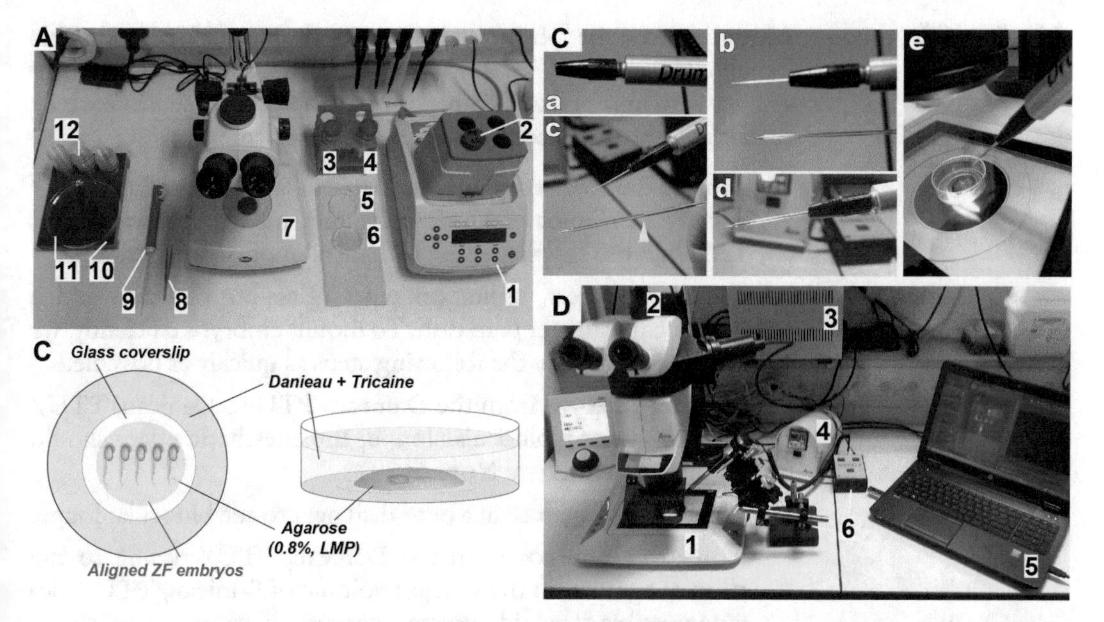

Fig. 2 Material for mounting and injection. (**A**) Example of bench organization used to mount the embryos: *1* Heating block to keep the LMP agarose at 40 °C. *2* LMP agarose 0.8% to immobilize the embryos. *3* Danieau/PTU/tricaine ready to use. *4* Danieau PTU to be added after agarose polymerization. *5* Petri dish filled with Danieau/PTU/tricaine. *6* Petri dish filled with Danieau/PTU and the embryos to be mounted. *7* Binoculars. *8* Tweezers to place the embryos within the LMP agarose drop. *9* Aspiration pipet used to transfer embryos from dish to dish. *10* Dark surface to be able to visualize the embryo. *11* Petri dish to mix the embryos in the drop of agarose. *12* Glass-bottomed petri dish compatible with imaging (MatTek). (**B**) Schematic view of the embryos aligned for injection and imaging in the MatTek dish. (**C**) Microneedle preparation and installation: (*a*) microinjector head, (*b*) microneedle (bottom) and microinjector with the piston out, (*c*) filling the microneedle with mineral oil, (*d*) microneedle mounted on the piston, (*e*) microinjector ready-to-use under the stereomicroscope. (**D**) Microinjection setup: *1* Stereomicroscope. *2* Camera. *3* Fluorescent illumination lamp. *4* Heating plate controller. *5* Computer with the Leica software. *6* Microinjector and its controller

microscope systems (Leica, either upright or inverted) which are used for confocal acquisition with an immersion 20× objective or a water-dipping 25× objective (HC FLUOTAR L 25×/0.95 W VISIR).

3 Methods

3.1 Zebrafish Embryos Handling

1. Zebrafish embryos should be maintained in 5–10 cm diameter petri dish, in Danieau 0.3× for the first 24 h post fertilization (hpf) and then transferred to Danieau/PTU for the rest of the experiments.
2. Embryos are carefully moved using plastic transfer pipet.
3. Dechorionate embryos at least two hours before mounting, using thin tweezers, to avoid any bending of the embryos during the mounting stage.

3.2 Zebrafish Mounting

Mounting can be performed using a simple and reproducible bench organization (Fig. 2a). The goal of the mounting step is to end up with immobilized embryos in a drop of LMP agarose, stuck on a glass-bottomed petri dish suitable for imaging.

1. Prepare the required solutions: Danieau/PTU, Danieau/PTU/tricaine (1 aliquot of 1 mL in 24 ml), melted and warm-kept LMP agarose supplemented with an aliquot of 1 mL tricaine.
2. Prepare tweezers, aspiration pipet, glass-bottom dish and a 10–15 cm diameter petri dish. To mount embryos efficiently, we realize sequentially the following steps as quickly as possible.
3. Transfer embryos from the Danieau/PTU to Danieau/PTU/tricaine under a binocular loupe, the anesthetic effect should be instantaneous (*see* **Note 6**).
4. Put a drop of agarose in a petri dish next to the binocular loupe.
5. Move the embryos from the Danieau/PTU/tricaine to the drop of agarose, in the smallest volume of Danieau/PTU/tricaine possible to avoid excessive agarose dilution.
6. Transfer the embryos (usually 8) with the agarose to the glass-bottomed dish.
7. Place the embryos as wished using tweezers, before the agarose solidifies (Fig. 2b, *see* **Note 7**).
8. After 1–2 min, the agarose should be solidified.
9. Add Danieau/PTU to keep the embryo in a liquid medium.
10. Store embryos in the incubator at 28 °C until injection and imaging.

3.3 Preparation of the Cells to Be Injected, at a Defined Concentration

1. Tumor cells (TCs) should be prepared for injection in their growing medium or in PBS.
2. Cells are usually concentrated at 10^8 cells per mL for injection, which roughly represents a final volume of 80 μL of cell suspension from a confluent 10 cm culture dish. These parameters need to be adjusted and adapted to your cell line of interest.

3.4 Injection Preparation

For the injection of TCs, the Nanoject II (Drummond) is used in association with a fluorescence stereomicroscope (Fig. 2c, d).

1. Microneedles are pulled from capillaries and microforged extemporaneously using a capillary puller (Shutter Instrument) and a microforge to have an inner diameter of 20–25 μm (*see* **Note 5**).
2. The microneedle is filled with mineral oil before inserting it around the Nanoject piston.
3. Five to ten microliters of cell suspension are deposited on a piece of Parafilm under the stereomicroscope, enabling a visual control during microneedle loading (Fig. 3a, *see* **Note 8**).

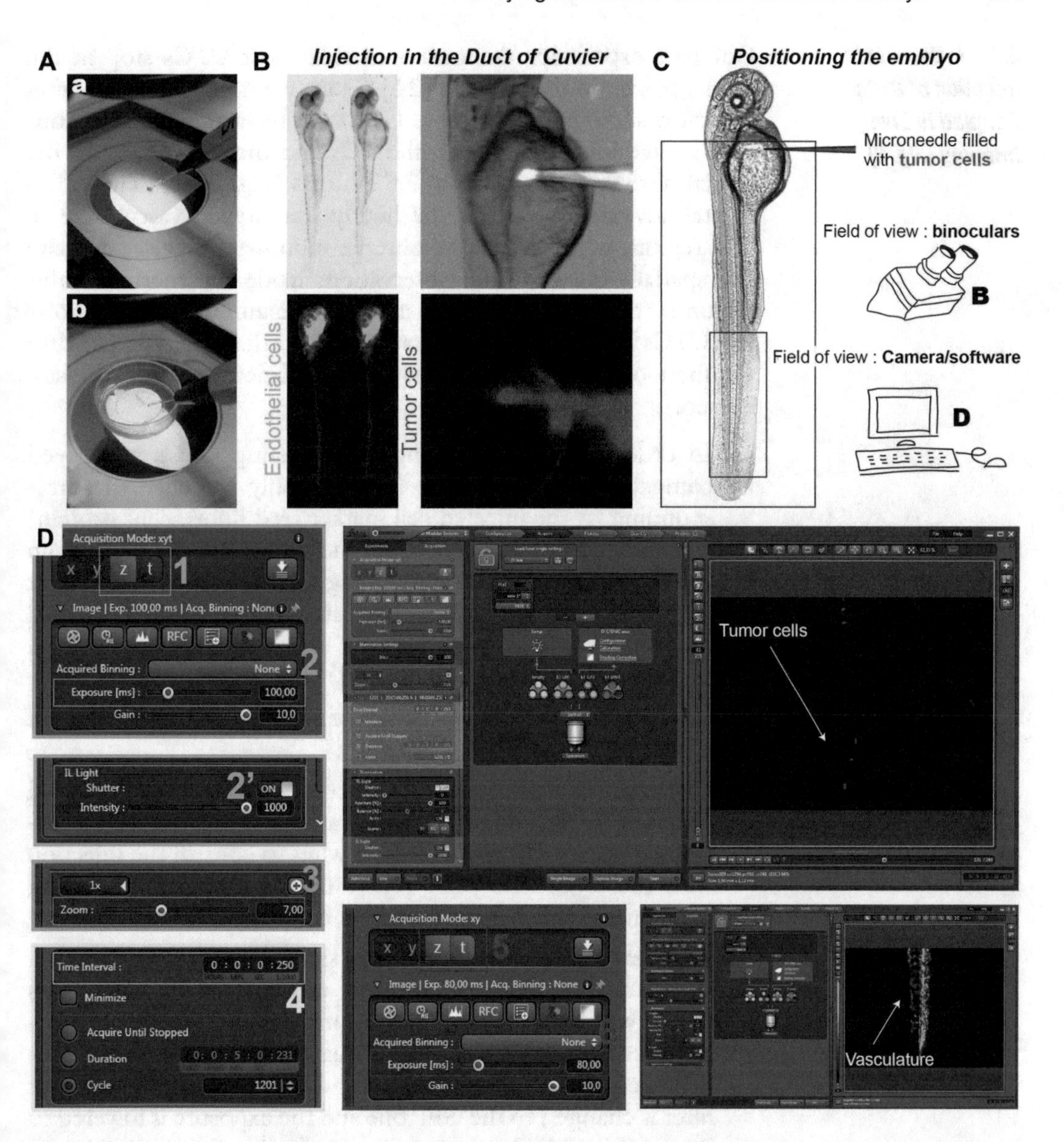

Fig. 3 Injection and live imaging setup. (**A**) Pictures showing the filling of the capillary with the solution containing tumor cells (*a*) and the rough positioning of the injector with a MatTek dish and zebrafish embryos (*b*). (**B**) Pictures showing the aligned fish, ready for injection under the stereomicroscope, with transmitted light and GFP filter (right); the position of the microneedle filled with cells for injecting in the duct of Cuvier with transmitted light and ET-C filter (left). (**C**) Schematic view of the situation for live injection imaging: with magnification optimized at ×7.0, one will see the point of injection and the caudal plexus in the oculars, and at the same time, camera field of view will cover the entire plexus. (**D**) Software setting for live injection imaging: First, to do the time-lapse imaging, ET-C filter, Time (*T*) dimension set to on (*1*), Adapted exposure with maximal intensity (*2,2′*), Zoom on ×7.00 (*3*), Time interval on 250 ms (*4*). Second, to record the endothelium (background image), GFP filter, Time dimension set to off (*5*), Adapted exposure (*6*)

3.5 Intravascular Injection of CTCs Coupled to Live Imaging (Fig. 3)

Our past experience demonstrated that most CTCs stop in the caudal plexus of the embryo [24]. Thus, we designed an imaging workflow such that injection of CTCs can be instantaneously coupled to live imaging of both the TCs and the vasculature in the caudal plexus of a developing ZF embryo (Fig. 3a–c). This offers several advantages such as the highly stereotyped vasculature of that region, which allows to observe multiple animals, cumulate and spatially correlate the observations made. Furthermore, this region is relatively thin, which allows to quantitatively track flowing CTCs in a vascular network that is rather flat. To perform simultaneous injection and imaging, the Leica LAS_AF software is used.

1. In order to capture flowing and arresting CTCs at a speed compatible with real time imaging, only the channel corresponding to the injected cell marker (red fluorescent protein) is recorded during injection. All the parameters are then set up to record this channel at the magnification of interest (light intensity, exposure, frame rate, etc.). In our case: maximal intensity, maximal gain, 100 ms exposure time allowing an acquisition at 4 frames per second) (Fig. 3d).
2. Place the microneedle at the site of injection within the duct of Cuvier (Fig. 3b).
3. Once the ZF embryo and the injector are properly positioned (Zoom ×7.00, Fig. 3c), image capture is started simultaneously with the injection itself allowing to control the injection with the binoculars while recording the behavior of flowing CTCs using the software (Fig. 3c, d).
4. At the end of the live recording, a single image of the zebrafish vasculature is acquired, without moving the embryo. This image will be used as a background for the registered time-lapse recording of CTCs. The acquisition parameters are modified accordingly: the time-lapse mode is turned off, the light filter is changed to the GFP one and the exposure is lowered to 80 ms (Fig. 3d). In our experience, 5 min of recording is sufficient to document the earliest events such as CTC slowing, transient arrest events, and definitive arrest.

3.6 Time-Lapse Analysis in Image J

Once the time-lapse recording of the behavior of CTCs has been performed, the data obtained offers several angles of analysis. Here, we first propose a method for optimal movie rendering of the earliest steps of metastatic behavior (Fig. 4).

1. We use the ImageJ software. The single image of the zebrafish vasculature is duplicated n-time to match the number of recorded images on the injected cell channel. The following Java script can be used:

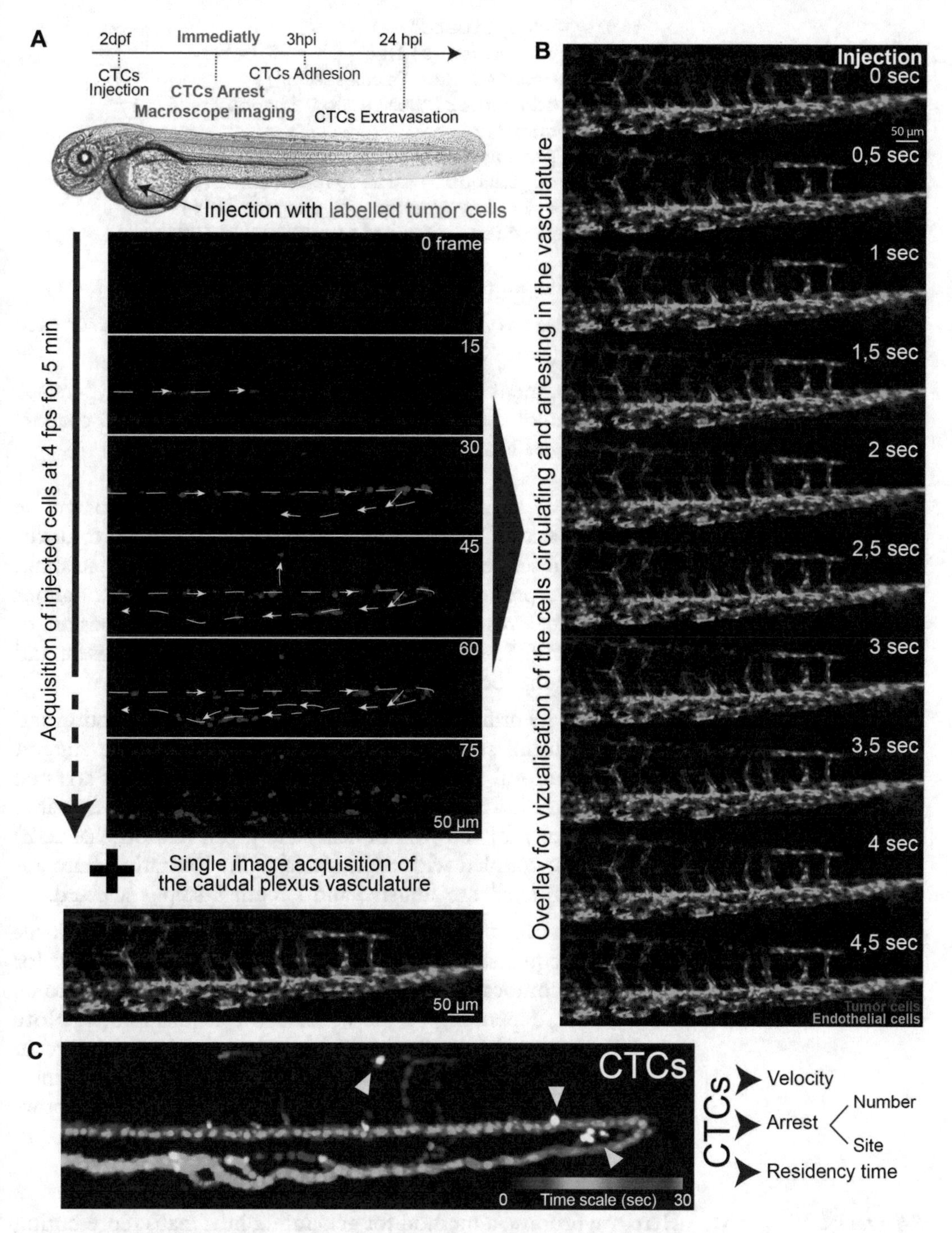

Fig. 4 Live injection imaging processing. (**a**) Timeline and schematic representation of the method: Single channel (red) acquisition is performed to live record the events with the maximal temporal resolution, followed by a single image acquisition in the green channel. (**b**) Processing with ImageJ allows the reconstruction of the injection movie. (**c**) Post-treatment with time color coding (ImageJ > Hyperstack > time color coding) allowing the qualitative and quantitative study of the movie

```
title = "duplicate";
Number_of_images=512;
Dialog.create("New Parameters");
Dialog.addNumber("Number_of_images:", 512);
Dialog.show();
title =Dialog.getString();
Number_of_images = Dialog.getNumber();;
for (i = 1; i < Number_of_images; i++){
run("Duplicate...", "title=ENDOTHELIUM");
}
run("Images to Stack", "name=endo title=[] use");
```

2. Merge the two stacks and optimize the color balance for each channel (Fig. 4b).
3. Post-treatment with time color coding (ImageJ > Hyperstack > time color coding) can be performed on the CTC channel (Fig. 4c; *see* **Note 9**).

3.7 Confocal Imaging to Record the Adhesion and Extravasation Pattern

1. Once early arrest of CTCs has been recorded, embryos can be placed back at 28 °C (incubator) and further studied to document subsequent important events of metastatic seeding. Here, we propose to further document, at high-throughput and spatial resolution, the stable intravascular adhesion of CTCs (Fig. 5a–c), as well as their capacity to extravasate and colonize the local parenchyma (Fig. 5d, e).
2. We have identified 3 hpi as an optimal timing for detecting and documenting stably attached intravascular TCs. We suggest using confocal imaging for studying the location of arrested TCs in relation to the vasculature. The caudal plexus is acquired sequentially for red (tumor cells) and green (endothelial cells) channels, coupled with transmitted light. Excitation/integration parameters are adjusted and 1.5 μm Z-step is selected.
3. For statistical analysis, a large batch of zebrafish caudal plexus can be acquired one after the other: We estimated the time for a single confocal stack (with a resolution of 1024 × 512) to be 5–8 min, depending on the width of the acquisition (*see* **Note 10**). Again, thanks to the highly stereotyped vasculature in that region, we have developed a method allowing the compilation of a high number of embryos, thereby providing a powerful statistical analysis of the behavior of metastatic cells in realistic conditions.

3.8 Image Preparation for High-Throughput Analysis Through Heat Mapping

Here, we propose a method for generating heat maps representing the density of TCs in a given region of the ZF embryo, over a high number of embryos. This method exploits the stereotyped vasculature of the embryo and uses a reference background image of the caudal plexus vasculature (Fig. 5b). The heat map is scaled from blue (lowest density of cells) to red (highest density). For display purposes,

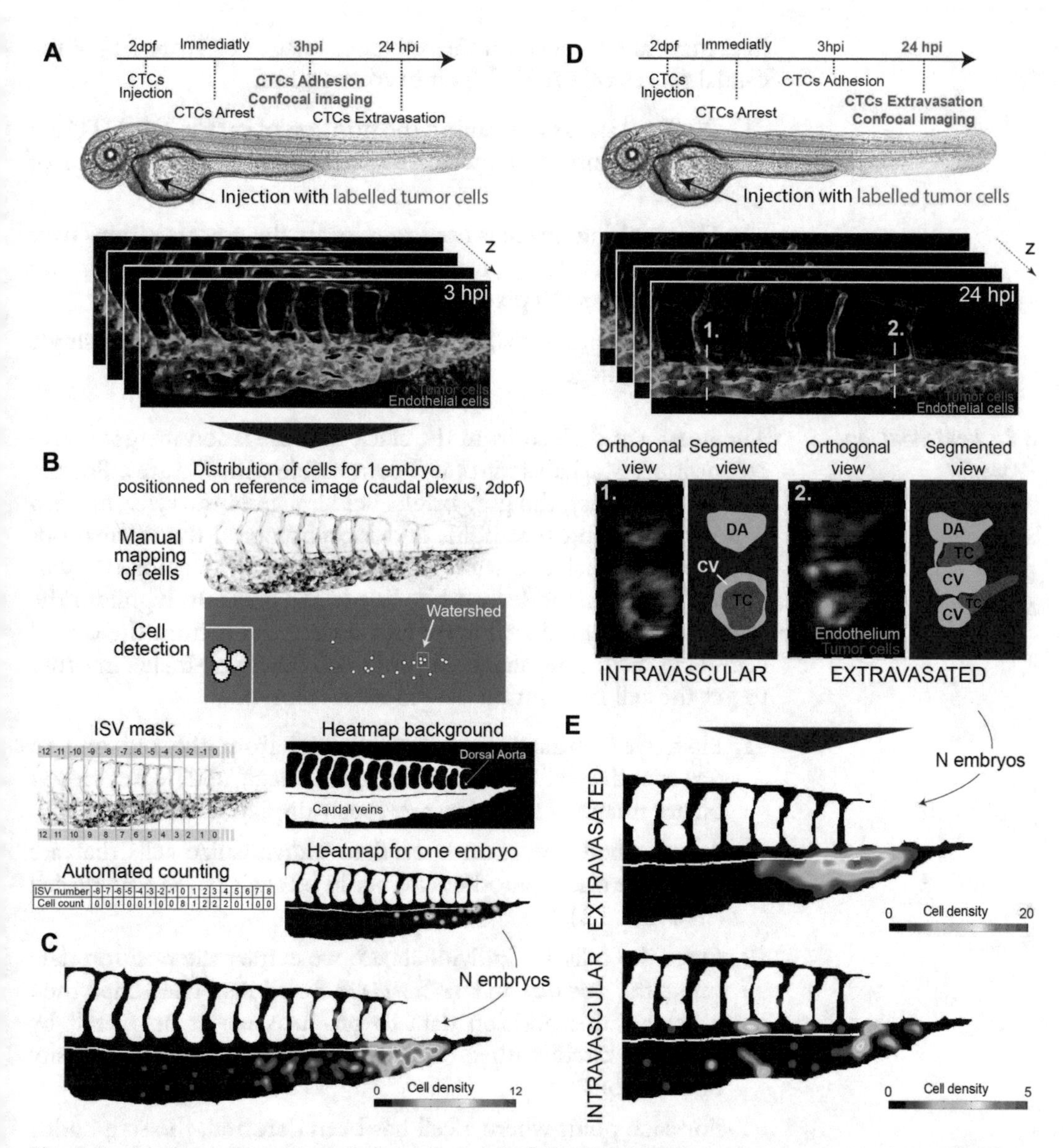

Fig. 5 Confocal imaging and analysis, at 3 and 24 hpi. (**a**) Schematic representation and timeline of the method: imaging at 3 hpi to record the stable adhesion pattern of tumor cells in the zebrafish caudal plexus. (**b**) Heat mapping protocol: From the manual mapping (top), to the automated counting of the cells per ISV and the heat map creation (bottom). Mapping automated recognition image is provided to illustrate the role of the watershed operation in the MatLab script (small insert show the watershed result, separating three cells). (**c**) Illustrative result from the heat mapping protocol at 3 hpi for several embryos. (**d**) Schematic representation and timeline of the method: imaging at 24 hpi for recording the extravasation pattern of tumor cells in the zebrafish plexus. Orthogonal views (ImageJ > Stack > Reslice) at different positions (yellow dashed lines) are provided to illustrate the best way to assess the cell situation (intravascular/extravasated). (**e**) Illustrative results from the heat-mapping protocol for intravascular versus extravasated cells

heat maps are revealed on the reference image (*see* **Note 11**) of the caudal plexus of a zebrafish embryo at 48 hpf.

1. Confocal stacks containing the position of each single CTC are projected into a single image (maximum projection tool of Image J).
2. The resulting image is used to relocate the exact positions over the reference image, using small red circle in ImageJ (Fig. 5b; brush tool, 5–10 pixels; *see* **Note 12**).
3. The localization images are saved as .tif and pooled in a single stack in ImageJ.

3.9 Heat Mapping Procedure

The heat map derived from the stack of localization images is created using a MatLab script (also compatible with Octave). For the sake of readability, the gray bright field image of the reference fish is processed using thresholds and morphological transformations to produce a black and white representation of the embryo vasculature. We further added a white line in the image to highlight the localization of the dorsal aorta and distinguish it from the caudal veins (Fig. 5b). The stacks are processed image-by-image in order to get the cell localization data in each embryo.

1. First, we subtract the green channel from the red one to remove the reference background image and leave only a binary image of the cells (previously displayed as red circles).
2. A watershed operation is used to individualize cells that are very close one to another and could be analyzed as a single cell (*see* **Note 12**).
3. Once the cells are individualized, we extract the position data using the *bwlabel* and *regionprops* functions. The script then processes this position data to produce a heat map built by adding a circle with a dimensional Gaussian shape intensity distribution.
4. For each point where a cell has been detected, the script adds a circle centered at the found position with a dimensional Gaussian intensity distribution with a maximal value of unity and a width of 6 (approximation of the real cell size at the resolution used here, *see* **Note 13**). The values of all the curves are added for each point in space, resulting in an image that represents the density of cells/pixel (Fig. 5c, e).
5. The color scale can then be adjusted to display a maximum (in red) where the largest cell density is observed (*see* **Note 14**).
6. The script also produces a table (saved as a .csv file) gives a quantitative analysis of cell locations data. The script then compares the position of a cell to the position of the ISVs and

gives its absolute value. This is automatically done for each fish and provides a file containing a line for each image/fish.

7. We propose here to divide the caudal plexus in several regions (Fig. 5c), based on the intersegmental vessels (ISV) stereotyped positions. The latter are registered from $-n$ to n: n being the number of ISVs in the reference image with negative values corresponding to positions within the dorsal aorta, positive values corresponding to positions within the caudal veins, and 0 representing the arteriovenous junction (*see* **Note 15**).

3.10 Tracking Tumor Metastasis at High Spatiotemporal Resolution

The experimental metastasis assay that we propose here is further compatible with a fine analysis of single CTC behavior. Here, we propose two angles of analysis that we have developed and used in our studies [24] for investigating various parameters of successful metastatic outgrowth (Fig. 6a). First, provided that one has access to fast confocal imaging (*see* **Note 16**), high-speed time lapse analysis of blood flow can be performed in vascular regions undergoing arrest or extravasation events of CTCs (Fig. 5b–e). While confocal imaging allows to capture, at high spatial resolution, the position and behavior of single CTC in relation to the vasculature, high-speed imaging (and subsequent tracking) of flowing red blood cells allows to document the hemodynamic environment of such an event, and study its potential influence on the extravasation process. We have recently shown that blood flow actively contributes to the arrest and extravasation of CTCs [24]. In addition, the ZF embryo is perfectly suited for tracking events such as arrest and extravasation at very high resolution using Correlative Light and Electron Microscopy methods [25, 26] (Fig. 5g). We have developed in the past protocols for achieving correlation of events imaged within living ZF embryos with high-resolution of Electron Microscopy [26]. When applied to single CTCs imaged within the ZF embryos, at different stages of the extravasation process, i.e., arrest, extravasation, or metastatic growth), this methodology allows to provide a high-resolution view of subcellular regions and organelles of cells, as well as of neighboring stromal cells such as endothelial or immune cells (Fig. 5g).

4 Notes

1. Many other lines (expressing fluorescent reporters in various cell types) are compatible with the following protocol. One is free to use any line of interest depending on the aims of the experimental approach. A long list of transgenic lines can be found on the ZFIN database (https://zfin.org/action/fish/search).

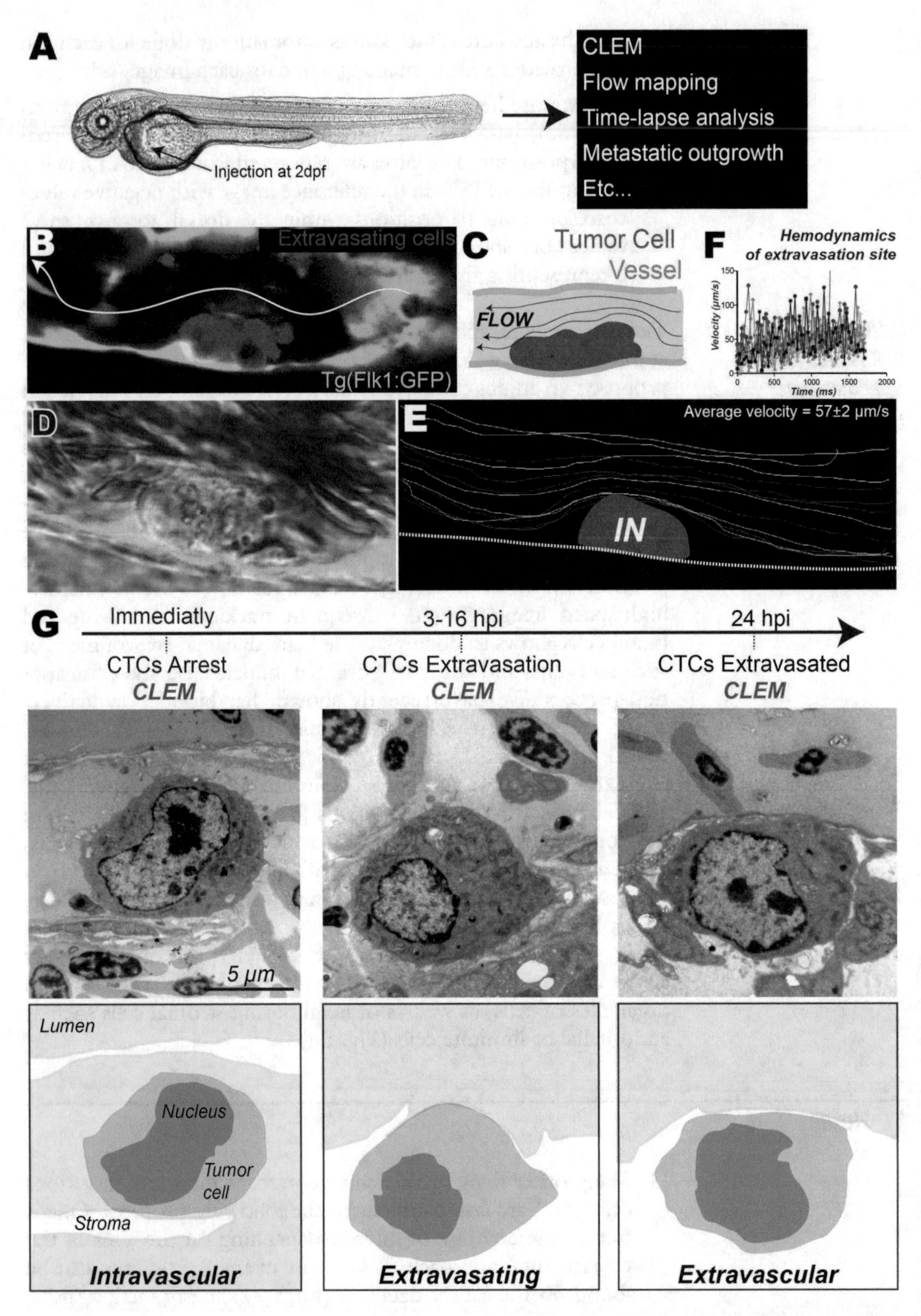

Fig. 6 Zebrafish embryo as a versatile experimental metastasis model. (**a**) Nonexhaustive list of the applications of the described method. (**b**) Zoom in a cluster of arrested tumor cells with confocal imaging. (**c**) Schematic of the arrested cluster of cells in the circulation from (**b**). (**d**) Confocal imaging with transmitted light allowing the visualization of the blood cells. (**e**) Representation of the blood cell tracking on the confocal images. (**f**)

2. PTU (phenylthiourea—used to prevent melanogenesis) is rather unstable at 28 °C. Store Danieau/PTU at room temperature for 2 weeks.
3. Tricaine (used to anesthetize the embryos) is unstable at room temperature and 40 °C. Prepare fresh solutions from your frozen stock 25× for each experiment.
4. Mineral oil is an incompressible fluid, nonmiscible with water that allows a perfect transmission of the piston movement during injection.
5. The microneedle tip can also be broken with tweezers, under a stereomicroscope, but we recommend to do this step with a prototype obtained with a microforge for careful comparison.
6. As tricaine gets slightly more diluted for each mounted dish, it progressively loses its anesthetic power. At some point, do not hesitate to change to fresh Danieau/PTU/tricaine.
7. The best option for injection and imaging is to have all the embryos as close as possible, with one single orientation, in the center of the glass coverslip. During the mounting, if you are not fast enough to arrange the embryos, you can redo the mounting easily by adding some Danieau/PTU/tricaine in the glass-bottomed dish. Then, pipet the embryos back to the Danieau/PTU/tricaine and start again the mounting.
8. This is the utmost critical step for a successful injection session. If at any moment during capillary filling, one can see any resistance, bubble formation, etc., it means that the filling step is compromised. We advise changing the capillary for a new one and restart the filling step.
9. Because the vasculature is mostly planar in that region of the ZF embryo, this method allows to reconstruct a fine time-lapse analysis of the very first events of intravascular CTCs seeding in perfused blood vessels. The highly stereotyped architecture of the ZF vasculature in that region further offers the possibility to analyze a high number of embryos upon 5 min of time-lapse recording. This method thus offers the first, unprecedented, protocol for studying the arrest, at high throughput and spatiotemporal resolution, of CTCs in realistic vascular environments.
10. Alternatively, one can start a time-lapse for several hours to follow the evolution of a single event (or multiple events using the multiposition tool). From our experience, zebrafish embryos can handle more than 18 h of time-lapse recording in a thermostatic chamber.

Fig. 6 (continued) Tracking results allowing a fine flow dissection around the tumor cells in the vasculature. (**g**) Experimental timeline with Correlative Light and Electron Microscopy (CLEM) allowing high-resolution dissection of the extravasation process. Transmitted electron microscopy images (top) and their segmentation (bottom) are provided to illustrate several steps of the metastatic cascade

11. This can be changed by entering the name of your own background image (line 26 of the Matlab script), keeping in mind that this image has to be the reference image for all the data (the same dimension is critical).
12. We advise to use a small dot diameter for the manual localization. Indeed, if the marks are overlapping too much, the watershed operation will not be able to distinguish and dissociate the cells. There is a risk that the script will produce errors by detecting one cell instead of a cluster of several cells. The script cannot work with smaller diameters than 5 pixels but this size is enough to prevent most of the critical overlaps.
13. This can be changed by changing the sigma value (line 42 of the Matlab script).
14. Any MatLab color scale can be used and modified for better readability (line 59 of the Matlab script).
15. This approach is further compatible with confocal acquisitions performed 24 hpi, aiming to document the extravasation behavior of TCs. Here, the intravascular (versus extravascular) location of cells is assessed by displaying the confocal acquisitions as orthogonal views (Fig. 5d). We suggest using this method for confidently attributing intravascular versus extravascular localization of recorded cells. Doing so and applying the previously described heat-mapping protocol, this will lead to two distinct heat maps for a single embryo: a heat map for intravascular cells, a heat map for extravascular cells (Fig. 5e). This step implies the creation of two heat maps for each embryo: one with the extravasated cells and one with the intravascular cells for each embryo.
16. Fast blood flow imaging can be performed using, for example, SP5/SP8 confocal microscopes equipped with resonant scanner. We usually record the displacement of red blood cells using transmitted light at a scanning speed of about 100 fps.

Acknowledgments

We thank all members of the Goetz Lab for helpful discussions throughout the development of this technology. We are grateful to Sofia AZEVEDO and Nina FEKONJA for their help in various aspects of this method. We are very much grateful to Francesca PERI (EMBL) and Kerstin RICHTER (EMBL) for providing zebrafish embryos. This work has been funded by Plan Cancer (OptoMetaTrap, to J.G. and S.H) and CNRS IMAG'IN (to S.H., J.G., and C.P.) and by institutional funds from INSERM and University of Strasbourg. G.F. is supported by La Ligue Contre le Cancer. N.O is supported by Plan Cancer. G.A. was supported by FRM (Fondation pour la Recherche Médicale).

References

1. Massagué J, Obenauf AC (2016) Metastatic colonization by circulating tumour cells. Nature 529:298–306
2. Katt ME, Placone AL, Wong AD, Xu ZS, Searson PC (2016) In vitro tumor models: advantages, disadvantages, variables, and selecting the right platform. Front Bioeng Biotechnol 4:12
3. Cheung KJ, Gabrielson E, Werb Z, Ewald AJ (2013) Collective invasion in breast cancer requires a conserved basal epithelial program. Cell 155:1639–1651
4. Karreman MA, Hyenne V, Schwab Y, Goetz JG (2016) Intravital correlative microscopy: imaging life at the nanoscale. Trends Cell Biol 26:848–863
5. Kienast Y et al (2010) Real-time imaging reveals the single steps of brain metastasis formation. Nat Med 16:116–122
6. Cheung KJ et al (2016) Polyclonal breast cancer metastases arise from collective dissemination of keratin 14-expressing tumor cell clusters. Proc Natl Acad Sci U S A 113: E854–E863
7. Cheon D-J, Orsulic S (2011) Mouse models of cancer. Annu Rev Pathol 6:95–119
8. Deryugina EI, Kiosses WB (2017) Intratumoral cancer cell intravasation can occur independent of invasion into the adjacent stroma. Cell Rep 19:601–616
9. Leong HS et al (2014) Invadopodia are required for cancer cell extravasation and are a therapeutic target for metastasis. Cell Rep 8:1558–1570
10. Amatruda JF, Shepard JL, Stern HM, Zon LI (2002) Zebrafish as a cancer model system. Cancer Cell 1:229–231
11. Stoletov K, Klemke R (2008) Catch of the day: zebrafish as a human cancer model. Oncogene 27:4509–4520
12. Stoletov K, Montel V, Lester RD, Gonias SL, Klemke R (2007) High-resolution imaging of the dynamic tumor cell–vascular interface in transparent zebrafish. Proc Natl Acad Sci U S A 104:17406–17411
13. Stoletov K et al (2010) Visualizing extravasation dynamics of metastatic tumor cells. J Cell Sci 123:2332–2341
14. White RM et al (2008) Transparent adult zebrafish as a tool for in vivo transplantation analysis. Cell Stem Cell 2:183–189
15. Heilmann S et al (2015) A quantitative system for studying metastasis using transparent zebrafish. Cancer Res 75:4272–4282
16. Kaufman CK et al (2016) A zebrafish melanoma model reveals emergence of neural crest identity during melanoma initiation. Science 351:aad2197
17. Tang Q et al (2016) Imaging tumour cell heterogeneity following cell transplantation into optically clear immune-deficient zebrafish. Nat Commun 7:10358
18. Kim IS et al (2017) Microenvironment-derived factors driving metastatic plasticity in melanoma. Nat Commun 8:14343
19. Corey DR, Abrams JM (2001) Morpholino antisense oligonucleotides: tools for investigating vertebrate development. Genome Biol 2:reviews1015
20. Zhang Y, Huang H, Zhang B, Lin S (2016) TALEN- and CRISPR-enhanced DNA homologous recombination for gene editing in zebrafish. Methods Cell Biol 135:107–120
21. De Santis F, Di Donato V, Del Bene F (2016) Clonal analysis of gene loss of function and tissue-specific gene deletion in zebrafish via CRISPR/Cas9 technology. Methods Cell Biol 135:171–188
22. Ablain J, Zon LI (2016) Tissue-specific gene targeting using CRISPR/Cas9. Methods Cell Biol 135:189–202
23. Lawson ND, Weinstein BM (2002) In vivo imaging of embryonic vascular development using transgenic zebrafish. Dev Biol 248:307–318
24. Follain G, Osmani N, Azevedo S, Allio G, Mercier L, Karreman MA, Solecki G, Garcia-Leon MJ, Lefebvre O, Fekonja N, Hille C, Chabannes V, Dollé G, Metivet T, Der Hovsepian F, Prudhomme C, Ruthensteiner B, Kemmling A, Siemonsen S, Schneider T, Fiehler J, Glatzel M, Winkler F, Schwab Y, Pantel K, Harlepp S, Goetz JG (2017) Hemodynamic forces tune the arrest, adhesion and extravasation of circulating tumor cells. Dev Cell. https://doi.org/10.1101/183046
25. Goetz JG et al (2014) Endothelial cilia mediate low flow sensing during zebrafish vascular development. Cell Rep 6:799–808
26. Goetz JG, Monduc F, Schwab Y, Vermot J (2015) Using correlative light and electron microscopy to study zebrafish vascular morphogenesis. Methods Mol Biol 1189:31–46

Chapter 16

Analysis of In Vivo Cell Migration in Mosaic Zebrafish Embryos

Arthur Boutillon, Florence A. Giger, and Nicolas B. David

Abstract

Being optically clear, the zebrafish embryo is a nice model system to analyze cell migration in vivo. This chapter describes a combination of injection and cell transplant procedures that allows creation of mosaic embryos, containing a few cells labeled differently from their neighbors. Rapid 5D confocal imaging of these embryos permits to simultaneously track and quantify the movement of large cell groups, as well as analyze the cellular or subcellular dynamics of transplanted cells during their migration. In addition, expression of a candidate gene can be modified in transplanted cells. Comparing behavior of these cells to control or neighboring cells allows determination of the role of the candidate gene in cell migration. We describe the procedure, focusing on one specific cell population during gastrulation, but it can easily be adapted to other cell populations and other migration events during early embryogenesis.

Key words Zebrafish, Live imaging, Cell migration, Cell transplantation, Mosaic embryos, Cell tracking

1 Introduction

Cell migration is key to build, shape, repair, and defend an organism. It has been extensively studied in vitro, providing invaluable knowledge on its cellular and molecular bases [1, 2]. It, however, clearly appeared in the past few years that a number of cells behave quite differently in vitro and in vivo [3, 4]. This likely stems from the more complex environment they encounter (3D extracellular matrix of varying stiffness, neighboring cells, guidance cues) leading, for instance, cells that would use only lamellipodia on a flat surface to use a wide array of cell processes (lamellipodia, pseudopodia of varying length and shapes, blebs, …) for their in vivo displacements [5]. Many cells, furthermore, display collective behaviors, their migration depending on interactions with neighboring cells [6, 7]. Understanding in vivo cell migration thus implies direct analysis of cells moving in their physiological environment.

Doing so obviously requires optically clear systems permitting cell observation without further manipulation or dissection. Cells also

Alexis Gautreau (ed.), *Cell Migration: Methods and Protocols*, Methods in Molecular Biology, vol. 1749, https://doi.org/10.1007/978-1-4939-7701-7_16, © Springer Science+Business Media, LLC 2018

need to be labeled, usually through the expression of fluorescent proteins. This can be achieved using transgenic lines expressing reporter constructs in specific cell types. However, such a strategy often leads to the labeling of a whole cell population, when labeling sparse, isolated cells, is key to good imaging, in particular for analyzing cell contours, or membrane protrusions, which cannot be seen if neighboring cells are similarly labeled [8, 9]. Creating mosaic embryos, containing only one or few cells differing from surrounding cells, is thus crucial for precise analysis of their dynamics. Mosaic embryos, furthermore, permit to test consequences of activation/repression of genes/pathways on the migration of individual cells. Comparing the tracks and behavior of these cells with surrounding control cells under the same experimental conditions allows determination of the role of a particular gene in cell migration [10]. Here we describe a procedure, based on injections and cell transplants, to create mosaic zebrafish embryos during gastrulation and to image migrating cells.

Gastrulation is the developmental stage during which cells first migrate, to organize the embryo in different germ layers and set up the basis of the future body plan [11]. This involves a wide array of migration events, including random walks, directed migrations and collective movements. The fish embryo, being optically clear, offers direct access to these processes [12]. Among the diverse movements taking place during gastrulation, cells of the organizer (termed "shield" in zebrafish) internalize and migrate collectively toward the animal pole of the embryo, forming the prechordal plate [9, 13, 14]. We combine injections of RNAs encoding fluorescently tagged proteins, cell transplants and in vivo imaging, to analyze mechanisms of cell migration. We have successfully used this protocol to describe behavior of wild-type cells, dissect the pathways controlling their motility and orientation, as well as assess the importance of candidate genes in regulating cell migration in vivo [14–16]. The proposed protocol can be used to test the potential implication of any candidate gene in controlling cell migration in vivo. It can furthermore be easily adapted to analyze other cell types, or other developmental stages.

2 Material

2.1 RNA Injection

1. Stereomicroscope (Nikon, model SMZ18).
2. Glass capillaries (Internal Diameter = 0.58 mm, External Diameter = 1 mm).
3. Needle puller (KOPF Vertical pipette puller, model 720, David Kopf Instruments, Tujunga, CA, USA).
4. Injection mold (Fig. 1a, 3D model available as a .stl file on request; commercially available from adaptive science tools, model TU1).

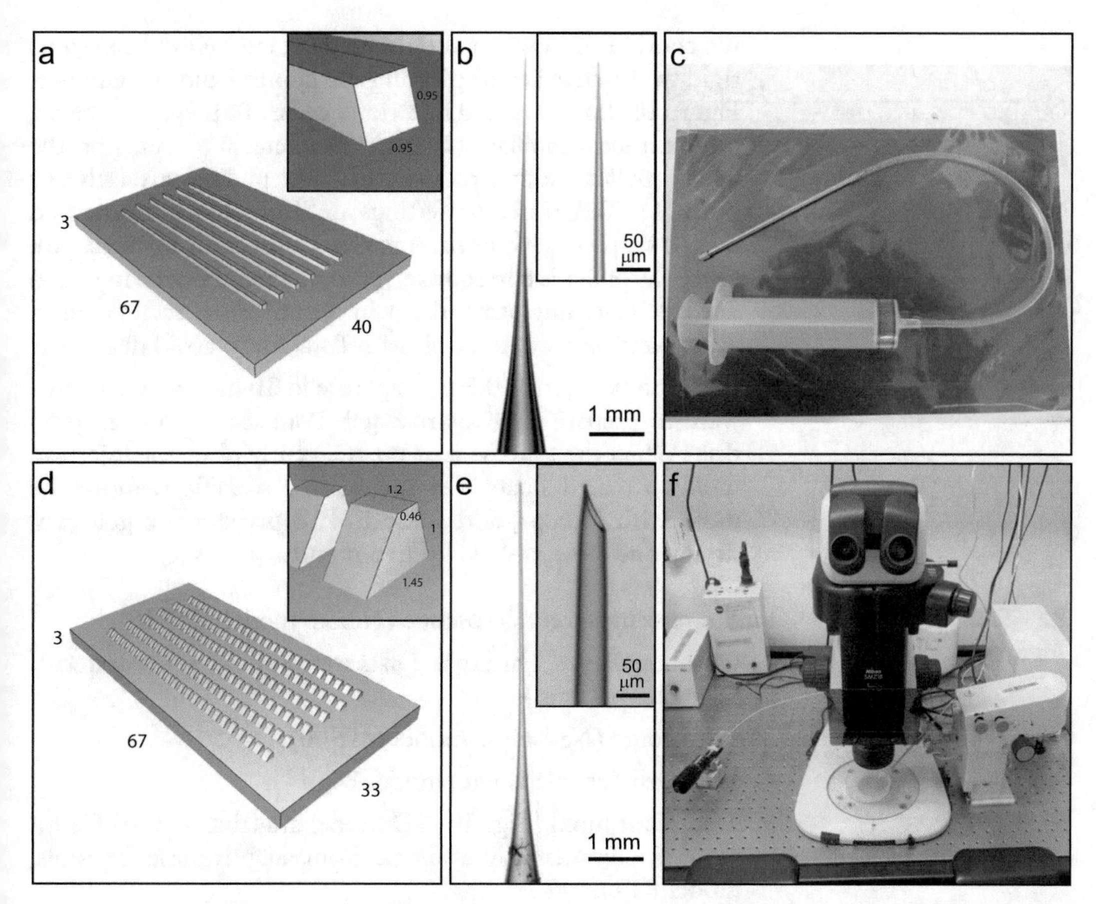

Fig. 1 Injection and transplant equipment. (**a**) 3D model of the injection mold used to prepare injection agarose plates. Dimensions are in mm. Model is available as a .stl file. (**b**) Typical injection needle. (**c**) Needle holder connected to a syringe, used to rinse transplant needles. (**d**) 3D model of the transplant mold used to prepare transplant agarose plates. Dimensions are in mm. Model is available as a .stl file. (**e**) Typical transplant needle. (**f**) General view of the transplant setup. Transplant needle is inserted in a needle holder, mounted on a mechanical micromanipulator. The needle is connected to a microinjector

5. Injector (Eppendorf, Transjector 5246).
6. Manipulator (Narishige, model M-152).
7. 20 µl microloader tips (Eppendorf) to load injection needles.
8. Plastic Pasteur pipettes for transferring embryos.
9. Embryo Medium (EM, Zebrafish book [17]): 15 mM NaCl, 0.5 mM KCl, 25 µM Na_2HPO_4, 45 µM KH_2PO_4, 1.3 mM $CaCl_2$, 1 mM $MgSO_4$•$7H_2O$, 4 mM $NaHCO_3$, adjust to pH 7.2 with NaOH.
10. RNAs: mRNAs are prepared in vitro, using the mMessage mMachine kit (Ambion) to produce fully capped RNAs.
11. Sterile-filtered water (Sigma) for diluting RNAs.
12. Injection needle: needles should be long enough to reach the embryo within its chorion, without creating too large a hole in

the chorion. It however should be short enough to keep some rigidity, which is key to piercing the chorion and the embryo. Figure 1b shows a typical injection needle. To prepare a needle, mount a glass capillary (Internal Diameter 0.58 mm) on the needle puller. Adjust temperature and pulling strength (*see* **Note 1**). Test different settings until obtaining the desired needle shape (check needles under a dissection microscope). Once the puller is correctly set, prepare several dozen injection needles. Carefully store them in a petri dish, secured on a putty band and seal the dish with Parafilm to avoid dust.

13. Injection plate: melt 0.5 g of agarose in 50 ml EM in a microwave to prepare a 1% agarose gel. Pour it in a 90-mm petri dish. When the agarose is at 60 °C, gently place the injection mold so that it floats. Let the agarose solidify. Remove the mold with forceps, add some EM to prevent the gel from drying and store at 4 °C up to a month.

2.2 Shield to Shield Transplant

1. Fluorescent stereomicroscope (Nikon, model SMZ18).
2. Glass capillaries (Internal Diameter = 0.78 mm, External Diameter = 1 mm).
3. Microforge (Narishige, model MF900).
4. Microgrinder (Narishige, model EG-44).
5. Transplant mold (Fig. 1d, 3D model available as a .stl file on request; commercially available from adaptive science tools, model PT1).
6. Eyelash mounted on a stick.
7. Mechanical micromanipulator (Leica).
8. Microinjector (Narishige, model IM-9B).
9. Fine tweezers for dechorionation (Dumont Fine Science Tools).
10. 35 mm petri dishes.
11. Fire-polished glass Pasteur pipettes, to manipulate dechorionated embryos: approach the tip of the pipette close to the flame of a lighter until the glass slightly melts. Be careful not to melt it too much, otherwise the pipette opening may end up too small and embryos will be damaged when drawn into the pipette.
12. Penicillin/Streptomycin (Thermo Fisher Scientific): for transplanted embryos, use at 100 unit/ml Penicillin and 100 μg/ml streptomycin in EM.
13. Transplant needle: mount a glass capillary (Internal Diameter 0.78 mm) on the needle puller and pull the capillary. The stretched part of the needle should be around 1.5 cm (Fig. 1e). Test different settings until obtaining the desired shape and pull several dozen needles (*see* **Note 1**). Mount a needle on the microforge. Position the needle very close to the microforge's filament, at the point where its inside diameter is 25 μm. Briefly heat the

filament: upon dilatation, the filament touches the needle, the glass locally melts then breaks upon retraction of the filament. Using the microgrinder, grind the tip of the needle for one minute at maximum speed to form a 35° bevel (Fig. 1e; *see* **Note 2**). To remove glass residues, mount the needle on a needle holder connected to a syringe (Fig. 1c) and wash the tip by briefly aspiring 2% hydrofluoric acid (toxic and highly corrosive, manipulate under a chemical fume hood) over 5 mm. Repeat three times. Immediately rinse three times by aspiring acetone over 1 cm. Store the needles on a putty band in a Parafilm-sealed petri dish.

14. Transplant plate: melt 0.5 g of agarose in 50 ml EM in a microwave to prepare a 1% agarose gel. Pour it in a 90 mm petri dish. When the agarose is at 60 °C, gently place the injection mold so that it floats. Let the agarose solidify and remove the mold with forceps.
15. Transplant setup: around the fluorescent stereomicroscope, install the microinjector, connected to a needle holder mounted on the micromanipulator (Fig. 1f).

2.3 Mounting and Imaging

1. MatTek chamber (35 mm Dish, No. 1.5 Coverslip, 7 mm Glass Diameter, MatTek Corporation, Ashland, MA).
2. Microscopy: we use a Ti PFS (Nikon) inverted microscope equipped with a CSUX1-A1 (Yokogawa) spinning disc module, a 40× water immersion objective (plan Apo, N.A. = 1.15, Nikon), a Cobolt Calypso laser (491 nm, 100 mW), and a Cobolt Jive laser (561 nm, 100 mW) (Cobolt, 04-01 Series). The microscope is caged and heated using The Cube (Life Imaging Services) to maintain constant temperature and the sample is placed in a The Brick (life imaging Services) chamber to maintain constant hygrometry. Metamorph is used for acquisitions.

2.4 Software for Imaging and Picture Analyzing

1. Metamorph (Molecular Devices, LLC).
2. ImageJ.
3. XuvTool [18].
4. IMARIS (Bitplane).
5. Matlab (Mathworks).
6. R (R Foundation for Statistical Computing).

3 Methods

3.1 Embryo Injection

1. Warm the injection plate by placing it at 28 °C at least 30 min before injection.
2. Once fishes have laid, collect the embryos. Rinse them well to remove all the droppings and scales from the adults and har-

vest in EM in a 90-mm petri dish. Transfer 50 embryos in an empty injection plate and squeeze them gently inside the grooves using forceps, without piercing the chorion. Orient the embryos with the animal pole up. Fill the plate with EM.

3. Wear gloves while manipulating RNA to prevent contamination with RNAse, keep all solutions on ice. Prepare 4 μl of 50 ng/μl Histone2B-mCherry RNA solution by diluting the stock RNA solution in sterile-filtered water. This solution will be injected in embryos used as hosts in the transplant. Prepare 4 μl of 50 ng/μl Histone2B-mCherry RNA and 70 ng/μl LifeAct-GFP RNA. This will be injected in embryos used as cell donors in the transplant. To affect a particular pathway, it is possible to add other compounds to this solution, which will be present only in the transplanted cells. In particular, morpholinos or RNAs encoding dominant-negative or constitutively active forms of proteins can be used.
4. Using microloader tips, fill the injection needle with 2 μl of RNA solution. Insert the needle in a needle holder, mounted on a manipulator and connected to the injector. Under the stereomicroscope, delicately open the needle using forceps, either by scraping the tip or by breaking the very tip by pinching it. To test whether the needle is opened, apply pressure with the injector and check that a drop of RNA solution forms at the tip. Then put the tip of the needle into the EM to prevent it from drying.
5. To prepare host embryos, inject half of the embryos at the 1-cell stage, with the H2B-mCherry solution (Fig. 2a) (*see* **Note 3**). To do so, place an embryo close to the injection needle. Using the manipulator, pierce through the chorion and through the cell membrane so that the needle tip is inside the cell (Fig. 2b). Avoid piercing the yolk and injecting into the yolk. Inject 4 nl (*see* **Note 4**) and remove the needle. The embryo should stay in the agarose groove. Repeat for 25 embryos and put the plate at 28 °C. To prepare donor embryos, let embryos reach the 4-cell stage (1 hpf), and inject 2 nl of the H2B-mCherry and LifeAct-GFP solution into one of the four cells (Fig. 2c). Repeat for 25 embryos and put the plate at 28 °C.

3.2 Dechorionate

1. Coat a 35-mm plate with 1 ml of 1% agarose in EM. When the agarose has solidified, fill the plate with warmed EM (28 °C). Transfer embryos to be dechorionated with a plastic Pasteur pipette.
2. With two fine tweezers, carefully remove the chorion. This is done by pinching the chorion with one tweezers and delicately pulling on the scratch with the second tweezers (*see* **Note 5**). Repeat until the embryo is freed from the chorion. Tear the

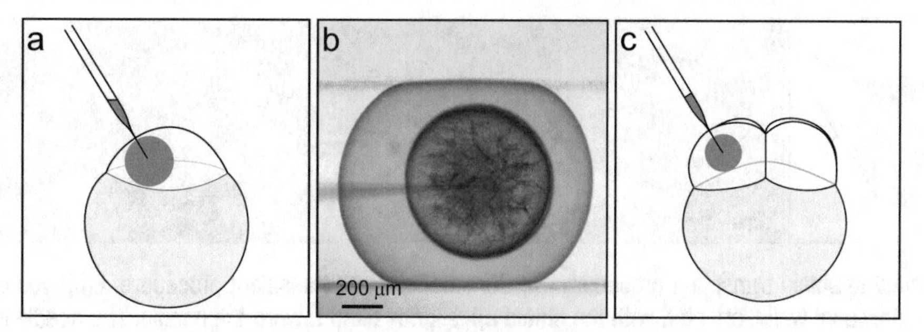

Fig. 2 Injection procedure. (**a**) Schematic of the injection procedure at the 1-cell stage. The injected volume on the schematic represents 4 nl. (**b**) 1-cell stage embryo maintained in an agarose groove. Injection needle is in the cell. (**c**) Schematic of the injection procedure at the 4-cell stage. The injected volume on the schematic represents 2 nl

chorion gently and progressively to avoid crushing the embryo. Dechorionated embryos are very fragile and will not survive contact with air or plastic: for further steps, use fire-polished glass pipettes and agarose-coated plates.

3. Clean the plate by removing torn chorions and damaged embryos with a glass pipette then carefully place the plate at 28 °C.

3.3 Shield to Shield Transplants

1. Fill a transplantation plate with Penicillin/Streptomycin EM and warm at 28 °C.
2. Wash the eyelash and transplant needle with 70% ethanol. For the transplant needle, mount it on a needle holder connected to a syringe and draw ethanol on 2 cm (Fig. 1c). Empty the needle and dry it by drawing air. Do not dry by blowing, as this may result in dust getting stuck in the needle.
3. When embryos have reached the shield stage (6 hpf), select donor and host embryos under a fluorescent stereo microscope. Pick embryos displaying bright and homogeneous fluorescence in the shield. Use the eyelash to manipulate and sort embryos. With a fire-polished glass pipette transfer selected embryos into the wells of the transplantation plate, aligning vertically and side by side hosts and donors. Rotate the embryos with the eyelash in order to place the shield up, slightly tilted toward the needle (Fig. 3a, b).
4. Install the transplant needle on the needle holder of the transplant setup (*see* **Note 6**). Under the fluorescent stereo microscope, orient correctly the bevel of the needle: if cells must be transplanted deep, turn the bevel downward; orient it upward to transplant cells superficially. Lower the needle into the EM and aspire EM over half of the stretched part of the needle (*see* **Note 7**).
5. Place a donor embryo in front of the needle and delicately insert the needle into the shield. Be careful not to pierce nor to be too

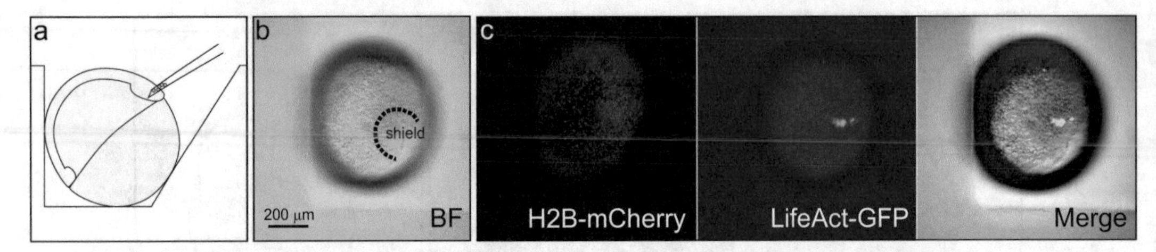

Fig. 3 Shield to shield transplant procedure. (**a**) Schematic of the transplant procedure. Embryos are maintained in individual wells, oriented with the shield up, slightly tilted toward the needle. The needle is used to draw cells from the shield of a donor embryo and inject them in the shield of a host embryo. (**b**) Bright field image of a host embryo in an agarose well. (**c**) This host embryo was injected at the 1-cell stage with RNAs encoding Histone2B-mCherry. All nuclei are thus labeled in red. A few cells expressing LifeAct-GFP were transplanted into the shield

close to the yolk. Gently draw up a few cells inside the needle (*see* **Note 8**). Using fluorescence, check that cells in the needle are labeled.

6. Place the corresponding host embryo in front of the needle and delicately insert the needle into the shield. Be careful not to approach the yolk too much. Gently blow the cells into the host, taking care not to add too much liquid with them. However, it may be necessary to keep expelling EM while moving the needle out of the embryo, in order for the cells not to stick to the needle. Be careful not to blow any air as this will damage the embryo. Use fluorescence to check that cells are now in the shield of the host embryo (Fig. 3c).
7. Repeat **steps 4** and **5** until all host embryos have been transplanted. Remove donor and damaged embryos. Carefully place the plate in a 28 °C incubator.
8. Clean transplant needle with water, as described in **step 2**.

3.4 Mounting Embryos

Embryos are mounted in a small volume of warm agarose, which cools and thus solidifies rapidly. Being fast is thus crucial: prepare workbench before mounting, so that all required equipment is at hand. Do not mount more than three embryos at a time.

1. Prepare 1 ml of 0.2% agarose in penicillin–streptomycin EM solution in a small glass vial and place it in a preheated 42 °C hot block.
2. Select 1–3 embryos and draw them in a fire-polished pipette. While the embryos are in the pipette, put the MatTek plate under the scope and focus on the bottom of the well.
3. Drop the embryos in the 0.2% agarose solution without adding too much EM. Discard the remaining EM and draw the embryos back into the pipette. Take care to draw enough agarose after the embryos in order to fill the well before the embryos fall out.

4. Blow a drop of agarose and the embryos into the MatTek well. Take care not to let the embryos touch air or the border of the well. Make sure the well is completely filled with agarose (*see* **Note 9**).
5. Depending on room temperature, the agarose will gel in about 1 min, during which embryos should be oriented using the eyelash. While orienting the embryo, be careful to touch it only on the blastoderm and not on the yolk, which is very fragile. Depending on the type of microscope used for imaging (upright or inverted), place the shield upward or downward, against the glass bottom. In the latter case, use fluorescence to spot transplanted cells and orient the embryo properly.
6. Wait a few minutes for the agarose to set completely, then add a drop of penicillin–streptomycin EM to prevent it from drying.

3.5 Imaging

1. Preheat the microscope cage at 28 °C at least 30 min before starting imaging.
2. Put the MatTek plate under the spinning-disc microscope and fill it with penicillin–streptomycin EM.
3. Set the hygrometry module to 80% relative humidity. In absence of hygrometry control, place a water-soaked tissue to keep constant moisture. This will prevent the EM from evaporating and the agarose from drying.
4. Using a 40× long range objective, spot an embryo and focus on labeled cells. Tune the laser power and exposure time to optimize image dynamics for both channels. Specified RNA doses should lead to bright GFP and mCherry signals, so that brief exposure times can be used (<100 ms; *see* **Note 10**). This limits photobleaching and toxicity, and allows for high frame rates, compatible with acquiring large z-stacks, at short time intervals (2 min). Using bright field, set the *X* and *Y* position, then the top and bottom of the z-stack, so that it encompasses the entire shield. Be careful to consider that the embryo is curved and that cells may therefore not migrate in a straight z plane. It may thus be useful to enlarge the z-stack in order to have the whole prechordal plate encompassed in the stack for the duration of the time lapse. To get a larger field of view, image several positions that will be stitched with the XuvTool software [18]. Neighboring stacks require a 5–10% overlap.
5. Repeat **step 4** for every embryo in the imaging plate then set the time interval to 2 min and z-step to 2 μm. Launch the time lapse for 2 h. Be careful that total acquisition time at each time step does not exceed 2 min.

3.6 Data Analysis

3.6.1 Quantification of Actin-Rich Protrusion Orientation

1. Load images in ImageJ using the Bioformat plugin (Fig. 4a).
2. Using the movement of labeled cells, determine the general direction of migration then rotate the movie using the "Rotate" function of ImageJ to set the direction angle to 0°.

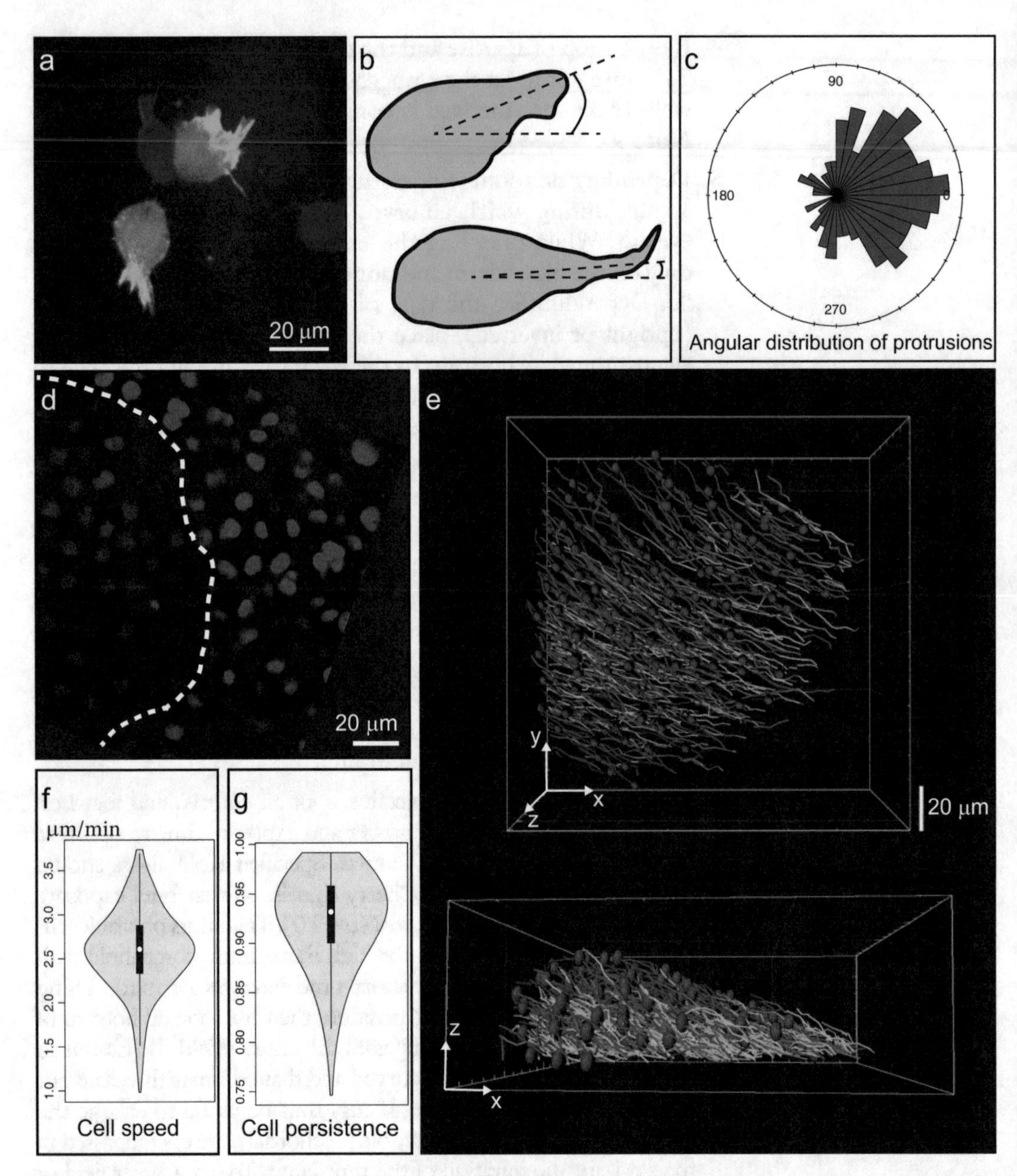

Fig. 4 Data analysis. (**a**) Transplanted cells expressing LifeAct-GFP. Large actin-rich protrusions are easily visible. (**b**) Measure of protrusion orientation. Image has been rotated so that the *x*-axis is aligned with migration movement. (**c**) Rose plot of the angular distribution of protrusions. (**d**) The host embryo expresses Histone2B-mCherry, labeling all nuclei in red. Dotted line delineates the limit between prechordal plate cells and ectodermal cells. (**e**) Tracks of prechordal plate cells, obtained in IMARIS. Upper part, *XY* view; lower part, *XZ* view. (**f, g**) Raw data from cell tracks are imported and processed in Matlab, to quantify migration properties. Here, distribution of average instant speed (**f**) of all tracked cells and distribution of cell persistence (**g**), calculated over five time steps

3. On the green channel, for each time step and each transplanted cell, look for actin-rich protrusions. We consider all protrusions exceeding 5 μm in length. If present, measure their 2D orientation using the "Angle" tool. Draw a straight line from the centroid of the cell toward the stem of the protrusion and measure the angle relative to the general direction of migration (Fig. 4b).
4. Use R and the circular package for data analysis and representation. If angles are measured over 0–180°, classic statistical tests can be used to compare two distributions, like the Kolmogorov-Smirnov test (ks.test). If angles are measured over 0–360°, circular statistical tests, like Watson's two sample test (watson.two.test), should be used. Data are best presented as a rose diagram (rose.diag) (Fig. 4c).

3.6.2 Cell Tracking

1. Open the time-lapse series with IMARIS. Set the pixel size, depending on the microscope and camera used for imaging (Fig. 4d).
2. Use the "Spot" function to locate and track nuclei. Set object size to 10 μm, which is the average nucleus size at this developmental stage. Due to lower axial resolution, nuclei may appear elongated in the axial dimension. It is therefore best to allow ellipsoidal shape, with 15 μm in the *Z* axis. Filter on spot quality to remove false positive. Filter on red intensity to discriminate between ectoderm cells and migrating prechordal plate cells, which, being deeper in the embryo, appear dimmer. Prechordal plate cells migrate at about 3 ± 0.8 μm/min [14], hence allow 9 μm as the maximal distance between consecutive time steps. Allowing gaps over one or two time points provides longer continuous tracks, without introducing too many track errors. Once automatic tracking is done, visually check tracks and manually correct them if necessary (Fig. 4e).
3. Export the results as a .csv file and process with MatLab. We use custom made Matlab routines to compute instant speed, persistence, coherence, orientation of movement, etc. for each cell. Routines are available on request (Fig. 4f, g).

4 Notes

1. Injections and transplants are technically demanding steps, which highly depend on needle quality. Pulling needles of the right shape (*see* Fig. 1b, e for reference) requires fine tuning of parameters but generally, to get short and sharp needles for injection, the temperature should be set low and the tension high. On the contrary, for long transplant needles, temperature should be high and tension low.

2. To ease penetration in embryos, which can be difficult at late developmental stages, it is possible to add a barb at the tip of transplant needles. Install a transplant needle on the microforge with the tip 50 μm above the filament. Heat the filament, lower the needle to briefly touch the filament and immediately pull back. By melting the very tip of the bevel, this will stretch a glass barb. If the barb is too long, break it with forceps (Fig. 1e, insert). Pay attention that this procedure may reduce the diameter of the needle opening.
3. Embryos meant to be hosts are injected at the 1-cell stage. Injected RNAs will diffuse in the cytoplasm, leading most cells to be labeled, but with varying intensities, depending on how much RNA they inherited. On the contrary, in donor embryos it is important that all labeled cells have inherited similar levels of RNAs, so that all transplanted cells are similar. Hence, donors are injected at the 4-cell stage, since the RNA will diffuse in a smaller volume and be distributed homogeneously into the daughter cells.
4. Some calibrate injection volume by injecting in an oil drop and measuring the diameter of the created droplet. Although theoretically precise, this technique implies that injection conditions are constant from one embryo to another. Experience proved this to be false, resistance to fluid flow in the needle varying largely from one embryo to another, due to differences in the position of the needle in the embryo, and partial, transient, clogging of the needle. To get reproducible injections, it is thus preferable to visually control the volume of injected liquid, which can be seen as a clearer droplet in the cytoplasm. By comparing the size of the droplet to the rest of the embryo, it is possible to have a good estimation of the injected volume (Fig. 2a, c). Phenol red may be added to the injection solution to better visualize the droplet.
5. While dechorionating, be very careful when first pinching the chorion, since this is when the embryo has the most chance to be crushed. In particular, for injected embryos, the chorion is pierced at the injection spot and embryos have a tendency to squeeze through this hole if chorion is brutally pinched.
6. Oil may be used instead of air in the transplant system. Being inelastic, oil provides a more reactive setup. Entirely fill the microinjector, the tubing and the needle with oil. Use the oil setup as described for the air setup. One drawback is that, to be efficient, the whole system must be purged of any air bubble, which may be difficult. Furthermore, needles filled with oil tend to get dirty and thus cannot be reused as many times as when used with air.
7. Be careful to have the right amount of liquid in your needle: enough to draw and blow cells without contact with air, but

not too much, otherwise the system will respond with a delay and sudden jolts. The interface between air and embryo medium must be kept in the stretched part of the needle.

8. Be careful not to draw up yolk in the needle, nor to draw up too much liquid after the cells, since injecting yolk or large volumes of liquid appear toxic to the host. Yolk can be seen as a transparent, nonfluorescent mass in the needle.
9. Once embryos have been deposited in the well, and before orienting them, briefly rinse the glass transfer pipette to avoid agarose solidifying into it. Otherwise, although not visible, agarose will partially clog the pipette and likely harm the next embryo that will be mounted.
10. Laser light and emitted fluorescent light are partially absorbed and diffracted by the sample, leading to signal attenuation when imaging deep. We use a custom Metamorph journal to compensate for this loss of signal. Exposure time is linearly increased while getting deeper in the sample, so that exposure time is doubled over the entire stack.

Acknowledgments

This work was supported by the grant PJA 20151203256 from Fondation ARC pour la Recherche sur le Cancer, and grant ANR-15-CE13-0016-02 from Agence Nationale de la Recherche.

References

1. Horwitz R, Webb D (2003) Cell migration. Curr Biol 13:R756–R759. https://doi.org/10.1016/B978-0-12-394447-4.20070-9
2. Vicente-Manzanares M, Horwitz AR (2011) Cell migration: an overview. Methods Mol Biol 769:1–24. https://doi.org/10.1007/978-1-61779-207-6_1
3. Lämmermann T, Sixt M (2009) Mechanical modes of “amoeboid” cell migration. Curr Opin Cell Biol 21:636–644. https://doi.org/10.1016/j.ceb.2009.05.003
4. te Boekhorst V, Preziosi L, Friedl P (2016) Plasticity of cell migration in vivo and in silico. Annu Rev Cell Dev Biol 3228361:1–28. https://doi.org/10.1146/annurev-cellbio-111315-125201
5. Ridley AJ (2011) Life at the leading edge. Cell 145:1012–1022. https://doi.org/10.1016/j.cell.2011.06.010
6. Haeger A, Wolf K, Zegers MM, Friedl P (2015) Collective cell migration: guidance principles and hierarchies. Trends Cell Biol 25:556–566. https://doi.org/10.1016/j.tcb.2015.06.003
7. Theveneau E, David NB (2014) Collective cell migrations. Med Sci (Paris) 30:751–757. https://doi.org/10.1051/medsci/20143008012
8. Row RH, Maître J-LL, Martin BL et al (2011) Completion of the epithelial to mesenchymal transition in zebrafish mesoderm requires Spadetail. Dev Biol 354:102–110. https://doi.org/10.1016/j.ydbio.2011.03.025
9. Montero J-A, Carvalho L, Wilsch-Bräuninger M et al (2005) Shield formation at the onset of zebrafish gastrulation. Development 132:1187–1198. https://doi.org/10.1242/dev.01667
10. Liang C-C, Park AY, Guan J-L (2007) In vitro scratch assay: a convenient and inexpensive method for analysis of cell migration in vitro. Nat Protoc 2:329–333. https://doi.org/10.1038/nprot.2007.30

11. Solnica-Krezel LL, Sepich DS (2012) Gastrulation: making and shaping germ layers. Annu Rev Cell Dev Biol 28:687–717. https://doi.org/10.1146/annurev-cellbio-092910-154043
12. Solnica-Krezel L, Stemple DL, Driever W (1995) Transparent things: cell fates and cell movements during early embryogenesis of zebrafish. BioEssays 17:931–939. https://doi.org/10.1002/bies.950171106
13. Kai M, Heisenberg C-P, Tada M (2008) Sphingosine-1-phosphate receptors regulate individual cell behaviours underlying the directed migration of prechordal plate progenitor cells during zebrafish gastrulation. Development 135:3043–3051. https://doi.org/10.1242/dev.020396
14. Dumortier JG, Martin S, Meyer D et al (2012) Collective mesendoderm migration relies on an intrinsic directionality signal transmitted through cell contacts. Proc Natl Acad Sci U S A 109:16945–16950. https://doi.org/10.1073/pnas.1205870109
15. Dumortier JG, David NB (2015) The TORC2 component, Sin1, controls migration of anterior mesendoderm during zebrafish gastrulation. PLoS One 10:e0118474. https://doi.org/10.1371/journal.pone.0118474
16. Dang I, Gorelik R, Sousa-Blin C et al (2013) Inhibitory signalling to the Arp2/3 complex steers cell migration. Nature 503:281–284. https://doi.org/10.1038/nature12611
17. Westerfield M (2000) The zebrafish book. A guide for the laboratory use of zebrafish (Danio rerio), 4th edn. University of Oregon, Eugene, OR
18. Emmenlauer M, Ronneberger O, Ponti A et al (2009) XuvTools: free, fast and reliable stitching of large 3D datasets. J Microsc 233:42–60. https://doi.org/10.1111/j.1365-2818.2008.03094.x

Chapter 17

Analysis of Cell Shape and Cell Migration of *Drosophila* Macrophages In Vivo

Marike Rüder, Benedikt M. Nagel, and Sven Bogdan

Abstract

The most abundant immune cells in *Drosophila* are macrophage-like plasmatocytes that fulfill central roles in morphogenesis, immune and tissue damage response. The various genetic tools available in *Drosophila* together with high-resolution and live-imaging microscopy techniques make *Drosophila* macrophages an excellent model system that combines many advantages of cultured cells with in vivo genetics. Here, we describe the isolation and staining of macrophages from larvae for ex vivo structured illumination microscopy (SIM), the preparation of white prepupae for in vivo 2D random cell migration analysis, and the preparation of pupae (18 h after puparium formation, APF) for in vivo 3D directed cell migration analysis upon wounding using spinning disk microscopy.

Key words *Drosophila*, Macrophages, Plasmatocytes, Cell shape, Cell motility, Wounding, Live cell imaging, Structured illumination microscopy (SIM), Laser ablation, Spinning disk microscopy

1 Introduction

The most abundant immune cells in *Drosophila* are plasmatocytes [1, 2]. They fulfill similar central roles to mammalian macrophages during animal development, in response to infections, tissue damage and cancer, henceforth referred to as macrophages [3]. Embryonic macrophages originate from the procephalic mesoderm, disperse throughout the embryo and proliferate from initially 600 to about 10,000 cells in hematopoietic pockets of the larval body wall [4–7]. The second major population of macrophages derives from a hematopoietic organ, the lymph gland, which develops during larval stages. Gland macrophages derive from undifferentiated progenitors and are released into the hemolymph at the onset of pupal metamorphosis [8, 9]. Both macrophage lineages, the self-renewing tissue macrophages from larval stages and progenitor-based monocyte-derived macrophages,

Marike Rüder and Benedikt M. Nagel contributed equally to this work.

Alexis Gautreau (ed.), *Cell Migration: Methods and Protocols*, Methods in Molecular Biology, vol. 1749, https://doi.org/10.1007/978-1-4939-7701-7_17, © Springer Science+Business Media, LLC 2018

persist through the pupal stage into the adult. They can easily be isolated from white prepupae by rupturing the cuticle. The morphology of these cells can be visualized at high spatial resolution using SIM. Compared to larval macrophages, which are nonpolarized, isolated monocyte-like pupal macrophages are highly polarized with very dynamic lamellipodial protrusions and are capable of migrating randomly at short distances ex vivo [10, 11]. Spinning disk confocal microscopy of macrophages in living prepupae, or mid-stage pupae, nicely complements the ex vivo analysis and allows to dissect the in vivo requirement of gene functions in cell motility and directed cell migration. Here, we describe the procedures for obtaining and analyzing larval and pupal *Drosophila* macrophages ex vivo and in vivo.

The isolation of macrophages from larvae/prepupae is a convenient technique for ex vivo cell morphology analyses. In comparison to larval cells, pupal macrophages have an increased cell size and exhibit a polarized cell morphology (with a lamellipodium and a trailing edge) when spread on uncoated glass surfaces or Concanavalin A (ConA)-coated surfaces [10]. Cell spreading on uncoated glass surfaces depends on integrin-mediated cell adhesion [12, 13]. To further stimulate integrin-mediated cell adhesion, macrophages can be plated on surfaces coated with the ECM protein vitronectin. Cells spread on vitronectin as on glass, but exhibit increased βPS-integrin-positive focal adhesion sites [13]. Isolated macrophages survive for several hours in culture, thus allowing standard cell analysis using immunostaining, flow cytometry or microarray.

Migrating fluorescent-marked macrophages (e.g., Lifeact-EGFP or any marker) can be imaged in vivo through the transparent prepupae (2–4 h APF). Macrophage-specific gene expression or knockdown (mediated by transgenic RNAi) can specifically be induced by the hmlΔ-Gal4 driver, which is expressed in almost all larval and pupal macrophages [14]. In white prepupae, subepidermal macrophages collectively initiate random integrin-dependent 2D single cell migration and disperse throughout the developing adult body. Cells exhibit a spread morphology with a clearly visible lamellipodium [13, 15, 16]. Cells can be tracked manually or automatically by using different software (e.g., Fiji, Imaris, and Bitplane). Different cell parameters including velocity, migration distance and contact inhibition of locomotion can be considered [13, 17].

Dispersed macrophages from the mid-pupal stage remain highly motile, patrolling for infection or dying larval cells. Their capacity to respond to wounds further increases with pupal age. Upon laser-induced single cell ablation or wounding of epithelia, wild type cells immediately switch from a random mode into a directed mode of 3D cell migration [13, 18]. Here, we describe a method to ablate a single cell in the pupae wing (16–18 h APF) and the subsequent computer-assisted analysis of directed cell migration using the Histogram-based Macrophage Migration Score (HMMS) [19]. Genetic tools, e.g., RNAi-based knockdown or MARCM technique [11, 16], are available to manipulate and analyze genes controlling cell migration.

2 Materials

2.1 Solutions and Imaging Chambers

Prepare all solutions using ultrapure water (prepared by purifying deionized water, to attain a sensitivity of 18 MΩ cm at 25 °C). Prepare and store at room temperature unless indicated otherwise.

1. M3++: Dissolve 39.36 g Shields and Sang M3 Insect Medium and 0.5 g $KHCO_3$ in 800 mL water (*see* **Note 1**) and adjust to pH 6.6. Add 1 g yeast extract, 2.5 g Bacto peptone and fill up to 1 L. Filter the medium with a sterile filter (0.2 μm) under the sterile hood. Remove 110 mL medium, add 100 mL heat-inactivated FBS and 5 mL penicillin (10,000 Units/mL)–streptomycin (10,000 μg/mL). Store at 4 °C.
2. 4% PFA: Preheat 300 mL water to 60 °C. Dissolve 20 g PFA and add 1 M NaOH until the solution gets transparent (*see* **Note 2**). Add 50 mL 10× PBS and adjust to approximately pH 7. Add water to a final volume of 500 mL. Aliquot and store at −20 °C or use immediately.
3. 10× PBS: 1.37 M NaCl, 100 mM Na_2HPO_4, 20 mM KH_2PO_4, 27 mM KCl. Dissolve in water, adjust to pH 7.4 with 1 M HCl, ultimately autoclave the solution.
4. 3% BSA: 3% (w/v) bovine serum albumin in PBS.
5. 0.1% (v/v) Triton X-100 in PBS.
6. Heptane glue: Add 1 m brown paracel tape to 100 mL heptane and shake until it is dissolved. Centrifuge 1 h at 12000 × *g* and decant the liquid.
7. Mowiol: Add 2.4 g Mowiol 4–88 to 6 g Glycerol and stir for 1 h. Add 6 mL water and stir for 1 h. Add 12 mL 0.2 M Tris–HCl (pH 8.5), incubate at 50 °C for 2 h and stir every 20 min for 2 min (*see* **Note 3**). Aliquot and store at −20 °C.
8. Phalloidin solution: Dilute fluorescent-dye-labeled phalloidin 1:100 in 3% BSA.
9. Imaging chamber: Cut out a 1.5 cm hole of tissue culture dishes (35 × 10 mm) and use a polyurethane-based adhesive sealing mass to stick coverslips under it.

2.2 Microscopy

2.2.1 Fluorescent Stereo Microscope

Larvae were imaged with a Leica Plan APO 1.0× objective using a Leica M165 FC stereo microscope equipped with a CCD camera (Leica DFC7000 T) and the laser line 488 nm. The Leica Application Suite (LAS) 4.7 was used for image acquisition and processing.

2.2.2 SIM

Cells were imaged through a 63× Plan-Apochromat (high-NA 1.4) oil-immersion objective using an inverted Zeiss Axio Observer Z1 SR microscope with a low light EMCCD camera (Andor iXon 885; 8 μm × 8 μm pixel size) and the laser lines 405, 488 and 561 nm. The ELYRA S.1 system and ZEN software 2010 (Zeiss) were used for image acquisition and processing. For image acquisition, a 5 grid rotation was used. SIM reconstructions of the cells were created using the following manual settings: Noise Filter −5.5, SR frequency weighting 1.0, sectioning zero order 98, first order 83, second order 83.

2.2.3 Spinning Disk Microscopy

Cells were imaged through a 63× Plan-Apochromat (NA 1.4) or 40× Plan-Apochromat (NA 1.3) oil-immersion objective using an inverted Zeiss Axio Observer Z1 microscope with a CCD camera (Axiocam MRm camera, 6.45 μm × 6.45 μm) and the laser lines 488 and 561 nm. The ZEN software 2012 (Zeiss) was used for acquisition and processing of the images.

2.2.4 Laser Ablation Setting

For wounding assays, the UV laser ablation system DL-355/14 direct from Rapp OptoElectronics was used on the spinning disk microscope.

2.3 Fly Strains

The following fly strains are used in this protocol (Table 1).

Table 1
Fly strains used

Genotype	Explanation	Reference
W1118	w, wild type X	Lindsey and Zimm (1992)
hmlΔGal4, UAS-EGFP	Expression of EGFP driven by the hemolectin promoter	[19]
FRT40A scarΔ37	*scar/wave* null allele	[20]
hsFLP; FRT40A,Gal80/CyO; tubGal4, CD8-GFP/TM6b	To induce cell mutant clones; used for MARCM analysis *X*, 2, 3	Bl 42725
UAS-scar/wave RNAi	RNAi transgene dsRNA *scar/wave*	NIG-Fly 4636R-1

3 Methods

Grow flies on standard *Drosophila* food at 25 °C (*see* **Note 26**).

3.1 Isolation of Pupal Macrophages for Ex Vivo SIM

1. Use a wet paintbrush to easily obtain at least ten white prepupae (*see* **Notes 4** and **5**; Fig. 1b) and transfer them into a dissection dish, containing water. Wash the prepupae by stirring with your paintbrush.
2. Transfer the animals with forceps into a new dissecting dish (*see* **Note 6**), containing 200 μL of M3++.
3. Take hold of the prepupa's head with a forceps and use a second forceps to grab directly behind the first one and tear open the prepupae (*see* **Note 7**; Fig. 1a). Once opened, slightly shake the prepupae to rinse out the hemolymph. Then remove the corpus and repeat for each prepupa (*see* **Note 8**).
4. Pipette those roughly 200 μL into a 1.5 mL tube, containing 300 μL M3++ and repeat **steps 1–3** for additional samples (*see* **Note 9**).
5. Put adequate ethanol cleaned coverslips into a 24-well plate and transfer your macrophage containing medium onto the coverslips (*see* **Note 10**).
6. Let the macrophages spread for 1 h at 25 °C (*see* **Notes 11** and **12**).
7. Subsequently remove the liquid carefully, cover the slip with 4% PFA (fixative) and incubate for 15 min at room temperature (*see* **Note 13**).
8. For antibody staining, remove the liquid, cover the slip with 0.1% Triton, directly remove it (*see* **Note 14**) and wash three times with 1× PBS (*see* **Note 15**).
9. Pipette 30 μL 3% BSA onto Parafilm in a light protected, humidified chamber (*see* **Note 16**), grab the coverslip with a forceps (*see* **Note 17**), place it upside down in the 3% BSA and close the lid.
10. Incubate for 30 min at room temperature.
11. Transfer the coverslip in a 30 μL drop (placed on Parafilm in a light protected chamber) of primary antibody solution (diluted in 3% BSA solution) and incubate for 2 h at room temperature.
12. Take the coverslip with forceps and remove any unbound antibodies by dipping it twice in deionized water and remove the adherent drop by dabbing the coverslip edge on paper tissue.
13. Transfer the coverslip in a 30 μL drop (placed on Parafilm in a light protected chamber) of secondary antibody solution (diluted in 3% BSA solution) and incubate for 45 min at room temperature.

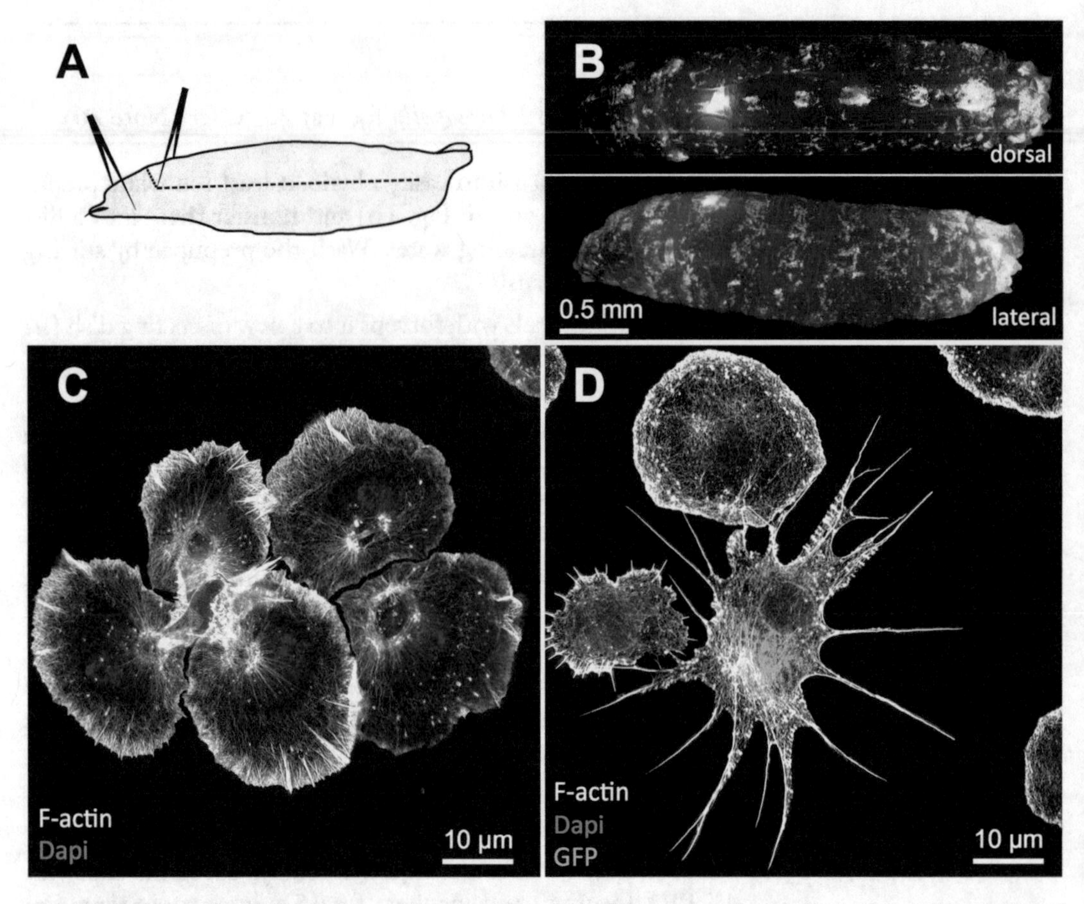

Fig. 1 Morphological analysis of *Drosophila* macrophages. (**a**) Preparation of a third instar larva, the position of the forceps at the anterior end is indicated (**b**) Montage image of a third instar larva with fluorescent marked macrophages in hematopoietic pockets of the larval body wall. (**c**, **d**) SIM images of isolated larval macrophages. Cells are stained with phalloidin (F-actin, white) and DAPI (nuclei, blue). Image (**c**) shows wild type macrophages. *scar/wave* mutant cells (**d**) were generated by the MARCM technique. Mutant cells are marked by GFP expression (green)

14. Again, take the coverslip with a forceps and remove any unbound antibodies by dipping it twice in deionized water and remove the adherent drop by dabbing the coverslip's edge on paper tissue.
15. Heat up Mowiol to 60 °C and put a drop (roughly 25 μL) on a microscope slide and mount the coverslip.
16. Let the Mowiol solidify for at least 1 h at 4 °C and afterward acquire images using SIM (Fig.1c, d).

3.2 Preparation of White Prepupae for In Vivo 2D Random Migration

1. Use a wet paintbrush to easily obtain a white prepupa and put it on a clean paper tissue and roll it using a paintbrush to dry the pupal case (*see* **Note 18**).
2. Use a thin layer of heptane glue to coat the coverslip of a small glass bottom culture plate (*see* **Note 19**).

3. Take the prepupa using the forceps and place it dorsal side down and slightly laterally tilted on the glue. Gently press on the prepupa (use a forceps or a paint brush) to get it firmly stuck.
4. Put a small wet piece of tissue paper in the culture plate and seal it with Parafilm to avoid prepupa dehydration. Wait until the prepupal case gets brownish (around 2 h) and start imaging using the spinning disk microscope.
5. Search for macrophages near the epithelium, since these cells migrate better and have a more pronounced lamellipodium. Use a Z-stack with a range of around 20 μm and record for at least 10 min with an interval of 20 s.
6. Make a maximum projection of the recorded Z-images of the time-lapse using Fiji. Save it afterward as AVI or single TIFF images (Fig. 2a–c).
7. Manually track cells and make the track visible using Fiji (Fig. 2b′, c′). Position information of cells over time can be used to extract parameters of cell migration.

3.3 Preparation of Pupae for In Vivo 3D Random or Directed Migration

1. Use a wet paintbrush to easily obtain white prepupae, put them in a petri dish laid with wet filter paper and seal with Parafilm.
2. Let the animals develop until 18 h APF (after puparium formation) at 25 °C (*see* **Note 20**).
3. Use the forceps to take hold of the pupa by grasping the operculum (*see* **Note 21**) and put it on a clean paper tissue and roll it using a paintbrush to dry the pupal case (*see* **Note 18**).
4. Prepare a microscope slide with double-sided sticky tape, put the pupa ventral side down, gently press on the pupa to get it firmly stuck and remove the operculum as depicted (Fig. 3a).
5. Cut along the lateral side, above the wing as a starting point toward the posterior end with spring scissors (*see* **Note 22**). Use the forceps to grasp the sliced pupal case and stick it to the tape (Fig. 3a).
6. Use a very thin layer of heptane glue to coat the coverslip of a small glass bottom culture plate (*see* **Note 23**).
7. Gently take the pupa with the forceps and stick one wing on the heptane glue (Fig. 3b).
8. Put a small wet piece of tissue paper in the culture plate and seal it with Parafilm to avoid dehydration.
9. Using the spinning disk microscope, the wing is easily identified by its disperse distribution of macrophages compared with the abdomen or thorax region.
10. Use a Z-stack with a range of around 20 μm and record for 30 min with an interval of 30 s.

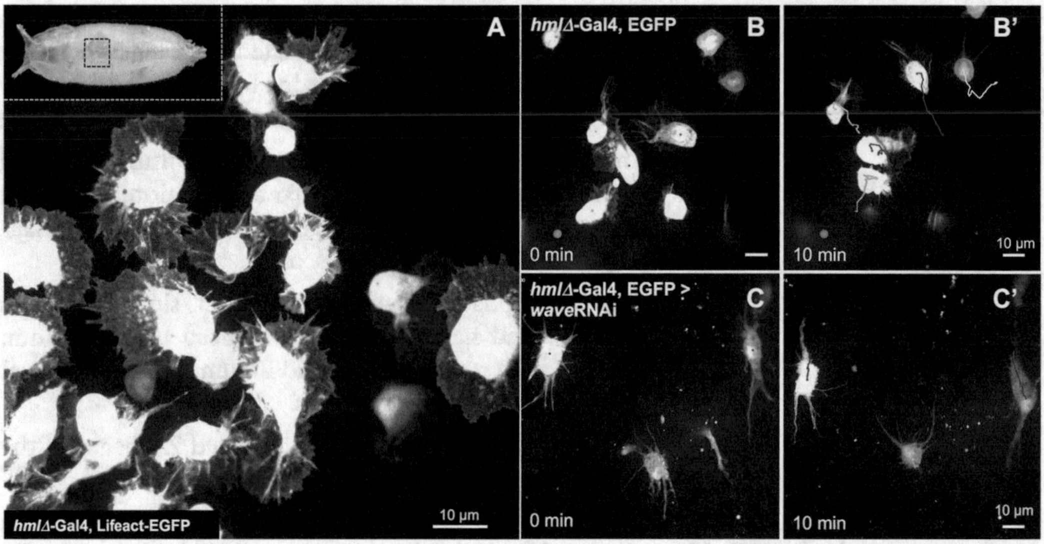

Fig. 2 In vivo spinning disk microscopy of prepupal macrophages. (**a**) Frame of a spinning disk microscopy video of highly motile wild type macrophages expressing an EGFP-Lifeact transgene imaged from a living prepupae (2 h APF). Single frames (0–10 min) of a time-lapse movie showing random 2D migration of wild type macrophages (**b**, **b′**) and RNAi-mediated knockdown of *scar/wave* (**c**, **c′**). In comparison to wild type cells, *scar/wave* depleted cells (**c**) show an impaired migratory behavior and exhibit defects in lamellipodia formation. (**b′** and **c′**) Cells were manually tracked using the Fiji Plugin "manual tracking" and resulting trajectories are shown

11. For directed migration analysis, acquire one Z-stack and pause the recording. Use the laser ablation system to define a region or cell of interest in all three dimension and ablate while still in the pause mode. After finishing the ablation sequence, resume recording (*see* **Note 24** and **25**).
12. Make a maximum projection of the recorded Z-images of the time-lapse movie using Fiji or equivalent microscope software. Save it afterward as an AVI movie or single TIFF images (Fig. 3d, e).
13. Manually track cells and make the track visible using Fiji (Fig. 3d, e, 30 min). Position information of cells over time can be used to extract parameters of cell migration.

4 Notes

1. The Shields and Sang M3 Insect Medium powder is toxic! Weighing-in should be done under the hood.
2. Paraformaldehyde is toxic! Wear a mask and gloves, when weighing PFA. Immediately after adding the paraformaldehyde to the water, transfer the beaker glass to a magnetic stirrer under the hood. You can use higher concentrations of NaOH

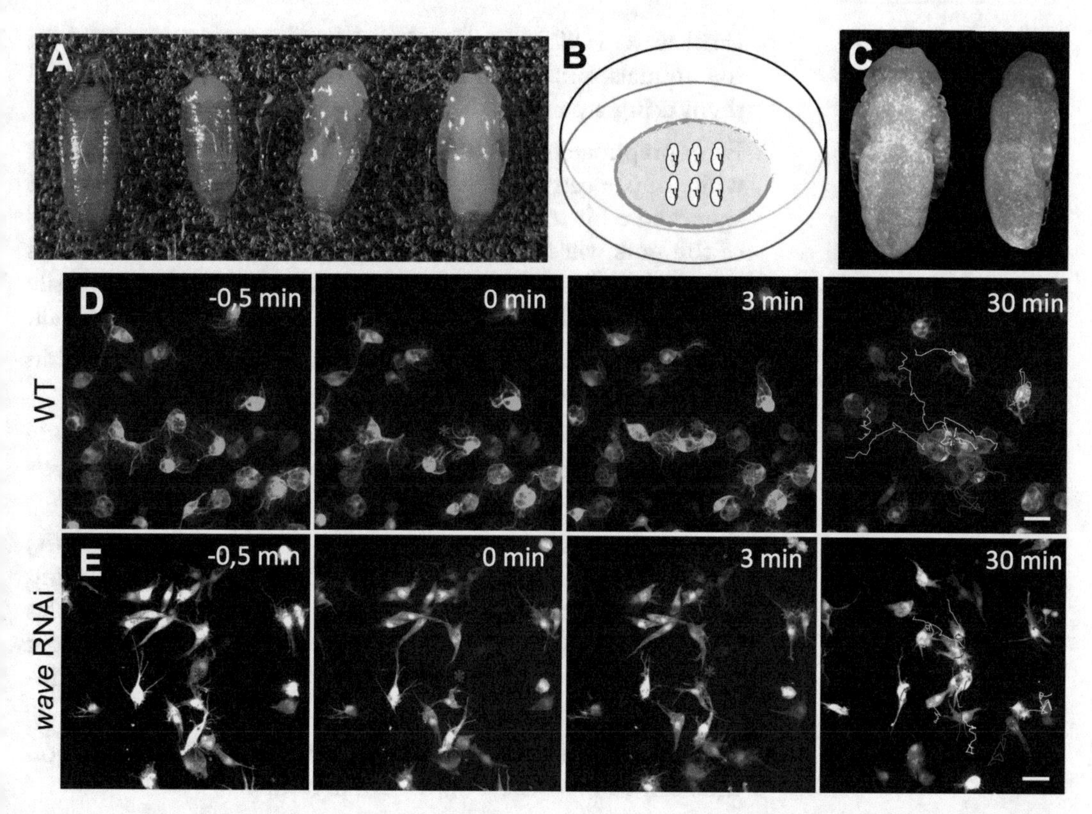

Fig. 3 In vivo spinning disk microscopy of mid-stage pupal macrophages. (**a**) Dissection of pupae 18 h APF. (**b**) Isolated pupae fixed on coverslip heptane glue. (**c**) Montage image of pupae with GFP marked macrophages 18 h APF. (**d**, **e**) Single frames of a time-lapse movie of wild type macrophages (**d**) and RNAi mediated knock-down of *scar/wave* (**e**) upon ablation of a single cell (red asterisk) in the pupal wing. Before wounding (−0.5 min) cells migrate randomly, after wounding (0 min) cells start to migrate toward the wound (3–30 min). *wave* depleted cells show an altered migratory behavior and exhibit defects in lamellipodia formation. Cells were manually tracked using Fiji Plugin "manual tracking" and resulting trajectories are shown

at the beginning, but it takes some time until the PFA dissolves, so add the NAOH slowly.

3. If the Mowiol 4–88 does not dissolve completely, centrifuge for 15 min at 3000 × *g*. Optionally add 25 mg DABCO (1,4-diazabicyclo-(2,2,2)octane) per mL solution until it is dissolved. DABCO additionally protects the samples from photobleaching.
4. Optionally, use at least ten third instar larvae and put them on water-moistened filter paper in a small petri dish. Use Parafilm to seal the petri dish so that the motile larvae cannot flee or get squashed between the dish and its lid.
5. Alternatively, use at least 20 third instar larvae and carry on with the next step. Pupal macrophages have an increased cell size and many of them are polarized [10].

6. If the water is highly polluted, or if too much water remains on the animals, put the animals on a clean paper tissue and roll them using a paintbrush.
7. Early prepupae have a soft cuticle and are easy to open. Older white pupae can be bled at the dorsal side. The animal can be squeezed a bit to increase the yield of macrophages, but parts of the guts will flow out. If using third instar larvae stated as described in **Note 4**, squeeze the larvae gently to loosen sessile macrophages from the hematopoietic pockets of the body wall.
8. Drop the animals at the dissecting dish's inner edge to drain off any remaining drops of medium.
9. While repeating this procedure with the other samples, already prepared samples can be stored on ice to minimize macrophage activity.
10. Optionally, coat the coverslips with Concanavalin A (ConA) for 30 min at room temperature to get an enhanced spreading effect. If the incubation time is over, but preparing the samples is not, remove ConA and coat with 1× PBS. Alternatively, coat with Vitronectin overnight at 4 °C to enhance integrin-dependent adhesion.
11. Use at least 1 h to get optimal cell spreading. The incubation time can be prolonged up to several hours.
12. If black dots appear on the coverslip, discard the sample. Crystal cells are responsible for this melanization reaction [21].
13. Incubation for a longer period of time could interfere with antibody binding due to over-cross-linking.
14. Triton is used to permeabilize cell membranes. Incubation for too long will dissolve cell membranes.
15. For incubation with Phalloidin only, move on to **step 13** and use in place of a secondary antibody, Phalloidin diluted 1:100 in 3% BSA solution. Incubation time decreases to 35 min.
16. Use wet filter paper stuck to the lid with adhesive tape.
17. Do not remove PBS from the last washing step, it is easier to take the coverslip with the well still filled.
18. Be sure to completely dry the pupa's case, since any residual water will prevent sticking it onto the double-sided sticky tape.
19. The glue's autofluorescence interferes with imaging, if the layer is too thick.
20. It is possible to use 16 h APF pupae.
21. If the operculum is not empty, but rather "mushy," the fly is not in the developmental stage it should have been. It is possible to keep them at 25 °C for a little longer, however it will then be harder to determine their exact age.

22. Using angled to side spring scissors might be helpful. Optionally, use fine point forceps.
23. Dip a coverslip into the glue and spread it with gentle pressure onto a microscope slide.
24. Suitable macrophages are found in the central or distal part of the wing. Proximal macrophages appear to be vacuolated and slow. In contrast, distal macrophages are livelier due to less prior phagocytosis activity.
25. Keep in mind that subepidermal macrophages need less laser intensity to be ablated than macrophages located further down.
26. The Gal4/UAS system is temperature sensitive. Using 29 °C increases expression and could lead to a stronger phenotype.

Acknowledgment

We thank L. Brüser for critical reading of the manuscript, and the Bloomington Stock Center and VDRC for fly stocks. This work was supported by a grant to S.B. from the cluster of excellence 'Cells in Motion' (CIM; Deutsche Forschungsgemeinschaft) (BO 1890/1, BO 1890/2, BO 1890/3, BO 1890/4).

References

1. Gold KS, Bruckner K (2016) Macrophages and cellular immunity in Drosophila melanogaster. Semin Immunol 27(6):357–368
2. Lemaitre B, Hoffmann J (2007) The host defense of Drosophila melanogaster. Annu Rev Immunol 25:697–743
3. Williams MJ (2007) Drosophila hemopoiesis and cellular immunity. J Immunol 178(8):4711–4716
4. Tepass U, Fessler LI, Aziz A, Hartenstein V (1994) Embryonic origin of hemocytes and their relationship to cell death in Drosophila. Development 120:1829–1837
5. Makhijani K, Alexander B, Tanaka T, Rulifson E, Bruckner K (2011) The peripheral nervous system supports blood cell homing and survival in the Drosophila larva. Development 138(24): 5379–5391
6. Holz A, Bossinger B, Strasser T, Janning W, Klapper R (2003) The two origins of hemocytes in Drosophila. Development 130(20): 4955–4962
7. Jung SH, Evans CJ, Uemura C, Banerjee U (2005) The Drosophila lymph gland as a developmental model of hematopoiesis. Development 132(11):2521–2533
8. Grigorian M, Mandal L, Hartenstein V (2011) Hematopoiesis at the onset of metamorphosis: terminal differentiation and dissociation of the Drosophila lymph gland. Dev Genes Evol 221(3):121–131
9. Gold KS, Bruckner K (2015) Macrophages and cellular immunity in Drosophila melanogaster. Semin Immunol 27(6):357–368
10. Sampson CJ, Williams MJ (2012) Real-time analysis of Drosophila post-embryonic haemocyte behaviour. PLoS One 7(1):e28783
11. Sander M, Squarr AJ, Risse B, Jiang X, Bogdan S (2013) Drosophila pupal macrophages—a versatile tool for combined ex vivo and in vivo imaging of actin dynamics at high resolution. Eur J Cell Biol 92(10–11):349–354
12. Jani K, Schock F (2007) Zasp is required for the assembly of functional integrin adhesion sites. J Cell Biol 179(7):1583–1597
13. Nagel BM, Bechtold M, Rodriguez LG, Bogdan S (2017) Drosophila WASH is required for integrin-mediated cell adhesion, cell motility and lysosomal neutralization. J Cell Sci 130(2):344–359
14. Sinenko SA, Mathey-Prevot B (2004) Increased expression of Drosophila tetraspanin, Tsp68C,

suppresses the abnormal proliferation of ytr-deficient and Ras/Raf-activated hemocytes. Oncogene 23(56):9120–9128

15. Brinkmann K, Winterhoff M, Onel SF, Schultz J, Faix J, Bogdan S (2016) WHAMY is a novel actin polymerase promoting myoblast fusion, macrophage cell motility and sensory organ development in Drosophila. J Cell Sci 129(3):604–620
16. Moreira CG, Jacinto A, Prag S (2013) Drosophila integrin adhesion complexes are essential for hemocyte migration in vivo. Biol Open 2(8):795–801
17. Davis JR, Luchici A, Mosis F, Thackery J, Salazar JA, Mao Y, Dunn GA, Betz T, Miodownik M, Stramer BM (2015) Inter-cellular forces orchestrate contact inhibition of locomotion. Cell 161(2):361–373
18. Stramer B, Wood W, Galko MJ, Redd MJ, Jacinto A, Parkhurst SM, Martin P (2005) Live imaging of wound inflammation in Drosophila embryos reveals key roles for small GTPases during in vivo cell migration. J Cell Biol 168(4):567–573
19. Lammel U, Bechtold M, Risse B, Berh D, Fleige A, Bunse I, Jiang X, Klambt C, Bogdan S (2014) The Drosophila FHOD1-like formin Knittrig acts through Rok to promote stress fiber formation and directed macrophage migration during the cellular immune response. Development 141(6):1366–1380
20. Zallen JA, Cohen Y, Hudson AM, Cooley L, Wieschaus E, Schejter ED (2002) SCAR is a primary regulator of Arp2/3-dependent morphological events in Drosophila. J Cell Biol 156(4):689–701
21. Vlisidou I, Wood W (2015) Drosophila blood cells and their role in immune responses. FEBS J 282(8):1368–1382
22. Lindsey DL, Zimm GG. San Diego: Academic Press; (1992) The genome of Drosophila melanogaster; pp. 804–1066

Check for updates

Chapter 18

Migration of Q Cells in *Caenorhabditis elegans*

Yongping Chai, Zhiwen Zhu, and Guangshuo Ou

Abstract

During *C. elegans* larval development, the Q neuroblasts produce their lineage by three rounds of divisions along with continuous cell migrations. Their neuronal progeny is dispersed from the pharynx to the anus. This in vivo system to study cell migration is appealing for several reasons. The lineage development is stereotyped; functional analysis and genomic screens are rendered easy and powerful thanks to powerful tools; transgenic manipulations and genome engineering are efficient and can be conveniently combined with live-cell imaging. Here we describe a series of protocols in Q cell migration studies, including quantifications of progeny position, genetic screening strategies, preparation of migration mutants or transgenic worms expressing related fluorescent proteins, multipositional time-lapse tracking of Q cell migration using confocal microscopy and image analyses of single cell movements and dynamics.

Key words *Caenorhabditis elegans*, Q neuroblast, CRISPR-Cas9, Time-lapse imaging

1 Introduction

At a late stage of *Caenorhabditis elegans* embryogenesis, about 60 min before the egg hatch, the two Q neuroblasts, QR and QL, are born in the posterior half of worm, mirroring each other on the right and left side of embryo, respectively (Fig. 1a) [1]. At birth, the Q neuroblasts are located within the two lateral rows of epidermal stem seam cells anterior to their V5 sister cells (Fig. 1b), from where they migrate intensively and derive identical cell lineages. Even though the two Q neuroblasts eventually generate the same types of neurons, their migration behavior is different. Shortly after larva hatch, the QR cell initiates polarization, moves out of the seam cell row toward the anterior, whereas QL migrates toward the posterior and end up crawling over its V5 sister (Fig. 1c). The initial migration of Q neuroblasts is slow and short in distance, but decisively steers the whole lineage migrations. For example, both daughters of QR cell, QR.a and QR.p, continue to migrate anteriorly, and QR.a moves faster than its sister, QR.p. On the other hand,

---

Yongping Chai and Zhiwen Zhu contributed equally to this study.

Alexis Gautreau (ed.), *Cell Migration: Methods and Protocols*, Methods in Molecular Biology, vol. 1749,
https://doi.org/10.1007/978-1-4939-7701-7_18,

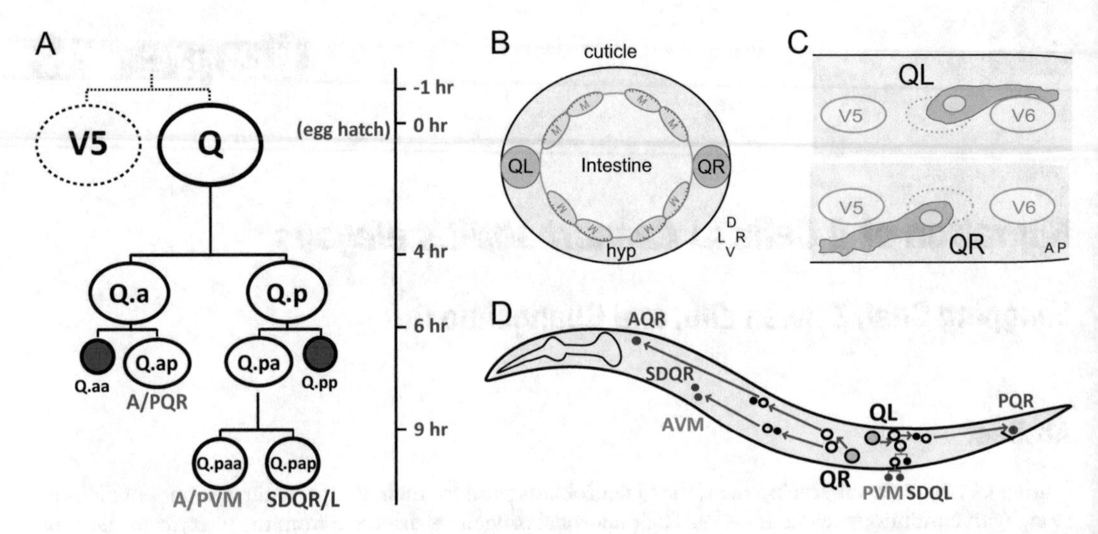

Fig. 1 *C. elegans* Q neuroblast lineages and Q cell migration. (**a**) Timeline of *C. elegans* Q neuroblast lineage development. After birth, during embryogenesis, Q neuroblasts undergo three rounds of asymmetric divisions to generate two apoptotic cells (black) and three different types of neurons: oxygen sensory neurons AQR (QR.ap) and PQR (QL.ap); mechanosensory neurons AVM (QR.paa) and PVM (QL.paa); interneurons SDQR (QR.pap) and SDQL (QL.pap). Time on the right indicates each cell division during Q lineage development, counting from egg hatching. (**b**) The position of Q neuroblasts in a cross section of the L1 animal. *M* muscle, *hyp* hypodermis, *D* dorsal, *V* ventral, *L* left, *R* right. (**c**) A cartoon showing the initial polarization of Q neuroblasts. *A* anterior, *P* posterior. Dashed circles represent the birth places of Q neuroblasts. (**d**) A diagram of the migration of Q neuroblasts and their descendants in the L1 larval stage. QR and its progeny migrate toward the anterior, and QL and its progeny migrate toward the posterior

when QL.p ceases moving, QL.a crawls over it to continue its journey toward the posterior. The migration speed and distance during this second stage is intermediate. The third stage of migration starts after the second round of asymmetric cell division. The QR.a cell divides to generate an apoptotic QR.aa cell and a fast-migrating QR.ap cell, which eventually stops migration near the pharynx and develops into AQR, an oxygen-sensory neuron. Similarly, the descendant cell from the posterior-directed QL lineage, QL.ap, migrates far and fast to position itself just behind the anus and differentiate into the oxygen-sensory neuron PQR. In the meanwhile, the nonapoptotic daughter of Q.p cells migrates a medium distance, and divides to give rise to a mechanosensory neuron, AVM from QR.p and PVM from QL.p, and an interneuron, SDQR from QR.p. and SDQL from QL.p (Fig. 1d) [1–3], which further migrates toward the dorsal or the ventral side respectively, before differentiation. The whole Q cell migration process is thus highly spatiotemporally stereotypical, and has been shown to be comprehensively orchestrated by multiple evolutionarily conserved mechanisms [3–11]. Thus, the multistaged Q cell migration can provide a good system to dissect migration transitions and interplays between different signaling and regulatory mechanisms.

Many of the crucial molecules regulating Q cell migration can also be involved in a variety of other fundamental cellular processes. Their mutants are mostly embryonic lethal, thus impeding their functional analyses in Q cell migration during larvae development. To overcome this problem, we have developed TALEN and CRISPR-Cas9-mediated gene editing platforms to construct somatic conditional TALEN or CRISPR-Cas9 mutants that can be induced in specific cell lineages or developmental stages of the nematode [12, 13]. We also developed live-cell imaging techniques to study Q cell migration with resolutions in the order of seconds for time and of micrometer order for space [14] and the technique of GFP knockin to achieve fluorescent labeling of endogenous target proteins. This combination of techniques greatly reduces the background and improves the observation of highly dynamic processes.

Using these genetic and imaging approaches, we discovered a series of embryonic lethal molecules modulating Q cell migration [13, 15, 16]. For example, we found that *C. elegans* embryonic essential gene *cor-1*, the homolog of Coronin, a molecule, which is associated with human neurobehavioral dysfunction, is involved in the regulation of actin organization and cell morphology during Q cell migration by promoting actin recycling [13]. Our work also uncovered that another embryonic essential gene encoding the multidomain protein Anillin, which is primarily considered to organize cytokinesis, can be recruited to the leading edge of migrating Q neuroblasts and can stabilize F-actin at this location to ensure coherent motility [15]. By combining conventional forward genetic screens with conditional gene editing, we were able to identify the signaling pathway that transduces input from the transmembrane protein MIG-13 to actin polymerization at the leading edge of Q cells. This pathway includes the embryonic essential genes *sem-5* and *abi-1* [16]. Further application of GFP knockin with high magnification objective lens enabled us to capture for the first time the punctum-like distribution of the actin nucleation promoting factors WASP and WAVE along the leading edge of migrating Q cells [16].

Here we will first describe the protocol to quantify the migration phenotype of Q cells, followed by conventional mutagenesis. Then we will outline how to apply the somatic CRISPR-Cas9 method to generate conditional knockout animal, and to label molecules of interest using both traditional transgenic strain construction and knockin techniques. Finally, we will describe the multipositional time-lapse confocal imaging procedure designed to image migration of individual descendants of Q cells in *C. elegans* L1 larvae using spinning disk confocal microscopy, and methods to analyze and characterize key subcellular properties that define the migration behavior.

2 Materials

2.1 Worm Maintenance and Culture

2.1.1 Equipment

1. Worms incubators with constant temperature of 20 °C.
2. Platinum pick for worm transferring.

2.1.2 Worm Strains

1. Wild-type strain—N2 Bristol strain.
2. Transgenic or mutant strains available from the *C. elegans* Genetics Center (CGC).
3. Mutant animals generated by EMS mutagenesis or CRISPR/Cas9 editing (knockout, conditional knockout, and knockin).
4. Transgenic animals carrying extrachromosomal arrays of appropriate fluorescent markers generated by microinjection.

2.1.3 Culture Media

1. Nematode growth medium (NGM) agar plates: For 1 L medium, dissolve 3 g NaCl, 2.5 g Bacto peptone, and 21 g agar in a flask containing 970 mL distilled H_2O. Stir to mix well and then autoclave. After the medium cools down to 60 °C, add 1 mL $CaCl_2$ (1 M stock), 1 mL $MgSO_4$ (1 M stock), 1 mL cholesterol (5 mg/mL stock) and 25 mL phosphate K_3PO_4 buffer (1 M stock, pH 6.0), then ddH_2O to the final volume of 1 L. Mix thoroughly, dispense the medium into the plastic plates with desired volumes.
2. Escherichia coli OP50 culture, to seed the NGM plate as food supply. Inoculate a single OP50 colony into 40 mL of sterile LB broth, culture overnight at 37 °C with 220 rpm shaking. Spot 100–200 μL OP50 cultured in the center of each NGM agar plate, spread and dry at room temperature. The plates are ready to use after the bacterial lawn has grown overnight.

2.2 Worm Manipulations and Transgenic Animal Constructions

2.2.1 Equipment

1. Dissecting microscope.
2. Fluorescence dissection scope.
3. Microinjection set.

2.2.2 Solution and Chemicals

1. M9 buffer: Add 5.8 g of Na_2HPO_4, 3.0 g of KH_2PO_4, 0.5 g of NaCl, 1.0 g of NH_4Cl into 1 L of ddH_2O. Autoclave for 20 min at 121 °C to sterilize.
2. Ethyl methanesulfonate (EMS) (99%, Sigma, cat. no. 62-50-0).
3. Bleach buffer: mix 2 mL NaClO, 5 mL NaOH (1 M stock), and 3 mL ddH_2O together, keep away from light.
4. Worm lyse buffer: mix 200 μL 10× Easy Taq buffer, 25 μL 20 mg/mL proteinase K, and 800 μL ddH_2O.

2.2.3 Molecular Reagents for Construction

1. Desired suitable PCR primers and vectors.
2. Phusion DNA polymerase premix (Phusion high-fidelity PCR master mix with HF Buffer, New England Biolabs, cat. no. M0531S), is used for all the PCR reactions.
3. In-Fusion enzyme (In-Fusion HD cloning kit, Clontech, cat. no. 639648).
4. Dpn-1 (NEB, cat. no. R0176L), a restriction enzyme to digest methylated DNA.
5. T7 endonuclease I (NEB, cat. no. M0302L).

2.3 Live-Cell Confocal Imaging

2.3.1 Chemical and Supplies

1. Imaging plate: leave regular 60 mm NGM plates to dry at room temperature, drop 40–60 μL OP50 bacteria culture on each plate the day before usage.
2. 3% "sandwich" agarose pad: place a 12-mm round coverslip on an even surface, use a Pasteur pipette to put one drop of 3% (weight/vol) agarose on the center of the coverslip and then immediately drop another 12-mm round coverslip on top of the agarose to make a "sandwich."
3. Levamisole (Sigma): store 10 mg/mL solution as the 10× stock, and prepare 1 mg/mL working solution freshly on the experimental day.
4. Immersion oil for microscope objective (Zeiss Immersol).

2.3.2 Equipment

Q cell migration can be sensitive to the phototoxicity, therefore a detector with high excitation efficiency and signal-to-noise ratio is favored to minimize the laser input. A frame rate of 1 min is adequate for a Q cell migration time-lapse movie. An automated stage is recommended to make the best of the interval time by recording multiple positions at a time. We use the system below:

1. Microscope: Zeiss Observer Z1 with "A-Plan" ×10/0.25 Ph1 (working distance (WD) = 4.2 mm) objective and the alpha "Plan-Apochromat" ×100/1.46 Oil DIC (ultraviolet) VIS-IR M27 (WD = 0.10 mm) objective.
2. MS-2000 flat-top xyz automated stage with MS-2000-WK multiaxis stage controller.
3. Solid-state lasers: Sapphire CW CDRH USB laser system with 405-nm (100 mW), 488-nm (150 mW) and 568-nm (150 mW) lines.
4. Spinning disk confocal scanner unit: CSU-X1 incorporating 405-, 488- and 561-nm dichroic filters suitable for the laser wavelengths.
5. Electron-multiplying charge-coupled device (EMCCD) digital camera: Andor iXon + DU-897D-C00#BV-500 (Andor Technology).

2.3.3 Software

1. For imaging acquisition: μ-Manager (http://valelab.ucsf.edu/~MM/MMwiki/) with suitable configurations to control the imaging units.
2. For imaging processing and data analyses: ImageJ (https://imagej.nih.gov/ij/).

3 Methods

3.1 Quantitative Evaluation of Q Cell Descendant Positions

3.1.1 Imaging of Q Cell Descendant Position in Grown Animals

1. Culture strain GOU246 worms on OP50-seeded NGM plates at 20 °C.
2. Under a dissecting microscope (*see* **Note 1**), transfer more than 100 individual L4 stage worms (*see* **Note 2**) onto a new OP50-seeded NGM plate for imaging.
3. Mount 10–20 transferred animals in a drop of M9 buffer on a 3% agar pad, add a drop levamisole solution beside the M9 drop (*see* **Note 3**), and cover the samples with coverslips immediately.
4. Put the sample on the stage of confocal microscope. Use the bright field to locate and focus animals on the coverslip under 10× objective.
5. Take Z-stacking images with both GFP and RFP channels for each animal (*see* **Note 4**).
6. Finish imaging of all 100 transferred animals.

3.1.2 Scoring of Q Cell Descendant Positions

1. Open one acquired image with imaging processing software, such as ImageJ.
2. Draw a segmented line from URXs to the PLMs along the midline of the animal body (Fig. 2a).
3. Measure the length of the line; document it as the full body length for reference.
4. Draw segmented lines and measure the lengths from URXs to each Q cell descendant AQR, PQR, AVM, and PVM, respectively (*see* **Note 5**).
5. Normalize the distance to the URXs of each Q cell descendant by the full body length of the animal as their position coordinates along the worm, where URXs are located at 0 while PLMs are located at 1 (Fig. 2a).
6. For each Q cell descendant neuron, pool their positioning data from 100 individual animals together for desired statistical analyses.

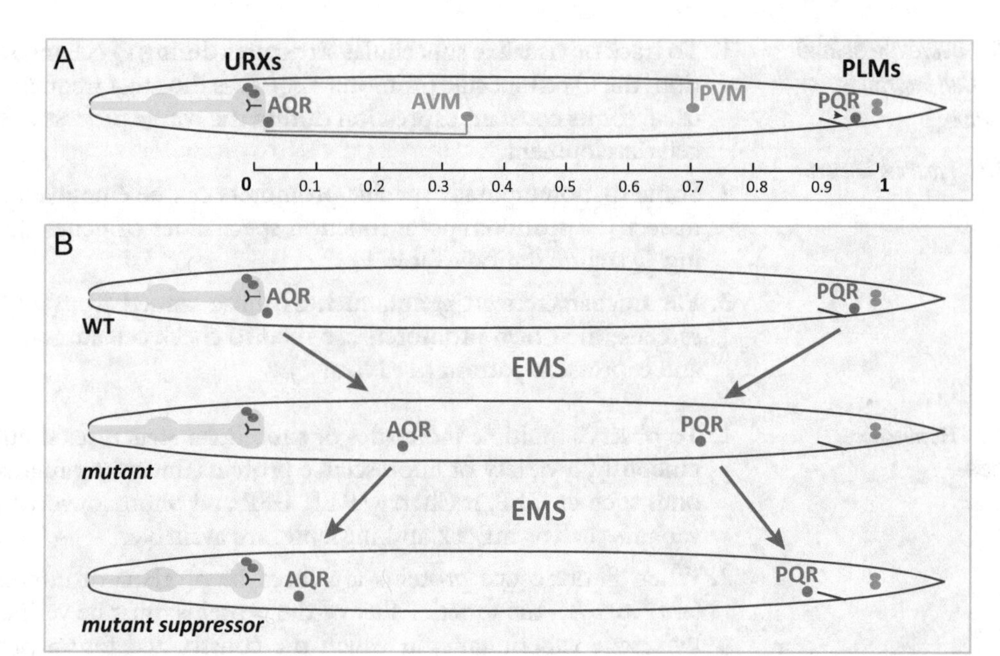

Fig. 2 Quantification and genetic study of Q cell descendant migration at the L4 stage. (**a**) Scheme showing the positions of Q neuroblast lineage descendants at L4 stage. The body length between URXs and PLMs are divided into ten blocks, which are used to score Q cell positions. Note that both the AVM and PVM project ventral-directed neurites, which distinguish them from ALMs and PLMs. (**b**) Diagram showing forward genetic strategies that can be used to screen for unknown modulators of Q cell migration

3.2 Forward Genetic Screen Using Ethyl Methane Sulfonate (EMS) (See Note 6, Fig. 2b)

1. Harvest adult animals from 20–50 NGM plates, bleach to get eggs, let hatch on un-seeded NGM plate to obtain synchronized L1 larva.
2. Collect the synchronized L1 worms, put on OP50-seeded NGM plates, let grow to L4 stage.
3. Gather the late L4 worms (P0s) from the culture plates and treat with 4 mL EMS (20 μl 99% EMS in 4 mL M9 buffer) in a 15 mL tube for 4 h with mild shaking at 20 °C.
4. Wash P0s three times with M9 and put worms on OP50-seeded NGM plates for recovery overnight at 20 °C.
5. Bleach the recovered P0s to get F1s, let grow on OP50-seeded NGM plates till L4-stage, then single 1000–5000 individual F1 animals.
6. When the next generation (F2) of the F1s grow to L4 stage, screen for the Q cell migration phenotype, single out 3–5 F2s with migration phenotype from each positive F1 plate (Fig. 2b).
7. Confirm the Q cell migration phenotype in the next generation, then characterize the mutant phenotype and clone the gene that contains the lesion.

3.3 Overexpressions in Q Cell Migration Studies

3.3.1 Promoter Choices

1. To track or visualize subcellular structures during Q cell migration, the Q cell specific promoter P*egl-17* is the most frequently used, for its constant expression during the whole process of Q cell development.
2. Some spatiotemporally specific promoters can be employed to dissect the spatiotemporal function specificities of genes during Q cell migration (Table 1).
3. For uncharacterized genes, such as those cloned from EMS screens, their own promoters are used to check cell autonomy and expression pattern (*see* **Note 7**).

3.3.2 Fluorescence Labels

1. To observe multiple molecules or subcellular structures simultaneously, a variety of fluorescence proteins, including regular ones such as GFP, mCherry/RFP, BFP and photoconvertible variants such as mEos2 and mMaple, are available.
2. When fluorescence proteins are fused to proteins of interest (*see* **Note 8**), the functionality of the proteins must be verified by rescue experiments, in which the constructed fusion proteins should restore the defect in Q cell migration associated with inactivation of the target gene (Table 2).
3. Sometimes, overexpression of particular protein causes Q cell migration defects even at the lowest dose. A knockin fusion strategy should then be considered.

3.3.3 Construction of Transgenic Worms Carrying Desired Overexpressions

1. A homologous recombination based protocol called In-Fusion cloning is used to construct the desired plasmids.
2. For In-Fusion cloning, 15–20 nt overlapping sequences at both ends of the vector and linearized insertion fragment are needed.
3. Design primers to amplify the gene of interest from cDNA or N2 genomic DNA, and the vector carrying proper promoter and fluorescence protein respectively with Phusion DNA polymerase, make sure that the 15 nt overlapping sequences was added to the 5′ end of the designed primers to insert the PCR products into the aimed construct.
4. Incubate the linear vector and insertion fragments together with the In-Fusion enzyme at 50 °C for 1 h (*see* **Note 9**), then transform the mixture into DH5α bacteria to obtain the desired plasmid.
5. Standard microinjection protocol for *C. elegans* germ line transformation is applied to introduce the transgene containing plasmids, together with a coinjection marker plasmid, such as P*odr-1::dsRed*, *rol-6(+)* or *unc-76(+)* into N2 or *unc-76* animals. Usually 20–50 young adult animals are injected with 2–20 ng DNA constructs (*see* **Note 10**).
6. At least two independent transgenic lines with reasonable transmission rate should be examined to obtain quantitative or qualitative data.

Table 1
Useful promoters in Q cell migration studies

Promoter	Q Cell specificity	Expression initiation during Q cell development
P*egl-17*	All descendants	All stages
P*hsp-16.2*	All descendants	Induced by 33 °C heat shock
P*egl-13*	Q.a lineage	After Q.a divisions, increasing afterward
P*egl-46*	Q.a lineage and Q.paa	After the birth of Q.a and Q.paa cells
P*mig-13*	QR lineage	Shortly after QR divides, increasing afterward
P*gcy-32*	AQR/PQR	To the end of Q.ap migration
P*mec-4*	AVM/PVM	To the end of Q.paa/p migration

Table 2
Frequently used markers in Q cell migration studies

Marker	Subcellular structure labeled	Recommended fluorescence protein	Construct
His-72	Chromosome	mCherry	P*egl-17::his-72::mCherry::unc-54 3'UTR*
Myri-	Plasma membrane	mCherry	P*egl-17::Myri-mCherry::unc-54 3'UTR*
–	Cytosol	BFP	P*egl-17::bfp::unc-54 3'UTR*
Moesin	F-Actin	GFP	P*egl-17::moesin::gfp::unc-54 3'UTR*

3.4 Generation of Knockout or Knockin Animals

3.4.1 CRISPR-Cas9 Based Knockout and Conditional Knockout

1. Design two or more sgRNA sites (*see* **Note 11**) within the first 1–2 exon of the target gene with a CRISPR design tool (http://crispr.mit.edu). The sequences of potential target sites should conform to G(N)N19NGG (N = A, C, G, or T) (*see* **Note 12**).
2. In-Fusion cloning system is employed to construct the CRISPR-Cas9 plasmid containing: Cas9 expression driven by *eft-3* promoter for mutant generation or heat shock promoter/tissue-specific promoter for conditional knockout; and sgRNA expression driven by P*U6* promoter (Fig. 3).

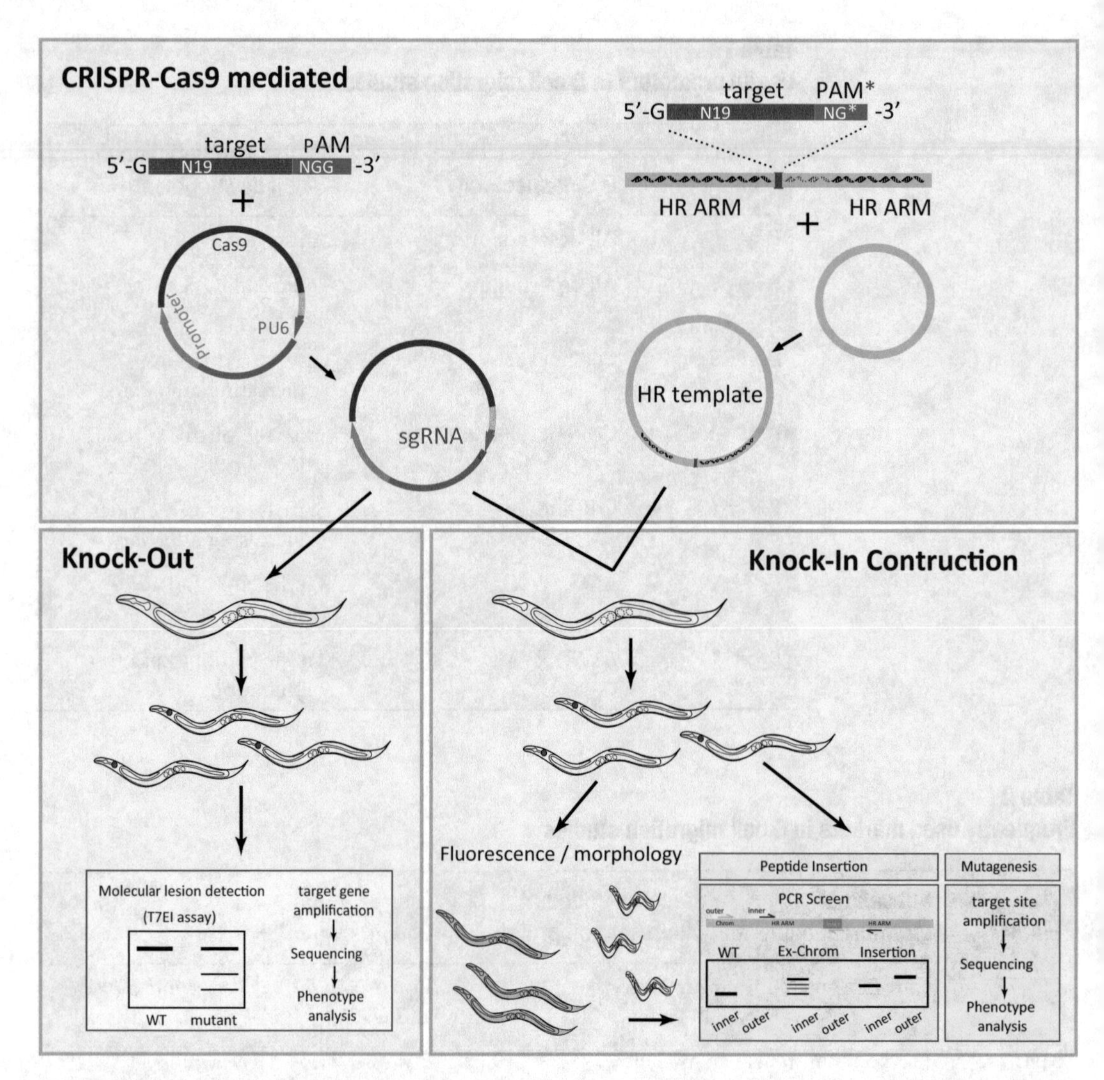

Fig. 3 Workflow of the CRISPR-Cas9 based protocol used to generate knockout or knockin animals. The CRISPR-Cas9 method includes the construction of a Cas9-sgRNA expression plasmid (Cas9 driven by P*eft-3*, P*hsp*, or any suitable promoter for conditional expression) and a plasmid carrying the homologous recombination template with a modified PAM site for knockin; the transgenesis by germ line injection; and the molecular and phenotypic screen of knockout or knockin animals. The red dot in the animal head represents the coinjected marker

3. Microinject the mixture of 10 ng CRISPR-Cas9 plasmid and 1 ng of a coinjection marker (e. g. rol-6(+) or Podr-1::dsRed) into the germ line of 50–100 young adult animals.
4. Multiple independent lines with the good transmission rate will be examined with the T7EI molecular assay (**steps 5**–7) to detect DNA lesions before phenotypic analyses.
5. From each individual line, 5–10 transgenic animals at the L4 stage with induced lesion are lysed to obtain genomic DNA,

from which a fragment flanking the sgRNA target site is amplified with Pfu DNA polymerase.

6. After purification, the PCR product is melted at 95 °C for 15 min and reannealed to form heteroduplexed DNA at room temperature, followed by digestion with T7 Endonuclease I (T7EI) at 37 °C for 30 min, to which the molecular lesion site is much more vulnerable than intact sequences.
7. The digested products are separated by 2% agarose gel electrophoresis for analysis, and ImageJ is used to measure the relative intensities of DNA bands. Indel is quantified by the formula $100 \times (1 - (b + c)/(a + b + c)1/2)$, in which a corresponds to the intact band intensity and b and c present the overall intensities of the digested small bands.
8. In the case of germ cell mutagenesis, the PCR fragment is further subjected to Sanger sequencing to identify lesion types.
9. In the case of conditional somatic knockouts, the PCR fragment is cloned into a pEASY-T1 Cloning Vector for Sanger sequencing.
10. Other than Q cell migration phenotypes, lethality phenotype is quantified for the lethal-gene conditional knockout (P*hsp*) animals: 4 h after being laid, 100 eggs are subjected to the 33 °C heat-shock treatment for 1 h and then culture at 20 °C until the L3 larval stage. The lethality is calculated as 100 minus the number of hatched worms (Fig. 3).

3.4.2 CRISPR-Cas9 Based Knockin to Generate In Situ Label or Designed Mutations

1. Design sgRNA site near the start codon to fuse tag to the N-terminus, the stop codon to fuse tag to the C-terminus, or at the desired mutagenesis site. Construct the CRISPR-Cas9 plasmid as described in (Subheading 3.4.1).
2. Construct the knockin homogenous recombination (HR) template plasmid containing the insertion fragment as follows: 1.5 kb HR arms upstream of the knockin site + tag in frame or mutagenesis sequence +1.5 kb HR arms downstream of the knockin site. For the site-directed mutagenesis, modify the template DNA sequence without altering the protein sequence to introduce a restriction enzyme recognition site in the template as a convenient way to identify mutants.
3. Microinject the mixture of 10 ng CRISPR-Cas9 plasmid, HR template plasmid and 1 ng of a coinjection marker (e. g. rol-6(+) or Podr-1::dsRed) into 50–100 young adult animals.
4. Select single F1s expressing coinjection marker.
5. To screen for successful genomic editing progenies, any predicted visible fluorescence, morphology or other phenotype in single F2s can be used for an initial candidate selection, but a molecular assay must be performed.

6. In the case of point-directed mutagenesis, lyse F2s without coinjection marker, design "outer" primer pairs (Fig. 3) to amplify a fragment covering the mutagenesis site, part of HR arms and some flanking sequence in the chromosome with Pfu DNA polymerase. Digest the PCR product with the enzyme whose recognition site is introduced to the HR template, pick out the samples that are susceptible to the enzyme treatment and then sequence the target site to confirm positive F2s.
7. In the case insertion or deletion, design two pairs of primers, inner and outer as illustrated in Fig. 3, to screen for positive F2s.

3.5 Live-Cell Imaging of Migrating Q Cells and Analyses

3.5.1 Live-Cell Imaging of Migrating Q Cells

1. 12–16 h before imaging (*see* **Note 13**), put 60–100 healthy young adult *C. elegans* (*see* **Note 14**) on an imaging NGM plate, allow them to lay eggs for 2 h then remove adult animals to get synchronized progenies.
2. Culture the synchronized eggs and hatched L1 at 20 °C.
3. Make fresh 3% "sandwich" agarose pads (Material) just before the imaging.
4. Keep the imaging room temperature at 20 °C, unless the worms are temperature-sensitive mutants.
5. To set up an imaging sample, drop 1 μL M9 buffer at the center of a microwell dish, transfer 20–30 synchronized L1 larvae into the droplet (*see* **Note 15**, Fig. 4).
6. Add 1 μL of 1 mg/mL levamisole solution in separate drops surrounding the animal containing puddle (Fig. 4).
7. Take one "sandwich" agarose pad, open it by removing either of the glass coverslips, and dry the agarose in the air for 10–30 s.
8. Gently cover the sample and levamisole droplets with the pad, make sure the agarose gel facing sample (*see* **Note 16**, Fig. 4).
9. Put wet tissue paper inside the microwell dish to keep sample moist, then cover the dish with a lid.
10. Secure the sample dish on the motorized XY stage of an inverted microscope (*see* **Note 17**), which is equipped with acquisition devices and software.
11. Under the 100× oil objective, look for L1 larva at desired developmental stages for imaging, mark their positions in the acquisition software.
12. Set the time-lapse interval to be 1 min, start the acquisition and keep the sample focused well throughout the recording (*see* **Note 18**).

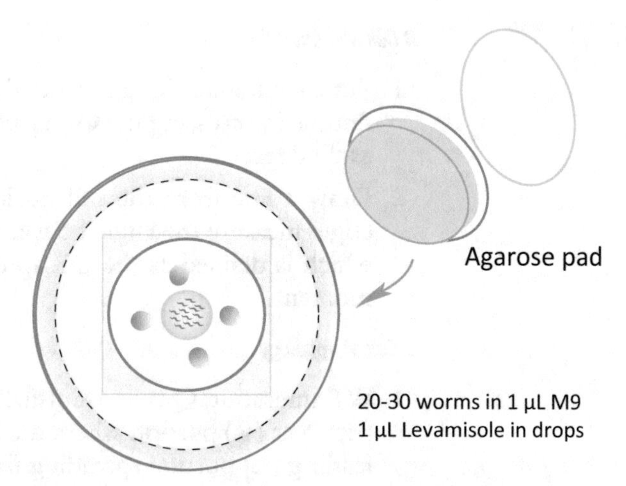

Fig. 4 Sample preparation of *C. elegans* larvae for live imaging. Pipette M9 buffer in the center of the glass bottom of a microwell dish, and transfer *C. elegans* larvae into the solution. Pipette 1 mg/mL levamisole in M9 buffer into drops, and place the agarose pad on top with agarose facing the sample

3.5.2 Data Analyses to Characterize Q Cell Migration

Explanations are provided on the example of QR.ap cell migration.

Raw Data Preparation

1. For manual live-cell imaging analyses, at least ten movies in high quality (well-focused, showing integral lamellipodium and cell body morphology from a top view, and no overexposure in any channel) are needed.
2. To ensure an unbiased analysis, correct genotypes and reliable movie quality should be absolute criteria to pool raw movies.
3. From each raw movie, crop a 40 min QR.ap chip with continued cell migration and no visible obstacle (*see* **Note 19**) in the path, usually starting from a time point of 30–45 min after QR.ap birth.
4. Put all the cropped chips in a folder to generate the movie set for the analysis.
5. For a given genotype, all kinetic parameters should be obtained from the same set of movie chips.
6. For each chip, data will be collected every 10 min. In other words, the 1st, 11th, 21th, 31th, and 41th frame (*see* **Note 20**) will serve as the data-gathering time points for data analyses.

Q Cell Migration Speed

1. Choose a reference point within the animal (*see* **Note 21**) which is relatively stationary throughout the movie.
2. Measure the distance from QR.ap to the reference point along animal body midline at data-gathering time points, calculate the relative movement during each 10 min interval, and obtain the mean velocity.

Q Cell Migration Direction

1. For each frame, along the anterior–posterior axis of the animal draw a line crossing the QR.ap cell and pointing to the anterior as "0 degree."
2. Draw a line from the cell nucleus to the middle of leading-edge, measure the angle between this line and the "0 degree," which is defined as the QR.ap cell migration direction at the moment.

Q Cell Morphology

1. WT migrating Q cells are usually composed of a round compact rear (R) portion where nucleus locates and a fan-like thin leading (L) portion spreading from the front of the nucleus.
2. The area ratio between the two portions can be measured and compared between genotypes, as long as the cells are imaged with top-view.
3. In general, the edge of either potion is relative smooth with small ruffles. However, filopodia-like growth or cracks in the edge can compromise cell mobility and therefore should be characterized.

Q Cell Polarity

1. In all data-gathering time points, identify the QR.ap cell for analysis without overlap onto adjacent cells.
2. If the cellular boundary is not whole due to single focal plane, try to integrate signal from multiple Z stacks into one imaging with the "Z-project" function in ImageJ.
3. In ImageJ, choose the fluorescence channel that is assigned for the F-actin marker acquisition, set the measurement parameter to be "mean gray value," and get a background reading (B) from any blank area in the imaging.
4. Draw segment lines along the plasma-membrane of the leading (L) and rear (R) portion (*see* **Note 22**) of QR.ap cell respectively, and then measure the fluorescence intensities of both lines.
5. Calculate the cortical F-actin fluorescent intensity ratio as $(L - B)/(R - B)$, which could be used as an indicator of the QR.ap cell polarity at the corresponding time point.
6. Data from all five time points can be averaged to represent the general QR.ap cell polarity of the movie.

Subcellular Distributions of Target Protein

1. For a cortically enriched target protein, such as integrin, the mean intensity along the plasma membrane over that within the cytosol can be characterized. When the mean intensity in cytosol is measured, the nucleus area should be excluded.

2. If a polarized subcellular distribution is observed, a similar strategy as the one described in the "Q cell polarity" section can be used to quantify the polarity of the target protein.
3. Some target proteins may be present in the cell with a particular pattern, in puncta for instance. This can be characterized in several ways. For example, the puncta number in a given area can be counted with the "Analyze Particles" function of ImageJ. The 2D intensity distribution of the particles can be plotted with the "Surface Plot" function in ImageJ. And the 1D "Plot Profile" function can be used to quantify the portion occupied by puncta along the leading edge or the overlap between different molecule puncta (*see* **Note 23**).

4 Notes

1. Animals for analysis should be collected without visualization of their fluorescence, otherwise the analysis might be biased.
2. Animals for assessment should be restricted to the L4 stage, since more and more bright puncta develop along neurites in aging animals.
3. The direct contact with levamisole solution will cause animal to curl, which is not favored for Q cell position measurement.
4. Acquisition with z-stacks to display neuronal process morphology can help to identify AVM from nearby ALMs.
5. The single neurite from ALMR/L grows straight toward the animal head, while the anterior pointing neurites of AVM and PVM usually project to the ventral side of body before turn to the anterior direction (Fig. 2a).
6. The strain used for genetic screen can be WT or mutant. Enhancers or suppressors of a particular mutant can be identified.
7. For the genes with multiple isoforms, the promoter length can vary among isoforms.
8. Theoretically the fluorescence protein can be fused to either N- terminus or C-terminus of the target protein. However, structural and functional aspects should be carefully considered for each protein.
9. To reduce contaminations from template vector residues, Dpn-1 digestion, which selectively removes the methylated template DNA, is frequently performed.
10. To minimize potential artifacts caused by overexpression, the amount of injected plasmid should be the lowest as possible, i.e., below which the fluorescence cannot be detected.

11. The comparison of efficiency and of phenotypes between mutants obtained using different sgRNAs helps to reveal potential off-target effects and to identify for the most efficient sgRNA.
12. The initial G optimizes the transcription process, and the NGG (PAM) is essential for Cas9 activity.
13. The synchronization is 12 h before imaging for QR/L migration imaging, and 16 h before imaging for imaging QR/L.a p migration.
14. For transgenic strains, the 100 animals should all carry extrachromosomal arrays.
15. Carry as little OP50 bacteria as possible to reduce their oxygen consumption in the restricted space and thus increase larva survival rate during imaging acquisition.
16. The M9 droplet containing animals and levamisole will be pushed to mix by the weight of agarose pad. The animals should be anesthetized within 1 min.
17. The speed of the automated stage should be set as intermediate, as too fast a speed can impede the focus in time-lapse imaging.
18. The laser intensity should be set as low as possible to reduce photobleaching and phototoxicity.
19. When QR.ap starts to migrate, it makes a detour to its apoptotic sister QR.aa that is born in the path. It usually takes 30–40 min for QR.ap to pass QR.aa.
20. The autofluorescent particles in the gut system can use as reference points.
21. The first frame is referred to as 0 min.
22. When drawing the line, the image should be zoomed in as much as possible to increase the accuracy.
23. The baseline that is used to define puncta from the fluorescence intensity plot should be reasonably set and fixed throughout the whole study.

Acknowledgements

This study was supported by the National Natural Science Foundation of China to G.O. and Y.C. (31671451, 31525015, 31561130153, and 31190063), the Junior Thousand Talents Program of China to G.O., and the National Basic Research Program of China to G.O. (973 Program, 2013CB945600).

References

1. Sulston JE, Horvitz HR (1977) Post-embryonic cell lineages of the nematode, Caenorhabditis elegans. Dev Biol 56(1):110–156
2. Chalfie M, Sulston J (1981) Developmental genetics of the mechanosensory neurons of Caenorhabditis elegans. Dev Biol 82(2):358–370
3. Ou G, Vale RD (2009) Molecular signatures of cell migration in C. elegans Q neuroblasts. J Cell Biol 185(1):77–85. https://doi.org/10.1083/jcb.200812077
4. Chapman JO, Li H, Lundquist EA (2008) The MIG-15 NIK kinase acts cell-autonomously in neuroblast polarization and migration in C. elegans. Dev Biol 324(2):245–257. https://doi.org/10.1016/j.ydbio.2008.09.014
5. Zinovyeva AY, Yamamoto Y, Sawa H, Forrester WC (2008) Complex network of Wnt signaling regulates neuronal migrations during Caenorhabditis elegans development. Genetics 179(3):1357–1371. https://doi.org/10.1534/genetics.108.090290
6. Dyer JO, Demarco RS, Lundquist EA (2010) Distinct roles of Rac GTPases and the UNC-73/Trio and PIX-1 Rac GTP exchange factors in neuroblast protrusion and migration in C. elegans. Small GTPases 1(1):44–61. https://doi.org/10.4161/sgtp.1.1.12991
7. Sundararajan L, Lundquist EA (2012) Transmembrane proteins UNC-40/DCC, PTP-3/LAR, and MIG-21 control anterior-posterior neuroblast migration with left-right functional asymmetry in Caenorhabditis elegans. Genetics 192(4):1373–1388. https://doi.org/10.1534/genetics.112.145706
8. Harris J, Honigberg L, Robinson N, Kenyon C (1996) Neuronal cell migration in C. elegans: regulation of Hox gene expression and cell position. Development 122(10):3117–3131
9. Maloof JN, Whangbo J, Harris JM, Jongeward GD, Kenyon C (1999) A Wnt signaling pathway controls hox gene expression and neuroblast migration in C. elegans. Development 126(1):37–49
10. Honigberg L, Kenyon C (2000) Establishment of left/right asymmetry in neuroblast migration by UNC-40/DCC, UNC-73/Trio and DPY-19 proteins in C. elegans. Development 127(21):4655–4668
11. Middelkoop TC, Korswagen HC (2014) Development and migration of the C. elegans Q neuroblasts and their descendants. WormBook:1–23. https://doi.org/10.1895/wormbook.1.173.1
12. Cheng Z, Yi P, Wang X, Chai Y, Feng G, Yang Y, Liang X, Zhu Z, Li W, Ou G (2013) Conditional targeted genome editing using somatically expressed TALENs in C. elegans. Nat Biotechnol 31(10):934–937. https://doi.org/10.1038/nbt.2674
13. Shen Z, Zhang X, Chai Y, Zhu Z, Yi P, Feng G, Li W, Ou G (2014) Conditional knockouts generated by engineered CRISPR-Cas9 endonuclease reveal the roles of coronin in C. elegans neural development. Dev Cell 30(5):625–636. https://doi.org/10.1016/j.devcel.2014.07.017
14. Chai Y, Li W, Feng G, Yang Y, Wang X, Ou G (2012) Live imaging of cellular dynamics during Caenorhabditis elegans postembryonic development. Nat Protoc 7(12):2090–2102. https://doi.org/10.1038/nprot.2012.128
15. Tian D, Diao M, Jiang Y, Sun L, Zhang Y, Chen Z, Huang S, Ou G (2015) Anillin regulates neuronal migration and neurite growth by linking RhoG to the actin cytoskeleton. Curr Biol 25(9):1135–1145. https://doi.org/10.1016/j.cub.2015.02.072
16. Zhu Z, Chai Y, Jiang Y, Li W, Hu H, Li W, Wu JW, Wang ZX, Huang S, Ou G (2016) Functional coordination of WAVE and WASP in C. elegans neuroblast migration. Dev Cell 39(2):224–238. https://doi.org/10.1016/j.devcel.2016.09.029

Check for updates

Chapter 19

Imaging the Molecular Machines That Power Cell Migration

Anika Steffen, Frieda Kage, and Klemens Rottner

Abstract

Animal cell migration constitutes a complex process involving a multitude of forces generated and maintained by the actin cytoskeleton. Dynamic changes of the cell surface, for instance to effect cell edge protrusion, are at the core of initiating migratory processes, both in tissue culture models and whole animals. Here we sketch different aspects of imaging representative molecular constituents in such actin-driven processes, which power and regulate the polymerisation of actin filaments into bundles and networks, constituting the building blocks of such protrusions. The examples presented illustrate both the diversity of subcellular distributions of distinct molecular components, according to their function, and the complexity of dynamic changes in protrusion size, shape, and/or orientation in 3D. Considering these dynamics helps mechanistically connecting subcellular distributions of molecular machines driving protrusion and migration with their biochemical function.

Key words Live cell imaging, Actin, Migration, WAVE complex, VASP, FMNL, Fluorescence imaging

1 Introduction

Over the years, we and others have accumulated evidence that the precise determination of localization and turnover of the molecular machines regulating cell migration will provide different aspects of insight. The technology employed will on the one hand provide information concerning the precise functions of the molecular players involved, but on the other hand also inform about the dynamics and complexities of protrusion behavior, teaching us about mechanisms and mechanics of migratory processes. The most prominent types of protrusions formed in cells adhering to and migrating on rigid, flat surfaces or along helical fibers built by extracellular matrix components are sheet- and

Electronic Supplementary Material: The online version of this chapter (https://doi.org/10.1007/978-1-4939-7701-7_19) contains supplementary material, which is available to authorized users.

Alexis Gautreau (ed.), *Cell Migration: Methods and Protocols*, Methods in Molecular Biology, vol. 1749,
https://doi.org/10.1007/978-1-4939-7701-7_19, © Springer Science+Business Media, LLC 2018

finger-like protrusions termed lamellipodia and filopodia, respectively [1–4]. Membrane blebbing [5], another protrusion mechanism gaining increasing attention for the migration in soft environments, as exemplified for instance by zebrafish primordial germ cells (PGCs) [6, 7], will not be covered in the examples shown here. Instead, here we focus on cells cultured on two-dimensional surfaces, the migration of which is accompanied by the formation of lamellipodia and filopodia, and which can be simply imaged with conventional live-cell imaging microscopy involving both phase-contrast and fluorescence optics. Due to the flatness of protrusions imaged during these migration events (with a lamellipodium rarely exceeding 150 nm in thickness), neither confocal microscopy nor any other method of contrast enhancement in the *Z*-direction such as TIRF (total interference reflection fluorescence) microscopy is required to reach the conclusions on subcellular distribution and dynamics of constituents of migration and actin polymerization machineries presented here. In spite of the overwhelming progress made in the discovery and/or further development of genetically encoded fluorescent proteins [8, 9] as well as on advanced imaging approaches including superresolution [10–12], allowing to extend our opportunities when dissecting protein function at the subcellular level, we here focus on simple general considerations concerning cell protrusion behavior and the dynamics of fluorescent protein-tagged actin regulatory factors using commercially available equipment available in virtually every cell biology laboratory. For instance, using fluorescently tagged variants of the protrusion marker VASP (Vasodilator-stimulated phosphoprotein, see Figs. 1 and 2; Movies 1 and 2) and the FMNL subfamily formin FMNL2 [13–16] (Fig. 3; Movies 3 and 4), both of which constitute actin assembly factors in vitro [13, 14, 17, 18], we show how we can image and control the forward protrusion of lamellipodia and filopodia versus their retraction, and how this correlates with the accumulation at the edges or tips of lamellipodia and filopodia, respectively. Using FMNL2 as example, we also illustrate how the choice of fluorescent protein tagging to the N- versus C-terminus of the protein of interest can modify the conclusion as to whether it localizes to lamellipodia or not (Fig. 3; Movies 3 and 4). In addition, using heterodimeric capping protein, we provide an example of a factor often displaying a partly distinct, i.e., more homogeneously distributed lamellipodial accumulation [19, 20] (Fig. 4 and Movie 5), likely explained last not least due to its highly distinct biochemical function [21]. We note though that we and others have previously also found instances of capping protein more concentrating toward the protruding edges of lamellipodia, depending perhaps on cell type and conditions [19, 22, 23], and also constituting an instance of comparably high variability concerning its precise localization and dynamics in these structures. Moreover, due to its prominent function as sub-

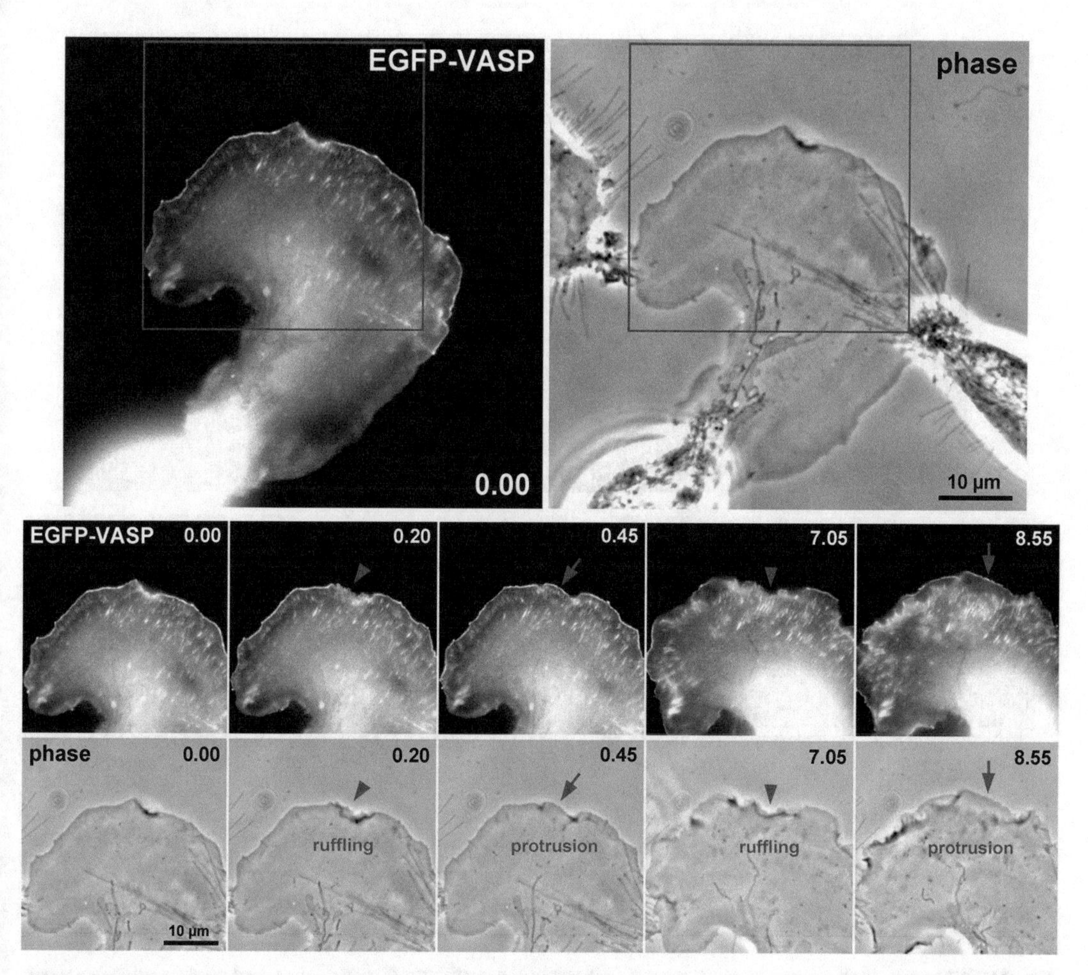

Fig. 1 EGFP-VASP specifically localizes to protruding lamellipodia tips. Top panel depicts epifluorescence and phase-contrast images of a B16-F1 cell transiently expressing EGFP-tagged VASP, as indicated. Note that EGFP-VASP localizes to lamellipodia tips as well as to focal adhesions within the cell. Red rectangles mark inset regions that are displayed in a time-lapse panel below. Bottom panels show phases of lamellipodial ruffling/backward-folding (red arrowheads, at some time points accompanied by retraction, see also Movie 1) and protrusion, as highlighted by red arrows. Note that at the cell edge, VASP localizes exclusively to protruding lamellipodia (red arrows), as described previously [16, 20], whereas it disappears from the focal plane upon ruffling and/or backward folding of the lamellipodium, or it delocalizes upon lamellipodial retraction. Time is in minutes and seconds

unit of the so-called ubiquitous WAVE regulatory complex (WRC) [1, 24, 25], required for activation of and hence actin filament generation and branching by Arp2/3 complex in lamellipodia [26–29], we are providing representative examples for subcellular dynamics during protrusion of the Abelson (Abl) interactor protein (Abi), which constituted the first protein indeed observed to be truly restricted to protruding lamellipodia and filopodia tips [30] (Figs. 5 and 6; and Movie 6), while absent for instance from focal adhesions, in contrast to VASP [20, 31–33] (Figs. 1 and 2, see also Movies 1 and 2). Finally, aside from distinguishing "non-specific" fluorescent protein accumulation in

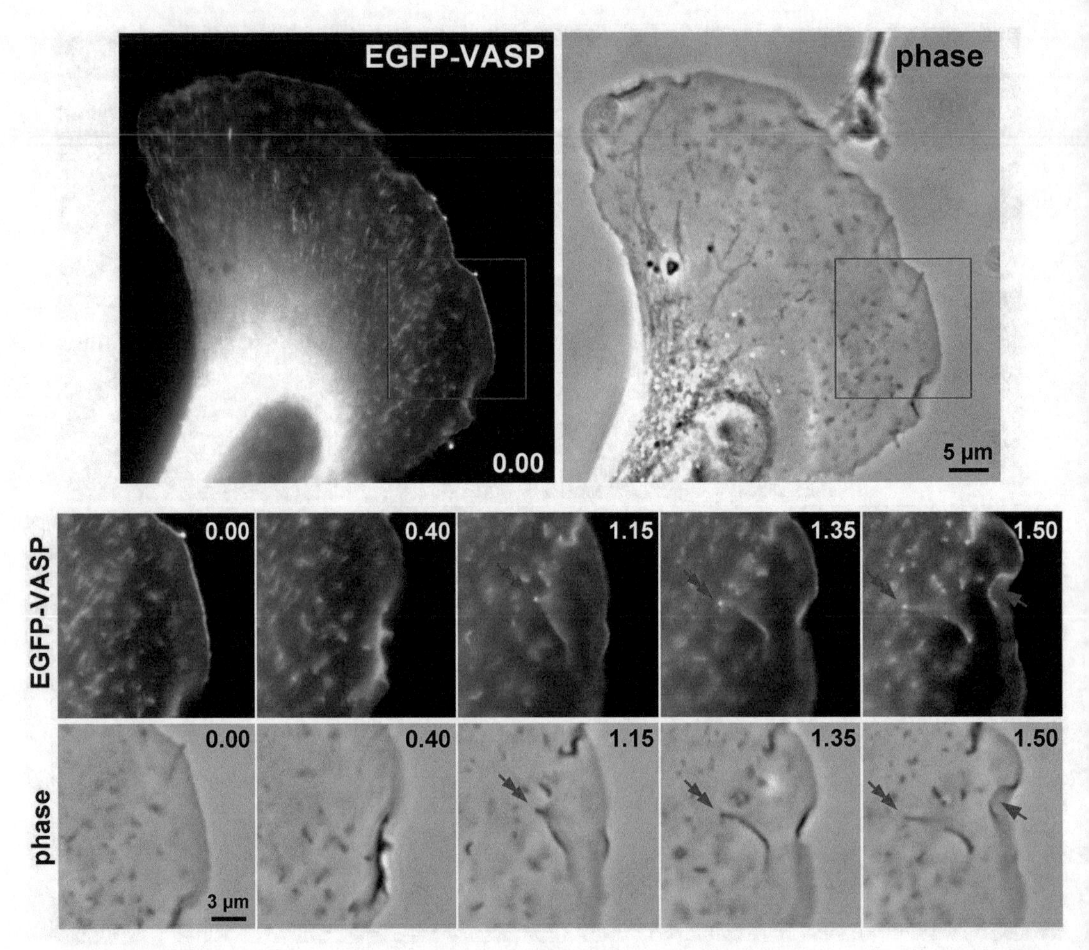

Fig. 2 VASP accumulation in a filopodium extending from a ruffle. Epifluorescence and corresponding phase-contrast images of a B16-F1 mouse melanoma cell transiently transfected with EGFP-VASP. Red rectangles depict inset regions that are shown in higher magnification in bottom panels. Bottom: Images show complexity of EGFP-VASP accumulation at the cell periphery with respect to lamellipodial and ruffling activity. In this example, a filopodium emerges from a lamellipodium after folding backward onto the dorsal cell membrane, and continues to protrude from the cell periphery, but now in the rearward direction (see double-headed arrowheads at time 1.15). In phase-contrast panels, the rearward-moving filopodium can be seen as a dark stick, tipped by the EGFP-VASP signal in fluorescence. Red arrow marks the tip of the lamellipodium after folding rearward onto the dorsal plasma membrane. Note that this tip localization can only be captured in these conditions of focusing onto the coverslip surface if the time-point of image acquisition precisely coincides with the backward folding of the lamellipodium. Time is in minutes and seconds

protrusions, as exemplified by the dynamics of expressed EGFP alone (Figs. 7 and 8; Movie 7), we compare two alternative cell model systems used in the past in our laboratories to study lamellipodia protrusion: the highly motile B16-F1 melanoma cells spreading and migrating nicely on laminin and first introduced as system for studying actin dynamics in vivo almost 20 years ago [34], and secondly, fibroblasts growing on fibronectin and transiently transfected with constitutively active Rac1. Fibroblasts treated in that manner, albeit nonmotile per definition due to the

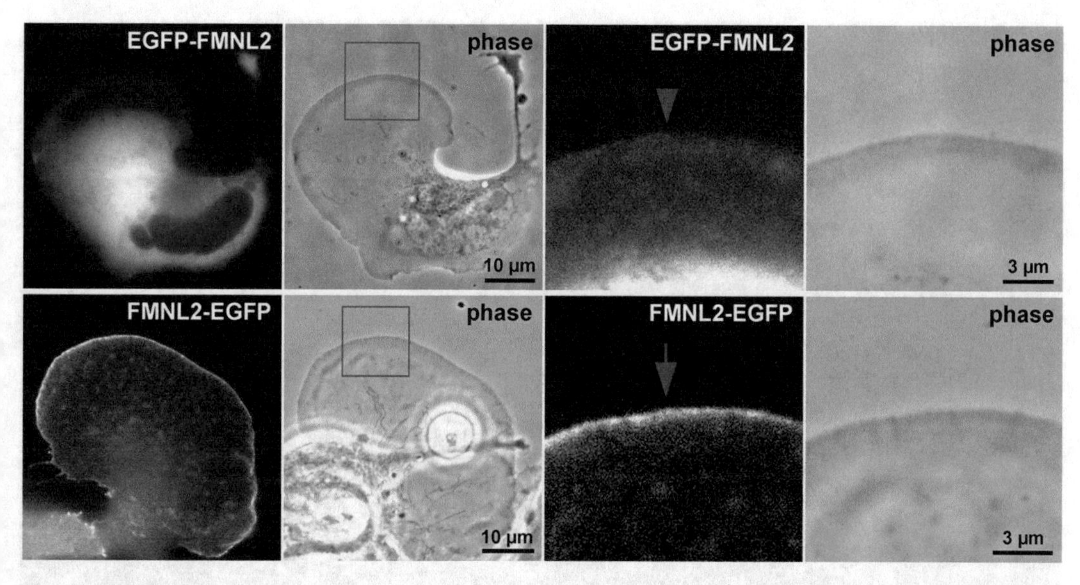

Fig. 3 Orientation of EGFP-tagging to full-length FMNL2 is decisive for proper subcellular localization. Epifluorescence and corresponding phase-contrast images of B16-F1 mouse melanoma cells transiently expressing EGFP-FMNL2 (top panel) or FMNL2-EGFP (bottom panel). Red rectangles mark regions that are shown in higher magnification on the right. Red arrow and arrowhead mark presence and absence from the tip region of the lamellipodium, respectively, in magnified insets on the right. Note prominent accumulation of this member of the formin family at the lamellipodium tip when tagged to the C-terminus (bottom), but not to the N-terminus (top). In the latter case, the formin appears mostly cytosolic, which can also be appreciated from its clear exclusion from the nucleus and from mitochondria appearing as sausage-like, dark stripes in the overview fluorescence image at the top left. As described previously [13], N-terminal EGFP tagging blocks N-terminal myristoylation of members of this formin family. Myristoylation contributes to form in activation and thus releases this formin member from autoinhibition in these conditions. In contrast, fusion of EGFP to the C-terminus of the formin enables N-terminal myristoylation, and hence activation of the formin driving its lamellipodial localization (bottom). Note though that myristoylation *per se* is not required for lamellipodial targeting, as the targeting domains are located in the center of the molecule [54] and release of formin autoinhibition by other means, such as removal of a C-terminal domain (DAD) mediating autoinhibition (not shown) will lead to indistinguishable lamellipodial accumulation [13, 14]

stimulation of lamellipodia protrusion all around the cell periphery [35, 36], can be micromanipulated in a much more flexible manner, for instance by microinjection, which we have previously employed to specifically but abruptly explore the effects on protrusion of instantly inhibiting the Arp2/3 complex [37, 38].

2 Materials

2.1 Preparation of Coverslips

1. Wash 15 mm-diameter coverslips No. 1.5 (0.17 mm thick) for 30 min in 60/40 Ethanol p.a.–HCl p.a. (37%) solution by gentle agitation.
2. Wash coverslips extensively with at least ten times 100 mL ddH_2O. Ensure that no Ethanol–HCl solution is left on the coverslips.

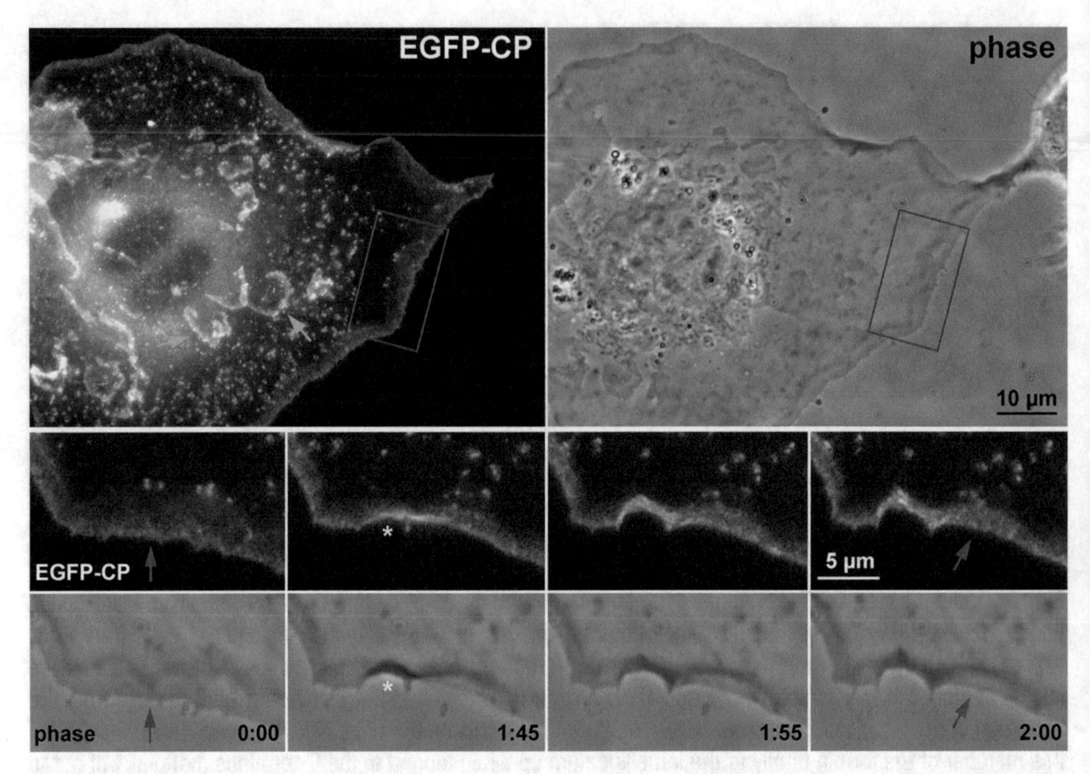

Fig. 4 Heterodimeric capping protein (CP) is localized to lamellipodia, ruffles, and endosomal structures. Top panel shows fluorescent (left) and phase-contrast (right) images of a time-lapse movie of a NIH3T3 cell expressing the beta-subunit of heterodimeric capping protein [55] fused to EGFP (EGFP-CP). Note that this cell is coexpressing Rac1-L61, resulting in a classical pancake-shape of the cell, and, furthermore, inducing multiple lamellipodia below the ventral side of the cell (green arrows; see also Fig. 6 illustrating how such structures are formed). Lower two panels show details of the time lapse of the region indicated in the top. Middle panels show EGFP-CP localization, bottom panels show corresponding phase-contrast images. Elapsed time is given in minutes and seconds. Red arrows point toward CP in the lamellipodium, with the highest intensity at the tip. Little dots might represent mini-ruffles at the ventral side of the plasma membrane and endosomal localization of CP, likely associated with the Arp2/3 activator WASH regulatory complex [56–58] (see also Movie 5). Yellow asterisks indicate plasma membrane ruffles, as can be appreciated in phase-contrast images. Note that in the corresponding fluorescent channel, EGFP-CP seems to localize in a thin line, indicative of a lamellipodium tip localization (compare also with EGFP-Abi1 in Fig. 5), which in this case, however, is due to the increase of specific localization derived from the ruffling lamellipodium

3. Separate coverslips with flat forceps (Dumont forceps, No. 5 or No. 7) and let them dry on a Whatman paper. Collect coverslips in a glass petri dish and heat sterilize in a dry heat sterilizer at 160 °C for 2 h.

2.2 Preparation of Media

1. Growth medium for NIH3T3 fibroblasts: add 50 mL FBS, 5 mL glutamine (200 mM), 5 mL sodium pyruvate (100 mM), and 5 mL MEM nonessential amino acids solution (100×) to 500 mL Dulbecco's Modified Eagle Medium (DMEM) containing 4.5 g/L glucose, and sterile filter medium (using sterile filters with 0.22 µm pore size).

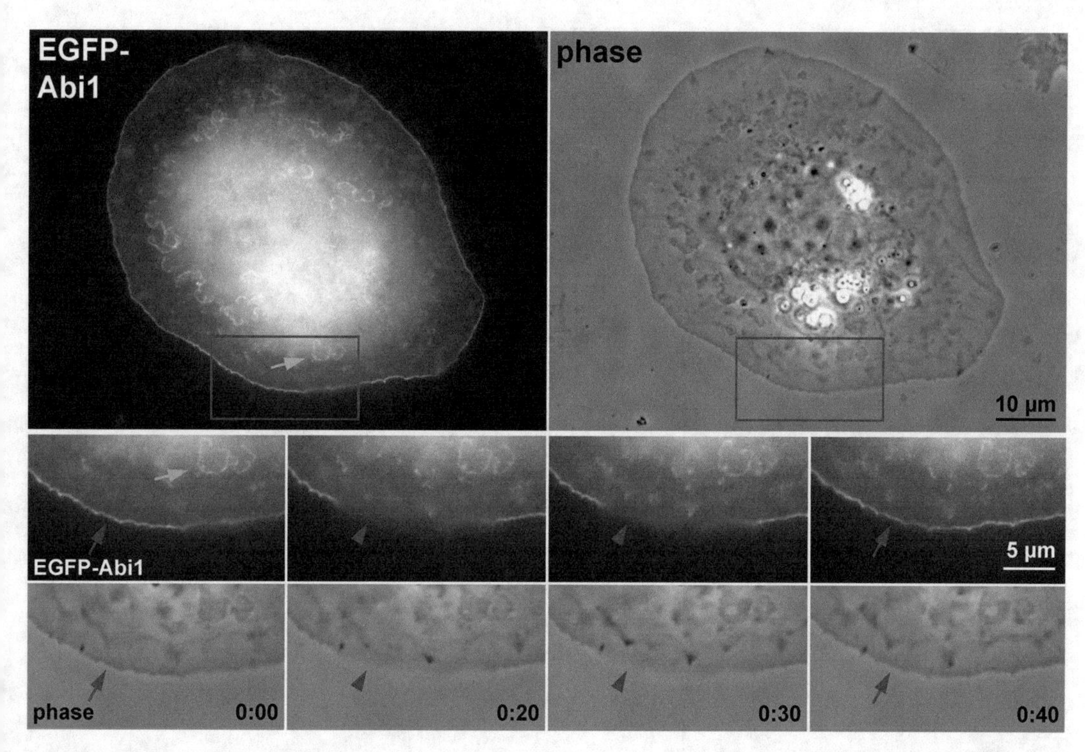

Fig. 5 Localization of a lamellipodial tip component, EGFP-Abi1. Top panel shows fluorescent (left) and phase-contrast (right) images of a time-lapse movie of a NIH3T3 cell expressing EGFP-Abi1. Note that this cell is coexpressing Rac1-L61, resulting in a classical pancake-shape of the cell, and, moreover, inducing multiple lamellipodia below the ventral side of the cell (green arrow; *see* also Fig. 6 for a scheme of how such structures are formed). Lower two panels show selected time-lapse frames of the region indicated in the top. Middle panels show EGFP-Abi1 localization, bottom panels show corresponding phase-contrast images. Elapsed time is given in minutes and seconds. Red arrows point toward Abi1 specifically accumulating at the lamellipodium tip [30], as has been shown as well for other WAVE complex components [28]. Note that Abi1 seems to locally disappear (red arrowheads) at 0:20 and 0:30 time points, which can be explained, however, by movement of the signal out of the focal plane, because the lamellipodium is lifting up from the substratum, as can also be appreciated from the blurred appearance of the periphery in phase-contrast images. At the same time, the Abi1 localization at ventral lamellipodia below the main cell body stays in the focal plane, confirming that these structures occur in the same focal plane as peripheral lamellipodia. *See* also Movie 6

2. Microscopy medium for NIH3T3 fibroblasts: add 50 mL FBS, 5 mL glutamine (200 mM), 5 mL sodium pyruvate (100 mM), 5 mL nonessential amino acids (100×), and 5 mL penicillin–streptomycin solution (10,000 U/mL) to 500 mL Nutrient mixture F-12 HAM, and sterile filter medium (using sterile filters with 0.22 μm pore size). It is important that the medium contains 25 mM HEPES, which buffers the pH value, thus allowing to work in the absence of a CO_2 atmosphere. This particular medium also lacks phenol red, which helps to avoid background fluorescence.

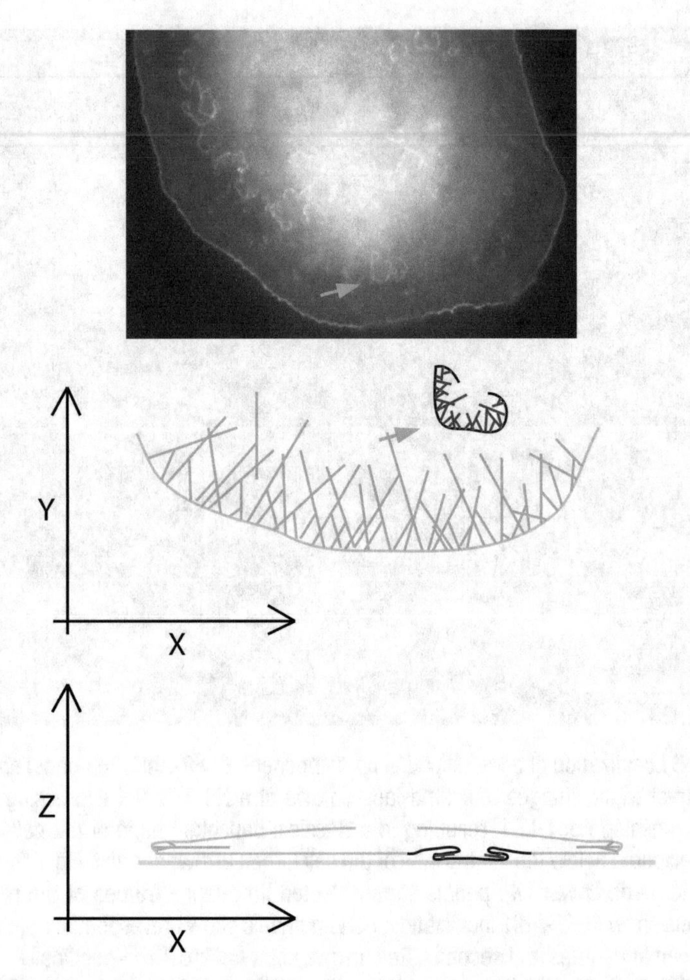

Fig. 6 Scheme illustrating the three dimensional appearance of lamellipodia at the ventral plasma membrane. Top image shows a NIH3T3 cell expressing EGFP-Abi1 and constitutively active Rac1 (*see* also Fig. 5). The green arrow highlights a ventral lamellipodium, which is frequently observed if constitutively active Rac is overexpressed. Middle panel illustrates the top view (*X*, *Y* coordinates) of a cell as exemplified in the top image. Actin filaments are illustrated as red lines, the plasma membrane is shown in black lines. The red and black transparent lines indicate the periphery of the cell, while red and black solid lines show a ventral lamellipodium appearing below the cell. The bottom panel shows the side view (*X*, *Z* coordinates) of the same cell with lines as depicted above. In addition, the bottom-limiting coverslip is drawn as grey bar. Note that this lamellipodium, as identified by the presence of protrusion-relevant proteins such as Abi family members and capping protein (*see* also Figs. 4 and 5), is a continuum of the plasma membrane evolving at the ventral plasma membrane

3. Growth medium for B16-F1 mouse melanoma cells: add 50 mL FBS (low endotoxin) and 5 mL glutamine (200 mM) to 500 mL Dulbecco's Modified Eagle Medium (DMEM) containing 4.5 g/L glucose, and sterile filter medium (using sterile filters with 0.22 μm pore size).

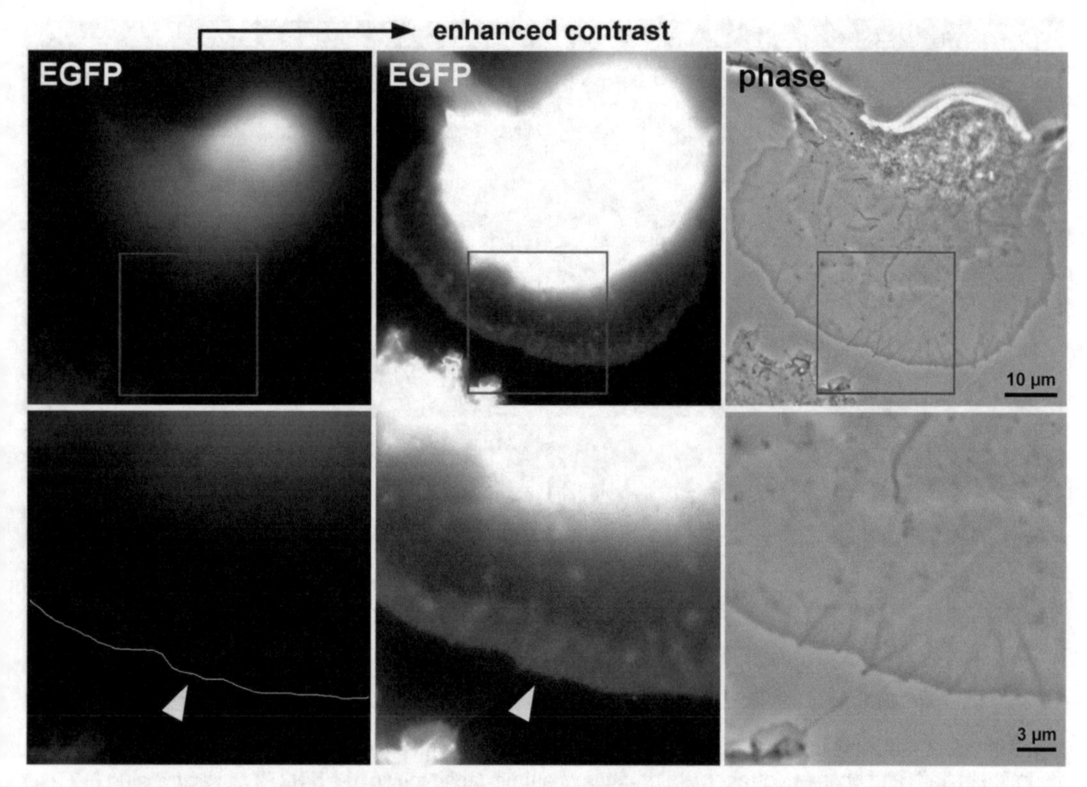

Fig. 7 Inappropriate contrast enhancement can cause erroneous lamellipodial EGFP-accumulation. B16-F1 mouse melanoma cell that was transiently transfected with EGFP and shown by epifluorescence and phase-contrast optics, as indicated. Bottom panel depicts zoom-images marked by red rectangles at the top. Modest image enhancement as shown on the left reveals strong EGFP-signal in and around the nucleus, as expected, but little fluorescence at the cell periphery. Note, however, that strong contrast enhancement as exemplified in the middle panel results in the erroneous impression of specific EGFP accumulation throughout the lamellipodium network (yellow arrowhead), which is not specific. The precise reason for the apparent increase in fluorescence intensity throughout the lamellipodium relative to the lamella region behind is not entirely clear, but it is reasonable to assume that the highly dense and branched lamellipodial actin network might constitute a certain diffusion barrier as compared to the rest of the cytoplasm and/or retain EGFP molecules in this region through molecular crowding

4. Microscopy medium for B16-F1 cells: add 50 mL FBS (low endotoxin), 5 mL glutamine (200 mM), and 5 mL penicillin–streptomycin solution (10,000 U/mL) to 500 mL Nutrient mixture F-12 HAM, and sterile filter medium (using sterile filters with 0.22 μm pore size). See also notes on F12-Ham medium in **item 2** above.

2.3 Cell Culture and Transfection Reagents

1. Autoclaved 1.5 mL microtubes.
2. Sterile transfer pipets.
3. Sterile 1× PBS pH 7.4.
4. 0.05% trypsin–EDTA.

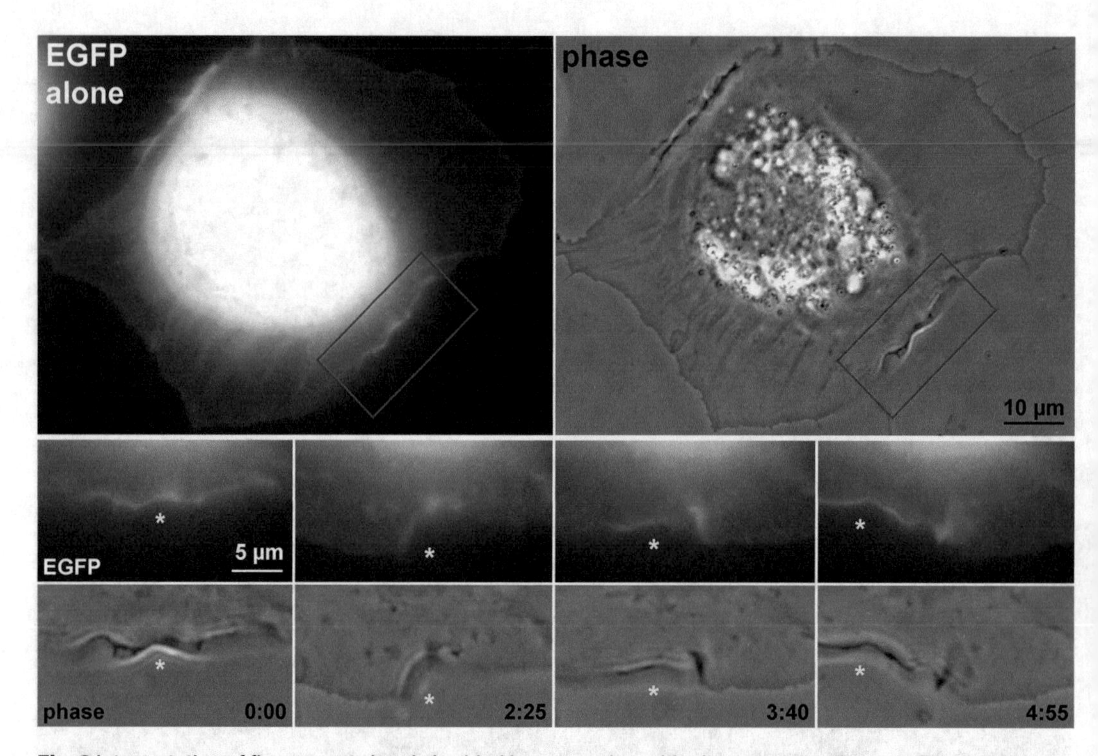

Fig. 8 Interpretation of fluorescent signals is aided by comparing with phase-contrast images. Top panel shows fluorescent (left) and phase-contrast (right) images of a time-lapse movie of a NIH3T3 cell expressing EGFP as control. Note that this cell was co-transfected with constitutively active Rac1-L61, leading to the well-known pancake-shape of cells treated in that manner. Lower two panels show selected time-lapse frames of the region indicated in the top (red rectangles). Middle panels show EGFP images, bottom panels show corresponding phase-contrast images. Elapsed time is given in minutes and seconds. Yellow asterisks highlight a fluorescence signal that appears to show a line, reminiscent of lamellipodial tip localization (*see* also Figs. 3 and 5). Note, however, that the periphery is ruffling, thus folding the lamellipodia that had initially protruded horizontally at the flat surface upward and backward, leading to a local increase in fluorescence due to cytoplasmic thickening at these subcellular locations. This can be appreciated when comparing the phase-contrast image of each corresponding time point. Darker areas in phase-contrast images correspond to denser and/or thicker structures—hence more contrast, in this case generated by the described ruffling activity. *See* also Movie 7

5. 3 cm diameter tissue culture plastic dishes.
6. 15 mL conical tubes.
7. Fibronectin (homemade [39] or commercially available for example from Roche, 1 05 407 001); dissolve 1 mg in 1 mL 2 M urea to generate a 1 mg/mL fibronectin stock solution, as described previously [39]. The stock solution can be kept at 4 °C for long-term storage, and small aliquots continuously removed for dilution into 1× PBS at required concentration (e.g., 10–25 µg/mL) prior to coating (*see* **Note 1**). For 2 M urea, dissolve 1.2 g urea in 10 mL ddH_2O and sterile filter (using sterile syringes and syringe filters, 0.22 µm pore size) into sterile 1.5 mL microtube.

8. Laminin, commercially available as powder or as 1 mg/mL stock solution in laminin coating buffer (*see* next following **item 9**).
9. Laminin coating buffer (LN buffer); 50 mM Tris, pH 7.4 in 150 mM NaCl; dissolve 121.4 g of Tris base and 58.44 g of NaCl in 900 mL ddH_2O, adjust pH to 7.4 with HCl, and fill up to 1000 mL with ddH_2O, then autoclave.
10. JetPei transfection reagent (Polyplus).
11. JetPrime (Polyplus).
12. Plasmid DNA encoding in frame-fusion of protein of interest with a fluorescent protein such as EGFP or mCherry, transfection grade purified, e.g., using commercially available plasmid DNA purification kits.
13. Centrifuge for conical tubes.
14. Water bath at 37 °C.

2.4 Microscope and Equipment

1. Any classical inverted epifluorescence microscope that is driven by software to control digital camera exposure times and to set up a time-lapse protocol is suitable. It is important, however, in particular for live cell imaging, to use a light source for fluorescence excitation that covers the entire visible spectrum, and that produces low heat such as metal halide and LEDs, which are superior over the first generation mercury arc lamps. Additionally required are appropriate fluorescence filters, reasonably fast electronic shutters (<20 ms switching time) or alternatively a TTL-trigger option of an electronically switchable light source, 63× or 100× oil immersion objectives with numerical apertures (NA) of at least 1.3, and a reasonably fast, sensitive and high resolution camera (e.g., a backilluminated, *c*harged-*c*oupling *d*evice [CCD] or scientific *c*omplementary *m*etal-*o*xide *s*emiconductor [sCMOS]). The experimental setups used in our labs are of variable generations, but have several features in common: in all setups, phase-contrast imaging employs a condenser with a PH3 phase ring. The corresponding phase-contrast ring is comprised in the objective lens. Phase contrast allows good visualization of extremely flat cellular structures as well as organelles in flat areas and hence delivers information on cell shape. Plan-Apochromat 100×/1.4NA oil and Neofluar 100×/1.3 NA oil immersion objectives (Zeiss) are commonly used. Apochromat objectives are strongly recommended when three colors (blue, green, and red) are being imaged or 405 nm lasers are used for photomanipulation experiments, since Apochromats are corrected for these spectra. Neofluar objectives are corrected for green and red and hence well suited for live cell imaging applications involving imaging of EGFP and red fluorescent dyes such as

mCherry. The images shown were captured using a Coolsnap HQ2 camera (Photometrics) with microscopes placed on vibration-isolation tables (e.g., Newport). In one setup, an inverted Axio Observer A1 (Zeiss) is equipped with an EGFP filter set (49002, ET-EGFP, Chroma), a DG-4 (Sutter instruments) xenon arc bulb light source and a Halogen lamp for transmitted light controlled by TTL shutters driven by VisiView software (Visitron Systems). As a second example, an inverted Axiovert 135 TV (Zeiss) is equipped with a FITC filter (Filter set 44, Zeiss), an HXP120 (Zeiss) light source, a halogen lamp for transmitted light and Uniblitz shutters (Model D122, Vincent Associates). In the experimental setups used here, no auto-focus system was used, but may be beneficial if available.

2. Warner chamber platform (PH-4, Warner instruments). This chamber offers the advantage of low cost use of coverslips, the use of little volumes of medium (750–1000 μL), which is advantageous for applications such as drug application, and does not require a complete incubation chamber enclosing the microscope stage (*see* **Note 2**).

   Alternatively, glass bottom dishes (e.g., ibidi μ dish, 81158, ibidi) can be used in combination with a temperature-, humidity-, and CO_2-controlled incubation chamber). Please note that optimal and stable temperature- and pH-value conditions are critical for experiments addressing any cellular function.

3. Open diamond bath imaging chamber for round 15 mm coverslips (RC25-F, Warner instruments).
4. Warner chamber temperature controller (TC-324B, SN:1176, Warner instruments).
5. Fill silicon grease (e.g., 111, Dow Corning) into a 10 mL syringe, put a 200 μL tip at the luer-lock and cut the first 2 mm of the tip with a scalpel.

3 Methods

3.1 Transfecting Cells

3.4.1 B16-F1 Transfection, Transfection of One 3 cm Dish

1. Cells should be 50–70% confluent at the time point of transfection. We routinely seed cells in a 3 cm dish 1 day prior to transfection by diluting cells 1:12.
2. Prepare micro tube labeled with A, pipet 50 μL 150 mM NaCl (Polyplus, supplied), add 0.75 μg plasmid DNA, when using two DNA constructs, use 0.375 μg each, vortex briefly.
3. Prepare micro tube labeled with B, pipet 50 μL 150 mM NaCl (Polyplus, supplied), add 1.5 μL JetPei, vortex briefly.
4. Add JetPei solution (B) to DNA solution (A), do not reverse order. Vortex briefly.

5. Incubate for 15–30 min at RT.
6. Add transfection mix dropwise to the cells. Incubate cells overnight at 37 °C, 7.5% CO_2, 95% humidity.

3.1.1 NIH3T3 Fibroblasts, Transfection of One 3 cm Dish

1. Cells should be 60–80% confluent at the time point of transfection. We normally seed cells 1:5 in a 3 cm dish 1 day prior to transfection.
2. Dilute 2 μg DNA into 200 μL jetPRIME buffer (Polyplus, supplied). Mix by vortexing.
3. Add 4 μL jetPRIME (Polyplus), vortex for 10 s, spin down briefly.
4. Incubate for 10 min at RT.
5. Add 200 μL of transfection mix per well dropwise onto the cells in serum containing medium, and distribute evenly. Incubate cells overnight at 37 °C, 7.5% CO_2, 95% humidity.

3.2 Preparing Transfected Cells on Coverslips

3.2.1 B16-F1 Cells

1. Put one or two 15 mm coverslips into a 3 cm dish using forceps.
2. Pipet 150 μL of 25 μg/mL laminin in LN coating buffer (1:40 dilution, *see* Subheading 2.3) carefully on top of the coverslip. The solution should spread well on the surface. Incubate for 60 min at RT. Avoid longer incubation times.
3. Wash transfected B16-F1 cells two times with PBS.
4. Aspirate PBS, add 0.5 mL trypsin–EDTA, and incubate at 37 °C.
5. Detach cells, add 3 mL of growth medium and resuspend cells, transfer cells into a 15 mL conical tube, pellet cells for 4 min at 300 × *g* in a centrifuge (*see* Subheading 2.3).
6. Aspirate medium, resuspend cells in 3 mL of growth medium.
7. Add 2 mL of PBS to laminin-coated coverslips, aspirate PBS and add growth medium.
8. Pipet 750 μL (1:4 dilution) of B16-F1 cell suspension in the 3 cm dish containing laminin-coated coverslips, gently shake the dish to ensure even distribution. Take care that if more than one coverslip is used, coverslips do not swim on top of each other.
9. Incubate B16-F1 cells for at least 3 h at 37 °C, 7.5% CO_2, 95% humidity to allow spreading. Upon spreading on laminin substrate, endogenous Rho GTPases become activated, leading to high endogenous Rac/Cdc42 activities in case of B16-F1 cells and promote a highly motile phenotype. B16-F1 cells on laminin generally show the migratory phenotype, which is not observable when plated on glass surfaces.

3.2.2 NIH3T3 Cells

1. Put one or two 15 mm coverslips into a 3 cm dish using forceps.
2. Pipet 150 μL of 10–25 μg/mL fibronectin in PBS (1:40–1:100 dilution, *see* Subheading 2.3) carefully on top of the coverslip. Incubate for at least 1.5 h at RT, or overnight at 4 °C.
3. Wash transfected NIH3T3 cells two times with PBS.
4. Aspirate PBS, add 0.5 mL trypsin–EDTA, and incubate at 37 °C.
5. Detach cells, add 3 mL of growth medium and resuspend cells, transfer cells into a 15 mL conical tube, pellet cells for 4 min at 300 × *g* in a centrifuge (see Subheading 2.3).
6. Aspirate medium, resuspend cells in 3 mL of growth medium.
7. Wash fibronectin-coated coverslips three times with PBS, aspirate and add growth medium.
8. Pipet 600 μL (1:5 dilution) of NIH3T3 cell suspension in the 3 cm dish containing fibronectin-coated coverslips, gently shake the dish to ensure even distribution. Take care that if more than one coverslip is used, coverslips do not move on top of each other.
9. Incubate NIH3T3 cells for at least 16 h at 37 °C, 7.5% CO_2, 95% humidity to allow spreading. NIH3T3 adhere better on fibronectin-coated substrates than on glass; moreover, if Rac is not overexpressed, fibronectin stimulates migration through integrin engagement and activation of Rho GTPases [40–42].

3.3 Mounting Coverslips in a Warner Chamber

1. Prewarm 10 mL of microscopy medium in a 15 mL conical tube to 37 °C in a water bath.
2. Put a ring of silicon grease (*see* Subheading 2.3) at the bottom gasket of the RC-25F transparent chamber (*see* also Fig. 9 for illustration).
3. Take out one 15 mm diameter coverslip with cells side toward the gasket and mount it concentrically to form the chamber bottom. Gently press it into place. Take care to prevent breakage of the coverslip.
4. Place the assembly into the Warner chamber platform.
5. Slide the platform side clamps above the chamber and use the four screws to tighten the assembly (in diagonal order to avoid breakage of the coverslip). Fill the chamber with 750–1000 μL of the respective microscopy medium.
6. Fit the thermistor into the hole of the platform; use a drop of immersion oil to ensure good thermal transfer.
7. Connect the heater elements to the temperature controller set to 37 °C.
8. Clean the coverslip bottom with ddH_2O-soaked tissue. Use a dry tissue to check leakage from the bottom side.

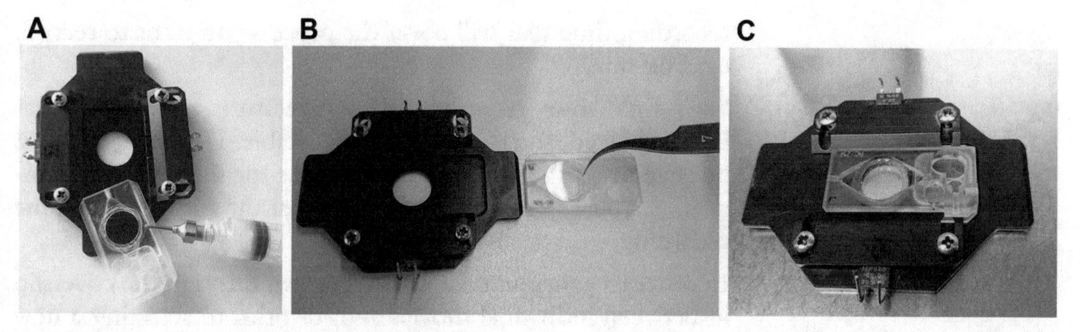

Fig. 9 Illustration of mounting a coverslip in a Warner chamber. (**a**) Put a ring of silicon grease at the bottom gasket of the RC-25F transparent chamber. (**b**) Put one 15 mm diameter coverslip with cells side toward the gasket and mount it concentrically to form the chamber bottom. Gently press it into place. Take care to prevent breakage of the coverslip and add growth medium immediately to avoid cell drying. (**c**) Correctly assembled chamber

3.4 Microscopy

1. Bring the coverslip into contact with the oil immersion objective. Wait approximately half an hour in order to achieve temperature equilibrium. Exchange microscopy medium every 30–60 min using a transfer pipette (also *see* **Note 2**).
2. Search the coverslip for low- to medium-level expressing cells. High-level expressing cells often show poor to no incorporation of the tagged protein of interest into appropriate structures and, in addition, often show perturbed behavior (*see* **Notes 3–5**).
3. Adjust Köhler illumination.
4. Adjust fluorescent exposure as follows: acquire one fluorescence image and read out the image depth (menu: image → image display, VisiView). Deselect "auto scale" to read out grey value intensities. Grey values of the fluorescence image, if possible, should exceed 1000 grey levels at the maximum. This can be achieved by adjusting lamp intensity and/or exposure time. If the fluorescence lamp intensity can be adjusted (e.g., in case of HXP120), set the lamp intensity to the lowest power and use between 400 and 800 ms exposure time in order to obtain intensity levels above 1000.
5. Adjust phase-contrast brightness by adjusting the lamp intensity and use exposure time of 100 ms or less.
6. Set up a time lapse with appropriate channels. In the examples shown here, select acquisition of phase-contrast and epifluorescence (EGFP) channel. Select a suitable interframe distance that will resolve changes in localization of proteins and dynamics (*see* **Note 6**). This strongly depends on the speed of the cellular process that is going to be analyzed. In the examples presented here, frames were captured every 5 s. Keep in mind that very fast acquisition cycles can potentially lead to photobleaching as well as photodamage (*see* **Note 7**). Select a total

recording time that will cover the process you want to record, e.g., 60 min.

7. Start time-lapse recording. If temperature equilibrium has been achieved, the focus will remain stable. However, it might be necessary to refocus. To do so, pause the time-lapse acquisition, use phase-contrast optics to adjust the focus and resume recording.
8. In order to generate reliable data, screen through the coverslip in between individual movies and/or prior to selecting a new cell to be recorded, using low magnification and thus light intensity to rapidly get an overview of respective cell behavior and/or localization of a given fluorescently tagged protein. Take as many images and movies as possible, always aiming at generating a dataset reflecting the "average" or representative behavior observed on a given coverslip. Moreover, record movies from several, independently prepared coverslips, and if possible on separate days. After finishing a given experiment, take sufficient time to inspect the raw data, and think of potential alternatives to confirm and solidify your conclusions (*see* **Notes 8** and **9**).

4 Notes

1. Dissolving fibronectin in 2 M urea keeps it soluble because in an unpolymerized state, while dissolving it into pH-neutral buffers such as PBS followed by aliquoting and freezing (as commonly recommended) will lead to polymerization, making it impossible to control precise concentrations of soluble protein after thawing.
2. Exchange the microscopy medium every 30–60 min, since water evaporates. You might notice too low medium level in phase contrast if the center of the viewing field is getting darker. This is due to misaligned Köhler illumination upon reduced medium levels.
3. The choice of cells overexpressing any fluorescently tagged protein requires careful consideration. Low expression levels are always preferred, as overexpression of specific components may lead to adverse effects on proper localization of the protein of interest or cause artifacts in cell behavior indirectly modifying protein localization and/or dynamics. In particular, proteins that are part of constitutive, multimeric protein complexes such as the pentameric WAVE regulatory complex (see Fig. 5 and Movie 6 for EGFP-Abi as example) or the heptameric Arp2/3 complex are difficult to image, since strong overexpression may affect protein complex stoichiometry and thus

interfere with the physiological function of endogenous complexes [43]. As specific example, preserving endogenous protein stoichiometry by engineering fluorescent proteins into the genomic loci of clathrin light chain A or dynamin-2 was found advantageous concerning dynamics and efficiency of endocytosis as compared to overexpression of these proteins [44]. This is of particular interest in times of highly efficient genome editing by CRISPR/Cas9 (ref. 45).

4. It may be beneficial to look at different cell types when addressing a particular question such as dynamic, subcellular localization. If available, probe cell lysates for the presence of proteins by Western Blotting to ensure that the protein of interest is also present endogenously. Also consider the same for known protein interactors.
5. It may sound trivial, but if exploring protein dynamics in a specific cell type or experimental conditions that disfavor the formation of a specific structure the protein of interest accumulates in, this will lead to irrelevant results that cannot be productively interpreted. As example, determination whether a protein of interest accumulates at the edge of a protruding lamellipodium essentially requires conditions in which protrusion of the edge is clearly confirmed and documented. In other words, the absence of given factor from the cell periphery cannot be taken as lack of localization in the lamellipodium, if these structures are retracted. Likewise, be careful not to overinterpret spatial fluctuations of fluorescence signals in specific subcellular locations if they could be simply explained by cytoplasmic thickening. Consider to use volume controls tagged to proteins of distinct fluorescence color. Always consider that the subcellular distribution of a given cytoskeletal factor comprises a cytoskeleton-bound fraction and a freely diffusive fraction, although relative quantities of the two distinct fractions may vary dependent on expression level.
6. When comparing different channels from time-lapse recordings (as for example EGFP fluorescence followed by phase-contrast optics as shown in all examples in this chapter), bear in mind that sequential channel acquisition necessarily includes a time shift in acquisition. This becomes particularly important if very fast processes are imaged and structures advance during the acquisition of different channels. In case of imaging two fluorescence colors, e.g., using a computer-controlled filter wheel, time-shifts can be overcome by using a dual filterset, allowing spectral splitting of the emission of simultaneously exited fluorophores either onto two halves of one camera or, alternatively, two separate cameras.

7. Phototoxicity or photodamage is a key issue in fluorescence live cell imaging, and caused by fluorophores converting to reactive states upon extended excitation followed by generation of free radicals, which in turn will modify cellular constituents if not eliminated by scavengers [46]. This can of course be avoided by low light intensities and by keeping exposure times at the minimum required for generating conclusive, interpretable data. Also keep in mind that the damage generated will correlate with the net energy introduced into a given cellular volume, so any method reducing the illumination volume per time, such as total internal reflection fluorescence (TIRF) microscopy [47], will also reduce phototoxicity. Hence, TIRF microscopy if appropriate for your sample and probe of interest—e.g., because the probe localizes in focal adhesions at the substratum layer—will not only increase contrast and resolution due to the excitation being restricted to a 150 nm region at the surface of the coverslip, but will also reduce the light dose per time and volume as compared to conventional widefield or confocal imaging. Moreover, if assuming that the damage correlates with the diffusion rate of scavengers to the illuminated region relative to the concentration of excited fluorophores in this volume, low lamp intensities in combination with long exposure times should be advantageous over high lamp intensities combined with short exposure times (Dr. Yu-Li Wang, Carnegie Mellon University, Pittsburgh, USA, personal communication).
8. Generate fusion constructs with fluorescent tags (e.g., EGFP) inserted at both the N- and C-terminus, since proteins might fold differently depending on where the fusion was engineered. In addition, fluorescent proteins with average molecular weights of roughly 30 kDa might mask interaction surfaces with other proteins, or interfere with specific posttranslational modifications that might serve as prerequisite of physiologic subcellular positioning (*see* also Fig. 3; Movies 3 and 4). If no convincing localization is observed, different linker lengths might improve protein folding in frame with fluorescent tag constructs (*see* for example [48]). For very small proteins, even intramolecular fusion proteins have been successfully employed in the past [49–52].
9. If possible, always consider to compare subcellular locations observed upon expressing fusions with fluorescent proteins with alternative localization studies, such as antibody staining of endogenous proteins, or localization of microinjected, purified protein or fluorescent antibody specific for the protein (*see* for example [53] for alternatives of studying zyxin distribution and dynamics).

Acknowledgments

This work was supported in part by the German Research Foundation (DFG) and by the Helmholtz Centre for Infection Research (HZI, Braunschweig, Germany). We would like to thank Dr. Yu-Li Wang (Pittsburgh, USA) for insightful discussions on fluorescence imaging.

References

1. Krause M, Gautreau A (2014) Steering cell migration: lamellipodium dynamics and the regulation of directional persistence. Nat Rev Mol Cell Biol 15(9):577–590. https://doi.org/10.1038/nrm3861
2. Ladwein M, Rottner K (2008) On the Rho'd: the regulation of membrane protrusions by Rho-GTPases. FEBS Lett 582(14): 2066–2074. https://doi.org/10.1016/j.febslet.2008.04.033
3. Mattila PK, Lappalainen P (2008) Filopodia: molecular architecture and cellular functions. Nat Rev Mol Cell Biol 9(6):446–454. https://doi.org/10.1038/nrm2406
4. Small JV, Stradal T, Vignal E, Rottner K (2002) The lamellipodium: where motility begins. Trends Cell Biol 12(3):112–120
5. Charras G, Paluch E (2008) Blebs lead the way: how to migrate without lamellipodia. Nat Rev Mol Cell Biol 9(9):730–736. https://doi.org/10.1038/nrm2453
6. Paksa A, Raz E (2015) Zebrafish germ cells: motility and guided migration. Curr Opin Cell Biol 36:80–85. https://doi.org/10.1016/j.ceb.2015.07.007
7. Paluch EK, Raz E (2013) The role and regulation of blebs in cell migration. Curr Opin Cell Biol 25(5):582–590. https://doi.org/10.1016/j.ceb.2013.05.005
8. Mishin AS, Belousov VV, Solntsev KM, Lukyanov KA (2015) Novel uses of fluorescent proteins. Curr Opin Chem Biol 27:1–9. https://doi.org/10.1016/j.cbpa.2015.05.002
9. Rodriguez EA, Campbell RE, Lin JY, Lin MZ, Miyawaki A, Palmer AE, Shu X, Zhang J, Tsien RY (2017) The growing and glowing toolbox of fluorescent and photoactive proteins. Trends Biochem Sci 42(2):111–129. https://doi.org/10.1016/j.tibs.2016.09.010
10. Demmerle J, Innocent C, North AJ, Ball G, Muller M, Miron E, Matsuda A, Dobbie IM, Markaki Y, Schermelleh L (2017) Strategic and practical guidelines for successful structured illumination microscopy. Nat Protoc 12(5):988–1010. https://doi.org/10.1038/nprot.2017.019
11. Kanchanawong P, Shtengel G, Pasapera AM, Ramko EB, Davidson MW, Hess HF, Waterman CM (2010) Nanoscale architecture of integrin-based cell adhesions. Nature 468(7323):580–584. https://doi.org/10.1038/nature09621
12. Schermelleh L, Heintzmann R, Leonhardt H (2010) A guide to super-resolution fluorescence microscopy. J Cell Biol 190(2):165–175. https://doi.org/10.1083/jcb.201002018
13. Block J, Breitsprecher D, Kuhn S, Winterhoff M, Kage F, Geffers R, Duwe P, Rohn JL, Baum B, Brakebusch C, Geyer M, Stradal TE, Faix J, Rottner K (2012) FMNL2 drives actin-based protrusion and migration downstream of Cdc42. Curr Biol 22(11):1005–1012. https://doi.org/10.1016/j.cub.2012.03.064
14. Kage F, Winterhoff M, Dimchev V, Mueller J, Thalheim T, Freise A, Bruhmann S, Kollasser J, Block J, Dimchev G, Geyer M, Schnittler HJ, Brakebusch C, Stradal TE, Carlier MF, Sixt M, Kas J, Faix J, Rottner K (2017) FMNL formins boost lamellipodial force generation. Nat Commun 8:14832. https://doi.org/10.1038/ncomms14832
15. Krause M, Dent EW, Bear JE, Loureiro JJ, Gertler FB (2003) Ena/VASP proteins: regulators of the actin cytoskeleton and cell migration. Annu Rev Cell Dev Biol 19:541–564. https://doi.org/10.1146/annurev.cellbio.19.050103.103356
16. Rottner K, Behrendt B, Small JV, Wehland J (1999) VASP dynamics during lamellipodia protrusion. Nat Cell Biol 1(5):321–322. https://doi.org/10.1038/13040
17. Breitsprecher D, Kiesewetter AK, Linkner J, Vinzenz M, Stradal TE, Small JV, Curth U, Dickinson RB, Faix J (2011) Molecular mechanism of Ena/VASP-mediated actin-filament elongation. EMBO J 30(3):456–467. https://doi.org/10.1038/emboj.2010.348

18. Hansen SD, Mullins RD (2010) VASP is a processive actin polymerase that requires monomeric actin for barbed end association. J Cell Biol 191(3):571–584. https://doi.org/10.1083/jcb.201003014
19. Mejillano MR, Kojima S, Applewhite DA, Gertler FB, Svitkina TM, Borisy GG (2004) Lamellipodial versus filopodial mode of the actin nanomachinery: pivotal role of the filament barbed end. Cell 118(3):363–373. https://doi.org/10.1016/j.cell.2004.07.019
20. Svitkina TM, Bulanova EA, Chaga OY, Vignjevic DM, Kojima S, Vasiliev JM, Borisy GG (2003) Mechanism of filopodia initiation by reorganization of a dendritic network. J Cell Biol 160(3):409–421. https://doi.org/10.1083/jcb.200210174
21. Edwards M, Zwolak A, Schafer DA, Sept D, Dominguez R, Cooper JA (2014) Capping protein regulators fine-tune actin assembly dynamics. Nat Rev Mol Cell Biol 15(10):677–689. https://doi.org/10.1038/nrm3869
22. Iwasa JH, Mullins RD (2007) Spatial and temporal relationships between actin-filament nucleation, capping, and disassembly. Curr Biol 17(5):395–406. https://doi.org/10.1016/j.cub.2007.02.012
23. Lai FP, Szczodrak M, Block J, Faix J, Breitsprecher D, Mannherz HG, Stradal TE, Dunn GA, Small JV, Rottner K (2008) Arp2/3 complex interactions and actin network turnover in lamellipodia. EMBO J 27(7):982–992. https://doi.org/10.1038/emboj.2008.34
24. Alekhina O, Burstein E, Billadeau DD (2017) Cellular functions of WASP family proteins at a glance. J Cell Sci 130(14):2235–2241. https://doi.org/10.1242/jcs.199570
25. Bisi S, Disanza A, Malinverno C, Frittoli E, Palamidessi A, Scita G (2013) Membrane and actin dynamics interplay at lamellipodia leading edge. Curr Opin Cell Biol 25(5):565–573. https://doi.org/10.1016/j.ceb.2013.04.001
26. Innocenti M, Zucconi A, Disanza A, Frittoli E, Areces LB, Steffen A, Stradal TE, Di Fiore PP, Carlier MF, Scita G (2004) Abi1 is essential for the formation and activation of a WAVE2 signalling complex. Nat Cell Biol 6(4):319–327. https://doi.org/10.1038/ncb1105ncb1105[pii]
27. Kunda P, Craig G, Dominguez V, Baum B (2003) Abi, Sra1, and Kette control the stability and localization of SCAR/WAVE to regulate the formation of actin-based protrusions. Curr Biol 13(21):1867–1875
28. Steffen A, Rottner K, Ehinger J, Innocenti M, Scita G, Wehland J, Stradal TE (2004) Sra-1 and Nap1 link Rac to actin assembly driving lamellipodia formation. EMBO J 23(4):749–759. https://doi.org/10.1038/sj.emboj.7600084
29. Stradal TE, Scita G (2006) Protein complexes regulating Arp2/3-mediated actin assembly. Curr Opin Cell Biol 18(1):4–10. https://doi.org/10.1016/j.ceb.2005.12.003
30. Stradal T, Courtney KD, Rottner K, Hahne P, Small JV, Pendergast AM (2001) The Abl interactor proteins localize to sites of actin polymerization at the tips of lamellipodia and filopodia. Curr Biol 11(11):891–895
31. Gertler FB, Niebuhr K, Reinhard M, Wehland J, Soriano P (1996) Mena, a relative of VASP and Drosophila enabled, is implicated in the control of microfilament dynamics. Cell 87(2):227–239
32. Haffner C, Jarchau T, Reinhard M, Hoppe J, Lohmann SM, Walter U (1995) Molecular cloning, structural analysis and functional expression of the proline-rich focal adhesion and microfilament-associated protein VASP. EMBO J 14(1):19–27
33. Rottner K, Hall A, Small JV (1999) Interplay between Rac and Rho in the control of substrate contact dynamics. Curr Biol 9(12):640–648
34. Ballestrem C, Wehrle-Haller B, Imhof BA (1998) Actin dynamics in living mammalian cells. J Cell Sci 111(Pt 12):1649–1658
35. Hall A (1998) Rho GTPases and the actin cytoskeleton. Science 279(5350):509–514
36. Nobes CD, Hall A (1995) Rho, rac, and cdc42 GTPases regulate the assembly of multimolecular focal complexes associated with actin stress fibers, lamellipodia, and filopodia. Cell 81(1):53–62. doi:0092-8674(95)90370-4[pii]
37. Koestler SA, Steffen A, Nemethova M, Winterhoff M, Luo N, Holleboom JM, Krupp J, Jacob S, Vinzenz M, Schur F, Schluter K, Gunning PW, Winkler C, Schmeiser C, Faix J, Stradal TE, Small JV, Rottner K (2013) Arp2/3 complex is essential for actin network treadmilling as well as for targeting of capping protein and cofilin. Mol Biol Cell 24(18):2861–2875. https://doi.org/10.1091/mbc.E12-12-0857
38. Steffen A, Koestler SA, Rottner K (2014) Requirements for and consequences of Rac-dependent protrusion. Eur J Cell Biol 93(5–6):184–193. https://doi.org/10.1016/j.ejcb.2014.01.008
39. Avnur Z, Geiger B (1981) The removal of extracellular fibronectin from areas of cell-substrate contact. Cell 25(1):121–132
40. Cailleteau L, Estrach S, Thyss R, Boyer L, Doye A, Domange B, Johnsson N, Rubinstein E, Boucheix C, Ebrahimian T, Silvestre JS,

Lemichez E, Meneguzzi G, Mettouchi A (2010) alpha2beta1 integrin controls association of Rac with the membrane and triggers quiescence of endothelial cells. J Cell Sci 123(Pt 14):2491–2501. https://doi.org/10.1242/jcs.058875

41. Mettouchi A, Klein S, Guo W, Lopez-Lago M, Lemichez E, Westwick JK, Giancotti FG (2001) Integrin-specific activation of Rac controls progression through the G(1) phase of the cell cycle. Mol Cell 8(1):115–127
42. Price LS, Leng J, Schwartz MA, Bokoch GM (1998) Activation of Rac and Cdc42 by integrins mediates cell spreading. Mol Biol Cell 9(7):1863–1871
43. Rottner K, Kaverina IN, Stradal TEB (2006) Cytoskeleton proteins. In: Celis JE (ed) Cell biology, a laboratory handbook, vol. 3, 3rd edn. Elsevier, London, pp 111–119
44. Doyon JB, Zeitler B, Cheng J, Cheng AT, Cherone JM, Santiago Y, Lee AH, Vo TD, Doyon Y, Miller JC, Paschon DE, Zhang L, Rebar EJ, Gregory PD, Urnov FD, Drubin DG (2011) Rapid and efficient clathrin-mediated endocytosis revealed in genome-edited mammalian cells. Nat Cell Biol 13(3):331–337. https://doi.org/10.1038/ncb2175
45. Hille F, Charpentier E (2016) CRISPR-Cas: biology, mechanisms and relevance. Philos Trans R Soc Lond Ser B Biol Sci 371(1707). https://doi.org/10.1098/rstb.2015.0496
46. Swedlow JR, Porter IM, Posch M, Swift S (2010) In vivo imaging of mammalian cells. In: Goldman RD, Swedlow JR, Spector DL (eds) Live cell imaging: a laboratory manual, 2nd edn. Cold Spring Harbor Laboratory Press, New York, NY, pp 317–332
47. Poulter NS, Pitkeathly WT, Smith PJ, Rappoport JZ (2015) The physical basis of total internal reflection fluorescence (TIRF) microscopy and its cellular applications. Methods Mol Biol 1251:1–23. https://doi.org/10.1007/978-1-4939-2080-8_1
48. Geese M, Schluter K, Rothkegel M, Jockusch BM, Wehland J, Sechi AS (2000) Accumulation of profilin II at the surface of Listeria is concomitant with the onset of motility and correlates with bacterial speed. J Cell Sci 113(Pt 8):1415–1426
49. Nejedla M, Li Z, Masser AE, Biancospino M, Spiess M, Mackowiak SD, Friedlander MR, Karlsson R (2017) A fluorophore fusion construct of human profilin I with non-compromised poly(L-proline) binding capacity suitable for imaging. J Mol Biol 429(7):964–976. https://doi.org/10.1016/j.jmb.2017.01.004
50. Okreglak V, Drubin DG (2007) Cofilin recruitment and function during actin-mediated endocytosis dictated by actin nucleotide state. J Cell Biol 178(7):1251–1264. https://doi.org/10.1083/jcb.200703092
51. Sakurai-Yageta M, Recchi C, Le Dez G, Sibarita JB, Daviet L, Camonis J, D'Souza-Schorey C, Chavrier P (2008) The interaction of IQGAP1 with the exocyst complex is required for tumor cell invasion downstream of Cdc42 and RhoA. J Cell Biol 181(6):985–998. https://doi.org/10.1083/jcb.200709076
52. Steffen A, Le Dez G, Poincloux R, Recchi C, Nassoy P, Rottner K, Galli T, Chavrier P (2008) MT1-MMP-dependent invasion is regulated by TI-VAMP/VAMP7. Curr Biol 18(12):926–931. https://doi.org/10.1016/j.cub.2008.05.044
53. Rottner K, Krause M, Gimona M, Small JV, Wehland J (2001) Zyxin is not colocalized with vasodilator-stimulated phosphoprotein (VASP) at lamellipodial tips and exhibits different dynamics to vinculin, paxillin, and VASP in focal adhesions. Mol Biol Cell 12(10):3103–3113
54. Dimchev G, Steffen A, Kage F, Dimchev V, Pernier J, Carlier MF, Rottner K (2017) Efficiency of lamellipodia protrusion is determined by the extent of cytosolic actin assembly. Mol Biol Cell 28(10):1311–1325. https://doi.org/10.1091/mbc.E16-05-0334
55. Schafer DA, Welch MD, Machesky LM, Bridgman PC, Meyer SM, Cooper JA (1998) Visualization and molecular analysis of actin assembly in living cells. J Cell Biol 143(7):1919–1930
56. Derivery E, Sousa C, Gautier JJ, Lombard B, Loew D, Gautreau A (2009) The Arp2/3 activator WASH controls the fission of endosomes through a large multiprotein complex. Dev Cell 17(5):712–723. https://doi.org/10.1016/j.devcel.2009.09.010
57. Gomez TS, Billadeau DD (2009) A FAM21-containing WASH complex regulates retromer-dependent sorting. Dev Cell 17(5):699–711. https://doi.org/10.1016/j.devcel.2009.09.009
58. Rottner K, Hanisch J, Campellone KG (2010) WASH, WHAMM and JMY: regulation of Arp2/3 complex and beyond. Trends Cell Biol 20(11):650–661. https://doi.org/10.1016/j.tcb.2010.08.014

Check for updates

Chapter 20

A Biologist-Friendly Method to Analyze Cross-Correlation Between Protrusion Dynamics and Membrane Recruitment of Actin Regulators

Perrine Paul-Gilloteaux, François Waharte, Manish Kumar Singh, and Maria Carla Parrini

Abstract

During mesenchymal cell motility, various actin regulators are recruited to the leading edge with exquisite precision in time and space to generate protrusion and retraction cycles. We present here an automated method, named CorRecD (from Correlation Recruitment Dynamics), which quantifies cell edge dynamics, protein recruitment and analyze their cross-correlation. The Wave Regulatory Complex (WRC), a master driver of protrusions, is used as a case-of-study. This biologist-friendly method relies on free software tools and can be applied to any fluorescently tagged protein of interest.

Key words Image analysis, Migration, Wave, Cell edge dynamics, Cross-correlation

1 Introduction

The dynamics of cell migration can be visualized by time-lapse imaging [1], which provides a great richness of quantitative digital information that is usually only partially exploited by biologists. One limiting step in extensive quantifications of migration parameters is the availability of automated computer tools that can be easily exploited by biologists. Interdisciplinary collaborations with imaging specialists, computer scientists and mathematicians are necessary to develop methods and software tools for image processing, statistical analysis, and modeling. Mathematical approaches, including cross-correlation, turned out to be extremely powerful to quantitatively interpret dynamical behaviors during cell migration, both at intra-cellular [2, 3] and supracellular levels [4, 5].

A recurrent problem in the study of cell migration is the measurement of the recruitment of regulator or effector proteins at the leading edge, where actin polymerization and protrusion activity occur. Several available approaches, in particular the QuimP

Alexis Gautreau (ed.), *Cell Migration: Methods and Protocols*, Methods in Molecular Biology, vol. 1749,
https://doi.org/10.1007/978-1-4939-7701-7_20,

software [6, 7], are powerful, but have a wide range of applications, leading to complex workflows. We describe here an alternative semiautomated method to quantify the recruitment of any fluorescently tagged protein of interest at the cell edge and to correlate the recruitment timing with the movements of the edge, following a strategy previously described in [2]. As example, this method is applied to study the Wave Regulatory Complex (WRC), a major effector of the active GTP-bound Rac1 GTPase [8, 9].

2 Materials

1. Cells expressing, transiently or stably (*see* **Note 1**), the protein of interest fused to a fluorescent tag.

   Our case-of-study: HEK-HT cells [10], normal human embryonic kidney cells, immortalized by ectopic expression of the telomerase catalytic subunit (hTERT) and of the SV40 large T antigen, and transiently transfected with pEGFP-Abi1 (pEGFP back-bone from Clontech).
2. Appropriate culture medium, with selection antibiotics when necessary.

   For HEK-HT cells: Dulbecco's modified Eagle's medium supplemented with 2 mM glutamine, penicillin, streptomycin, 10% fetal bovine serum, hygromycin (100 μg/mL), geneticin (400 μg/mL) and its variant without phenol red (ThermoFisher Scientific, # 21063029).
3. Glass-bottom 3.5 cm dishes (MatTek, # P35G-0.170-14-C).
4. Transfection reagents, such as jetPRIME (Polyplus) or Lipofectamine Plus reagents (Invitrogen).
5. Time-lapse spinning-disk confocal microscope, equipped with a heating chamber (37 °C) and a CO_2 controller (5% CO_2), to achieve fast 2D + t imaging with high sensitivity and optical sectioning on living cells.

   The system used for data presented here is composed of a TiE Nikon microscope with a 60× oil immersion objective, an Yokogawa CSU-X1 spinning-disk module, a 4 color laser bench (Errol), an iXon EM-CCD camera (Andor, pixel size 16 μm), a microscope incubator and CO_2 controller from Life Imaging Service. The imaging system is under control of the MetaMorph software (Universal Imaging).
6. A computer (*see* **Note 2**) with the ImageJ open-source software installed (https://imagej.nih.gov/ij/)
7. Homemade ImageJ macro: "MacroIdentifyandSegment".

   The macro is downloadable from http://minilien.curie.fr/ykjseg.

For installation, go to ImageJ menu Plugins > Macros/ Install. The macro will then appear under Plugins > Macros.

For expert users, *see* **Note 3**.

8. Homemade ImageJ plugin: "Recruitment Edge Analysis".

   The plugin is downloadable from http://minilien.curie.fr/u9zfc9.

   For installation, copy the .jar file in your ImageJ/plugins directory, restart ImageJ. Recruitment Edge Analysis should now appear in your list of plugins.

   For expert users, *see* **Note 3**.

9. Homemade Matlab open-source code: "CorrelationHeatMap".

   The program is downloadable for Windows 64 from http://minilien.curie.fr/sympfw.

   For installation on Windows 64 bits, uncompress correlation.zip, launch CorrelationHeatMapInstaller_mcr.exe. The installer will then be downloaded and will install the MatLab emulator, so that you do not need a MatLab license to run the software. Keep the default parameters, and click on install. The installation can take up to 10 min, but is needed only once. After correct installation, an icon should be available on the desktop, or you can find CorrelationHeatMap from the Windows Start up menu.

   For expert users, *see* **Note 3**.

3 Methods

3.1 Cell Preparation and Imaging

1. Day 0. Plate cells on glass-bottom 3.5 cm dishes in 2 mL medium. For HEK-HT cells, typically 60,000 cells per well.
2. Day 1. Transfect cells with plasmid expressing the fluorescent fusion protein of interest. When working with stable cell lines, this step is not necessary (*see* **Note 1**). Several routine transfection protocols are applicable. We are currently using jetPRIME (Polyplus) according to manufacturer's instructions. Briefly:

   Make sure cells are at 60–70% confluence.

   Dilute 2 μg DNA into 200 μL jetPRIME buffer, mix by vortexing.

   Add 4 μL jetPRIME, vortex for 10 s, spin down briefly.

   Incubate for 10 min at RT.

   Add 200 μL of transfection mix per well drop-wise onto the cells in 2 mL medium and distribute evenly by gently rocking the plates.

   Return the plates to the incubator.

   Replace transfection medium after 4 h by cell growth medium.

3. Day 3. Change the medium for the medium without phenol red (to avoid interferences with fluorescence imaging). Wait at least 3 h before imaging.
4. Image cells with the spinning-disk confocal microscope, at 37 °C and 5% CO_2, with a 60× or 100× objective. Acquire images every 1 or 2 s (*see* **Note 4**). Importantly, be sure that metadata, such as pixel size and time interval between two frames in a movie, are carefully noted on laboratory notebook or easily accessible from the image file metadata.

3.2 Image Analysis

"CorRecD" analysis at a glance, recapitulated in Fig. 1, is divided in three parts: segmentation of the protrusion, generation of edge velocity and recruitment maps, cross-correlation calculation. Users can practice on a sample data set available at http://minilien.curie.fr/5w2izs.

The first part is the identification and the segmentation of the protrusion (Fig. 1a, **steps 1** and **2**). We detail here a method that should be applicable to most fluorescence images. This first preprocessing part can be achieved by using the proposed macro MacroIdentifyandSegment.ijm. or can be replaced by any segmentation method. The only requirement is to have ImageJ ROIs (Region Of Interest) contouring protrusion at the end of this step, with one ROI per time frame.

1. Convert the images sequence to a stack in order to generate a video. Keep the images at 16-bit grayscale; do not convert to 8-bit.
2. Launch the macro called MacroIdentifyandSegment.ijm (in ImageJ, go to PluginsMacros/Run… and select the file). The macro will:
   - Crop the video to select the region to be analyzed (such as a protrusion).
   - Apply the smooth filter.
   - Achieve the cell edge segmentation by creating a threshold-based time-lapse binary mask.

Fig. 1 (continued) (**b**) Generation of edge velocity and recruitment maps using the ImageJ plugin "Recruitment Edge Analysis." (**c**) Calculation of cross-correlation between recruitment and edge velocity using the MatLab code "CorrelationHeatMap." The red line represents the fitting of the correlation coefficients as function of time lag. The vertical green line represents the maximum value of cross-correlation. In this case, the maximum value of cross-correlation is close to 0, indicating that Abi recruitment is synchronous with edge movement, as we reported in [9]

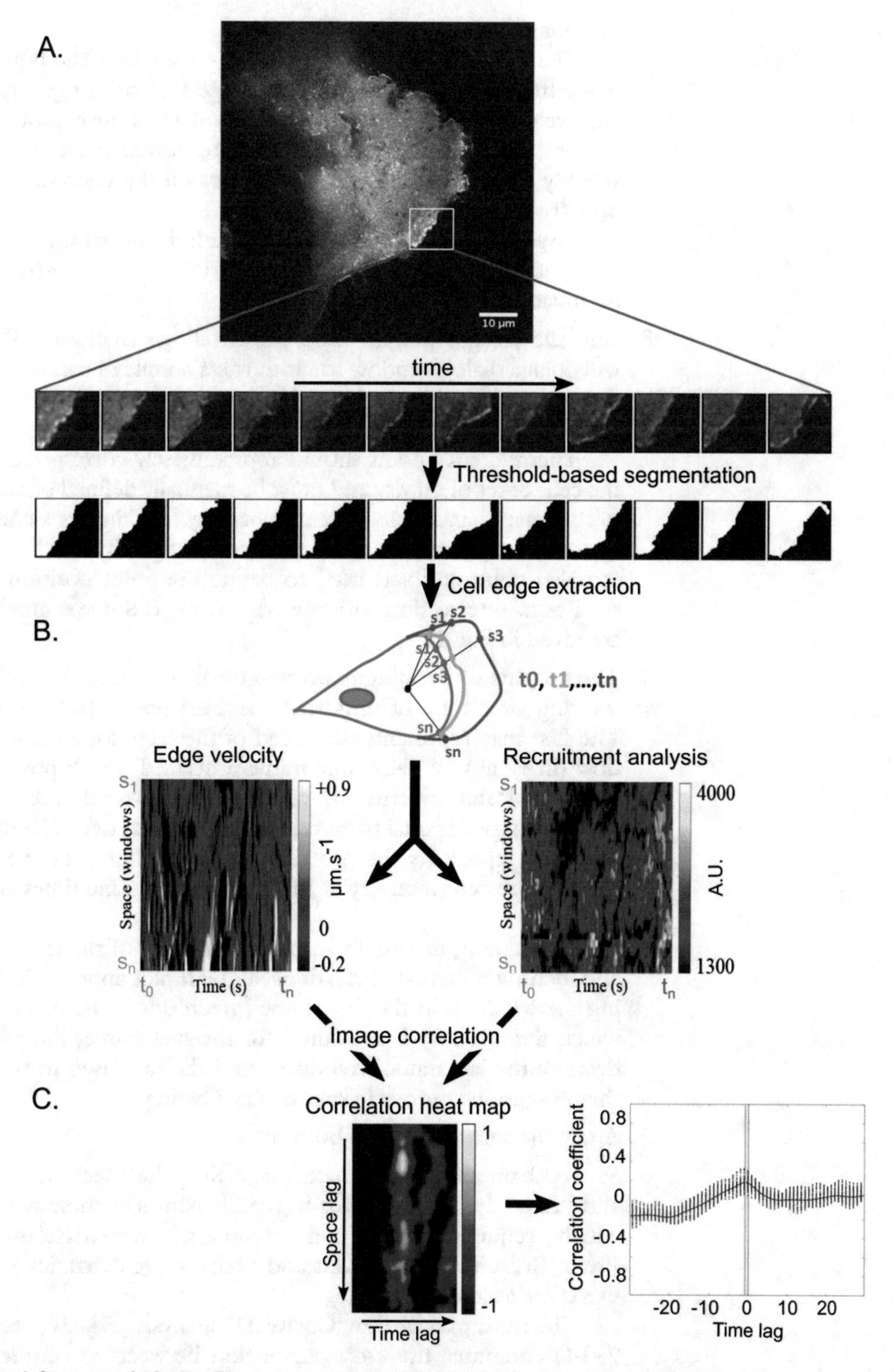

Fig. 1 Strategy of "CorRecD" image analysis method. (**a**) Selection and segmentation of a representative protrusion. The entire video sequence can be downloaded from http://minilien.curie.fr/5w2izs (time length 5 min, time interval 1 s, objective 60×, exposure time of 400 ms, and a binning factor of 2).

For a video tutorial, *see* **Note 3**.

The second part (Fig. 1b, **steps 3–6**) exploits the home-made ImageJ plugin "Recruitment Edge analysis" to generate edge velocity and recruitment maps. It will track the contour of the protrusions over time, and measure how it moves (edge velocity map) and how intensity evolves on the contour over time (recruitment map).

Be sure to have two windows opened: the protrusion stack file and the ROI Manager containing all protrusion boundaries.

3. Run the ImageJ plugin "Recruitment Edge Analysis". This will open a dialog window to modulate a couple of parameters that are explained in the legend of Fig. 2.

   For a video tutorial, *see* **Note 3**.

   The reference point should approximately correspond to the cell center of gravity and must be manually defined outside of the image, since cells were cropped, by selecting its x and y coordinates. A line passing by this center of gravity and by the sampled point at t was used to define the point position at $t + 1$ as its intersection with the edge at $t + 1$. Some examples are given in Fig. 2.

4. The outputs of the plugin are two maps that result from the tracking over time of uniformly sampled points (windows). The first map represents the speed of the edge for each window (in y) and for each time frame processed (in x); positive speed indicates protrusion, while negative speed indicates retraction; speed equal to zero indicates a static edge. The second map represents the protein recruitment, i.e., the maximum fluorescent intensity at the edge for the same times and windows.

   In addition, to visually inspect the quality of the tracking and identify potential aberrations, a graph plot appears showing the windows in the first frame (green dots), the band on which the intensity is measured for the first frame, the windows of the last frame (red dots), and the rays used to track the protrusion contour in case of Ray Casting.

5. Apply the smooth filter to both maps.

6. Save both smoothed map as text image. Keep the "Recruitment" and "Edge dynamics" words in the file names as these words will be required for the code computing cross-correlation. Prefix_Recruitment_suffix.txt and Prefix_Edge dynamics_suffix.txt are acceptable names.

   The third part of the "CorRecD" analysis (Fig. 1c, **steps 7–14**) computes the cross-correlation between recruitment and edge velocity by generating a curve of correlation values where x is the time lag tested, and y the value of the correlation between the edge velocity map and the recruitment map

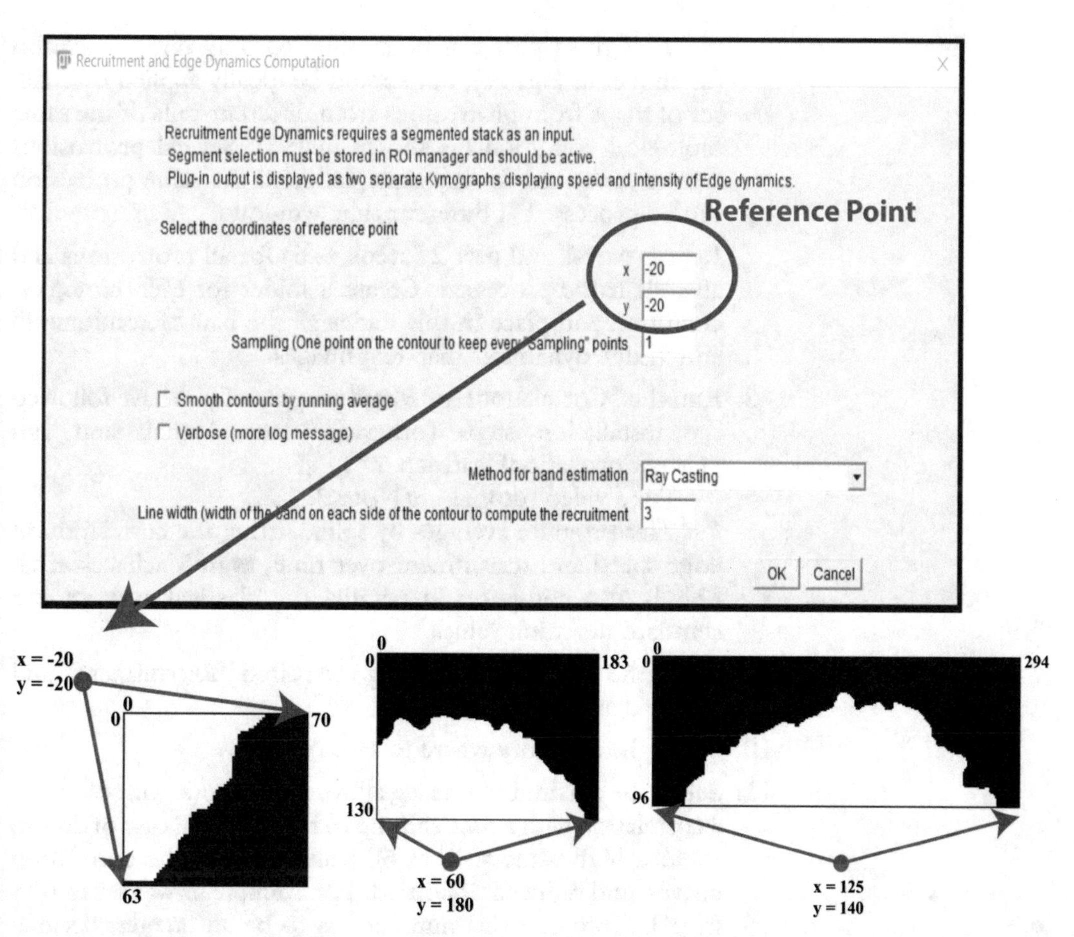

Fig. 2 User interface of the ImageJ plugin "Recruitment Edge Analysis". The following parameters should be selected: *Reference point*. The *x* and *y* coordinate values of the reference point, approximately corresponding to the cell center of gravity outside of the image, needs to be entered. We give below some examples with different protrusion positioning in the cropped stack. *Sampling* value represents the number of pixels at the perimeter of cell contour, it is recommended to keep it 1. This value may allow to under-sample the contour if needed (for faster computation for example). *Smooth contours by running average* box. If checked, this option will further smooth the contour before measurements. *Verbose* box. If checked, all processes will be displayed on a log file during the measurement. *Method for band estimation*. We recommend the Ray casting method because it is more stable. *Line width*. We typically work with line width = 3, that means that the maximal fluorescent intensity is computed over a band with a width of 7 pixels across the segmented edge, three pixels each side on the defined boundary pixel

with this given lag for each window (uniformly sampled point on the protrusion contour). Curves (correlation by windows against lag) will then be averaged to compute the lag corresponding to the maximum correlation value. If the value is positive, the event recruitment happened in average before the event on edge dynamics, if negative, the event on edge dynamics happened in average after the event on recruitment. If 0, the two events tend to be simultaneous.

This third part can be applied to only one protrusion (as shown in Fig. 1c), but should be ideally applied on a full set of maps from protrusions from different cells of the same biological condition (as shown in [9]). Several protrusions can be processed for the same cell, and the same protrusion can be processed at different time windows.

7. Repeat part 1 and part 2 (**steps 1–6**) for all protrusions and all cells to be processed. Create a folder for each biological condition and place in this folder all the pair "Recruitment" and "Edge dynamics" map text images.
8. Run the "CorrelationHeatMap" program if you have followed the installation steps (otherwise Run Matlab and run "CrossCorrelationHeatMap.m").

   For a video tutorial, *see* **Note 3**.

   This program averages by spline fitting the correlation of edge speed and recruitment over time, as in Machacek et al. [2]. It also computes in parallel the classical average and standard deviation values.
9. Select the directory containing the paired "Recruitment" and "Edge dynamics" maps.
10. Select the directory where to save results.
11. Select the maximum time lag allowed. The code will search for a correlation with a time shift up to this value. In case of doubt, select a high value such as 60 s and observe the correlation curves, and reduce it if needed. For example, it was set to 60 s in [9]. Note that this number has to be an integer, decimal values are not allowed.
12. For each protrusion to be analyzed, a window will pop up asking for the time interval between two frames for this particular movie. This is expressed in seconds, and does not have to be an integer, decimal values are allowed.
13. The "CorrelationHeatMap" program considers that the correlation is significant if it passes the two following criteria: (a) the averaged maximum value of the cross correlation coefficient is superior to the 90% confidence level of the correlation value, depending on the number of sampled points, as defined by Pearson; (b) the confidence interval of the correlations values obtained by spline fitting of the correlation curves, computed by bootstrapping 2000 samples from the residuals, does not include 0. A pop up will appear for each protrusion, letting the user decide to keep or discard this protrusion from the analysis (*see* **Note 5**).

14. As output of the program, for each pair of "Recruitment" and "Edge dynamics" maps, two files are saved:

    The first file is named ProtrusionName_CorrelationHeatMap.tif and shows the value of correlation for each window (sampled point) in *Y* (as many *Y* as sampled point) against the shift in time (lag) in *X* (*X* will go from +/− the maximum value of lag allowed).

    The second file is called ProtrusionName_CorrelationFitting, and shows three plots from top to bottom:

    Average curve of cross-correlation with Standard deviation.

    Average curve of cross-correlation with 90% of confidence level by spline fitting and bootstrapping.

    Average curve with a *y*-axis rescaled to the maximum correlation value.

    When the code is run on several protrusions, it merges correlation curves for individual protrusion to obtain the averaged lag leading to a maximum of correlation, and calculates a confidence interval for this value.

    An additional file called Correlationfitting.jpg displays the correlation fitting merge of all protrusions correlation curves, and gives the final result.

4 Notes

1. Comparison between transient and stable expression of the protein of interest. In the example described in this chapter, we exploited HEK-HT cells transiently transfected with the vector pEGFP-Abi1 and imaged 1–2 days after transfection. By lentiviruses infection, we also generated HEK-HT cells stably expressing mCherry-Abi1. Indeed, even though more time-consuming, the use of stable cell lines offers several important advantages: lower (i.e., more physiological) level of expression, homogeneity of cell population and consequent facility in cell selection for imaging.
2. Requirements for computer. If the computer is running on Windows 64 bits, a free executable version of the homemade "CorrelationHeatMap" code with an installer is provided. On other operative systems, license for MatLab is required, with the Signal Processing and the Statistics Matlab toolboxes. The software has been tested with Matlab versions from 2015A.
3. Supplementary information. All source codes (for ImageJ and Matlab), datasets and video tutorials are available on github: https://github.com/PerrineGilloteaux/migrationChapter.

4. Warning about photobleaching and phototoxicity. As always in cell imaging, the undesirable effects of illumination should be critically evaluated. In general, the laser power should be maintained as low as possible. The use of a very sensitive EM-CCD camera allowed us to perform image acquisition of weak fluorescent signals with moderate laser intensity, thus without excessive photobleaching (estimated to 20% for 277 frames in the selected movie).
5. Tips to discard or not a cell or protrusion from the analysis. In some cases, the user may choose to keep a cell advised to be skipped. For example, if the maximum correlation value is very similar to the 90% confidence level (0.2 for 0.21 for example), or if the spline confidence interval is very similar to 0 (dotted red horizontal lines crossing green vertical line in protrusionname_CorrelationFitting.jpg). Conversely, a user may choose to discard a protrusion, which passed the tests, in particular if the lag found corresponds to the maximum lag entered as a parameter (i.e., green vertical lie at the border of the curve in protrusionname_CorrelationFitting.jpg).

Acknowledgments

This work was supported by Fondation ARC pour la Recherche sur le Cancer (PJA 20151203371 to M.C.P.), Ligue National contre le Cancer (RS14/75-54 to M.C.P.) and Institut national de la Santé et de la Recherche médicale (Inserm, PC201530 to M.C.P.). The authors acknowledge the France Bio-Imaging infrastructure supported by the French National Research Agency (ANR-10-INBS-04, "Investments for the future").

References

1. Coutu DL, Schroeder T (2013) Probing cellular processes by long-term live imaging—historic problems and current solutions. J Cell Sci 126:3805–3815. https://doi.org/10.1242/jcs.118349
2. Machacek M, Hodgson L, Welch C, Elliott H, Pertz O, Nalbant P, Abell A, Johnson GL, Hahn KM, Danuser G (2009) Coordination of Rho GTPase activities during cell protrusion. Nature 461:99–103. https://doi.org/10.1038/nature08242
3. Lee K, Elliott HL, Oak Y, Zee C-T, Groisman A, Tytell JD, Danuser G (2015) Functional hierarchy of redundant actin assembly factors revealed by fine-grained registration of intrinsic image fluctuations. Cell Syst 1:37–50. https://doi.org/10.1016/j.cels.2015.07.001
4. Deforet M, Parrini MC, Petitjean L, Biondini M, Buguin A, Camonis J, Silberzan P (2012) Automated velocity mapping of migrating cell populations (AVeMap). Nat Methods 9:1081–1083. https://doi.org/10.1038/nmeth.2209
5. Hayer A, Shao L, Chung M, Joubert L-M, Yang HW, Tsai F-C, Bisaria A, Betzig E, Meyer T (2016) Engulfed cadherin fingers are polarized junctional structures between collectively migrating endothelial cells. Nat Cell Biol 18:1311–1323. https://doi.org/10.1038/ncb3438
6. Dormann D, Libotte T, Weijer CJ, Bretschneider T (2002) Simultaneous quanti-

fication of cell motility and protein-membrane-association using active contours. Cell Motil Cytoskeleton 52:221–230. https://doi.org/10.1002/cm.10048

7. Tyson RA, Epstein DBA, Anderson KI, Bretschneider T (2010) High resolution tracking of cell membrane dynamics in moving cells: an electrifying approach. mathematical modelling of natural phenomena. Math Model Nat Phenom 5:34–55. https://doi.org/10.1051/mmnp/20105102
8. Derivery E, Gautreau A (2010) Generation of branched actin networks: assembly and regulation of the N-WASP and WAVE molecular machines. BioEssays 32:119–131. https://doi.org/10.1002/bies.200900123
9. Biondini M, Sadou-Dubourgnoux A, Paul-Gilloteaux P, Zago G, Arslanhan MD, Waharte F, Formstecher E, Hertzog M, Yu J, Guerois R, Gautreau A, Scita G, Camonis J, Parrini MC (2016) Direct interaction between exocyst and wave complexes promotes cell protrusions and motility. J Cell Sci 129:3756–3769. https://doi.org/10.1242/jcs.187336
10. Hahn WC, Counter CM, Lundberg AS, Beijersbergen RL, Brooks MW, Weinberg RA (1999) Creation of human tumour cells with defined genetic elements. Nature 400:464–468. https://doi.org/10.1038/22780

Chapter 21

Using Single-Protein Tracking to Study Cell Migration

Thomas Orré, Amine Mehidi, Sophie Massou, Olivier Rossier, and Grégory Giannone

Abstract

To get a complete understanding of cell migration, it is critical to study its orchestration at the molecular level. Since the recent developments in single-molecule imaging, it is now possible to study molecular phenomena at the single-molecule level inside living cells. In this chapter, we describe how such approaches have been and can be used to decipher molecular mechanisms involved in cell migration.

Key words Cell migration, Single-protein tracking, Superresolution microscopy, Integrin-dependent adhesion, Lamellipodium

Abbreviations

AOTF	Acousto-optic tunable filter
Cas9	CRISPR associated protein 9
CRISPR	Clustered regularly interspaced short palindromic repeats
D	Diffusion coefficient
DMEM	Dulbecco's modified Eagle medium, high glucose
EDTA	Ethylenediaminetetraacetic acid
EMCCD	Electron-multiplying charge-coupled device
FBS	Fetal bovine serum
FWHM	Full width at half maximum
GFP	Green fluorescent protein
MEF	Mouse embryonic fibroblast
mEos2	Monomeric Eos2
MSD	Mean square displacement
NA	Numerical aperture
PALM	Photoactivation localization microscopy
PBS	Phosphate buffered saline
sCMOS	Scientific complementary metal-oxide semiconductor

Thomas Orré and Amine Mehidi contributed equally to this work.

Alexis Gautreau (ed.), *Cell Migration: Methods and Protocols*, Methods in Molecular Biology, vol. 1749, https://doi.org/10.1007/978-1-4939-7701-7_21,

SRR	Superresolved reconstruction image
STL	Superresolution time-lapse movie
TIRF	Total internal reflection fluorescence
uPAINT	Universal point accumulation imaging in nanoscale topography

1 Introduction

Cell migration is a complex process which requires the spatiotemporal coordination of molecular machineries, such as force generating modules (myosins, actin polymerizing filaments, ...), cell anchoring structures (integrin-dependent, cadherin-dependent adhesions, ...), and force transmitting modules (anchor-cytoskeleton linkages, actin cross-linkers, ...).

To break down the coordination of these modules and understand the involved mechanisms, it is necessary to study the dynamics of those molecular machines in living cells at the protein level. In fact, mobility of proteins is often correlated to their activity. This is particularly true for proteins inside adhesive structures that assemble into molecular linkages transmitting forces between the extracellular matrix (ECM) and the actin cytoskeleton. For instance, structural molecules of focal adhesions exhibit distinct levels of connection to the actin retrograde flow, suggesting hierarchical mechanical coupling to the actin cytoskeleton [1]. Furthermore, the immobilization of integrins, core components of ECM adhesive structures, is correlated to their activation [2, 3]. Correlations between protein mobility and activity has also been reported for other families of proteins including synaptic neurotransmitter receptors [4] and small G-proteins. For instance, the small G-protein Ras is mainly undergoing fast diffusion along the plasma membrane in its basal state, whereas it shows slower diffusion and increased occurrence of immobilization after activation by EGF stimulation [5]. This correlation between level of activation and mobility of proteins was also demonstrated for Rho GTPases, molecular switches controlling the actin cytoskeleton [6, 7].

Besides, living systems show important spatial and temporal heterogeneities, which, most of the time, are only probed partially by ensemble imaging techniques. Yet, probing the full distribution of protein behaviors rather than an average parameter can reveal the existence of multiple subpopulations and enable to observe intermediate states or rare events. Moreover, to study the molecular mechanisms underlying cell migration, it is particularly informative to probe the spatiotemporal distribution of the involved proteins, as movements at the cellular scale are largely generated by mechanical interplays between structural elements. Furthermore, specific spatiotemporal distribution of signaling proteins (e.g., Cdc42, Rho, and Rac), is required for the establishment of cell polarity, which is critical for cell migration [8].

Previous single-particle tracking studies in living cells used 20–40 nm gold nanoparticles [9, 10], micron-sized beads functionalized with ligands of adhesion receptors [11], quantum dots [12, 13], or antibodies coupled to organic fluorophores [14]. Tracking integrins with these approaches provided important insights in the molecular understanding of integrin-mediated adhesion and force generation, sensing and response [9, 15, 16]. However, these approaches mostly reported the behavior of groups of integrin molecules [9, 15, 16], though single-molecule tracking was achievable using low degrees of bead functionalization [17]. Moreover, most of these probes cannot cross the plasma membrane unless using invasive approaches (microinjection, electroporation, osmotic shock, ...). Therefore, they are not convenient to study the dynamic behavior of intracellular proteins or of transmembrane proteins inside adhesive and protrusive structures of living cells, which are confined and crowded multimolecular assemblies mostly localized on the ventral side of cells [18, 19].

These problems were solved thanks to developments of photoactivatable/photoconvertible fluorescent proteins, enabling straightforward single-molecule tracking in living cells [20]. Single-molecule imaging approaches, such as PALM (photoactivation localization microscopy) [21] were initially developed to bypass Abbe diffraction limit and are therefore usually referred to as superresolution imaging techniques (for extensive reviews of superresolution techniques [22, 23]. These techniques rely on the detection of fluorescent emission of sparse, isolated objects at a single time. By enabling the detector to record nonoverlapping fluorescent signals, the signal emitter can then be computationally localized with a precision down to 10 nm [24]. Low densities of emitting fluorophores in a biological sample can be obtained by exposing the sample to a highly diluted labeling solution, or by using photoactivatable/photoconvertible labels. Photoactivatable /photoconvertible fluorophores have photophysical properties that can be modified by light-induced chemical reactions. For instance, the Eos photoconvertible protein is converted from a green fluorescent form to a red fluorescent form following UV irradiation, a property found in numerous photoconvertible proteins [25]. By irradiating the sample with very low intensity UV light, only a few Eos molecules will undergo photoconversion per unit of time, enabling the experimentalist to image sparse and individual red-emitting fluorophores.

In this chapter, we focus on single-particle tracking rather than on superresolution microscopy. Nevertheless, it is important to note that single-particle tracking with ultrahigh densities of trajectories can also be used to generate superresolved images or density/diffusion maps below the diffraction limit [6, 26, 27].

Labeling of the target protein can be achieved by strategies falling into three categories:

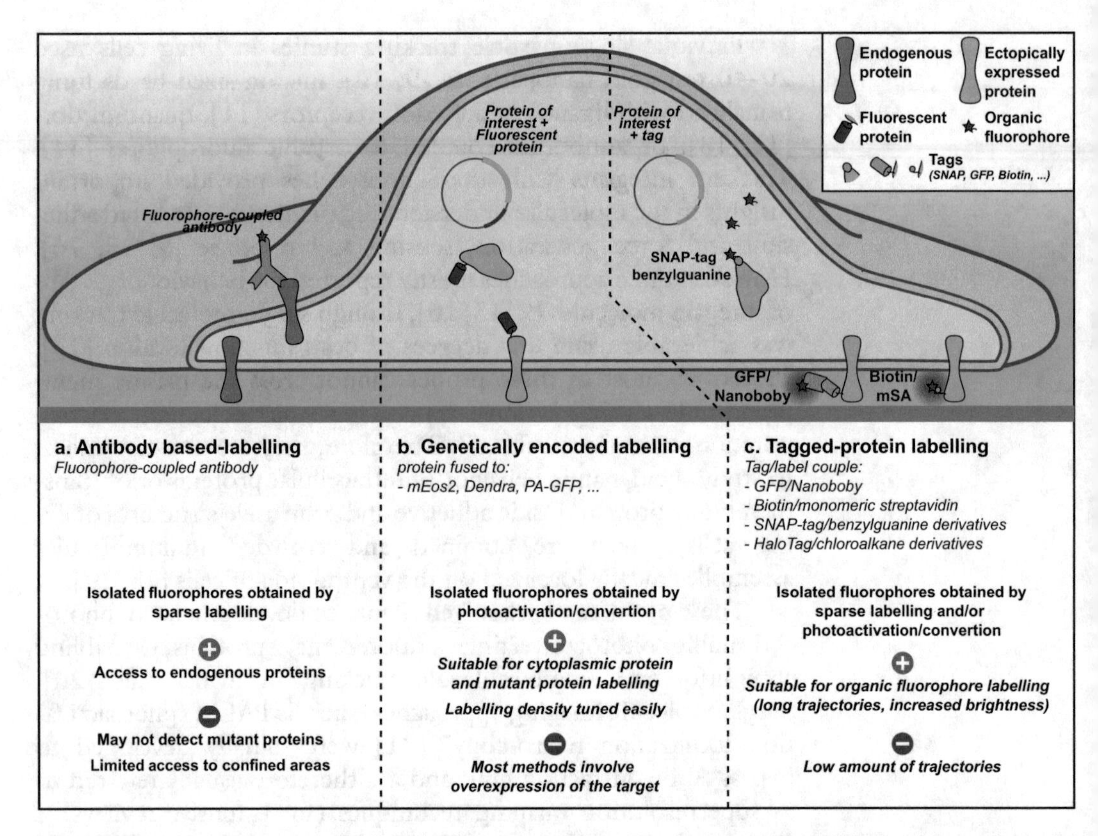

Fig. 1 Different labeling strategies for single-protein tracking. The three main labeling strategies suitable for single-protein tracking are shown here, along with the advantages and drawbacks of each approach

(a) Antibody-based labeling [14, 26] (Fig. 1a): sample cells are treated with a fluorophore-coupled antibody recognizing the target protein. However, at the exception of few studies [18, 28], this approach is limited to the study of transmembrane or extracellular targets as antibodies cannot cross the plasma membrane. Furthermore, the use of multivalent antibodies can prevent quantitative imaging and induce clustering artifacts.

(b) Genetically encoded labeling (Fig. 1b): sample cells are transfected with a DNA encoding for the targeted protein genetically fused with a fluorescent protein. Cells transfected with a such construct then produce the target protein with a label attached to it. In this labeling approach, the endogenous, nonlabeled protein is expressed concomitantly to the labeled protein.

(c) Tagged-protein labeling (Fig. 1c), which combines aspects of the two first approaches. In this case, sample cells are transfected with DNA encoding for the targeted protein fused with the sequence of a “tag” able to chemically react with a given labeling agent (SNAP-tag/benzylguanine derivatives [29–31], HaloTag/chloroalkane derivatives [7, 31], GFP/Nanoboby [32], Biotin/monomeric streptavidin [19, 33]).

Compared to the antibody-based labeling, the tagged-protein approach provides more specific labeling, with controlled valency, enabling quantitative labeling of the protein of interest. It is also an easily achievable labeling approach when antibodies targeting the protein of interest are not available.

Nevertheless, both antibody and tagged-protein based approaches enable the use of organic fluorophores as labeling agents (e.g., Alexa Fluor dyes, Atto dyes), which present several advantages in term of imaging. In fact, compared to fluorescent proteins, these probes show increased brightness and photostability, thereby facilitating single-molecule tracking. Moreover, some of these probes are membrane permeant, enabling intracellular protein labeling [31]. Application of single-molecule imaging techniques consists in three steps: (1) Sample preparation, (2) Sample imaging, and (3) Data treatment and analysis, each step involving critical aspects to take into account to get relevant data. In this chapter, we describe those steps in order to study molecular processes occurring in cellular structures involved during migration, such as the lamellipodium and integrin-dependent adhesions.

2 Materials

2.1 Solutions and Reagents

Use ultrapure water and cell culture tested products.

1. Ringer medium: 150 mM NaCl, 5 mM KCl, 2 mM $CaCl_2$, 2 mM $MgCl_2$, 10 mM HEPES 11 mM glucose at pH 7.4 (*see* **Notes 1** and **2**).
2. Cell culture medium: Dulbecco's modified Eagle medium, high glucose (DMEM), 10% fetal bovine serum (FBS), GlutaMAX supplement, 100 U/ml penicillin–streptomycin, 1 mM sodium pyruvate, 15 mM HEPES.
3. Commercial 0.05% trypsin, 0.02% EDTA solution.
4. Trypsin inactivation medium: DMEM, 1 mg/ml soybean trypsin inhibitor, GlutaMAX supplement, 100 U/ml penicillin–streptomycin, 1 mM sodium pyruvate, 15 mM HEPES.
5. Phosphate buffered saline (PBS).
6. Purified human fibronectin solution (10 μg/ml) in PBS.
7. Nitric acid solution (65% wt/wt in water).

2.2 Imaging Setup

1. Inverted microscope (Nikon TiE or equivalent).
2. High NA (1.45 or 1.49) oil immersion objectives designed for TIRF microscopy.
3. System to maintain the focus (e.g., Perfect Focus System, Nikon).

4. Microscope temperature control system (The Cube, The Box, Life Imaging Services).
5. Continuous wave (cw) lasers for excitation depending on the fluorophore to be imaged, e.g., 23 mW HeNe laser (Thorlabs), frequency doubled Nd:Yag (Coherent) or solid state lasers.
6. Low noise highly sensitive electron-multiplying charge-coupled device (EMCCD) camera (e.g., Evolve, Photometrics), or scientific complementary metal-oxide semiconductor (sCMOS) camera (e.g., Orca Flash4 by Hamamatsu).
7. Several optical and optomechanical components including mirrors, dichroic filters, and lenses.
8. Fast shutter (Uniblitz) or AOTF (AA optoelectronic).
9. For each fluorophore, an appropriate set of filters (Chroma or Semrock) is required.
10. Ludin opened sample holder (Life Imaging Services).
11. Computer for image acquisition, image processing, and visualization.
12. Software for image acquisition (e.g., Metamorph, Molecular Devices).
13. Software for image processing and visualization: Wavetracer commercial software [34] or free alternatives: ThunderSTORM [35] for superresolution reconstruction, ICY [36] with spot detector and spot tracker plugins for single particle detection and tracking [37]. Kymotoolbox ImageJ plug-in from Fabrice Cordelière [38].

2.3 Others

1. Nucleofactor™ transfection kit for MEF-1 and Nucleofactor™ 2b device (Amaxa™) (*see* **Note 3**).
2. #1.5H coverslips (Marinfield).
3. 75 cm^2 flasks for cell culture.
4. Multicolor fluorescent microbeads (Tetraspeck, Invitrogen).
5. Incubator at 37 °C supplied with humidified air containing 5% CO_2.

3 Methods

Most protocol steps are identical to achieve single-molecule imaging using photoactivatable/convertible fluorescent proteins or labeling of a tagged protein with organic dyes. Therefore, we present below a single procedure in which the protein of interest, either coupled to a fluorophore or a tag recognized by organic dyes, is referred to as "fusion protein". For a comparison of these different labeling strategies, *see* **Note 4**.

3.1 Cell Preparation

3.1.1 Cell Electroporation (24 h Before Imaging)

1. Actively dividing immortalized Mouse Embryonic Fibroblasts (MEFs, *see* **Note 3**) cultured in DMEM with 10% FBS in 75 cm² flasks are detached with trypsin/EDTA solution (1.5 ml). After 1–3 min, trypsin is inactivated by adding serum-containing DMEM (5 ml). Cells are then counted and cell suspension volume is adjusted to keep 1–2 million cells per tube per experimental condition.
2. After centrifugation at 300 × *g* for 5 min, cell pellet is resuspended in transfection reagent and mixed with the DNA. Cells are usually cotransfected with DNAs encoding for the fusion protein, (3–5 μg per condition, e.g., Eos-protein or tag-protein, *see* **Notes 5–7**) and for a GFP-coupled reporter of the structure of interest (1–2 μg per condition, e.g., GFP-paxillin for adhesive structures, GFP-alpha-actinin for lamellipodia). Cells are then electroporated with the Nucleofector™ 2b Device using the MEF T-020 program (Lonza Nucleofactor protocol) (*see* **Note 3**).
3. After electroporation, cells are replated in a 6-well plate (about 0.3 million cells/well) in preheated culture medium and placed in a 37 °C incubator with humidified air containing 5% CO_2.

3.1.2 Coverslip Cleaning, Sterilization, and Coating

Before coating, coverslips are cleaned to ensure absence of nonspecific fluorescent materials.

1. Coverslips are placed in ceramic racks.
2. Racks are placed in concentrated nitric acid (65% wt/wt) bath in staining glass boxes overnight.
3. The racks are moved in ultrapure water bath in other staining glass boxes to rinse the coverslips. Six changes of ultrapure water bath every 30 min (or more) are required.
4. Coverslips are quickly rinsed in absolute ethanol.
5. Ceramic racks are placed in a glass beaker, covered with aluminum foil and sterilized in an oven at 240 °C for 8 h.

Coverslip coating (1 day before, or same day of the microscopy experiment, *see* **Note 2**).

1. Cleaned coverslip are covered with a 10 μg/ml fibronectin solution (minimal volume: 150 μl per coverslip) and placed at 37 °C for 60–90 min.
2. After incubation, fibronectin solution is aspirated and coverslips are washed three times in PBS.
3. Coverslips are stored in PBS medium at 4 °C until the experiment.

3.1.3 Sample Preparation (Day of the Experiment, See Note 8)

1. Culture medium is removed and cells are washed twice with PBS to remove trypsin-inactivating proteins. Cells are then detached with trypsin–EDTA solution (0.3 ml per well) for 1–3 min at 37 °C.

2. Trypsin is inactivated with trypsin inactivation medium (1 ml per well) and are counted using any conventional cell counting method.
3. After centrifugation at 300 × *g* for 5 min, cells are resuspended in Ringer medium (1–2 ml).
4. *For PALM experiments (see* **Note 9**):

   Particle tracking inside focal adhesions: right after resuspension, cells are plated on fibronectin coated coverslips in Ringer medium (30,000–50,000 cells per coverslip). After 3 h, coverslips can be mounted on an open chamber with 0.8–1.6 ml of Ringer medium for imaging.

   Particle tracking inside lamellipodia: after resuspension, cells are incubated at 37 °C and 5% CO_2 for 30 min before plating (to allow integrin turnover after trypsin degradation). After incubation cells are stored at room temperature and 10,000–20,000 cells are loaded directly on fibronectin coated coverslips mounted on an open chamber containing ~800 μl of Ringer medium. Imaging is started when cells start spreading and forming lamellipodia, typically 5–10 min after loading.

   For exogenous labeling with nonphotoactivatable dyes (see **Note 9**):

   To avoid cell detachment during ulterior washing steps, labeling is started at least 30 min after loading the cells on the coverslip. The coverslip is then incubated 10 min at 37 °C in a solution containing the labeling agent, typically at a concentration of 1 nM. The coverslip is then washed three times in PBS and placed in the imaging chamber with Ringer medium.

3.2 Acquisition

1. Cells chosen for single-molecule imaging are typically identified by observing the GFP-fused reporter of the structure of interest.
2. To measure the mobility of a protein of interest within a given region of the cell, a subregion of the camera field of view is selected for recording at high frequency (typically 20–50 Hz, *see* **Note 10**). Several sequences of single-molecule imaging are then continuously recorded (for a total of 5000–20,000 images, *see* **Notes 11–13**), interlaced with images of the fluorescent reporter to monitor possible displacement of the structure of interest (e.g., focal adhesion, lamellipodium, growth cone, …). Acquisitions are made in the Total Internal Reflection Fluorescence (TIRF) mode, using an inverted motorized microscope (Nikon Ti) equipped with a 1.49 NA 100× oil immersion objective and a Perfect Focus System, placed in a thermostatic enclosure at 37 °C. Excitation of the fluorophore coupled to the protein of interest at high laser power can induce rapid photobleaching, thereby preventing tracking of molecules over time. In the case of mEos2 imaging, we typically

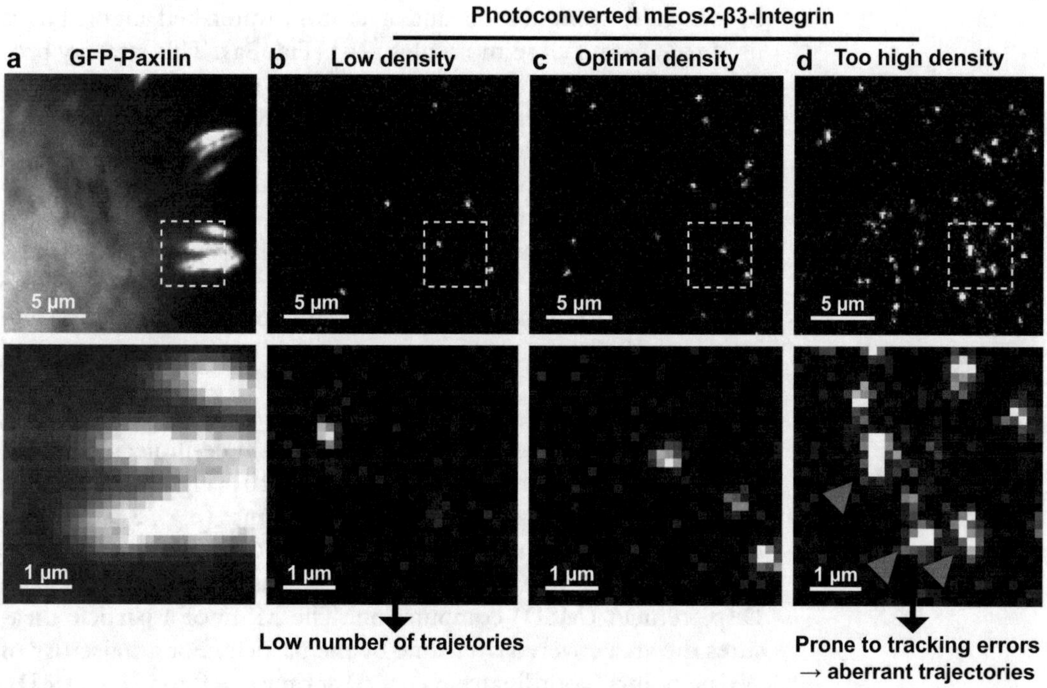

Photoconverted mEos2-β3-Integrin

a GFP-Paxilin b Low density c Optimal density d Too high density

5 μm 5 μm 5 μm 5 μm

1 μm 1 μm 1 μm 1 μm

Low number of trajetories

Prone to tracking errors → aberrant trajectories

Fig. 2 Raw images of single-molecule imaging. Images of a MEF coexpressing GFP-paxillin (**a**) for focal adhesion labeling, and mEos2-β3-integrin for single-protein tracking of integrins using sptPALM. (**b–d**) Raw images of sptPALM of mEos2-β3-Integrin. Increasing power of photoactivating 405 nm irradiation correlates with increasing density of photoconverted mEos2-β3-integrin. Representations of suboptimal (**b**), adequate (**c**) or tracking error prone (**c**) fluorophore densities for single-molecule tracking. Orange arrow heads highlight areas where photoconverted mEos2-β3-integrin are overlapping. Lower panels correspond to zoomed-in insets of upper panels

use a 561 nm beam power ranging from 4 to 8 mW at objective output to allow single fluorophores to emit on several consecutive recorded planes before photobleaching.

For photoactivatable/convertible fluorophore imaging (e.g., mEos2), fluorophores are continuously photoconverted and excited by TIRF illumination at corresponding wavelengths. (i.e., 405 and 561 nm for mEos2). Photoactivation/conversion laser power is adjusted to maintain fluorophore densities compatible with single-molecule tracking (Fig. 2).

3.3 Data Treatment and Analysis

3.3.1 Detection and Superresolved-Reconstruction Image (Fig. 3b)

Treatment and analysis of single-molecule images are computer resource demanding and are therefore usually performed after the experiment. However, software tools allowing real-time single-particle detection and tracking have been recently developed (e.g., WaveTracer [34]). In the single-molecule images, individual emitting fluorophores appear as diffracted bright spots (Fig. 2). After decomposition of the original images into wavelet maps using custom algorithms, each fluorophore is localized on segmented images by centroid computation and a custom-watershed algorithm is used to separate close molecules [24] (Fig. 3a). This strategy provides similar localization precision as more conventional Gaussian fitting, but with a tenfold decreased computation time. Those single-molecule detections can then be plotted on a single image, forming a superresolved reconstruction (SRR) image (Fig. 3b). To measure the pointing accuracy of our setup, we repetitively imaged purified mEos2 proteins adsorbed on a glass coverslip. As measured with the Full Width at Half Maximum (FWHM) of the obtained distribution of localizations, the resolution of our system is around 50 nm.

3.3.2 Fast Dynamics: Tracking and Superresolved-Tracks Image (Fig. 3c)

Single-molecule trajectories are then generated by reconnecting single molecule detections on consecutive single-molecule images using simulated annealing algorithms [39, 40] (Fig. 3c). There are several ways to analyze single-particle dynamics (e.g., 2D displacement [41], Mean Square Displacement [2, 6, 7], Bayesian analysis [42]). Here we will briefly present the analysis via Mean Square Displacement (MSD) computation. The MSD of a particle measures the area covered over time by the particle. For a trajectory of N data points (coordinates $x(t)$, $y(t)$) at times $t = 0$ to $(N-1) \times Dt$, with for example $Dt = 20$ ms, the MSD for time intervals $t = n \times Dt$ is calculated using the formula (Eq.1):

$$\mathrm{MSD}(t = n \cdot \Delta t) = \frac{\sum_{i=1}^{N-n} \left(x_{i+n} - x_i\right)^2 + \left(y_{i+n} - y_i\right)^2}{N - n}. \quad (1)$$

Particles undergoing free-diffusion show linear MSD over time, whereas the MSD of particles diffusing in a confined area will reach a plateau over time. Besides, the MSD over time of immobile particles is constant and arises from the localization error of the system used. Therefore, molecule diffusion characteristics can be determined by fitting the MSD curves obtained from the single-molecule trajectories (Fig. 3b, *see* **Note 14**). Arising from the above formula, the last points of the MSD curve are computed using a small number of values. Therefore, these poorly averaged

Fig. 3 (continued) Fractions of particles undergoing free diffusion, confined diffusion or immobilization for mEos2-Actin (orange) and mEos2-Integrin β3 (blue). (**d**) Superresolution time-lapse movies and kymographs to study slow dynamics. A superresolution time-lapse movie is generated by merging 25 consecutive planes of the single-molecule detection movie. Kymographs are then produced from this movie. The horizontal axis displays the spatial dimension and the vertical axis represents elapsed time. Kymographs perpendicular to the cell edge (highlighted by a violet line in (b), middle), generated from mEos2-Actin sptPALM time-lapse. The mEos2-Actin motions are highlighted by dashed violet lines with the corresponding velocity in nm/s (middle). Kymographs obtained from mEos2-β3-integrin superresolution time-lapse images along a focal adhesion (highlighted by violet lines in (b), bottom). The mEos2-β3-integrin motions are highlighted by dashed violet lines with the corresponding velocity in nm/s (bottom)

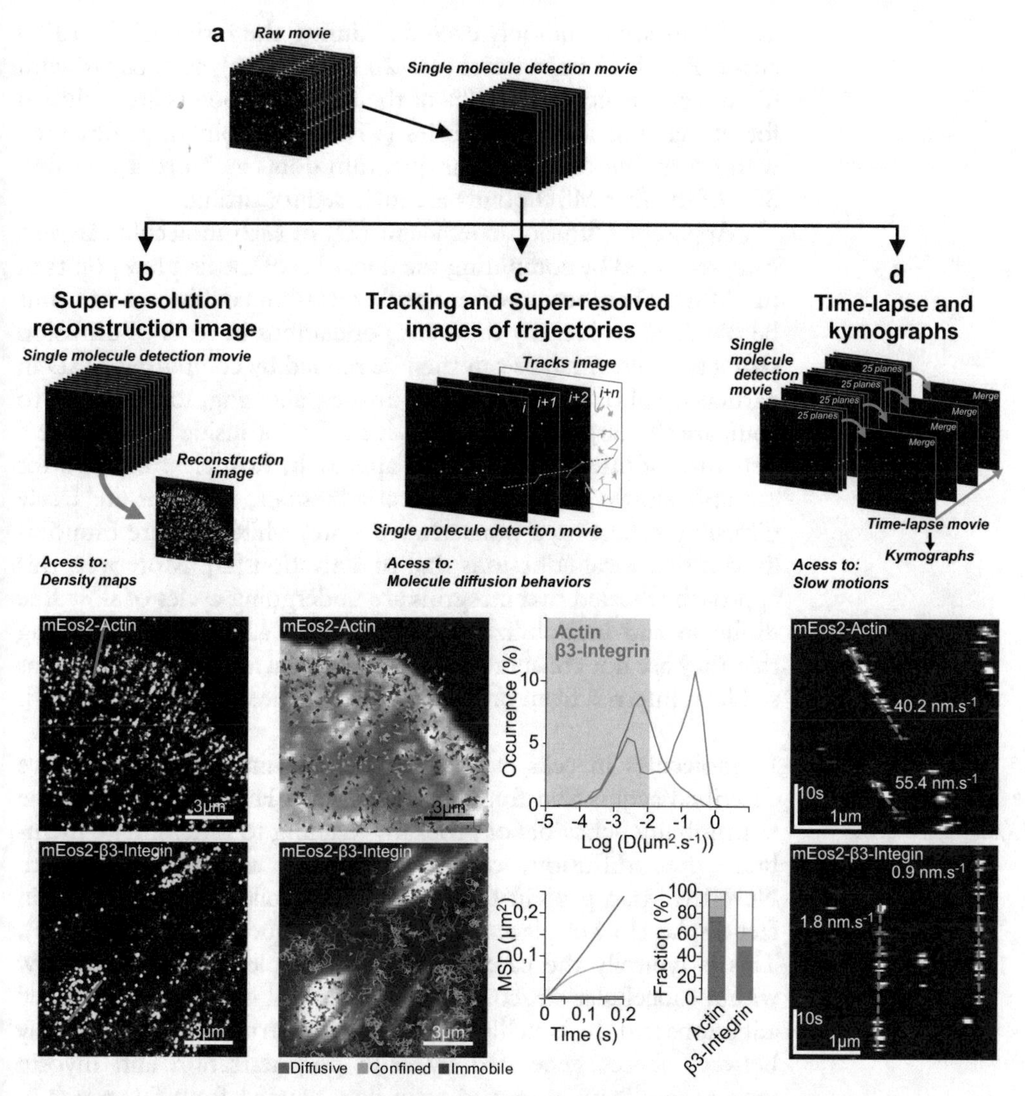

Fig. 3 Single-molecule images: treatment and analysis. (**a**) Single molecules are detected with a 50 nm spatial resolution from a stream recording of sptPALM images. (**b–d**) Various ways to extract information from single-molecule detection movies. (**b**) Superresolved-reconstruction image and distribution of localizations. The superresolved-reconstruction image is obtained by superimposing the single-molecule detection images (top). Middle panel shows the superresolved-reconstruction image of mEos2-Actin in the lamellipodium of a spreading MEF, obtained from a sptPALM recording of 150 s at 50 Hz. Bottom panel shows the superresolved-reconstruction image of mEos2-β3-integrin in focal adhesions of a MEF, obtained from a sptPALM recording of 320 s at 50 Hz. (**c**) Superresolved trajectories image and molecule diffusion behaviors. The superresolved trajectories image is obtained by reconnecting single-molecule detections on consecutive planes (top). The trajectories from mEos2-Actin (middle left) and mEos2-Integrin β3 (bottom left) are color-coded to show their diffusion modes: free diffusion (green), confined diffusion (yellow), and immobile (red). The lamellipodium and focal adhesions are labeled by GFP-α-actinin (middle left) and GFP-Paxilin (bottom left), respectively. Distributions of the Diffusion coefficient D computed from the obtained trajectories with mEos2-Actin (orange) and mEos2-Integrin β3 (blue) are shown in a logarithmic scale (middle right). The grey area including D values inferior to 0.011 μm²/s corresponds to immobile trajectories. Variation of theoretical values of MSD over time for particles undergoing free diffusion (green), confined diffusion (yellow), or immobilization (red) (bottom, middle).

last points are commonly excluded during the fitting of the MSD curve. For short trajectories (10–20 time points), as obtained with fluorescent proteins, 60–80% of the first MSD points are included for fitting. For long trajectories (>100 time points), as obtained with organic fluorophores or quantum dots (*see* **Note 4**), around 30% of the first MSD points are included for fitting.

Apparent Diffusion coefficient (D) of each molecule can also be determined by computing the linear fits of the first few points of the MSD. Distribution of molecule diffusion coefficients can thus be obtained to identify different populations in term of diffusion characteristics. Analysis can then be refined by computing MSD in particular subregions of the raw images, allowing, for instance, to compare the target protein diffusive behavior inside and outside a structure of interest. Using this approach, Rossier *et al.* have for example shown that outside focal adhesions, integrins are freely diffusing, reflecting a nonactivated state, while they are immobilized inside focal adhesions due to activation [2]. Moreover, this approach revealed that integrins are undergoing cycles of slow free diffusion and immobilization within focal adhesions, suggesting that they are not constantly activated in focal adhesions, and not as stable as inferred from previous ensemble measurements [43, 44].

3.3.3 Slow Dynamics: Time-Lapse and Kymograph (Fig. 3d)

Biomolecules in cells evolve at different time scales. So far we described acquisition frequencies (20–300 Hz) used to study the fast mobility behaviors of proteins, enabling to differentiate membrane free diffusion, confined diffusion and immobilization. Nevertheless, a protein that appears immobile at 50 Hz could in fact display slow displacements at lower frequencies (e.g., 0.5 Hz). This is typically the case for proteins coupled to the actin flow within subcellular structures involved in cell migration, e.g., focal adhesions and the lamellipodium. In those structures, the interplay between forces generated by actin polymerization and myosin motors results in an inward actin flow ranging from few nm/s in focal adhesions up to 100 nm/s in lamellipodia or growth cones [2, 6, 45–48]. The actin flow is the engine powering cell migration, since its connection to adhesive proteins via a myriad of adaptors transmits internal forces to the ECM and triggers membrane protrusion [49]. Originally, those slower directed movements were visualized and quantified using speckle microscopy, which combines long exposures and low expression of proteins fused to fluorescent proteins [1, 48, 50]. The use of photoactivatable/photoconvertible fluorescent proteins is an alternative way to obtain isolated emitters (*see* **Note 4**).

Superresolution time-lapse movies of protein slow movements could be generated in two ways. One can use long exposures (e.g., 250 ms) and low acquisition frequencies (e.g., 2 Hz), as in speckle microscopy. In that scheme, photoactivation/conversion and excitation laser power should be decreased to prevent excessive photoactivation and bleaching (e.g., 1.5 kw/cm^2 for the excitation laser) [6, 51]. Another way to study protein slow movements is to merge several superresolved reconstruction images (e.g., 25 images) obtained from 50 Hz single-molecule imaging, generating a superresolution time-lapse movie (STL) (e.g., 2 Hz) (Fig. 3d). In fact, a single fluorophore can undergo reversible on/off fluorescent state switching ("blinking") for tens of seconds. Therefore, it is possible to track a slowly moving fluorophore for a prolonged duration by reconnecting the different "on" periods. Directed movements of proteins (rather than Brownian motion) can thereby be analyzed by tracking the position of individual proteins over time [6, 51].

Slow protein behavior can also be analyzed by generating kymographs (see Kymotoolbox ImageJ plugin developed by Fabrice Codelière from the Bordeaux Imaging Center, available on demand: fabrice.cordelieres@u-bordeaux.fr). In a region of interest, a straight line or a segmented line is typically drawn directly on the STL-movie (Fig. 3d), to generate a kymograph (*X* position over time) and a kymostack (*Y* position over *X* position). It is then possible to extract different parameters such as the speed, direction and dwell time of proteins.

Using this approach, it was demonstrated that F-actin motions within focal adhesions are transmitted differentially to specific classes of α/β integrins. β3-class integrins are stationary and enriched in focal adhesions, whereas β1-class integrins are less enriched and display rearward movements [2]. Thus, specific classes of α/β integrins act as distinct "nanoscale adhesion units" within an individual focal adhesion, with specific dynamics, organization, and force transmission of F-actin motion to the ECM. F-Actin and actin binding protein (e.g., Arp2/3 complex) flow could also be quantified in the lamellipodium of migrating cells [52], neuronal growth cones [6, 51], but also in much smaller motile structures such as neuronal dendritic spines [6, 45, 52, 53]. Importantly, unlike the fast concerted flow of actin and Arp2/3 in the lamellipodium and growth cones, actin elongation from immobile Arp2/3 in dendritic spines does not generate a fast and concerted rearward flow of actin [6]. Since actin regulators in those structures are almost identical, this suggests that the shape and dynamics of protrusive structures are determined by the nanoscale organization of F-actin regulators rather than their nature.

The developments of superresolution microscopy techniques revealed that proteins, which seem colocalized using conventional light microscopy, are spatially segregated at the nanoscale into distinct functional layers or centers [54–56]. The ability to track proteins at the nanoscale, which comes in handy with single-molecule localization microscopy techniques, is a powerful tool to study the spatiotemporal regulation of specific protein interactions within those distinct functional areas. Thus, we can now zoom into subcellular structures involved in cell migration (focal adhesions, lamellipodium, …) and understand the mechanisms locally controlling protein activities.

4 Notes

1. Other physiological imaging buffers than the one proposed here can be used. Imaging medium should, however, have low background fluorescence level, as signal/noise ratio is critical for single-molecule imaging. Components increasing background fluorescence such as phenol-red should therefore be avoided as much as possible.
2. To promote the formation of a given type of cell adhesions, specific molecules (fibronectin, collagen, …) can be adsorbed on the coverslips before cell seeding. In that case, as serum contains ECM proteins (fibronectin, vitronectin), serum-free media can be used so that adhesion formation is strictly promoted by the chosen adsorbed protein.
3. To our knowledge, there is no restriction on the transfection method or cell type to use.
4. *Comparison of the different labeling strategies*:

   Genetically encoded labeling, i.e., via the expression of the protein of interest fused to a photoactivatable/convertible fluorescent protein, enables quick and easy adjustment of the fluorophore density during single-molecule imaging experiments. One can therefore compensate for photobleaching and maintain a constant fluorophore density for a long time, allowing to accumulate thousands molecule trajectories within the same region of interest. Moreover, such ease of fluorophore density adjustment makes it possible to compensate for differences regarding the target protein density within the cell. For a same cell, it is therefore possible to sequentially perform single-molecule tracking in a region where the target protein is highly concentrated (e.g., Paxillin in focal adhesions), and then in a region containing low levels of the target protein (e.g., Paxillin outside focal adhesions), which can reveal the existence of different states of the target protein (e.g., freely diffusing vs confined). On the other hand, the vast majority of these trajectories typically last only few hundreds of milliseconds, due to frequent blinking. Using our setup to continuously image membrane-anchored mEos2 with a 561 nm laser power of around 5 mW (objective output), about 75% of the trajectories obtained last less than 200 ms due to blinking.

   Compared to fluorescent proteins, organic dyes and quantum dots present increased brightness and photostability. Increased brightness improves the signal/noise ratios during single-molecule imaging, and higher spatial resolution-pointing accuracy can be obtained with such probes. These probes can thereby enable to measure smaller confinement areas than achieved with fluorescent proteins. Besides, increased photostability enables to record longer trajectories (>10 s), which can be required to measure the duration of given single-molecule behaviors (immobilization duration, diffusion duration) and observe transitions between different molecular behaviors [12, 13, 18] (Fig. 4). On the other hand, unless working with photoactivatable organic dyes, [31], or using the uPAINT approach (c.f. below), fluorophore density cannot be adjusted after labeling with this approach. As a consequence, one cannot restore the pool of emitting fluorophores after photobleaching. Therefore, fewer trajectories are obtained in a given region compared to photoactivatable/convertible label imaging, making it more time-demanding to obtain statistically representative data samples with organic fluorophores/quantum dots.

When studying a membrane protein, the uPAINT approach can be applied. This consists in recording thousands of single-molecule trajectories appearing sequentially on the cell surface while labeling of the protein of interest is ongoing [26, 27]. In this case, the pool of labeling molecules is constantly replenished due to the continuous and/or reversible binding of a labeling agent to the protein of interest. This enables to combine the use of organic fluorophores with the accumulation of a high number of trajectories. However, this method is not optimal to study proteins located under the cell, because of reduced accessibility of those proteins to labeling molecules. Besides, if the labeling molecule binds irreversibly to the target, and if the target is expressed at a low level, it may be difficult to accumulate a sufficient number of trajectories with uPAINT. This can be solved

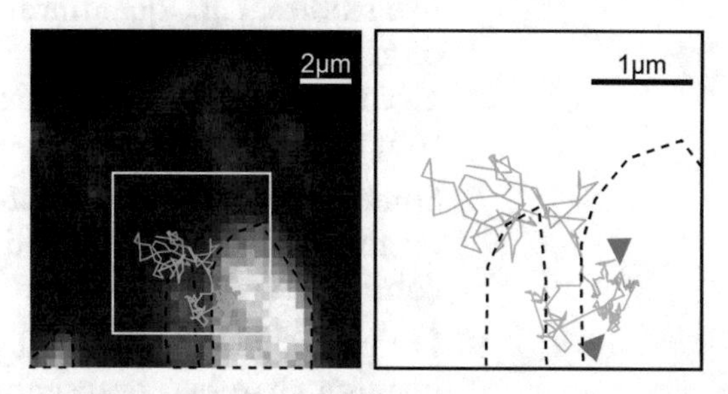

Fig. 4 Long single-molecule trajectories using organic fluorophores. Integrins immobilize after entering focal adhesions. Single-protein tracking of SNAP-tag-β3-Integrin using benzylguanine-ATTO 565 in MEFs cotransfected with GFP-Paxilin for focal adhesion labeling (Left). The outlined area in the left panel is shown at a higher magnification in the right panel. The orange arrow heads highlight the immobilization of β3-Integrin in focal adhesions

by using the DNA-Paint approach, which relies on the transient binding of short fluorescently labeled oligonucleotides [57–59]. In fact, as the binding of labeling molecules to the target is strictly transient in this approach, photobleached molecules are quickly replaced by unbleached ones.

5. Other fluorophores than mEos2 can be used to perform PALM experiments (Dendra [60], mEos3.2 [61], ...) and other fluorophores than GFP can be used to localize the structures of interest. However, strictly avoid overlap of emission wavelengths between photoactivated/converted state of the tracked protein and the reporter of the structure of interest.
6. Fusion of a protein of interest with a fluorescent protein or a tag can alter its function due to steric hindrance or oligomerization. This may be solved by changing the position of the tag/fluorophore in the fusion protein (e.g., N-terminal instead of C-terminal). Besides, some fluorophores seem more likely to oligomerize than others [61, 62]. Therefore, functions of fusion proteins used for such approaches should be checked. For instance, functionality of a fusion protein can be tested by inducing its expression in cells showing specific defects due to the absence or low expression of the protein of interest (rescuing assay). Expression of the fusion protein should have the same effect on cells as reexpression of the nonlabeled equivalent protein.
7. To perform quantitative structural studies, i.e., quantitative composition of given structures, or to circumvent potential artifacts introduced by the overexpression of the protein of interest, it is possible to express a genetically encoded labeled protein at endogenous level by using the CRISPR-Cas9 approach [63–65]. With this approach, the native, unlabeled version of the protein of interest is completely removed and replaced by the labeled fusion protein. This enables quantitative studies, e.g., quantification of the molecular composition of given structures, as the amount of protein of interest in a given region can be measured based on the fluorescent signal [66, 67].
8. Some fluorophores are photoactivated/converted or bleached by ambient light and should be manipulated in reduced room lighting.
9. For prolonged imaging (>1 h) of the sample mounted on the imaging chamber, evaporation can be prevented by adding a glass cap on top. Note that smaller volumes are more sensitive to evaporation-induced concentration increase of buffer constituents. Due to the minimal composition of Ringer imaging buffer, it is not adapted to long cell incubation/

imaging (>12 h). In such cases, specific nutrients or growth factors may be added to Ringer medium.

10. Studying the fast motion of single proteins requires high frequency imaging. In this case, only the most relevant subregion in the camera field of view is selected for recording, as most setup cannot generate whole field 50 Hz movie due to limited data transfer/writing speed. Increased acquisition speed (100–500 Hz) may be obtained by reducing the size of the camera field of view subregion chosen for recording or using fast cameras (sCMOS).
11. Typically, 10,000–20,000 planes are recorded for single-molecule imaging, so as to ensure proper sampling of the fusion protein molecular behavior. For proteins showing low levels of expression/labeling, this amount may be increased to obtain a sufficient number of single-molecule trajectories.
12. Fluorescent beads (e.g., 100 nm Tetraspeck, Invitrogen) can be adsorbed to the coverslip before image recording to enable post-experiment acquisition registration (lateral drift correction). Typically, a drop of 100 μl of diluted bead solution (1:10,000) is loaded on Parafilm and the coverslip is put upside down on the drop for 15 min at room temperature. After incubation, the coverslip is rinsed twice with PBS before imaging.
13. When using photoconvertible fluorophores (and some photoactivatable ones), nonspecific background signal can be decreased by photobleaching the sample before photoactivation/photoconversion, improving signal/noise ratio in the following single-molecule recordings. This can be achieved by illuminating the sample during few seconds at the excitation wavelength of the fluorophore used for single-molecule imaging.
14. For the MSD analysis presented above, the average duration of the trajectories selected for computing should be below the transition time between the different single-molecule behaviors of the protein of interest (e.g., transition time between free diffusion and immobilization). In fact, whereas the MSD analysis is relevant for trajectories reflecting the same behavior over time (i.e., free-diffusion or confined diffusion or immobilization), it is biased for particles shifting between different behaviors (e.g., free diffusion then immobilization). For instance, a trajectory reflecting free diffusion for half of its length and immobilization for the other half, would often be scored as a confined behavior with this approach, which is inaccurate. Moreover, the part of this trajectory corresponding to immobilization would interfere with the Diffusion coefficient computing method mentioned above. This would induce an underestimation of the Diffusion coefficient compared to the

value that would be obtained when treating the free-diffusion part alone. Nevertheless, for a given persistent molecular behavior (i.e., free diffusion or confined diffusion), the longer the trajectory, the more accurate is the MSD computing. Therefore, to study trajectories representing a unique molecular behavior while using the maximal trajectory duration available in the data set, it is possible to split each trajectory into multiple ones upon transition between different molecular behaviors. These transitions can be detected by computing *rolling* MSDs along the trajectory, enabling to measure a confinement probability used to segment the trajectories [68, 69].

References

1. Hu K, Ji L, Applegate KT et al (2007) Differential transmission of actin motion within focal adhesions. Science 315:111–115. https://doi.org/10.1126/science.1135085
2. Rossier O, Octeau V, Sibarita J-BB et al (2012) Integrins β1 and β3 exhibit distinct dynamic nanoscale organizations inside focal adhesions. Nat Cell Biol 14:1057–1067. https://doi.org/10.1038/ncb2588
3. Paszek MJ, DuFort CC, Rossier O et al (2014) The cancer glycocalyx mechanically primes integrin-mediated growth and survival. Nature 511:319–325. https://doi.org/10.1038/nature13535
4. Constals A, Penn AC, Compans B et al (2015) Glutamate-induced AMPA receptor desensitization increases their mobility and modulates short-term plasticity through unbinding from stargazin. Neuron 85:787–803. https://doi.org/10.1016/j.neuron.2015.01.012
5. Murakoshi H, Iino R, Kobayashi T et al (2004) Single-molecule imaging analysis of Ras activation in living cells. Proc Natl Acad Sci 101:7317–7322. https://doi.org/10.1073/pnas.0401354101
6. Chazeau A, Mehidi A, Nair D et al (2014) Nanoscale segregation of actin nucleation and elongation factors determines dendritic spine protrusion. EMBO J 33:2745–2764. https://doi.org/10.15252/embj.201488837
7. Shibata ACE, Chen LH, Nagai R et al (2013) Rac1 recruitment to the archipelago structure of the focal adhesion through the fluid membrane as revealed by single-molecule analysis. Cytoskeleton 70:161–177. https://doi.org/10.1002/cm.21097
8. Charest PG, Firtel RA (2007) Big roles for small GTPases in the control of directed cell movement. Biochem J 401:377–390. https://doi.org/10.1042/BJ20061432
9. Felsenfeld DP, Choquet D, Sheetz MP (1996) Ligand binding regulates the directed movement of β1 integrins on fibroblasts. Nature 383:438–440. https://doi.org/10.1038/383438a0
10. De Brabander M, Geuens G, Nuydens R et al (1985) Probing microtubule-dependent intracellular motility with nanometre particle video ultramicroscopy (nanovid ultramicroscopy). Cytobios 43:273–283
11. Suter DM, Forscher P (2001) Transmission of growth cone traction force through apCAM–cytoskeletal linkages is regulated by Src family tyrosine kinase activity. J Cell Biol 155:427–438. https://doi.org/10.1083/jcb.200107063
12. Dahan M (2003) Diffusion dynamics of glycine receptors revealed by single-quantum dot tracking445. Science 302:442. https://doi.org/10.1126/science.1088525
13. Groc L, Heine M, Cognet L et al (2004) Differential activity-dependent regulation of the lateral mobilities of AMPA and NMDA receptors. Nat Neurosci 7:695–696. https://doi.org/10.1038/nn1270
14. Tardin C, Cognet L, Bats C et al (2003) Direct imaging of lateral movements of AMPA receptors inside synapses. EMBO J 22:4656–4665. https://doi.org/10.1093/emboj/cdg463
15. Choquet D, Felsenfeld DP, Sheetz MP (1997) Extracellular matrix rigidity causes strengthening of integrin–cytoskeleton linkages. Cell 88:39–48. https://doi.org/10.1016/S0092-8674(00)81856-5
16. Giannone G, Jiang G, Sutton DH et al (2003) Talin1 is critical for force-dependent reinforcement of initial integrin-cytoskeleton bonds but not tyrosine kinase activation. J Cell Biol 163:409–419. https://doi.org/10.1083/jcb.200302001

17. Jiang G, Giannone G, Critchley DR et al (2003) Two-piconewton slip bond between fibronectin and the cytoskeleton depends on talin. Nature 424:334–337. https://doi.org/10.1038/nature01805
18. Leduc C, Si S, Gautier J et al (2013) A highly specific gold nanoprobe for live-cell single-molecule imaging. Nano Lett 13:1489–1494. https://doi.org/10.1021/nl304561g
19. Chamma I, Rossier O, Giannone G et al (2017) Optimized labeling of membrane proteins for applications to super-resolution imaging in confined cellular environments using monomeric streptavidin. Nat Protoc 12:748–763. https://doi.org/10.1038/nprot.2017.010
20. Lippincott-Schwartz J, Patterson GH (2009) Photoactivatable fluorescent proteins for diffraction-limited and super-resolution imaging. Trends Cell Biol 19:555–565. https://doi.org/10.1016/j.tcb.2009.09.003
21. Manley S, Gillette JM, Patterson GH et al (2008) High-density mapping of single-molecule trajectories with photoactivated localization microscopy. Nat Methods 5:155–157. https://doi.org/10.1038/nmeth.1176
22. Cognet L, Leduc CC, Lounis B (2014) Advances in live-cell single-particle tracking and dynamic super-resolution imaging. Curr Opin Chem Biol 20:78–85. https://doi.org/10.1016/j.cbpa.2014.04.015
23. Huang B, Bates M, Zhuang X (2009) Super-resolution fluorescence microscopy. Annu Rev Biochem 78:993–1016. https://doi.org/10.1146/annurev.biochem.77.061906.092014
24. Izeddin I, Boulanger J, Racine V et al (2012) Wavelet analysis for single molecule localization microscopy. Opt Express 20:2081–2095. https://doi.org/10.1364/OE.20.002081
25. Shcherbakova DM, Sengupta P, Lippincott-Schwartz J, Verkhusha VV (2014) Photocontrollable fluorescent proteins for superresolution imaging. Annu Rev Biophys 43:303–329. https://doi.org/10.1146/annurev-biophys-051013-022836
26. Giannone G, Hosy E, Levet F et al (2010) Dynamic superresolution imaging of endogenous proteins on living cells at ultra-high density. Biophys J 99:1303–1310. https://doi.org/10.1016/j.bpj.2010.06.005
27. Nair D, Hosy E, Petersen JD et al (2013) Super-resolution imaging reveals that AMPA receptors inside synapses are dynamically organized in nanodomains regulated by PSD95. J Neurosci 33:13204–13224. https://doi.org/10.1523/JNEUROSCI.2381-12.2013
28. Courty S, Luccardini C, Bellaiche Y et al (2006) Tracking individual kinesin motors in living cells using single quantum-dot imaging. Nano Lett 6:1491–1495. https://doi.org/10.1021/nl060921t
29. Keppler A, Kindermann M, Gendreizig S et al (2004) Labeling of fusion proteins of O6-alkylguanine-DNA alkyltransferase with small molecules in vivo and in vitro. Methods 32:437–444. https://doi.org/10.1016/j.ymeth.2003.10.007
30. Keppler A, Pick H, Arrivoli C et al (2004) Labeling of fusion proteins with synthetic fluorophores in live cells. Proc Natl Acad Sci 101:9955–9959. https://doi.org/10.1073/pnas.0401923101
31. Grimm JB, English BP, Chen J et al (2015) A general method to improve fluorophores for live-cell and single-molecule microscopy. Nat Methods 12:244–250. https://doi.org/10.1038/nmeth.3256
32. Ries J, Kaplan C, Platonova E et al (2012) A simple, versatile method for GFP-based super-resolution microscopy via nanobodies. Nat Methods 9:582–584. https://doi.org/10.1038/nmeth.1991
33. Chamma I, Letellier M, Butler C et al (2016) Mapping the dynamics and nanoscale organization of synaptic adhesion proteins using monomeric streptavidin. Nat Commun 7:10773. https://doi.org/10.1038/ncomms10773
34. Kechkar A, Nair D, Heilemann M et al (2013) Real-time analysis and visualization for single-molecule based super-resolution microscopy. PLoS One. https://doi.org/10.1371/journal.pone.0062918
35. Ovesný M, Křížek P, Borkovec J et al (2014) ThunderSTORM: a comprehensive ImageJ plug-in for PALM and STORM data analysis and super-resolution imaging. Bioinformatics 30:2389–2390. https://doi.org/10.1093/bioinformatics/btu202
36. de Chaumont F, Dallongeville S, Chenouard N et al (2012) Icy: an open bioimage informatics platform for extended reproducible research. Nat Methods 9:690–696. https://doi.org/10.1038/nmeth.2075
37. Chenouard N, Bloch I, Olivo-Marin J-C (2013) Multiple hypothesis tracking for cluttered biological image sequences. IEEE Trans Pattern Anal Mach Intell 35:2736–3750. https://doi.org/10.1109/TPAMI.2013.97
38. Zala D, Hinckelmann MV, Yu H et al (2013) Vesicular glycolysis provides on-board energy for fast axonal transport. Cell 152:479–491. https://doi.org/10.1016/j.cell.2012.12.029
39. Racine V, Hertzog A, Jouanneau J, et al (2006) Multiple-target tracking of 3D fluorescent objects based on simulated annealing. In: Third IEEE international symposium on biomediocal

imaging: macro to nano, 2006. IEEE, pp 1020–1023

40. Racine V, Sachse M, Salamero J et al (2007) Visualization and quantification of vesicle trafficking on a three-dimensional cytoskeleton network in living cells. J Microsc 225:214–228. https://doi.org/10.1111/j.1365-2818.2007.01723.x. JMI1723 [pii]
41. Das S, Yin T, Yang Q et al (2015) Single-molecule tracking of small GTPase Rac1 uncovers spatial regulation of membrane translocation and mechanism for polarized signaling. Proc Natl Acad Sci U S A 112:E267–E276. https://doi.org/10.1073/pnas.1409667112
42. El BM, Dahan M, Masson J-B (2015) InferenceMAP: mapping of single-molecule dynamics with Bayesian inference. Nat Methods 12:594–595. https://doi.org/10.1038/nmeth.3441
43. Cluzel C, Saltel F, Lussi J et al (2005) The mechanisms and dynamics of αvβ3 integrin clustering in living cells. J Cell Biol 171:383–392. https://doi.org/10.1083/jcb.200503017
44. Brown CM, Hebert B, Kolin DL et al (2006) Probing the integrin-actin linkage using high-resolution protein velocity mapping. J Cell Sci 119:5204–5214. https://doi.org/10.1242/jcs.03321
45. Frost NA, Shroff H, Kong H et al (2010) Single-molecule discrimination of discrete perisynaptic and distributed sites of actin filament assembly within dendritic spines. Neuron 67:86–99. https://doi.org/10.1016/j.neuron.2010.05.026
46. Honkura N, Matsuzaki M, Noguchi J et al (2008) The subspine organization of actin fibers regulates the structure and plasticity of dendritic spines. Neuron 57:719–729. https://doi.org/10.1016/j.neuron.2008.01.013
47. Lai FPL, Szczodrak M, Block J et al (2008) Arp2/3 complex interactions and actin network turnover in lamellipodia. EMBO J 27:982–992. https://doi.org/10.1038/emboj.2008.34
48. Watanabe N, Mitchison TJ (2002) Single-molecule speckle analysis of actin filament turnover in Lamellipodia. Science 295:1083–1086. https://doi.org/10.1126/science.1067470
49. Giannone G, Mège RM, Thoumine O et al (2009) Multi-level molecular clutches in motile cell processes. Trends Cell Biol 19:475–486. https://doi.org/10.1016/j.tcb.2009.07.001
50. Ponti A, Machacek M, Gupton SL et al (2004) Two distinct actin networks drive the protrusion of migrating cells. Science 305:1782–1786. https://doi.org/10.1126/science.1100533
51. Garcia M, Leduc C, Lagardère M et al (2015) Two-tiered coupling between flowing actin and immobilized *N*-cadherin/catenin complexes in neuronal growth cones. Proc Natl Acad Sci 112:6997–7002. https://doi.org/10.1073/pnas.1423455112
52. Tatavarty V, Kim EJ, Rodionov V, Yu J (2009) Investigating sub-spine actin dynamics in rat hippocampal neurons with super-resolution optical imaging. PLoS One. https://doi.org/10.1371/journal.pone.0007724
53. Tatavarty V, Das S, Yu J (2012) Polarization of actin cytoskeleton is reduced in dendritic protrusions during early spine development in hippocampal neuron. Mol Biol Cell 23:3167–3177. https://doi.org/10.1091/mbc.E12-02-0165
54. Kanchanawong P, Shtengel G, Pasapera AM et al (2010) Nanoscale architecture of integrin-based cell adhesions. Nature 468:580–584. https://doi.org/10.1038/nature09621
55. Case LB, Baird M, Shtengel G et al (2015) Molecular mechanism of vinculin activation and nanoscale spatial organization in focal adhesions. Nat Cell Biol 17:880–892. https://doi.org/10.1038/ncb3180
56. Xu K, Zhong G, Zhuang X (2013) Actin, spectrin, and associated proteins form a periodic cytoskeletal structure in axons. Science 339:452–456. https://doi.org/10.1126/science.1232251
57. Jungmann R, Avendaño MS, Woehrstein JB et al (2014) Multiplexed 3D cellular super-resolution imaging with DNA-PAINT and exchange-PAINT. Nat Methods 11:313–318. https://doi.org/10.1038/nmeth.2835
58. Jungmann R, Steinhauer C, Scheible M et al (2010) Single-molecule kinetics and super-resolution microscopy by fluorescence imaging of transient binding on DNA origami. Nano Lett 10:4756–4761. https://doi.org/10.1021/nl103427w
59. Schnitzbauer J, Strauss MT, Schlichthaerle T et al (2017) Super-resolution microscopy with DNA-PAINT. Nat Protoc 12:1198–1228. https://doi.org/10.1038/nprot.2017.024
60. Gurskaya NG, Verkhusha VV, Shcheglov AS et al (2006) Engineering of a monomeric green-to-red photoactivatable fluorescent protein induced by blue light. Nat Biotechnol 24:461–465. https://doi.org/10.1038/nbt1191
61. Zhang M, Chang H, Zhang Y et al (2012) Rational design of true monomeric and bright photoactivatable fluorescent proteins. Nat Methods 9:727–729. https://doi.org/10.1038/nmeth.2021
62. Cranfill PJ, Sell BR, Baird MA et al (2016) Quantitative assessment of fluorescent proteins. Nat Methods 13:557–562. https://doi.org/10.1038/nmeth.3891

63. Guan J, Liu H, Shi X et al (2017) Tracking multiple genomic elements using correlative CRISPR imaging and sequential DNA FISH. Biophys J 112:1077–1084. https://doi.org/10.1016/j.bpj.2017.01.032
64. Ratz M, Testa I, Hell SW, Jakobs S (2015) CRISPR/Cas9-mediated endogenous protein tagging for RESOLFT super-resolution microscopy of living human cells. Sci Rep 5:9592. https://doi.org/10.1038/srep09592
65. Truong Quang BA, Mani M, Markova O et al (2013) Principles of E-cadherin supramolecular organization in vivo. Curr Biol 23:2197–2207. https://doi.org/10.1016/j.cub.2013.09.015
66. Karathanasis C, Fricke F, Hummer G, Heilemann M (2017) Molecule counts in localization microscopy with organic fluorophores. ChemPhysChem 18:942–948. https://doi.org/10.1002/cphc.201601425
67. Levet F, Hosy E, Kechkar A et al (2015) SR-Tesseler: a method to segment and quantify localization-based super-resolution microscopy data. Nat Methods 12:1065–1071. https://doi.org/10.1038/nmeth.3579
68. Giannone G, Mondin M, Grillo-Bosch D et al (2013) Neurexin-1β binding to neuroligin-1 triggers the preferential recruitment of PSD-95 versus gephyrin through tyrosine phosphorylation of neuroligin-1. Cell Rep 3:1996–2007. https://doi.org/10.1016/j.celrep.2013.05.013
69. Simson R, Sheets ED, Jacobson K (1995) Detection of temporary lateral confinement of membrane proteins using single-particle tracking analysis. Biophys J 69:989–993. https://doi.org/10.1016/S0006-3495(95)79972-6

Chapter 22

Optogenetic Control of Cell Migration

Xenia Meshik, Patrick R. O'Neill, and N. Gautam

Abstract

Subcellular optogenetics allows specific proteins to be optically activated or inhibited at a restricted subcellular location in intact living cells. It provides unprecedented control of dynamic cell behaviors. Optically modulating the activity of signaling molecules on one side of a cell helps optically control cell polarization and directional cell migration. Combining subcellular optogenetics with live cell imaging of the induced molecular and cellular responses in real time helps decipher the spatially and temporally dynamic molecular mechanisms that control a stereotypical complex cell behavior, cell migration. Here we describe methods for optogenetic control of cell migration by targeting three classes of key signaling switches that mediate directional cellular chemotaxis—G protein coupled receptors (GPCRs), heterotrimeric G proteins, and Rho family monomeric G proteins.

Key words Cell migration, Optogenetics, Subcellular, Signaling, Opsin, GPCRs, Light induced dimerization, Fluorescence microscopy

1 Introduction

Cell migration is central to biological processes such as the immune response, inflammation, morphogenesis, wound healing, and tumor invasion [1]. Genetic and pharmacological perturbations have helped identify many of the molecules that play a role in cell migration. However, such perturbations are typically limited to time scales that are much longer than processes that drive migration, and do not offer subcellular control. Such control is required in order to develop a better understanding of how these proteins interact to form a spatiotemporally dynamic network capable of polarizing the cell and driving migration in a directionally sensitive manner. To overcome these limitations, our laboratory has been using optogenetic methods to initiate and control cell migration [2, 3]. We do this by modulating the activation state of three classes of proteins present at various points in the migration pathway: (a) G-protein coupled receptors (GPCRs) [4, 5], (b) Gαi family proteins [6], and (c) small Rho family GTPases, Cdc42 and Rac [7].

Alexis Gautreau (ed.), *Cell Migration: Methods and Protocols*, Methods in Molecular Biology, vol. 1749,
https://doi.org/10.1007/978-1-4939-7701-7_22, © Springer Science+Business Media, LLC 2018

In the case of GPCRs, we transfect cells with a G protein coupled opsin and locally activate one side of the cell to create a gradient of GPCR activity. In the case of G proteins, we transfect cells with a light-inducible dimerization pair, CRY2-CIBN [8], and optically recruit a protein that accelerates Gα hydrolysis or trap Gβγ on one side of a cell. When combined with uniform exposure to a chemoattractant, this leads to a gradient of heterotrimeric G protein or Gβγ activity. Lastly, to create a gradient of Cdc42 or Rac activity, we transfect cells with another light inducible dimerization construct, iLID [9], to recruit a guanine nucleotide exchange factor (GEF) for activating Cdc42 or Rac on one side of the cell (Fig. 1). All these methods induce rapid, reversible, and directionally sensitive migration in macrophage cells. Here we describe the methods for transfecting macrophages like RAW 264.7 cells with these constructs and for driving the migration of these cells by photoactivation of subcellular regions. Since opsin activation also controls the migration of a neutrophil-like cell line and optical activation of Cdc42 also directs the migration of an epithelial cell line, the optogenetic methods described here may be widely applicable to a variety of important cell types (Fig. 2).

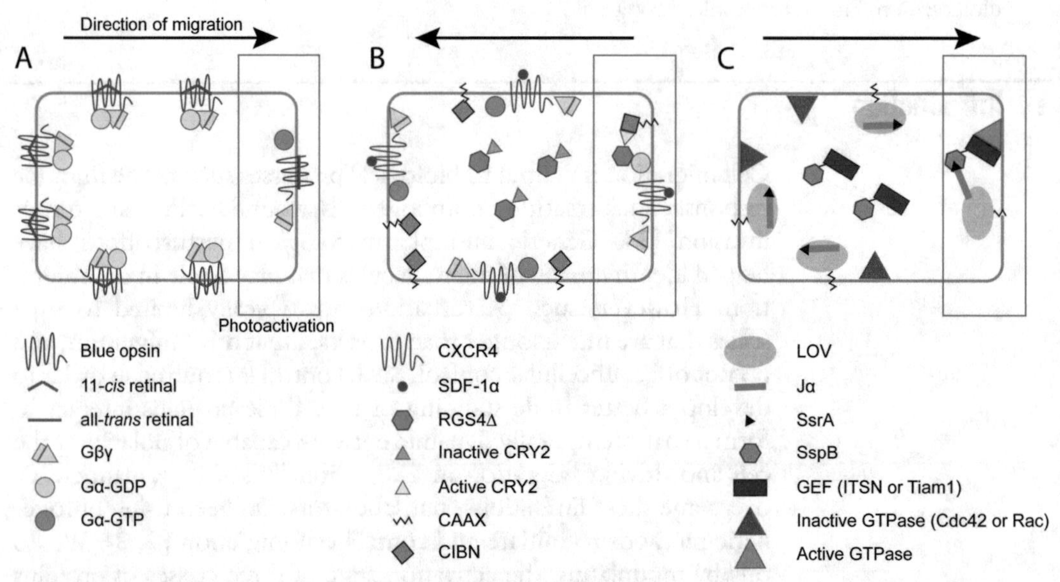

Fig. 1 Diagram of optogenetic methods for directing cell migration. (**a**) Localized activation of blue opsin creates a gradient of heterotrimeric G protein activity. (**b**) Localized activation of CRY2 in CXCR4-expressing cells causes CRY2 it to dimerize with CIBN, recruiting the RGS4Δ to one side of the cell and deactivating Gα proteins on that side. Simultaneous global activation of CXCR4 with SDF-1α creates a gradient of Gα activity. (**c**) Localized activation of iLID causes dimerization of SspB and SsrA, recruiting a GEF for Cdc42 or Rac to one side. This creates a gradient of Cdc42 or Rac activity. Blue box represents the area of photoactivation. Arrows indicate the direction of migration

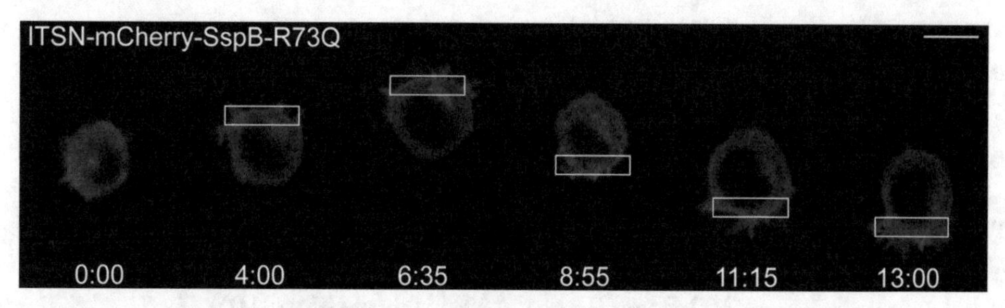

Fig. 2 Example of a RAW macrophage cell undergoing optogenetically driven cell migration through Cdc42 activation. Note the accumulation of mCherry-tagged GEF at the leading edge. Scale bar is 10 μm. Time is in min:s

2 Materials

2.1 Cell Culture

1. RAW 264.7 cells (*see* **Note 1**).
2. RAW cell culture media (*see* **Note 2**): Dulbecco's modified Eagle's medium (DMEM) containing 4500 mg/L glucose, L-glutamine, sodium pyruvate, and sodium bicarbonate, fetal bovine serum dialyzed using a 12–14 kDa cutoff membrane (DFBS) (*see* **Note 3**), penicillin/streptomycin (10,000 U/mL penicillin, 10,000 mg/mL streptomycin). Combine 445 mL DMEM, 50 mL DFBS, and 5 mL penicillin/streptomycin to make 500 mL of media.
3. Versene: 0.2 mg/mL ethylenediaminetetraacetic acid (EDTA).
4. T75 tissue culture flasks.
5. Cell scraper.
6. 10 mL serological pipettes.
7. Serological pipettor.
8. 15 mL conical tubes.
9. Hemocytometer.
10. Standard micropipettes with corresponding tips.
11. Cell culture incubator maintained at 37 °C and 5% CO_2.
12. Vacuum aspiration system.
13. Hank's Balanced Salt Solution (HBSS) containing 1 g/L glucose.

2.2 Transfections

1. Lonza Nucleofector 2b (*see* **Note 4**).
2. Lonza Cell Line Nucleofector Kit V (Lonza VCA-1003): Nucleofector Cell Line Solution V, Supplement 1, 0.5 μg/μL GFP in 10 mM Tris (for troubleshooting), cuvettes for nucleofection, plastic pipettes (*see* **Note 5**). Add Supplement 1 to Solution V and store at 4 °C.
3. Glass bottom dishes with 10-mm diameter well, #1 glass (Cellvis) (*see* **Note 6**).

2.3 DNA Constructs (See Note 7)

2.3.1 GPCR-Driven Migration

1. 3 μg Blue opsin-mCherry in pcDNA3.1 (Addgene #62932) (*see* **Note 8**).

2.3.2 Gα-Driven Migration

1. 2.5 μg CRY2-mCherry-RGS4Δ in pcDNA3.1 (Addgene #64207) (*see* **Note 9**).
2. 2.5 μg CIBN-CAAX [10] (*see* **Note 10**).
3. 2 μg CXCR4 [11] (*see* **Note 11**).

2.3.3 Cdc42-Driven Migration

1. 4 μg ITSN-mCherry-SspB-R73Q in pcDNA3.1 (Addgene #85220) (*see* **Note 12**).
2. 2 μg iLID-CAAX in pcDNA3.1 (Addgene #85220) (*see* **Note 13**).

2.3.4 Rac-Driven Migration

1. 4 μg Tiam1-mCherry-SspB-R73Q in pcDNA3.1 (*see* **Note 14**).
2. 2 μg iLID-CAAX in pcDNA3.

2.4 Imaging

1. Automated inverted fluorescence microscope with camera (*see* **Note 15**).
2. 63× oil immersion objective.
3. 445-nm wavelength laser for photoactivation (*see* **Note 16**).
4. Lasers for fluorescence excitation (*see* **Note 17**).
5. Light-proof chamber around microscope stage (*see* **Note 18**).
6. Localized photoactivation unit with appropriate control software (*see* **Note 19**).

2.5 Additional Reagents

2.5.1 GPCR-Driven Migration

1. 11-*cis* retinal. Working in the dark, dissolve 11-*cis* retinal in 100% ethanol to a stock concentration of 50 mM and store in light-proof vials at −80 °C until use (*see* **Note 20**).

2.5.2 Gα-Driven Migration

1. SDF-1α. Dissolve SDF-1α to a concentration of 10 μg/mL in 1× phosphate buffered saline (PBS) containing 0.1% bovine serum albumin (BSA) and store at −20 °C until use (*see* **Note 21**).

3 Methods

3.1 Cell Culture

1. Grow RAW cells in a T75 flask in the cell culture incubator at 37 °C and 5% CO_2 until confluent (*see* **Note 22**).
2. Aspirate media and add 6 mL of Versene (EDTA).
3. Leave flask with EDTA in the incubator for 10 min.

4. Remove cells from the bottom of the flask with a cell scraper or by repeatedly triturating with a 10 mL serological pipette.
5. Transfer the cell suspension to a 15 mL conical tube and centrifuge at 210 × *g* for 90 s. Cell pellet should be visible at the bottom of the tube after centrifugation.
6. Aspirate the EDTA and add 3 mL of 37 °C media. Triturate gently with 1 mL micropipette to resuspend the cells.
7. Place a drop of cell suspension on a hemocytometer and count to find cell density (*see* **Note 23**).
8. Calculate the volume corresponding to three million cells and transfer it to a 1.5 mL vial (*see* **Note 24**).
9. Centrifuge at 0.1 × *g* for 10 min.

3.2 Transfections (See Note 25)

1. Combine 100 μL of supplemented Nucleofector Solution V with the appropriate amount of DNA (*see* Subheading 2.3) in a 1.5 mL vial.
2. Put 500 μL of media in a 1.5 mL vial and place uncapped in the incubator.
3. After cells finish centrifuging, aspirate the media and replace with the Nucleofector-DNA mix.
4. Triturate the mix gently using a 1 mL micropipette until cells are resuspended (*see* **Note 26**) and transfer to a Lonza cuvette for nucleofection.
5. Electroporate with the Lonza Nucleofector 2b using program T-20 (*see* **Note 27**).
6. Remove media-containing vial from the incubator and transfer some of the warm media into the cuvette. Then immediately move the entire volume of cell suspension from the cuvette into the vial with the remaining media (*see* **Note 28**).
7. Place the uncapped vial with cells in the incubator for 10 min (*see* **Note 29**).
8. Pipette cells into 10-mm well glass bottom dishes, putting 50–100 μL of cell suspension per well.
9. Leave the dishes in the incubator for 1 h.
10. After 1 h, add 2 mL of warm media to each dish (*see* **Note 30**).
11. Incubate dishes for 4 h before imaging.
12. Prior to imaging, aspirate the cell media from dishes and replace with 1 mL of fresh warm media.

3.3 Imaging (See Note 31)

3.3.1 GPCR-Driven Migration

1. Working in the dark, dilute 11-*cis* retinal in cold HBSS to a concentration of 100 μM.
2. Add 100 μL of 11-*cis* retinal to the cell dishes for a final concentration of 10 μM and leave the dishes in the incubator for at least 20 min prior to imaging.

3. Move cell dishes to the microscope stage (*see* **Note 32**) and verify expression of fluorescently tagged blue opsin in the cells (*see* **Note 33**).
4. Find a cell that expresses blue opsin and designate a rectangular photoactivation box on one side of the cell (*see* **Note 34**).
5. Set up and run an automated imaging protocol that exposes the cell to localized 445-nm wavelength photoactivation pulses every 5 s (*see* **Note 35**) and acquire an image between photoactivation pulses.
6. The cell should begin to polarize and extend lamellipodia on the photoactivated side within 5 min. Continue moving the box to induce further leading edge extension and migration (*see* **Note 36**).

3.3.2 Gα-Driven Migration

1. Run an automated protocol that photoactivates the cell and acquires an image every 5 s.
2. Begin photoactivating the cell on one side to induce localized inhibition of Gαi activity (*see* **Note 37**).
3. While maintaining localized photoactivation, add SDF-1α to the cell dish for a final concentration of 50 ng/mL to globally activate CXCR4 receptors (*see* **Note 38**). This will create a gradient of Gαi activity.
4. The cell should begin to polarize and migrate in the direction opposite from the photostimulus. Continue moving the photoactivation box to follow the trailing edge of the cell.

3.3.3 Cdc42- and Rac-Driven Migration

1. Run an automated protocol that photoactivates the cell and acquires an image every 5 s.
2. Begin photoactivating the cell on one side to generate a gradient of Cdc42 or Rac activity (*see* **Note 39**).
3. The cell should polarize within minutes and begin migrating in the direction of the photostimulus. Continue moving the photoactivation box as the leading edge advances (Fig. 2).

4 Notes

1. Other cell types can be used for migration studies. However, we found RAW 264.7 cells to be optimal because they are capable of migration, yet display low levels of basal movement, unlike commonly used *Dictyostelium discoideum* amoeboid cells and HL-60 neutrophils.
2. This is the suggested media for RAW cells. If using another cell line, use the recommended media for that particular line.

3. We keep DFBS at −80 °C for long-term storage. Prior to use, thaw the DFBS in 37 °C water bath, then put in 56 °C water bath for 30 min to inactivate the serum complement. Then put in ice water bath for 5 min. We then make 50 mL aliquots and store them at −20 °C until use.
4. We use nucleofection, an electroporation-based method, to transfect RAW cells. We find that this method yields much higher expression levels of our desired constructs in RAW cells than Lipofectamine-based transfections. We often transfect as many as four or five different constructs simultaneously for experiments that combine optical control with imaging of multiple molecular responses.
5. Lonza Nucleofector Kit V is made specifically for use with the Lonza Nucleofector 2b and is the recommended kit for RAW cells. If using a different cell line with the Lonza Nucleofector 2b, refer to Lonza protocols (http://www.lonza.com/products-services/bio-research/transfection/nucleofector-kits-for-cell-lines.aspx) for the recommended kit. If using a different electroporator, use the suggested protocols and reagents for that particular electroporator.
6. The #1 cover glass has a thickness of 0.13–0.16 mm and is convenient for imaging with 63× and other short working distance oil objectives. We find that the small (10 mm) well size helps keep cells at a higher density, making it easier to locate cells while imaging.
7. All DNA amounts listed in this work have been optimized for use with RAW cells and the Lonza Nucleofector 2b electroporator. The DNA amounts may need to be adjusted if using another cell line or transfection method. Additionally, transfection results may vary if using another viral vector.
8. Blue opsin is a GPCR that couples to heterotrimeric Gi proteins. When exposed to blue light, it undergoes a conformational change that results in phosphorylation and activation of its associated Gα protein. The GTP-Gα and βγ complexes initiate signaling cascades that leads to activation of downstream regulators of migration, such as Rho GTPases, resulting in actin remodeling and lamellipodia formation (Fig. 1a) [12].
9. RGS4 (Regulator of G protein Signaling 4) is a protein that inactivates Gαi by accelerating its conversion to the GDP-bound form. RGS4Δ is a truncated mutant lacking amino acids 1–33, which prevents the protein from being localized to the plasma membrane at all times. This mutation is necessary because for our application RGS4 must be in the cytosol and translocate to the plasma membrane only upon photoactivation. CRY2 (cryptochrome 2) is a domain of a protein found

in *Arabidopsis thaliana* that dimerizes with its binding partner, CIBN, upon photoactivation (Fig. 1b) [8].

10. CAAX is the C-terminal plasma-membrane targeting domain of the protein KRas. By fusing CAAX to CIBN, we ensure that CIBN is localized to the plasma membrane, which allows the CRY2 complex to be recruited to the plasma membrane upon photoactivation (Fig. 1b).
11. CXCR4 is a chemoattractant receptor that is not expressed in RAW 264.7 cells. If using a different cell type that is known to express this receptor endogenously, then it is not necessary to transfect cells with CXCR4. Alternatively, one could utilize another chemoattractant receptor, such as C5a, and its corresponding chemoattractant instead of CXCR4 and SDF-1α (Fig. 1b).
12. The DH/PH domain of the protein ITSN (Intersectin) has GEF activity for Cdc42, a small GTPase known to control filopodia formation and actin remodeling in migrating cells [13]. Here the DH/PH domain of ITSN is fused to SspB, which is a bacterial protein with high binding affinity for the peptide SsrA. SspB is mutated (R73Q) to give it a lower affinity for SsrA, since we found that the unmutated version displays high levels of interaction with SsrA even in the absence of photoactivation (Fig. 1c).
13. iLID is a complex consisting of SsrA fused to the C terminus of LOV-Jα. LOV undergoes a conformational change upon blue light absorption, rearranging the Jα helix and making SsrA available for binding to SspB. Since CAAX keeps iLID anchored to the plasma membrane, ITSN-mCherry-SspB can thus be recruited to the plasma membrane with blue light exposure (Fig. 1c). During the initial optimization steps, it may be beneficial to use a fluorescently tagged version of iLID-CAAX, such as Venus-iLID-CAAX (Addgene #60411) [9] to optically verify its expression. After the optimal DNA concentrations and transfection conditions have been found, we typically use the untagged iLID-CAAX in order to image other fluorescently tagged proteins of interest.
14. Tiam1 is a GEF for Rac, a small GTPase that is involved in actin polymerization at the leading edge of migrating cells [13].
15. We use a Leica DMI6000 inverted fluorescence microscope with a Yokogawa spinning disk confocal unit and Andor iXon EM-CCD camera. We find this system very effective for combining subcellular optogenetic control of cell migration with fluorescent imaging experiments due to its ability to capture images at high speed and using relatively low laser power. The system is run through Andor iQ software.

16. We have a 50 mW solid state 445-nm laser. For photoactivation, the laser is attenuated using an AOTF to achieve a power of approximately 5 μW at the sample. The focused laser spot is scanned across the selected photoactivation region at a rate of 0.9 ms/μm^2. The 445-nm wavelength is convenient because it is capable of photoactivating all optogenetic tools described in this work: blue opsin, CRY2, or iLID. All of these can also be activated with 488-nm wavelength green light, which is important to consider in experiments where photoactivation is combined with fluorescent imaging. For imaging proteins of interest in migrating cells, we therefore use fluorophores that excite in the yellow or red range (e.g., YFP, Venus, and mCherry) to avoid unintentionally photoactivating the optogenetic constructs.
17. We use 50 mW solid state 515- and 594-nm wavelength lasers for confocal imaging. For brightfield illumination we use a 660-nm LED to avoid globally photoactivating the sample.
18. We use a light-proof chamber that also provides environmental control by maintaining the temperature at 37 °C and CO_2 level at 5%. This requires a heating unit and CO_2 tank with controller to be incorporated into the imaging setup.
19. We use an Andor FRAPPA (Fluorescence Recovery After Photobleaching and Photoactivation) unit for localized photoactivation, controlled by Andor iQ3 software. We have also found that LED illumination with a DMD (micro-mirror array) such as the Andor Mosaic works well for this purpose. When considering any other localized photoactivation system, one must ensure that it can be integrated with the imaging software and allow for the photoactivation area to be selected and repositioned in real time.
20. 11-*cis* retinal is a chromophore required for activation of blue opsin. It is light-sensitive and undergoes photoisomerization to all-*trans* form when exposed to blue light (Fig. 1a). When working with retinal, we therefore use a red light lamp and minimize any other light present in the room. We make 20 μL aliquots of 11-*cis* retinal to avoid repeated freeze–thaw cycles. 9-*cis* retinal can also be used in place of 11-*cis* retinal.
21. SDF-1α is a chemoattractant that is a ligand for CXCR4 receptors. We make 25 μL aliquots to avoid repeated freeze–thaw cycles.
22. We find that RAW cells plated at a density of 8 million per T75 flask reach confluency in around 24 h, whereas cells plated at three million per flask reach confluency in 48 h. We avoid using cells that have been plated more than 48 h prior to the transfection, as we find them to be less healthy. We also avoid using cells that have been passaged more than 15 times.

23. We typically dilute a small volume (10 μL) of cell suspension by a factor of 10 and use it for counting.
24. Lonza protocols recommend using two million cells per transfection, but we find that three million cells gave better results and did not impact transfection efficiency.
25. We use transient transfections for all experiments. We have not been able to generate stably transfected lines for migration experiments, since we find that optogenetic constructs in lentiviral vectors do not express well in RAW cells.
26. Triturate only the minimum number of times necessary to resuspend the cells. Try to minimize formation of air bubbles, as those can be problematic during the electroporation step.
27. Program T-20 is recommended for RAW cells. Parameters such as voltage and pulse duration are not known for Lonza programs. If using a different cell type, refer to the Lonza protocols for the optimal program. While testing various electroporation settings, one can use the GFP plasmid provided with the Nucleofector Kit for transfections, since GFP expresses very well in cells and is therefore convenient for optimizing protocols.
28. We often find a white aggregate of cells floating in the cuvette after electroporation. Avoid transferring these aggregates into the vial with media, as they are likely dead cells.
29. **Steps 6** and 7 should be performed quickly.
30. We keep the media in a T25 flask in the incubator for an hour prior to adding it to the dishes. This ensures that the media has the desired CO_2 level as well as temperature. After adding media, we find it convenient to move the dishes to a small incubator kept in the microscope room. This makes it easier to keep the samples in darkness once the cells start to express the optogenetic constructs of interest. This is especially useful for minimizing unwanted photoactivation in experiments using opsins. We find that the requirements for minimizing unwanted photoactivation by ambient light are less stringent when using CRY2 and iLID optogenetic tools.
31. We image cells 4–10 h after transfection. We find that RAW cells do not appear healthy 1 day after electroporation.
32. When moving retinal-containing samples from the incubator to the light-proof chamber on microscope, we turn off computer monitors and cover electronics displays to minimize retinal's exposure to light.
33. If the cells do not express a particular construct, we increase the amount of DNA used for the transfection, though we typically do not exceed 4 μg of DNA per sample. We also frequently transfect constructs into HeLa cells to check their

expression prior to trying them in RAW cells. HeLa cells are large, robust, and can be transfected using Lipofectamine rather than electroporation, leaving them healthier for a longer period of time. Lipofection protocols can be found at https://tools.thermofisher.com/content/sfs/manuals/Lipofectamine_2000_Reag_protocol.pdf.

34. Our photoactivation regions are typically around 5 × 15 μm. When working with opsin-driven migration, we place the photoactivation box next to the edge of the cell, rather than on top of it. We find this induces migration more efficiently, possibly due to a steeper gradient of opsin activity between the front and the back of the cell.
35. We find that a laser power of around 5 mW and exposure time of 0.9 ms/μm^2 (1.5–3% laser power with our 445-nm laser) provides sufficient photoactivation.
36. Not all cells expressing blue opsin will migrate in response to localized photoactivation. However, if multiple cells consistently fail to polarize, more troubleshooting may be necessary. We use the βγ translocation assay in HeLa cells to confirm that opsins are activating G proteins as intended. When Gαi assumes its active GTP-bound form, the Gβγ complex dissociates from Gα and translocates from the plasma membrane to intracellular membranes. We therefore use fluorescently tagged γ_9 subunits, which exhibit fast translocation, to monitor the activity state of G proteins. We transfect HeLa cells with blue opsin-mCherry and YFP-γ_9, incubate with 11-*cis* retinal, locate a cell in which γ_9 is localized on the plasma membrane in the basal state, and expose the cell to multiple pulses of blue light to induce maximal activation. Translocation of γ_9 to intracellular membranes confirms that opsins are being photoactivated and subsequently activate their associated G protein [14]. The βγ translocation assay can be used to find the optimal photoactivation settings, such as pulse duration and intensity, for generating opsin-induced G protein activity.
37. If the CRY2-CIBN system is working as intended, there will be an increase in red fluorescence on the plasma membrane in the photoactivated region due to increased binding of CRY2-mCherry-RGS4Δ to CIBN-CAAX.
38. Localized inhibition of Gαi without SDF-1α addition is not expected to induce polarization or migration, since basal levels of Gαi activity are low.
39. Accumulation of ITSN-mCherry-SspB-R73Q or Tiam1-mCherry-SspB-R73Q at the plasma membrane in the photoactivated region will confirm activation of the iLID system and the resulting dimerization of SspB and SsrA.

Acknowledgment

This work was funded by the National Institutes of Health through National Institute of General Medical Sciences Grants GM069027, GM107370, and GM122577.

References

1. Devreotes P, Horwitz AR (2015) Signaling networks that regulate cell migration. Cold Spring Harb Perspect Biol 7(8):a005959. https://doi.org/10.1101/cshperspect.a005959
2. Karunarathne WK, O'Neill PR, Gautam N (2015) Subcellular optogenetics – controlling signaling and single-cell behavior. J Cell Sci 128(1):15–25. https://doi.org/10.1242/jcs.154435
3. O'Neill PR, Gautam N (2015) Optimizing optogenetic constructs for control over signaling and cell behaviours. Photochem Photobiol Sci 14(9):1578–1585. https://doi.org/10.1039/c5pp00171d
4. Karunarathne WK, Giri L, Patel AK, Venkatesh KV, Gautam N (2013) Optical control demonstrates switch-like PIP3 dynamics underlying the initiation of immune cell migration. Proc Natl Acad Sci U S A 110(17):E1575–E1583. https://doi.org/10.1073/pnas.1220755110
5. Karunarathne WK, Giri L, Kalyanaraman V, Gautam N (2013) Optically triggering spatiotemporally confined GPCR activity in a cell and programming neurite initiation and extension. Proc Natl Acad Sci U S A 110(17):E1565–E1574. https://doi.org/10.1073/pnas.1220697110
6. O'Neill PR, Gautam N (2014) Subcellular optogenetic inhibition of G proteins generates signaling gradients and cell migration. Mol Biol Cell 25(15):2305–2314. https://doi.org/10.1091/mbc.E14-04-0870
7. O'Neill PR, Kalyanaraman V, Gautam N (2016) Subcellular optogenetic activation of Cdc42 controls local and distal signaling to drive immune cell migration. Mol Biol Cell. https://doi.org/10.1091/mbc.E15-12-0832
8. Kennedy MJ, Hughes RM, Peteya LA, Schwartz JW, Ehlers MD, Tucker CL (2010) Rapid blue-light-mediated induction of protein interactions in living cells. Nat Methods 7(12):973–975. https://doi.org/10.1038/nmeth.1524
9. Guntas G, Hallett RA, Zimmerman SP, Williams T, Yumerefendi H, Bear JE, Kuhlman B (2015) Engineering an improved light-induced dimer (iLID) for controlling the localization and activity of signaling proteins. Proc Natl Acad Sci U S A 112(1):112–117. https://doi.org/10.1073/pnas.1417910112
10. Idevall-Hagren O, Dickson EJ, Hille B, Toomre DK, De Camilli P (2012) Optogenetic control of phosphoinositide metabolism. Proc Natl Acad Sci U S A 109(35):E2316–E2323. https://doi.org/10.1073/pnas.1211305109
11. Zhao M, Discipio RG, Wimmer AG, Schraufstatter IU (2006) Regulation of CXCR4-mediated nuclear translocation of extracellular signal-related kinases 1 and 2. Mol Pharmacol 69(1):66–75. https://doi.org/10.1124/mol.105.016923
12. Wang F (2009) The signaling mechanisms underlying cell polarity and chemotaxis. Cold Spring Harb Perspect Biol 1(4):a002980. https://doi.org/10.1101/cshperspect.a002980
13. Raftopoulou M, Hall A (2004) Cell migration: Rho GTPases lead the way. Dev Biol 265(1):23–32
14. Saini DK, Kalyanaraman V, Chisari M, Gautam N (2007) A family of G protein betagamma subunits translocate reversibly from the plasma membrane to endomembranes on receptor activation. J Biol Chem 282(33):24099–24108. https://doi.org/10.1074/jbc.M701191200

Chapter 23

Electrotaxis: Cell Directional Movement in Electric Fields

Jolanta Sroka, Eliza Zimolag, Slawomir Lasota, Wlodzimierz Korohoda, and Zbigniew Madeja

Abstract

Electrotaxis plays an important role during embryogenesis, inflammation, wound healing, and tumour metastasis. However, the mechanisms at play during electrotaxis are still poorly understood. Therefore intensive studies on signaling pathways involved in this phenomenon should be carried out. In this chapter, we described an experimental system for studying electrotaxis of *Amoeba proteus*, mouse embryonic fibroblasts (MEF), Walker carcinosarcoma cells WC256, and bone marrow adherent cells (BMAC).

Key words Electric field, Electrotaxis, Galvanotaxis, Electrotactic chamber, Time-lapse videomicroscopy

1 Introduction

Cell migration is a basic property of animal cells and fundamental in many biological processes, including embryogenesis, inflammation, wound healing and tumor metastasis. Regulation of cell motility requires activation of several signaling pathways, which enable the cell to respond to various external signals. Among the extracellular factors that orient the animal cell locomotion, there are chemoattractants, chemorepellents, tissue architecture, adhesion site gradient, matrix topography, matrix stiffness, or electric fields [1].

Although endogenous direct current electric fields (dcEFs) were first recorded more than 150 years ago, their significance has remained poorly understood and largely ignored. Recently, the existence of these endogenous wound currents/EFs has been confirmed by several modern techniques [2]. The mechanism of EF generation in wounded epithelia is based on transepithelial potential (TEP) vanishing after injury. TEP is generated in intact skin tissue by polarized ion transport across the epithelial layer and maintained by tight junction presence. After wounding, ions begin to leak out at the wound site and TEP

Alexis Gautreau (ed.), *Cell Migration: Methods and Protocols*, Methods in Molecular Biology, vol. 1749,
https://doi.org/10.1007/978-1-4939-7701-7_23, © Springer Science+Business Media, LLC 2018

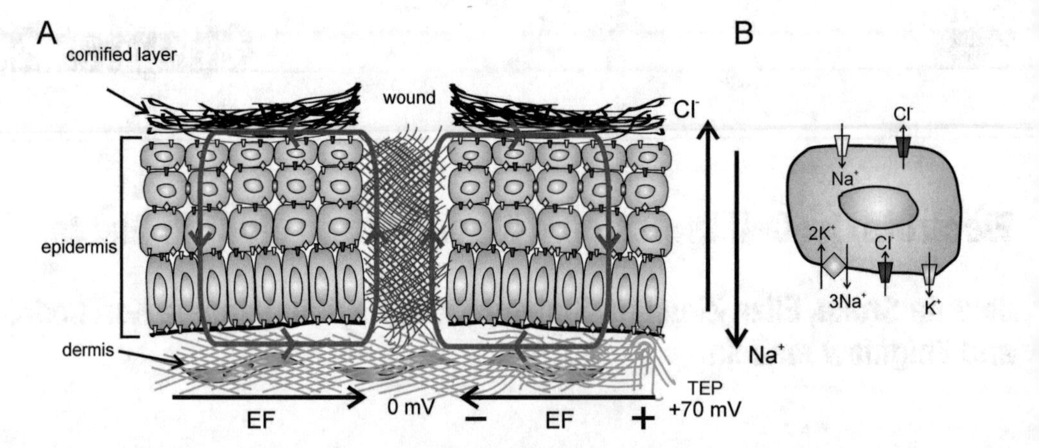

Fig. 1 Electric field in damaged skin after transepithelial potential (TEP) collapse. (**a**) Damage of epithelial layer results in ion leak through the wound and injury current flow (red line). (**b**) Uneven distribution of ion channels and sodium potassium pump in apical and basal side of the epidermal cell that are responsible for the TEP. Modified from [3, 4]

reduces to 0 mV. The difference between potential at the area of intact epithelium (+70 mV) and wound site (0 mV) results in injury current and electric field generation lateral to the epithelium plane. The return path for the current is in the layer between the dead, cornified tissue, and the living epidermis. The electric field persists until ion leakage is prevented by reepithelialization with the negative pole (cathode) located at the center of the wound bed (Fig. 1).

It is known that many cell types respond to applied dcEFs. Most cells such as human keratinocytes [5], fish epidermal cells [6], fibroblasts [7], highly metastatic rat prostate MAT-LyLu cancer cells [8], Walker carcinosarcoma cells [9, 10], and mesenchymal stem cells [11] migrated toward the cathode. However, human granulocytes [12], macrophages [11], rabbit corneal endothelial cells [13], human vascular endothelial cells (HUVECs) [14], and weakly metastatic rat prostate AT-2 cancer cells [8] show anodal response.

Several hypotheses concerning the reaction of cells to dcEFs have been suggested. Moreover, numerous molecules and signalling pathways were suggested to be involved in the electrotactic movement, but it is not clear how electric migration cues are transformed into cellular responses. One of the commonly suggested mechanism is asymmetrical distribution of membrane receptors, for example epidermal growth factor receptors (EGFR), concanavalin A receptor (ConA), acetylcholine receptor (AchR), or integrins [15–20]. A number of other mechanisms were also proposed, although less precisely documented [8, 21, 22]. At the intracellular level, some signaling pathways were considered as involved in electrotaxis, such as signaling through EGF receptors/ERK1/2, through integrin-Rac, cAMP/PKA, Rho GTPases, phosphatidylinositol 3-kinase (PI3K) [16, 23–25].

In conclusion, it is well documented that endogenous dcEFs exist and are likely to be involved in cellular physiology and pathophysiology. What is more, many cell types exhibit electrotaxis. However, the precise mechanisms by which cells detect and respond to endogenous or exogenous dcEFs are still unknown. We believe that further studies reveal the signaling pathways of the responses to dcEFs in different cells.

2 Materials

2.1 Cell Culture

1. Glass crystallizers (diameter 10 cm); Glass Pasteur pipette (Dom Handlowy Nauki, Krakow, Poland) (*see* **Note 1**).
2. Plastic tissue culture vessels. Tissue culture flasks and Primaria™ tissue culture flasks 25 or 75 cm^2 (BD Falcon).
3. Medium for *Tetrahymena pyriformis:* 1% Bact®™ Proteose-Peptone (Difco Laboratories), 0.1% Bact®™ Yeast Extract (Difco Laboratories) in distilled water; Pringsheim solution: solution I: 1.2 g $Na_2HPO_4 \cdot 12H_2O$, 0.25 g KCl, 500 ml H_2O; solution II: 2 g $Ca(NO_3)_2 \cdot 4H_2O$, 0.2 g $MgSO_4 \cdot 7H_2O$, 0.02 g $Fe_2(SO_4)_3 \cdot 7H_2O$, 500 ml H_2O (keep at 4 °C). For cell culture mix: 100 ml solution I + 100 ml solution II + 1800 ml distilled H_2O (keep at 4 °C) (*see* **Note 2**).
4. Medium for *Amoeba proteus (Princeton strain):* Pringsheim solution; simplified Chalkley's solution: (81.9 mg NaCl, 1.94 mg KCl, 1.11 mg $CaCl_2$, 1000 ml H_2O); *Tetrahymena pyriformis* (*see* **Note 2**).
5. Medium for Mouse embryonic fibroblasts (MEFs): Dulbecco's modified Eagle's medium (DMEM) with 4 mM L-glutamine, 15 mM HEPES, 3.7 g/l sodium bicarbonate, supplemented with 50 U/ml penicillin, 50 μg/ml streptomycin, and 10% fetal bovine serum (FBS) (Sigma-Aldrich); phosphate buffered saline (Sigma-Aldrich), 0.25% trypsin–EDTA solution (Gibco, Thermo Fisher Scientific).
6. Medium for Rat Walker carcinosarcoma cells WC 256: RPMI-1640 medium with 1.2 g/l sodium bicarbonate, supplemented with 100 U/ml penicillin, 10 μg/ml streptomycin, and 5% fetal bovine serum (FBS) (Sigma-Aldrich); phosphate buffered saline (Sigma-Aldrich), 0.25% trypsin–EDTA solution (Gibco).
7. Medium for Bone marrow adherent cells (BMACs): Dulbecco's modified Eagle's medium (DMEM) F12 HAM with 2.5 mM/ml L-glutamine, 15 mM HEPES, 1.2 g/l sodium bicarbonate, supplemented with 50 U/ml penicillin, 50 μg/ml streptomycin, and 10% fetal bovine serum (FBS) (Sigma-Aldrich); phosphate buffered saline (Sigma-Aldrich), 0.25% trypsin–EDTA solution (Gibco).

2.2 Electrotaxis

1. Agar.
2. High vacuum silicone grease (Dow Corning) or double-sided tape.
3. Heat conductive silicone Paste H (Termopasty, Sokoly, Poland).
4. Cover glass slides (size 60 mm × 35 mm; 60 mm × 10 mm) (*see* **Note 1**).
5. Cover glass strips (size 60 mm × 10 mm) (*see* **Note 1**).
6. Silver chloride (Ag/AgCl) electrodes (Sigma-Aldrich).
7. U-shape glass bridges (~8 cm long with the inner diameter ~7 mm) (Pejko-Szklo, Krakow, Poland) (*see* **Note 1**).
8. Direct current power supply.
9. Voltmeter with a high resistance.
10. Ammeter (Velleman).
11. Plexiglass electrotactic chamber (custom made, dimensions: 14 cm × 7 cm × 2 cm).
12. 0.5 M KCl solution (37 g KCl, 1000 ml distilled H_2O).
13. Labanowski's buffers solutions [26]: Cathode buffered solution: 3.3 mM Hepes and 6.7 mM triethanolamine (TEA) in 153 mM NaCl (or PBS); Anode buffered solution: 10 mM Hepes in 153 mM NaCl.

2.3 Time-Lapse Video Microscopy System

Fully automatic epifluorescence microscopes Leica DM IRE2 and Leica AF7000 system equipped with cooled, digital CCD cameras designed for time-lapse image acquisition according to the user-designed algorithm (including, among others, definition of camera basic settings, shutter opening and closing parameters (time of acquisition), sequence of excitation and emission filter changes, acquisition intervals, change of stage position in *xy* as well as *z* plane, sequential registration from predefined regions of the specimen and specimen dynamic sectioning). Leica AF7000 system is especially dedicated for live cell investigation and is equipped with the CO_2 and temperature incubator system. This system effortlessly performs fast 3D time-lapse experiments. It is also fully capable of applying the FRET method, fast Fura2 and ROS measurements.

3 Methods

3.1 Cell Culture

Tetrahymena pyriformis

1. Prepare medium for *Tetrahymena pyriformis* and autoclave.
2. Place 20 ml of culture medium into a sterile crystallizer and add 1×10^6 *Tetrahymena pyriformis.* Culture at room temperature in the dark.

3. Make a passage twice a week by transferring 1×10^6 of cells to the fresh culture medium.

Amoeba proteus

1. Prepare the Pringsheim solution.
2. Wash twice *Tetrahymena pyriformis* with Pringsheim solution by centrifuging at $600 \times g$ for 2 min.
3. Place 50 ml of Pringsheim solution into a sterile crystallizer and add few thousands clean *Amoeba proteus* and a few drops of washed *Tetrahymena pyriformis.* Culture at room temperature in the dark.
4. Add new food organisms (*Tetrahymena pyriformis*) to the culture every 3–5 days or as frequently as necessary [27].
5. If the population of Amoeba proteus increase renew the culture by transferring a few hundreds of cells to a new glassware using Pasteur pipette with Pringsheim solution and *Tetrahymena pyriformis.* Do not centrifuge Amoeba proteus.

MEF

1. Passage every third day in order to avoid exceeding confluency of 80%. Wash cells gently with PBS and subsequently add 0.25% trypsin–EDTA solution.
2. If necessary, the flask can be shaken a few times in order to accelerate detachment of cells. It is strongly recommended to avoid overtrypsinization.
3. Inactivate trypsin by washing with complete medium. Centrifuge cells at $200 \times g$ for 5 min.
4. Dilute cells 1:10 and propagate to a new culture flask.

WC256

1. To obtain weakly adherent, spontaneously blebbing WC256 cells (BC) from cells initially growing in a suspension culture, keep cells for about 1–2 weeks in the same flask (25 cm^2) [28].
2. Change the medium every 2 days by means of a pipette without shaking the culture flask.
3. After about 2 weeks, harvest cells by repeated clapping (about 2–3×) with the palm and transfer into a new culture flask. Repeat every 2 days.
4. To obtain lamellipodia forming cells strongly adhering to plastic flasks (LC), keep cells in the same flask for about 3–4 weeks (as described above) [9].
5. Change medium every 2 days by means of a pipette without shaking the culture flask.

6. When confluence of adherent cell culture reaches almost 90%, dissociate cells using 0.25% trypsin–EDTA solution and dilute 1:5 to a new culture flasks.

BMAC (mixture of bone marrow mesenchymal stem cells and macrophages)

1. Harvest tibias and femurs of 4- to 6-week-old C57Bl/6 mice immediately after animal euthanasia.
2. Flush cavities of femurs and tibias with DMEM/F12 medium. Remove large portions of bone marrow tissue by passing through a 40-G needle.
3. Centrifuge cell suspension at 200 × *g* for 5 min. Resuspend the cell pellet in complete medium [11].
4. Plate cells into a Primaria culture flask at the density of 3×10^5 nucleated cells/cm^2. Place flasks in a humidified incubator with 5% CO_2 at 37 °C for 72 h.
5. Remove nonadherent cell population and wash adherent cell layer once with fresh medium.
6. When a confluence of cell culture reaches almost 90%, passage cells with 0.25% trypsin–EDTA solution and dilute by 1:2 to new culture flasks.

3.2 Preparation of Cells for Electrotaxis Assay

Amoeba proteus

1. Starve *Amoebae proteus* for 2–3 days in a simplified Chalkley's solution.
2. Transfer cells into the fresh simplified Chalkley's solution.
3. Transfer about 100 cells into the observation chamber made of cover glasses.
4. After 15 min, when the amoebae attach to the bottom surface of the chambers, place the observation chamber in the plexiglass electrotactic chamber and mount it with silicone grease.

MEF

1. Passage cells from the culture flask with 0.25% trypsin–EDTA.
2. Centrifuge cells at 200 × *g* for 5 min. Resuspend cells in fresh complete medium.
3. Plate cells onto a bottom glass of the observation chamber at the density of 2.5×10^3 cells/cm^2 and incubate in humidified atmosphere with 5% CO_2 at 37 °C overnight.
4. Immediately before the experiment, assemble the complete observation chamber using a double-sided tape. Subsequently mount it to the plexiglass electrotactic chamber with silicone paste.

WC256

1. Passage cells from the culture flask by clapping (BC cells) or with 0.25% trypsin–EDTA solution (LC cells).
2. Centrifuge cells at 200 × *g* for 5 min. Resuspend cells in a fresh complete medium.
3. Plate cells into the observation chamber made of cover glasses at the density of 5.5×10^4 cells/cm^2 (BC cells) or 3.5×10^4 cells/cm^2 (LC cells) (*see* **Note 6**) and incubate in humidified atmosphere of 5% CO_2 at 37 °C for 2 h.
4. After 2 h, place the observation chamber in the plexiglass electrotactic chamber and mount it with silicone grease.

BMACs

1. Passage cells from the culture flask with 0.25% trypsin–EDTA solution.
2. Centrifuge cells at 200 × *g* for 5 min. Resuspend cells in a fresh complete medium.
3. Plate cells into the observation chamber made of cover glasses at the density of 2×10^4 cells/cm^2 and incubate in humidified atmosphere with 5% CO_2 at 37 °C for 2 h (*see* **Note 7**).
4. After 2 h, place the observation chamber in the plexiglass electrotactic chamber and mount it with silicone grease.

3.3 Preparation of the Electrotactic Chamber

1. Sterilize cover glass slides, and silicone grease or double-sided tape (*see* **Note 8**).
2. Prepare the observation chamber by sealing cover glass strips (60 mm × 10 mm × 0.2 mm) to the bottom and upper cover glasses (60 mm × 35 mm × 10 mm) using high vacuum silicone grease or double-sided tape (*see* **Note 9**).
3. Introduce the suspension of cells (*see* above in Subheading 3.2) into the observation chamber using micropipette (Fig. 2a) (*see* **Note 10**).
4. Dissolve 2% agar by boiling in 0.5 M KCl solution and fill the glass bridges with it. Avoid air bubbles formation to prevent disturbing of current running or unstable voltage across the observation electric chamber during experiment. Wait until agar solidifies in bridges.
5. Mount the electrotactic chamber with heat conductive silicone Paste H in the plexiglass apparatus avoiding wetting both of them (Fig. 2b) (*see* **Note 11**).
6. Add culture medium (the same as in the observation chamber) to the inner wells (red wells in Fig. 2b) of plexiglass apparatus.

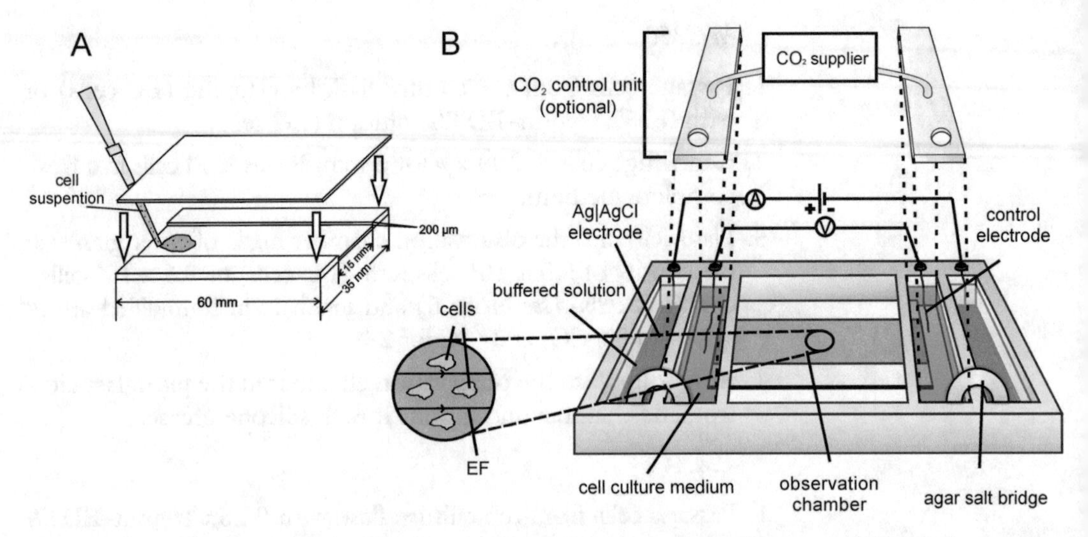

Fig. 2 Experimental chamber for electrotaxis. (**a**) Two-part observation chamber constructed of cover glasses of dimensions: 60 × 35 × 0.2 mm (*see* **Note 3**). (**b**) Complete plexiglass apparatus. Ag/AgCl electrodes for applying dcEF are placed in external wells filled with buffered solution and linked by agar bridges to internal wells filled with culture medium. Ag/AgCl probe electrodes (control electrodes) for real-time monitoring of dcEF are placed in the internal wells (*see* **Notes 4** and **5**). The CO_2 control unit comprises two plexiglass covers that can be fixed to the internal wells. CO_2 is provided from the CO_2 supplier through silicone wires. The observation chamber with cells inside is mounted between internal wells. Cells within chamber can be monitored with an inverted microscope (both dry or oil objectives). The enlargement presets the scheme of microscopic picture of cells in the chamber with respect to the dcEF lines

7. Add buffered solutions or PBS to the outer wells (green wells in Fig. 2b) to immerse Ag/AgCl reversible electrodes of 6 cm^2. Connect the electrodes in saline-filled wells by agar bridges to the neighboring wells, to which the observation chamber is attached.
8. If necessary, put the plexiglass cover on the apparatus and connect to the CO_2 supply to keep the pH stability.

3.4 Recording Cell Movements

Any inverted microscope can be used for time lapse imaging. The characteristics of our setup are useful (*see* **Note 12**). An economical setup can easily be put together if necessary (*see* **Note 13**).

1. Turn on the heater and set the temperature to 37 °C if necessary.
2. Stabilize the atmosphere to 5% CO_2, if necessary.
3. Place the chamber under the microscope so that the anode and cathode is on the left and right side, respectively (*see* **Note 14**).
4. Connect the electrotactic chamber to the direct current power supply, voltmeter, and ammeter (Fig. 2b) (*see* **Note 15**). Choose the optimal field(s) of view with cells in the electrotactic chamber.

5. Select the time interval between successive images and the total length of recording.
6. Start the image acquisition.
7. Turn on the EF.

3.5 Quantification of Cell Migration

1. Construct cell trajectories from cell centroid positions.
2. Under each experimental condition, analyze the trajectories of at least 50 cells.
3. Use the program Hiro (written by Wojciech Czapla) (*see* **Note 16**) to transfer all cell trajectories to the circular diagrams as previously reported [29, 30] to illustrate the orientation of cell movement. Instead of final cell displacements, we use full cell trajectories with the starting positions of cells at the origin of the coordinate axes. The *x* axis is the axis of electric field direction (Fig. 3).
4. Calculate for each cell or cell population (*see* **Note 17**) the following parameters:
   - Total length of cell displacement from the starting point to the final cell position.
   - Average rate of cell displacement, i.e., the distance from the starting point direct to the cell's final position/time of recording [31].
   - Total length of cell trajectory, i.e., the "true" length of cell trajectory; the trajectory of a cell was considered as a sequence of "n" straight-line segments, each corresponding to cell centroid translocation within one time interval between two successive positions.
   - Average speed of cell movement, i.e., the total length of cell trajectory/time of recording.
   - Coefficient of movement efficiency (CME), i.e., the ratio of cell displacement to cell trajectory length; CME would

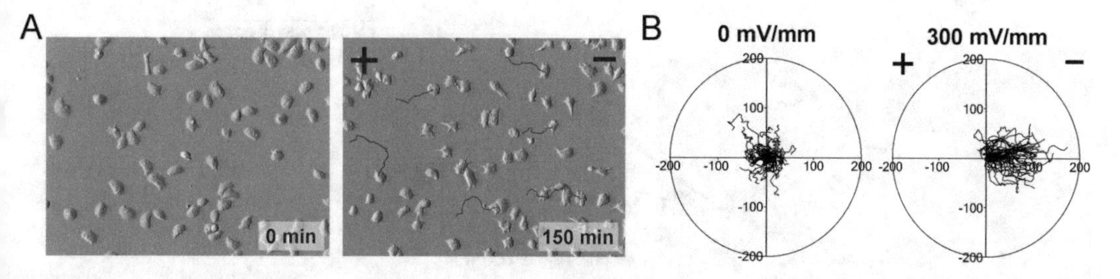

Fig. 3 Migration of rat Walker carcinosarcoma WC256 cells in electric field of 300 mV/mm. (**a**) Photographs of WC256 cells migrating in the EF of 300 mV/mm at the indicated time points. Red lines represent the trajectories of cells migrating toward the cathode. (**b**) Diagrams presenting trajectories of WC256 cells migrating in the EF of 0 and 300 mV/mm for 2.5 h. Scale in μm, $n = 50$ cells

equal 1 for the cell moving persistently along one straight line in one direction and 0 for a random movement [32].

- McCutcheon coefficient, i.e., the ratio between displacement parallel to the electric field direction and the total length of trajectory [33].
- Intersegmental angles α; α is defined as an angle between subsequent segments of cell trajectory [34, 35] (Fig. 4a);
- Average directional cosine β. The angle β is defined as the directional angle between the x axis (parallel to the electric field) and a vector AB, BC, CD; and A, B, C, D being successive positions of the cell, respectively. This parameter would equal +1 for a cell moving toward the cathode (the right side of the chamber), −1 for a cell moving in the direction of the anode (the left side of the chamber) and 0 for random movement [34, 35] (Fig. 4b).
- Average directional cosine γ. The angle γ is defined as the directional angle between the x axis (parallel to the electric field) and a vector AB, AC, AD; and A, B, C, D being the original and each subsequent positions of the cell, respectively. This parameter would equal +1 for a cell moving toward the cathode, −1 for a cell moving in the direction of the anode and 0 for random movement [34, 35]. This parameter is used generally to quantify the directionality of movement (Fig. 4c).

5. To detect the reaction of cells to changes in EF direction, cells are recorded for 30 min in EF and when all cells are polarized, the direction of the field is changed [11].

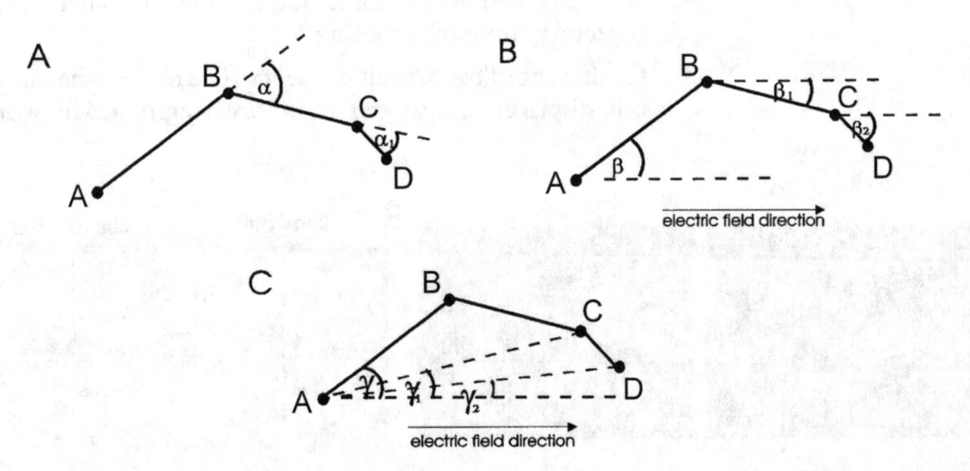

Fig. 4 Three types of angles characterizing cell locomotion. (**a**) intersegmental angles α; (**b**) directional segmental angles defined as the angle between the x axis (parallel to the electric field) and subsequent segments; (**c**) directional angles defined as the angle between the x axis (parallel to the electric field) and a vector AB, AC, AD and A, B, C, D being the original and each subsequent positions of the cell, respectively defined as in [34, 35]

6. For a quantitative description of cell reaction, the projections of new surface area of right and left sides of cells are measured with ImageJ.
7. The right and left sides of cells are estimated as follows: cell surface projection is divided into three parts along with the x axis (Fig. 5). The right side of the cell corresponds to the 1/3 fragment of the cell surface facing the cathode and, after field reversal, the anode. The left side corresponds to the opposite surface. The center of the cell is excluded from the analysis. The difference in the area of the right or left sides between two positions of a cell at different times t_0 and t is calculated according to the formula [11]:

$$S = (S_t - S_0) / S_0.$$

where S_t—surface area at time t, S_0—surface area at time t_0, t_0—5 min before field reversal.

8. Quantitative analysis is performed for 5 min before and 5 min after field reversal with a 30 s time interval (*see* **Note 18**).
9. Visualization of cell surface projection changes is carried out with Corel PaintShop Pro X6 software (Corel Corporation). Cell contours at the moment of field reversal and 1, 3, or 5 min later are merged. The zone of lamellipodium expansion is marked green, whereas the zone of cell retraction red (*see* **Note 18**).

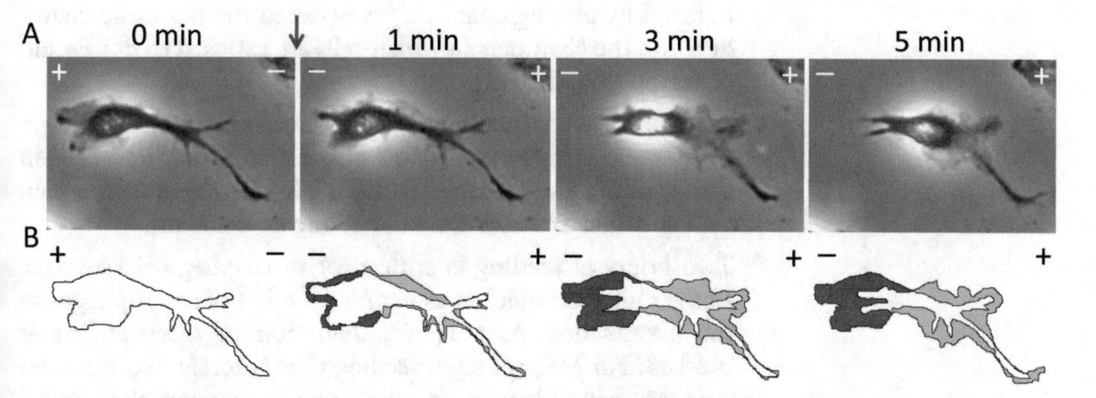

Fig. 5 Analysis of BMAC return upon electric field reversal. (**a**) Phase-contrast image sequence of a BMAC cell changing cell polarity. The time of EF reversal is indicated with a red arrow. (**b**) Analysis of changes in the cell surface projection between time of EF reversal and 1, 3, or 5 min after it. Red represents the retracted cell area and green represents the protrusive area

4 Notes

1. Glass crystallizers, glass Pasteur pipettes, glass cover slips and glass bridges can be obtained from various manufacturers.
2. *Tetrahymena pyriformis* and *Amoeba proteus* are very sensitive to impurities in the media, so high purity distilled water should be used in preparing all culture media. It is also critical to make sure that no detergent remains on the surface of the glass crystallizers after washing. For alternative culture methods, maintenance, and storage of *Tetrahymena pyriformis*, *see* [36].
3. The depth of the chamber is only 0.2 mm. The high ratio (about 10) of the surface area of upper and bottom walls of the chamber to the chamber depth and volume assures good dispersion of heat produced by the electric current due to the Joule's effect and thermal stabilization of the system [37].
4. The application of reversible Ag/AgCl electrodes instead of platinum, aluminum or stainless steel electrodes reduces the formation of several cytotoxic products. In addition, the reversible electrodes prevent uneven distribution of electric fields in solution between electrodes. When irreversible electrodes are used, the electric field changes over time due to polarization of the electrodes. As time passes, voltage drop begins to occur at the electrodes and reduces the electric field. Moreover, because in long-lasting experiments PBS or single buffer for both electrodes do not effectively stabilize the pH of solutions, in which the current is applied, cathode and anode buffered solutions according to Labanowski's formula [26] are used to decrease pH changes in solutions in which electrodes are immersed [38].
5. The harmful side effects of cytotoxic substances produced by the electrodes (metal ions, bubbles of gas, pH changes) may be reduced by placing agar bridges between the electrode chambers and the chambers in which cells are exposed to dcEFs and observed (Fig. 2) [38].
6. We use different plating densities because BC cells have a smaller surface area and it was possible to plate more cells in the electrotactic chamber without physical contact between cells and analyze more BC cells in one experiment.
7. Two hours of seeding is critical for macrophage electrotaxis. Longer incubation leads to strong adhesion of macrophages to the glass slides. As a result, migration of macrophages is blocked. For MSCs, longer seeding time is acceptable, however it is recommended to use the same incubation time in all experiments in order to obtain consistent results.
8. We recommend the use of a dry heat sterilization method in a hot air oven at 120 °C.

9. It is better to use silicone grease instead of double side tape to stick cover glasses. It is necessary if you want to use cells after dcEFs stimulation for other purposes such as immunostaining and Western blot which require opening of electrotactic chamber.
10. For experiments concerning the influence of some factors on the electrotaxis of cells, stick the glass strips only to the bottom cover glass then introduce the suspension of cells onto the space between the strips. When cells adhere to the surface aspirate the medium gently using a micropipette, add the medium with examined substance and seal the upper coverslip. Take care not to wet the strips to avoid the lack of tightness and medium leaking. If strips are wet, dry them with filter paper.
11. Take care of not wetting the electrotactic chamber and the plexiglass apparatus to avoid the lack of tightness and medium leaking resulting in unstable voltage or loss of current. If they are wet before sticking, dry them with filter paper.
12. Cell electrotaxis can be imaged with various microscope systems. We use a fully automated Leica DMI6000B inverted fluorescence microscope equipped with a broad range of objectives. The system is characterized by:
    - A motorized stage in combination with custom insert compatible with our electrotaxis chamber which enables recording of multiple fields of view, with XYZ coordinates specified individually.
    - Integrated modulation contrast, differential interference contrast or phase contrast optics.
    - The possibility of fluorescence imaging with a few filter cubes changed sequentially. Alternatively, application of external fast filter wheels provides extremely fast filter switching in just tens of milliseconds. It is crucial in the case of ratiometric calcium level measurements, FRET signal imaging or very rapid cell morphology changes.
    - A digital DFC360FX CCD camera (Leica Microsystems). For weak fluorescence signals, the use of an EM-CCD camera is recommended.
    - A Leica EL6000 metal-halide external light source (Leica Microsystems) which is suitable for long term excitation of fluorochromes in a stable manner.
    - An environmental control system (PeCon GmbH Erbach, Germany) set at 37 °C and 5% CO_2. Additionally, it is useful to humidify air infused to the chamber in order to reduce medium evaporation. To minimize optics and stage drift it is strongly recommended to stabilize the temperature overnight.

- A control software, Leica Application Suite X, allowing for long-term time-lapse image acquisition with appropriate time intervals and further processing of the obtained data.

13. Time-lapse imaging of electrotaxis research can be performed with an economical setup using any inverted microscope equipped with optics that enable imaging of unstained living cells (integrated modulation contrast, differential interference contrast or phase contrast). Introduction of a green light filter to the light path can be beneficial for cells condition during bright-field imaging (this color of incident light seems to be most harmless for living cells). If the microscope is not motorized with shutter control, fluorescence imaging is impossible. An inexpensive CMOS camera can be mounted on a separate port or in the place of an eyepiece. Camera control can be achieved with Micro-Manager, an open source microscopy software. Time series can be assembled into a movie with open source ImageJ software and can be quantitatively analyzed. The microscopy system described above is only suitable for imaging of organisms living at room temperature (*D. discoideum,* A. proteus). In the case of vertebrate cells, maintaining a higher temperature is necessary. Temperature control can be achieved by covering the microscope, by providing a source of hot air coupled with a temperature controller. When slow-migrating cells are recorded, pH control can be achieved by supplementing the medium with 15 mM HEPES. If medium needs to be replaced, pay a special attention to avoid changing the field of view and the focus.
14. Note the proper position of the chamber, i.e., place it parallel to the bottom edge of the microscope table.
15. Electric current is measured continuously with an ammeter. The voltage is measured directly with a high input impedance voltmeter and Ag-AgCl electrodes inserted at the ends of the experimental chamber (Fig. 2b). Alternatively, voltage gradient is calculated according to the Ohm's rule. DcEF strength is defined as follows: Field strength = I/dX, where I is the current in A, d is the medium conductivity in S/m, and X is the area of the cross section of the chamber in m^2.
16. Other programs for cell migration quantification can be used: Adapt, CellCognition, CellTracker, iTrack4U, MTrackJ, Pathfinder, SpotTracking, or TrackMate. Most of them enable automatic analysis of fluorescent images and manual tracking of cells on time-series obtained by various imaging methods [39].
17. Since the observed cell populations may be heterogeneous, quantitative estimation of cell movement parameters can benefit from histograms and/or dot-plot diagrams [40], in addition to the circular diagrams.

18. The best time points may differ depending on the cell type and should be determined experimentally in pilot experiments. Other programs to calculate cell surface area or to draw cell contours can be used.

Acknowledgments

This work was supported by a grant from the National Science Centre 2012/07/B/NZ3/02909, Poland. Faculty of Biochemistry, Biophysics, and Biotechnology of Jagiellonian University is a partner of the Leading National Research Center (KNOW) supported by the Ministry of Science and Higher Education.

References

1. Bear JE, Haugh JM (2014) Directed migration of mesenchymal cells: where signaling and the cytoskeleton meet. Curr Opin Cell Biol 30:74–82
2. Nuccitelli R, Nuccitelli P, Li C, Narsing S, Pariser DM, Lui K (2011) The electric field near human skin wounds declines with age and provides a noninvasive indicator of wound healing. Wound Repair Regen 19:645–655
3. McCaig CD, Rajnicek AM, Song B, Zhao M (2005) Controlling cell behavior electrically: current views and future potential. Physiol Rev 85:943–978. https://doi.org/10.1152/physrev.00020.2004
4. Martin-Granados C, McCaig CD (2013) Harnessing the electric spark of life to cure skin wounds. Adv Wound Care 3:127–138
5. Sheridan DM, Isseroff RR, Nuccitelli R (1996) Imposition of a physiologic DC electric field alters the migratory response of human keratinocytes on extracellular matrix molecules. J Invest Dermatol 106:642–646
6. Cooper MS, Schliwa M (1986) Motility of cultured fish epidermal cells in the presence and absence of direct current electric fields. J Cell Biol 102:1384–1399
7. Kim MS, Lee MH, Kwon B-J, Koo M-A, Seon GM, Park J-C (2015) Golgi polarization plays a role in the directional migration of neonatal dermal fibroblasts induced by the direct current electric fields. Biochem Biophys Res Commun 460:255–260
8. Djamgoz MBA, Mycielska M, Madeja Z, Fraser SP, Korohoda W (2001) Directional movement of rat prostate cancer cells in direct-current electric field: involvement of voltagegated Na+ channel activity. J Cell Sci 114:2697–2705
9. Sroka J, Krecioch I, Zimolag E, Lasota S, Rak M, Kedracka-Krok S, Borowicz P, Gajek M, Madeja Z (2016) Lamellipodia and membrane blebs drive efficient electrotactic migration of rat walker carcinosarcoma cells WC 256. PLoS One 11:e0149133
10. Krecioch I, Madeja Z, Lasota S, Zimolag E, Sroka J (2015) The role of microtubules in electrotaxis of rat Walker carcinosarcoma WC256 cells. Acta Biochim Pol 62:401–406
11. Zimolag E, Borowczyk-Michalowska J, Kedracka-Krok S, Skupien-Rabian B, Karnas E, Lasota S, Sroka J, Drukala J, Madeja Z (2017) Electric field as a potential directional cue in homing of bone marrow-derived mesenchymal stem cells to cutaneous wounds. Biochim Biophys Acta, Mol Cell Res 1864:267–279
12. Rapp B, De Boisfleury-Chevance A, Gruler H (1988) Galvanotaxis of human granulocytes – dose-response curve. Eur Biophys J 16:313–319
13. Chang PC, Sulik GI, Soong HK, Parkinson WC (1996) Galvanotropic and galvanotaxic responses of corneal endothelial cells. J Formos Med Assoc 95:623–627
14. Zhao M, Bai H, Wang E, Forrester JV, McCaig CD (2004) Electrical stimulation directly induces pre-angiogenic responses in vascular endothelial cells by signaling through VEGF receptors. J Cell Sci 117:397–405
15. Li L, El-Hayek YH, Liu B, Chen Y, Gomez E, Wu X, Ning K, Li L, Chang N, Zhang L, Wang Z, Hu X, Wan Q (2008) Direct-current electrical field guides neuronal stem/progenitor cell migration. Stem Cells 26:2193–2200

16. Pullar CE, Baier BS, Kariya Y, Russell AJ, Horst BAJ, Marinkovich MP, Isseroff RR (2006) Beta4 integrin and epidermal growth factor coordinately regulate electric field-mediated directional migration via Rac1. Mol Biol Cell 17:4925–4935
17. Poo M, Robinson KR (1977) Electrophoresis of concanavalin A receptors along embryonic muscle cell membrane. Nature 265:602–605
18. Zhao M, Pu J, Forrester JV, McCaig CD (2002) Membrane lipids, EGF receptors, and intracellular signals colocalize and are polarized in epithelial cells moving directionally in a physiological electric field. FASEB J 16:857–859
19. Orida N, Poo MM (1978) Electrophoretic movement and localisation of acetylcholine receptors in the embryonic muscle cell membrane. Nature 275:31–35
20. Fang KS, Ionides E, Oster G, Nuccitelli R, Isseroff RR (1999) Epidermal growth factor receptor relocalization and kinase activity are necessary for directional migration of keratinocytes in DC electric fields. J Cell Sci 112(Pt 12):1967–1978
21. Hart FX (2006) Integrins may serve as mechanical transducers for low-frequency electric fields. Bioelectromagnetics 27:505–508
22. Ozkucur N, Perike S, Sharma P, Funk RHW (2011) Persistent directional cell migration requires ion transport proteins as direction sensors and membrane potential differences in order to maintain directedness. BMC Cell Biol 12:4
23. Rajnicek AM (2006) Temporally and spatially coordinated roles for Rho, Rac, Cdc42 and their effectors in growth cone guidance by a physiological electric field. J Cell Sci 119:1723–1735
24. Zhao M, Dick A, Forrester JV, McCaig CD (1999) Electric field-directed cell motility involves up-regulated expression and asymmetric redistribution of the epidermal growth factor receptors and is enhanced by fibronectin and laminin. Mol Biol Cell 10:1259–1276
25. Meng X, Arocena M, Penninger J, Gage FH, Zhao M, Song B (2011) PI3K mediated electrotaxis of embryonic and adult neural progenitor cells in the presence of growth factors. Exp Neurol 227:210–217
26. Labanowski J (1979) Analysis of conditions for preparative electrophoresis of cells and subcellular fractions. Dissertation, Jagiellonian University
27. Prescott DM, James TW (1955) Culturing of Amoeba proteus on Tetrahymena. Exp Cell Res 8:256–258
28. Sroka J, von Gunten M, Dunn GA, Keller HU (2002) Phenotype modulation in non-adherent and adherent sublines of Walker carcinosarcoma cells: the role of cell-substratum contacts and microtubules in controlling cell shape, locomotion and cytoskeletal structure. Int J Biochem Cell Biol 34:882–899
29. C a E, Nuccitelli R (1984) Embryonic fibroblast motility and orientation can be influenced by physiological electric fields. J Cell Biol 98:296–307
30. Gruler H, Nuccitelli R (1991) Neural crest cell galvanotaxis: new data and a novel approach to the analysis of both galvanotaxis and chemotaxis. Cell Motil Cytoskeleton 19:121–133
31. Sroka J, Antosik A, Czyż J, Nalvarte I, Olsson JM, Spyrou G, Madeja Z (2007) Overexpression of thioredoxin reductase 1 inhibits migration of HEK-293 cells. Biol Cell 99:677–687
32. Sroka J, Kamiński R, Michalik M, Madeja Z, Przestalski S, Korohoda W (2004) The effect of triethyllead on the motile activity of walker 256 carcinosarcoma cells. Cell Mol Biol Lett 9:15–30
33. McCutcheon M (1946) Chemotaxis in leukocytes. Physiol Rev 26:319–336
34. Korohoda W, Madeja Z, Sroka J (2002) Diverse chemotactic responses of Dictyostelium discoideum amoebae in the developing (temporal) and stationary (spatial) concentration gradients of folic acid, cAMP, Ca2+ and Mg2+. Cell Motil Cytoskeleton 53:1–25
35. Korohoda W, Golda J, Sroka J, Wojnarowicz A, Jochym P, Madeja Z (1997) Chemotaxis of Amoeba proteus in the developing pH gradient within a pocket-like chamber studied with the computer assisted method. Cell Motil Cytoskeleton 38:38–53
36. Cassidy-Hanley DM (2012) Tetrahymena in the laboratory: strain resources, methods for culture, maintenance, and storage. Methods Cell Biol 109:237–276
37. Korohoda W, Mycielska M, Janda E, Madeja Z (2000) Immediate and long-term galvanotactic responses of Amoeba proteus to dc electric fields. Cell Motil Cytoskeleton 45:10–26
38. Grys M, Madeja Z, Korohoda W (2017) Avoiding the side effects of electric current pulse application to electroporated cells in disposable small volume cuvettes assures good cell survival. Cell Mol Biol Lett 22:1
39. Masuzzo P, Van Troys M, Ampe C, Martens L (2016) Taking aim at moving targets in computational cell migration. Trends Cell Biol 26:88–110
40. Waligórska A, Wianecka-Skoczeń M, Nowak P, Korohoda W (2007) Some difficulties in research into cell motile activity under isotropic conditions. Folia Biol (Praha) 55:9–16

Check for updates

Chapter 24

Analysis of Random Migration of *Dictyostelium* Amoeba in Confined and Unconfined Environments

Christof Litschko, Julia Damiano-Guercio, Stefan Brühmann, and Jan Faix

Abstract

Dictyostelium discoideum has proven to be an excellent model to study amoeboid cell migration. During their life cycle, *Dictyostelium* cells exhibit distinct modes of motility. Individual growth-phase cells explore new territories by random cell migration using the core cell motility machinery, but they can also hunt bacteria by detection and chemotaxis toward the by-product folate. After depletion of nutrients, the cells initiate a developmental program allowing streaming of the cells into aggregation centers by chemotaxis toward cAMP and by cell-to-cell adhesion. Subsequent development is associated with complex rotational movement of the compacted aggregates to drive cell type specific sorting, which in turn is necessary for terminal culmination and formation of fruiting bodies. Here we describe a protocol for the analyses of cell motility of vegetative *Dictyostelium* cells in unconfined and mechanically confined settings.

Key words Cell migration, *Dictyostelium*, Confinement, Cell tracking, Agar overlay, Random migration, Amoeboid motility

1 Introduction

A substantial fraction of our knowledge on cellular functions is based on studies with a few model organisms that were established during the second half of the last century. The social amoeba *Dictyostelium discoideum* is one such model, and has been particularly beneficial for the study of a wide range of fundamental cellular processes including actin cytoskeleton dynamics and cell motility, chemotaxis, phagocytosis, endocytosis, cell adhesion, pattern formation, multicellular development, and more recently, autophagy as well as social behavior and evolution [1]. *Dictyostelium* is accessible to genetic manipulations allowing the expression of fluorescent proteins or generation of knockout strains lacking one or multiple genes [2]. Due to its haploid genome, phenotypes can easily be detected. Moreover, mammalian cells frequently express multiple paralogs of a given protein family [3],

Alexis Gautreau (ed.), *Cell Migration: Methods and Protocols*, Methods in Molecular Biology, vol. 1749,
https://doi.org/10.1007/978-1-4939-7701-7_24, © Springer Science+Business Media, LLC 2018

which may complicate analyses of depleted cells or knockout mutants whereas in *Dictyostelium* most proteins are encoded by a single gene.

In this chapter of the book, we describe a protocol for the analyses of cell motility of vegetative *Dictyostelium* cells in unconfined and mechanically confined settings using the agar overlay technique initially established to study cytoskeletal components at high resolution during mitotic cell division [4]. Although it seems somewhat artificial at the first glance to compress cells with a sheet of agar, confinement is highly physiological as the cells have to frequently squeeze through narrow spaces in their natural habitats. Moreover, recent work has clearly revealed that the motile behavior of cytoskeletal mutants in unconfined versus confined scenarios can be highly diverse [5]. We describe an image analysis method for semiautomatic and manual tracking of migrating cells and include user-friendly tools to quantify important parameters such as cells speed, directionality ratio, turning angles and mean square displacement.

2 Materials

2.1 Preparation of Unconfined Cell Samples

1. Culture medium: HL5-C medium with glucose (Formedium) supplemented with 10 μg/mL Ampicillin/Streptomycin. Prepare according to manufacturer's instructions.
2. *Dictyostelium* cells: 10-cm Ø culture plate with *Dictyostelium* cells to be analyzed in 10–13 mL culture medium at approx. 50% confluence.
3. Microscopy dish: standard 35 mm microscopy glass bottom dishes (e.g., μ-Dish 35 mm, high Glass Bottom from Ibidi).
4. Sterile 10 mL serological pipettes.
5. 1× PB: 17 mM Na-K-phosphate buffer, pH 6.0. For a 20× stock solution dissolve 7.12 g of Na_2HPO4 and 39.8 g of KH_2PO4 in 1 L deionized water and autoclave. Store at 4 °C. Prepare 1× working solution from the 20× stock with deionized water. Autoclave. Store the working solution at room temperature.
6. Cell culture microscope.

2.2 Preparation of Confined Cell Samples (Agar Overlay Assay)

1. Culture medium: *see* Subheading 2.1.
2. *Dictyostelium* cells: *see* Subheading 2.1.
3. Microscopy dish: *see* Subheading 2.1.
4. 1× PB: *see* Subheading 2.1.
5. 76 × 26 mm microscope slides (e.g., Menzel, *see* **Note 1**).
6. 99% ethanol.
7. 24 × 24 mm glass coverslips (e.g., Menzel #1 (0.13–0.16 mm)).
8. Agarose gel: to make a 1.5% (w/v) gel dissolve 0.6 g agarose in 40 mL 1× PB in a laboratory glass bottle by careful heating using a microwave. Shake from time to time and avoid

excessive boiling of the liquid. Store for up to 1 week at room temperature (*see* **Note 2**).

9. Water bath at room temperature.
10. Scalpel.
11. Standard polystyrene petri dishes (Ø10 cm).
12. Distilled water.
13. Whatman filter paper GB005 (or similar) cut in small pieces of approx. 0.5 × 0.5 cm.
14. Cell culture microscope.
15. Laboratory wipes.

2.3 Imaging

1. Microscope equipped with 10× phase-contrast optics and a digital camera.
2. Microscopy software allowing time-lapse imaging (e.g., Metamorph 7, Molecular Devices).

2.4 Cell Tracking and Migration Analysis

1. Fiji/ImageJ [6] (https://fiji.sc/ or https://imagej.net/Fiji).
2. Cell tracking software, e.g., Dicty Tracking [7] (our semiautomatic tracking tool is available at (https://doi.org/10.6084/m9.figshare.5024552) or MTrackJ [8] (manual tracking tool for ImageJ available at https://imagescience.org/meijering/software/mtrackj/).
3. Excel (Microsoft Office).
4. Visual Basic for Applications (VBA)-based Excel workbooks *Dicty Tracking Evaluation.xslm*, *MTrackJ Evaluation.xslm* and *MSD Calculation.xslm* for further analysis. The workbooks are included in the Dicty Tracking package file (*see* **item 2**). For a more sophisticated analysis including velocity and direction autocorrelation we recommend the Excel macros of Gorelik and Gautreau [9].
5. Standard graphing software (e.g., SigmaPlot, GraphPad Prism, or Origin).

3 Methods

3.1 Preparation of Unconfined Cell Samples

1. Fill 3 mL culture medium in a glass bottom dish.
2. Wash off *Dictyostelium* cells from the 10-cm culture plate using a 10 mL serological pipette and add 1–2 drops of the cell suspension to the microscopy dish containing 3 mL of culture medium. Carefully shake the dish to homogenously distribute the cells.
3. Allow cells to adhere for at least 30 min. Subsequently, control adhesion of the cells to the glass bottom of the microscopy dish using a cell culture microscope (*see* **Note 3**).

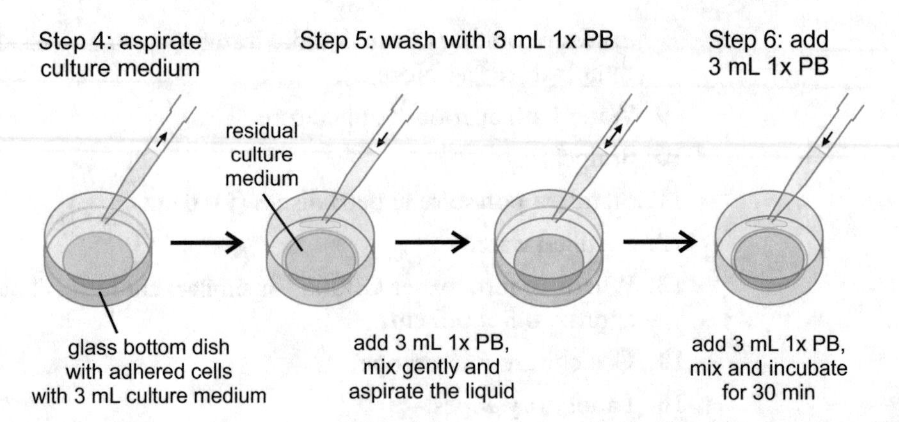

Fig. 1 Replacement of culture medium by 1× PB. Schemes shown refer to **steps 4–6** in Subheading 3.1

4. Carefully aspirate the culture medium using a 1000 μL pipette. Leave a small amount of medium on the glass bottom to prevent detachment and dessication of the cells (Fig. 1).
5. To wash out the remaining culture medium, slowly add 3 mL 1× PB to the microscope dish using a 1000 μL pipette, mix by pipetting cautiously up and down and aspirate the liquid as described above. Again, leave a small amount of liquid around the glass bottom (Fig. 1).
6. Add 3 mL 1× PB and mix as described above (Fig. 1). Incubate for 30 min at room temperature prior to time-lapse imaging.

3.2 Preparation of Confined Cell Samples (Agar Overlay Assay)

1. Perform **steps 1–6** of Subheading 3.1. During the 30 min incubation in 1× PB (**step 6** in subheading 3.1) perform the following steps.
2. Clean four microscope slides per sample with ethanol (*see* **Note 4**).
3. Place two glass coverslips on a microscope slide as shown in Fig. 2a.
4. Carefully melt the agarose gel (*see* Subheading 2.2, **step** 7) using a microwave. Shake the solution from time to time and avoid superheating. Cool down the agarose solution to 45–60 °C using the water bath.
5. Add 500 μL of the agarose solution to the center of the prepared microscope slide using a 1000 μL pipette (Fig. 2b).
6. Instantly, place the second microscope slide on top to build a slide-agarose-slide "sandwich" (Fig. 2c). Wait for 30 s to allow hardening of the agarose.
7. If necessary, remove excess of agarose from the edges of the slide-agarose-slide "sandwich" using the scalpel. Then carefully remove the upper slide with help of the scalpel as a lever arm (Fig. 2d).

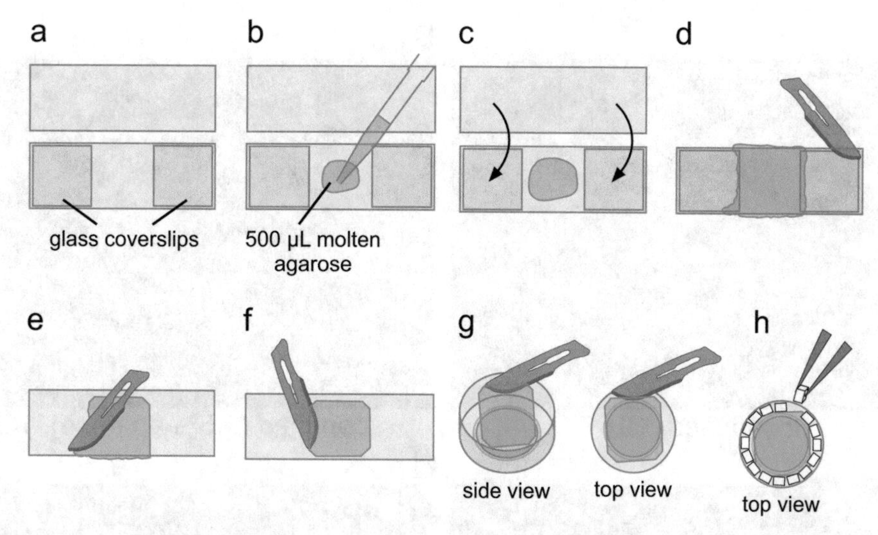

Fig. 2 Preparation of a confined cell sample using the agar overlay technique. (**a**) Setup for the generation of a 0.17 μm thin agarose slice consisting of two microscope slides and two glass coverslips. The glass coverslips function as spacers. (**b**) Addition of 500 μL molten agarose (*see* Subheading 3.2, **step 5**). (**c**) Generation of the slide-agarose-slide "sandwich". (**d**) Removal of the upper microscope slide using a scalpel. (**e**) Trimming of the agarose gel using a scalpel. (**f**) Inserting the scalpel blade between the agarose gel and the lower microscope slide allows easy separation of the sandwich. (**g**) Positioning of the trimmed agarose gel into a glass bottom dish with adhered cells in residual 1× PB (*see* Subheading 3.2, **step 10**). (**h**) Placing of distilled water saturated filter paper pieces along the wall of microscopy dish using tweezers

8. Trim the agarose gel on the corners for better fit into the microscope dish using the scalpel (Fig. 2e).
9. After 30 min of incubation aspirate 1× PB from the side of the microscopy dish containing the adhered cells using a 1000 μL pipette. Leave a small amount of liquid around the glass bottom (*see* Subheading 2.1, **step 4**).
10. Carefully place the trimmed agarose gel on top of the cells adhered to the glass bottom using the scalpel (Fig. 2f, g). Try to avoid wrinkling of the agarose slice as well as sliding with the agarose sheet over the cells. If necessary, remove excess of liquid from the edges of the glass bottom using a 200 μL pipette.
11. Control confinement (pancake-like appearance) of the cells using a cell culture microscope (*see* **Note 5**). If necessary, remove excess of liquid as described in **step 10**.
12. Fill some distilled water in a petri dish. Saturate the cut filter paper pieces by dipping them into the water droplet using tweezers.
13. Place the moistened filter paper pieces along the wall of the microscopy dish. They will serve as a humidity reservoir to prevent desiccation of the sample during imaging.

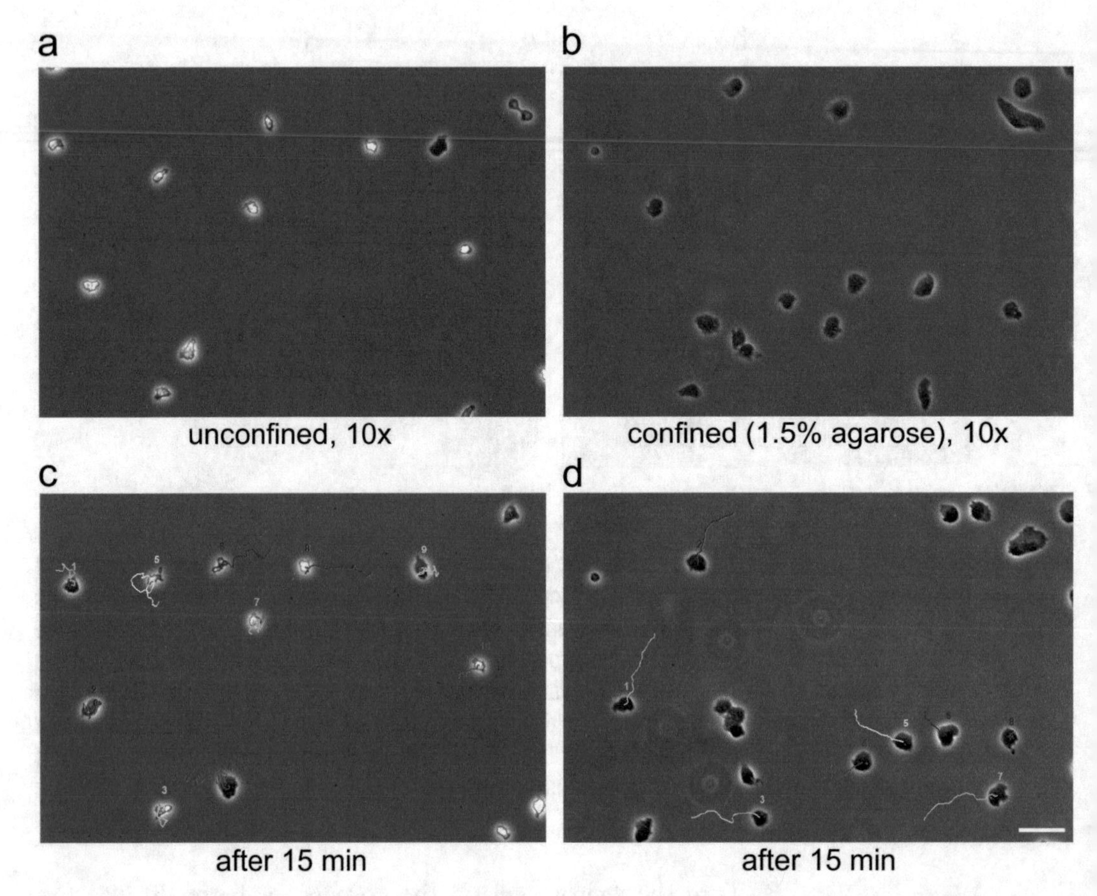

Fig. 3 Phase-contrast images of randomly migrating Ax2 wild-type cells in unconfined and confined settings. (**a**) Typical phase-contrast micrograph of unconfined Ax2 wild-type cells in 1× PB at 10×. (**b**) Typical phase-contrast micrograph of Ax2 wild-type cells in 1× PB confined through by a 1.5% agarose slice as described in Subheading 3.2. (**c**) Trajectories of unconfined Ax2 cells shown in (**a**) after 15 min. (**d**) Trajectories of confined Ax2 cells shown in (**b**) after 15 min. Note, that confined cells migrate much more directional. Cells in (**c**) and (**d**) were tracked with the Dicty Tracking tool (*see* Subheading 2.4). Scale bar, 50 μm

14. Close the microscopy dish with a punctured lid (*see* **Note 6**).
15. Prior to imaging, incubate the sample for at least 1 h at room temperature allowing the cells to adapt to increased stress in the physical confinement.
16. During incubation, control confinement from time to time using a cell culture microscope.

3.3 Imaging

1. Image unconfined cells with a time interval of 10 s for 15 min and confined cells every 20 s for 30 min at room temperature (*see* **Note 7**). Typical examples of phase-contrast images of unconfined and confined Ax2 wild-type cells at 10× magnification are shown in Fig. 3a, b, respectively.

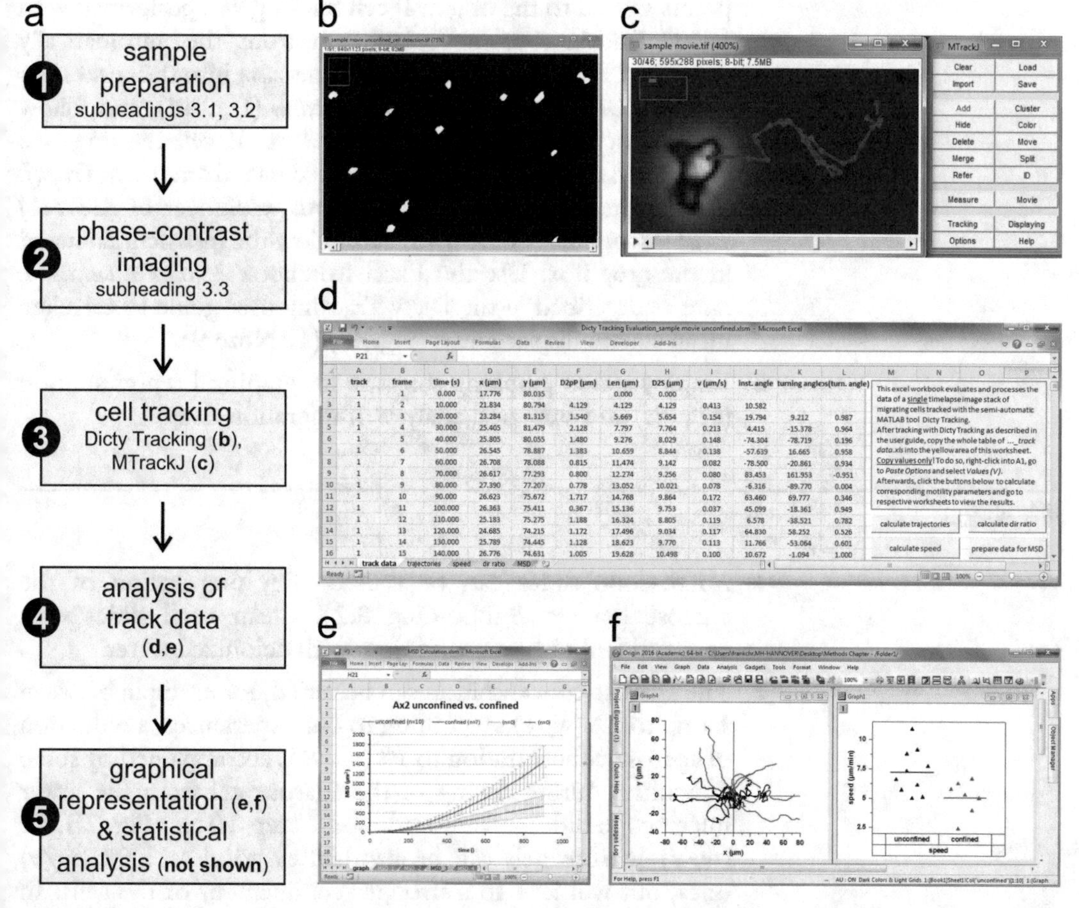

Fig. 4 Cell tracking and subsequent migration analysis. (**a**) Flowchart showing the different steps of migration analysis as described in this chapter. (**b**) Automatic detection of migrating cells from phase-contrast images by Dicty Tracking. The example shown corresponds to the phase-contrast image depicted in Fig. 3a. (**c**) Manual tracking of a migrating *Dictyostelium* cell using MTrackJ. (**d**) *Track data* sheet of the Excel workbook *Dicty Tracking Evaluation.xlsm* allowing calculation of mean speed, directionality ratio and trajectories shifted to the origin. (**e**) Calculation and graphical representation of the MSD of unconfined (blue) and confined cells (red) shown in Fig. 3c, d, respectively, using the Excel workbook *MSD Calculation.xlsm*. (**f**) Graphical representation of migration data using Origin 2016. Left, *x*-*y*-plot of trajectories shifted to the origin. Right, scatter plot of mean cell speeds. Both representations refer to unconfined (black) and confined cells (red) shown in Fig. 3c, d, respectively. Note, that confined cells cover a larger area within the same observation time (15 min) due to their highly directional migration behavior

3.4 Cell Tracking and Migration Analysis

1. Perform semiautomatic or manual cell tracking using Dicty Tracking (Fig. 4b) and MTrackJ (Fig. 4c), respectively. See corresponding user manuals for detailed instructions (*see* **Note 8**).
2. Use the VBA-based Excel workbook *Dicty Tracking Evaluation.xlsm* and *MTrackJ Evaluation.xlsm*, respectively, to calculate mean speed, directionality ratio and trajectories with starting

points shifted to the origin. If cell tracking was performed with Dicty Tracking, paste copied data from the automatically generated Excel workbook *_track data.xlsm* into the *track data* sheet of *Dicty Tracking Evaluation.xlsm* (Fig. 4d), and follow the instructions stated in the grey box. If cell tracking was performed using MTrackJ paste copied data from the *MTrackJ: points* window to the *MTrackJ-Points* worksheet of *MTrackJ Evaluation.xlsm* (not shown) and follow the instructions stated in the grey box. Use the Excel workbook *MSD Calculation.xlsm* as described in the Dicty Tracking user guide to calculate mean square displacement (Fig. 4e) (*see* **Note 9**).

3. Employ stated graphing software for graphical representation (Fig. 4f) and statistical analysis of migration data.

4 Notes

1. Microscope slides can be reused after preparation of the agarose gel (*see* Subheading 3.2). Clean used slides with conventional dish soap and rinse with deionized water.
2. The agarose concentration can be varied, for example between 1% up to 2% (w/v). According to our experiences, a reduction of agarose concentration to 1% (w/v) is accompanied by some difficulties during flaying of the agarose gel from the lower microscope slide (*see* Subheading 3.2 **step 10** and Fig. 2f). 2% (w/v) agarose gels can be handled as good as 1.5% (w/v) ones, but will lead to a stronger confinement of the cells. In some cases, for example when analyzing *Dictyostelium* mutants with strong defects of the actin cytoskeleton, higher agarose percentages can cause rupture of cells.
3. The procedure described in **steps 2** and **3** of Subheading 3.1 typically results in a cell density of 50–60 cells/cm^2 which turned out to be optimal for migration assays of unconfined Ax2 wild-type cells. The cell density should be decreased for migration analysis of confined cells, as these cells tend to migrate much more directional (*see* Fig. 3c, d) and, hence, will collide more frequently.
4. It has proven to be helpful to prepare two agarose gels for a confined cell sample, as the gel get damaged occasionally during **steps** 7, **8** and **10** of Subheading 3.2. Additionally, it is helpful to prepare two microscopy dishes with adhered cells in 1× PB according to Subheading 3.1 in parallel to have a backup dish, as overlaying of the cells by the agarose gel in some cases results in nonsuitable samples, e.g., due to formation of wrinkles or too many broken cells.
5. Sinking of the agarose gel onto the cells can occur very slowly due to the residual amount of 1× PB. Hence, in some cases it

takes 5–10 min until the cells are fully confined (visible under the microscope).

6. Clean lids are important for imaging. If the lid of the microscopy dish gets steamy due to the high humidity inside, it may help to open the lid for a short time or to puncture some holes at the periphery of the lid using a spatula tip heated up by a gas burner (perform under a laminar flow).
7. In cases of slowly migrating *Dictyostelium* mutants it is sufficient to image unconfined cells every 20 s to reduce noise resulting from tracking errors. Alternatively, frames can be removed from an existing image stack recorded at a 10 s interval using the *Slice Remover* function of ImageJ (Image ➔ Stack ➔ Tools ➔ Slice Remover).
8. Do not track cells that divide or collide during the experiment (as these processes interfere with the migratory behavior of the cells) as well as cells that leave and enter the field of view, respectively. Tracking of cells that leave or enter the observation area will result in different track lengths within the sample population.
9. Dicty Tracking as well as MTrackJ additionally calculates the turning angles for each cell at every time point. The probability distribution of these turning angels is another measure of directional persistence of migrating cells that is not biased by increasing observation time such as the directionality ratio [8], and can be easily generated using the histogram functions of many contemporary graphing programs.

Acknowledgments

This work was supported by the Deutsche Forschungsgemeinschaft (DFG grants FA 330/9-1 and FA330/11-1).

References

1. Bozzaro S (2013) The model organism Dictyostelium discoideum. In: Eichinger L, Rivero F (eds) Dictyostelium discoideum protocols, Methods in molecular biology, vol 983. Springer, New York, NY, pp 17–37
2. Faix J, Linkner J, Nordholz B et al (2013) The application of the Cre-loxP system for generating multiple knock-out and knock-in targeted loci. In: Eichinger L, Rivero F (eds) Dictyostelium discoideum protocols, Methods in molecular biology, vol 983. Springer, New York, NY, pp 249–267
3. Derivery E, Lombard B, Loew D, Gautreau A (2009) The wave complex is intrinsically inactive. Cell Motil Cytoskeleton 66:777–790. https://doi.org/10.1002/cm.20342
4. Yumura S, Mori H, Fukui Y (1984) Localization of actin and myosin for the study of ameboid movement in Dictyostelium using improved immunofluorescence. J Cell Biol 99:894–899
5. Ramalingam N, Franke C, Jaschinski E et al (2015) A resilient formin-derived cortical actin meshwork in the rear drives actomyosin-based motility in 2D confinement. Nat Commun 6:8496. https://doi.org/10.1038/ncomms9496
6. Schindelin J, Arganda-Carreras I, Frise E et al (2012) Fiji: an open-source platform for

biological-image analysis. Nat Methods 9:676–682. https://doi.org/10.1038/nmeth.2019
7. Litschko C (2017) Dicty Tracking: A stand-alone tool for fast and easy tracking of migrating Dictyostelium cells. figshare. https://doi.org/10.6084/m9.figshare.5024552.v3
8. Meijering E, Dzyubachyk O, Smal I (2012) Methods for cell and particle tracking. In: Conn PM (ed) Imaging and spectroscopic analysis of living cells. Methods Enzymol 504:183–200
9. Roman Gorelik, Alexis Gautreau, (2014) Quantitative and unbiased analysis of directional persistence in cell migration. Nature Protocols 9 (8):1931-1943

Chapter 25

Neutrophil Chemotaxis in One Droplet of Blood Using Microfluidic Assays

Xiao Wang and Daniel Irimia

Abstract

Neutrophils are the most abundant leukocytes in blood serving as the first line of host defense in tissue damage and infections. Upon activation by chemokines released from pathogens or injured tissues, neutrophils migrate through tissues toward sites of infections along the chemokine gradients, in a process named chemotaxis. Studying neutrophil chemotaxis using conventional tools, such as a transwell assay, often requires isolation of neutrophils from whole blood. This process requires milliliters of blood, trained personnel, and can easily alter the ability of chemotaxis. Microfluidics is an enabling technology for studying chemotaxis of neutrophils in vitro with high temporal and spatial resolution. In this chapter, we describe a procedure for probing human neutrophil chemotaxis directly in one droplet of whole blood, without neutrophil isolation, using microfluidic devices. The same devices can be applied to the study the chemotaxis of neutrophils from small animals, e.g., mice and rats.

Key words Chemotaxis, Microfluidics, Neutrophils, Speed, Persistence, Blood

1 Introduction

Neutrophils are the first line of host defense in tissue injury and infections [1]. Upon activation by chemokines released from pathogens or injured tissues, neutrophils transmigrate from peripheral blood and then migrate toward sites of infection along gradients of chemokines. This process is named chemotaxis. Impaired neutrophil chemotaxis, due to genetic defects [2] or to pathological conditions such as major burn [3] or drug treatment such as chemotherapy [4], weakens the host defense against pathogens and may further lead to severe infections. On the opposite side, excessive activation and recruitment of neutrophils in an uncontrolled manner may lead to chronic inflammation [5] and organ failure [6]. Thus, understanding neutrophil chemotaxis in homeostasis and pathology can provide important insights in clinical diagnosis, treatment, and prognosis of diseases.

Alexis Gautreau (ed.), *Cell Migration: Methods and Protocols*, Methods in Molecular Biology, vol. 1749,
https://doi.org/10.1007/978-1-4939-7701-7_25,

During the past four decades, a variety of assays including Boyden chamber [7], Zigmond chamber [8], Dunn chamber [9], and micropipette [10] have been developed to study chemotaxis in vitro. Although these techniques enabled numerous discoveries and elevated our understanding of chemotaxis, they have several limitations that hinder them from being used in clinical research. These assays lack the ability to control the gradient with precision over time and only use isolated neutrophils. Therefore, they require large volume of blood (>1 mL) and significant amount of work and time to isolate homogenous subpopulations of cells from the blood. This sample preparation process has the potential to alter neutrophil phenotype, interfering with the chemotaxis measurements in subsequent experiments.

Microfluidics is a technology using microfabricated valves, channels, and chambers to manipulate microscale fluids and organisms with high spatial and temporal resolution [11]. The application of microfluidics to the study of leukocyte biology has facilitated several insights into neutrophil chemotaxis [12]. Compared to the conventional chemotaxis assays, microfluidic assays enable the use of stable gradients, of sophisticated spatial profiles [13]. Microfluidic assays also enable precise mechanical confinement of the migrating cells, which resemble the interstitial spaces present in vivo in tissues [14]. Moreover, multifunctional microfluidic assays allow for seamless integration of sample preparation with downstream chemotaxis assay on a single chip [15–22]. A majority of these chips accomplish the on-chip blood purification first, and then probe neutrophil chemotaxis [15, 19–21]. Although these devices significantly simplify and reduce the time for sample preparation, they rely on P-selectin functionalized surfaces, which could activate the neutrophils and interfere with the chemotaxis measurements [15].

In this chapter, we present the complete work flow (Fig. 1) used in our group to measure neutrophil chemotaxis directly in a droplet of whole blood [16–18]. The device allows investigation of neutrophil chemotaxis, with high temporal and spatial resolutions, in microchannels that mimic the interstitial spaces. The moving neutrophils emerge from the droplet of blood at the time when they enter the channels, in a self-sorting process that does not interfere with their original activation status. The assay also preserves the serum components which are well known to be important for the functional status of the neutrophils [23].

2 Materials and Equipment

Soft lithography is the most widely used method for prototyping microfluidic devices. It includes two major steps: fabrication of a master mold in SU-8 and replication of devices in Polydimethylsiloxane (PDMS) from the mold. The fabrication of

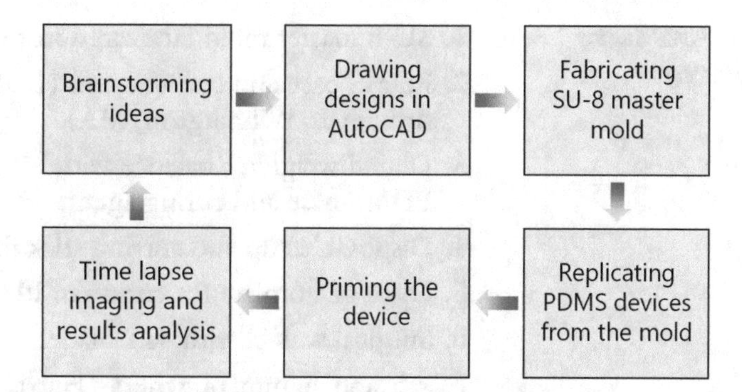

Fig. 1 The work flow of developing microfluidic devices for chemotaxis studies is an iterative process that may include several cycles of design, fabrication, and testing. A microfluidic design is first captured as an AutoCAD drawing. The design is implemented in SU-8 photoresist on silicon wafers using photolithography masks printed from the AutoCAD drawings. Devices are then fabricated in PDMS that is cast on the SU-8-silicon mold. The devices are then tested with whole blood samples. Depending on the initial results, the design may need to be optimized further

the SU-8 mold is conducted in class 100 clean room using standard microfabrication equipment. Basic tools and containers such as tweezers and beakers are broadly available in research labs, thus they are not mentioned in the section.

2.1 Soft Lithography

2.1.1 Fabrication of the SU-8 Master Mold

1. Photomasks with microfluidic designs (Front Range Photo Mask, Palmer Lake, CO).
2. 10 cm silicon wafers (Desert Silicon, Grandale, AZ).
3. SU-8 5 negative photoresist (SU-8 5, Microchem, Newton, MA) for patterning the first layer of the mold, which contains the migration channels.
4. SU-8 100 negative photoresist (SU-8100, Microchem) for patterning the second layer of the mold, which contains the chemokine chambers.
5. SU-8 developer (BTS-220, J.T. Baker, Center Valley, PA).
6. Dehydration oven at 200 °C.
7. Plasma treatment system (March PX-2527 Plasma System, Nordson March, Concord, CA).
8. Spin coater for uniformly coating layers of SU-8 with desired thickness.
9. Hot plates for baking SU-8 before and after exposure.
10. Mask aligner for multilayer alignment and exposure.
11. Profilometer (Contour Elite K, Bruker BioSpin Corporation, Billerica, MA) for precision metrology and quality control of the fabricated mold.

2.1.2 PDMS Casting and Bonding

1. SU-8 master mold fabricated from the SU-8 process.
2. PDMS base and curing agent (PDMS, Sylgard, 184, Elsworth Adhesives, Wilmington, MA).
3. Digital weighing balance with >0.1 g resolution for weighing PDMS base and curing agent.
4. Disposable cup and stirring stick for mixing PDMS.
5. Oven or hotplate for curing of PDMS at 65–80 °C.
6. Surgical scalpel with #11 blade.
7. 1.5 and 5 mm punchers (Harris Uni-Core, Ted Pella Inc., Reading, CA).
8. Clean-room adhesive tape.
9. Glass-bottom well plates (MatTek Co., Ashland, MA).
10. Plasma treatment system.

2.2 Preparation of Chemoattractant

1. Chemokine stock purchased from vendors, such as Sigma-Aldrich, Cayman Chemicals, R&D Systems Inc. (*see* **Note 1**).
2. Medium: Iscove's Modified Dulbecco's Medium (IMDM) containing 20% FBS. Mix 400 mL (IMDM) with 100 mL FBS. Filter the mixture and store at 4 °C. Warm up to 37 °C in a bead or water bath before use.
3. Human fibronectin (R&D systems, Inc., Minneapolis, MN): the stock is aliquoted into small vials and stored at 4 °C.
4. 500 μL snap vials.

2.3 Preparation of Human Whole Blood Sample

1. Lancing devices and disposable lancets (BD Contact-Activated Lancet 30G, Becton Dickinson, Franklin Lakes, NJ).
2. Media: IMDM containing 20% FBS.
3. Nucleus staining solution: Hoechst 33342 (10 mg/mL solution in water). Store at 4 °C.
4. 500 μL snap vials.

2.4 Time-Lapse Imaging

1. Nikon Eclipse TiE microscope equipped with Nikon perfect focus system.
2. Biochamber compatible with the microscope stage, set at 37 °C temperature and 5% CO_2 content.

3 Methods

3.1 Design and Fabrication of the Microfluidic Device

3.1.1 Designing the Microfluidic Device

1. Draw the microfluidic design in AutoCAD software (*see* **Note 2**).
2. Send the AutoCAD file to a vendor to fabricate the photomasks with desired resolutions (*see* **Note 3**).

3.1.2 Fabrication of the SU-8 Master Mold

1. Dehydrate two 4″ wafers at 200 °C in the dehydration oven for 90 min (*see* **Notes 4** and **5**).
2. Spin-coat SU-8 5 on the wafers with a thickness of 5 μm (*see* **Notes 6** and 7).
3. Bake the wafers on a 75 °C hotplate and then on a separate 100 °C hotplate.
4. Load the first photomask and the wafer on the mask aligner and expose the wafer under UV light.
5. Bake the wafers on a 75 °C hotplate and then on a separate 100 °C hotplate.
6. Fill a 15 cm-diameter glass container with ~30 mL SU-8 developer (the level of the developer should be ~1.5 cm above the bottom). Develop the SU-8 until unexposed SU-8 is completely removed (*see* **Note 8**).
7. Dry the wafer under nitrogen flow (*see* **Note 9**).
8. Repeat the similar process to fabricate the second layer, 50 μm thick, on top of the first layer (*see* **Note 10**).
9. Place the SU-8 master mold in a 15 cm diameter petri dish and tape the edge of the wafer to fix it.

3.1.3 PDMS Casting and Bonding

1. Weigh PDMS base and curing agent at a ratio of 10:1 (*see* **Note 11**).
2. Mix the base and curing agent homogeneously with a disposable stirring stick or a fork and pour the mixture on the mold.
3. Place the master mold into a vacuum chamber and degas until the PDMS is bubble-free.
4. Cure the PDMS in an oven or on a hot plate at 75 °C overnight.
5. After curing, cut PDMS along the outline of the wafer with a scalpel and peel out the entire PDMS layer.
6. Punch the inlet of the device with a 1.25 mm puncher and then remove the device from the PDMS layer with a 5 mm puncher.
7. Clean the surface of the device multiple times with clean-room adhesive tape.

Fig. 2 The microfluidic device for neutrophil chemotaxis in a droplet of whole blood. (**a**) A macrophotography of the microfluidic device fabricated in a single well plate. (**b**) The device immersed in IMDM media after priming. (**c**) Loading of 1 μL whole blood in the device using a fine pipette tip. The scale bars are 5 mm

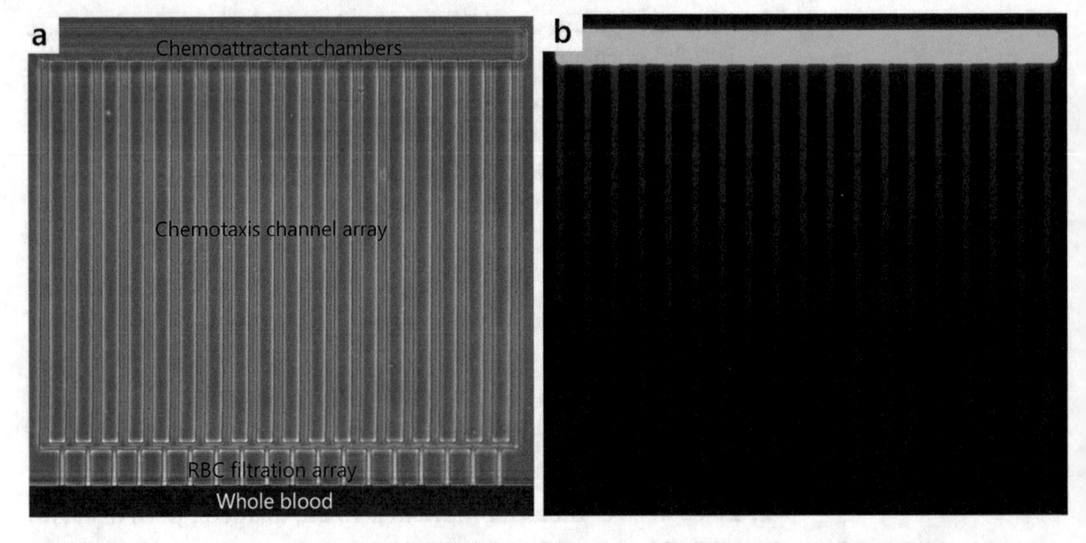

Fig. 3 Geometrical details of the microfluidic device for neutrophil chemotaxis. (**a**) Bright-field microscopic image showing the microfluidic unit for characterizing neutrophil trafficking from a drop of whole blood. (**b**) Fluorescence microscopic image showing the chemogradient established in the unit. The scale bars are 40 μm

8. Treat a glass bottom dish or multi-well plate and the device with O_2 plasma for 35 s (*see* **Note 12**).
9. Bond the PDMS device on top of the plate. Press the device gently with tweezers to ensure the PDMS and glass contact seamlessly.
10. Place the dish or plate on a hotplate for 20 min at 65 °C to enhance the bonding. Figure 2a is a photo of the device that is ready to use. Figure 3a shows a microscopic image of the microchannels and chambers.

3.2 Device Priming

1. Dilute chemokine to desired concentration in IMDM containing 20% FBS with 100 nM fibronectin. In this experiment, we use 100 nM LTB4 as the chemoattractant.

2. Pipetting 5 μL chemoattractant from the inlet into the device and degas the device in a vacuum chamber for 10 min.
3. Remove the device from the vacuum chamber and wait for 20 min until the microchannels and chambers are completely filled with chemoattractant.
4. To establish the gradient, flush the central cell-loading chamber with chemoattractant-free media using a 1 mL a pipette.
5. Immerse the device under 4 mL media (Fig. 2b).
6. Wait for 30 min to allow the gradient to stabilize (*see* **Note 13**) (Fig. 3b).

3.3 Loading Human Whole Blood Sample

1. Acquire 5 μL human whole blood by finger prick and transfer it into a 0.5 mL tube containing 5 μL of media, heparin anticoagulant, and ~4 μM of Hoechst stain.
2. After gentle mixing with pipette, incubate the sample at 37 °C and 5% CO_2 for 15 min to allow proper staining of the nuclei.
3. Dilute the sample further with media in a ratio of 1:3.
4. Load 1 μL diluted human blood per device, gently, from the inlet using a gel loading tip (Fig. 2c).

3.4 Time-Lapse Imaging

1. Place the dish or multi-well-plate on the microscope stage.
2. Cover the plate with the biochamber which maintains the temperature and CO_2 content at 37 °C and 5%.
3. Set up imaging positions, filter cubes, time duration of each cycle and total imaging time in the software (*see* **Note 14**) and start the imaging process.
4. After the experiment, import the time-lapse images into ImageJ or other imaging processing software for further analysis (*see* **Note 15**). For example, cell migration trajectories can be tracked and analyzed using Trackmate module in ImageJ. Figure 4 shows a neutrophil migrating up the LTB4 gradient in the microchannel.

4 Notes

1. We recommend thawing chemokine solutions right before priming the device. Repetitive thawing and freezing should be avoided.
2. The presented microfluidic design has two layers: a 5 μm thick layer containing channels for cell migration and a 50 μm thick layer containing chemokine chambers. Fabrication of the two-layer SU-8 master mold requires two photomasks with proper alignment marks.

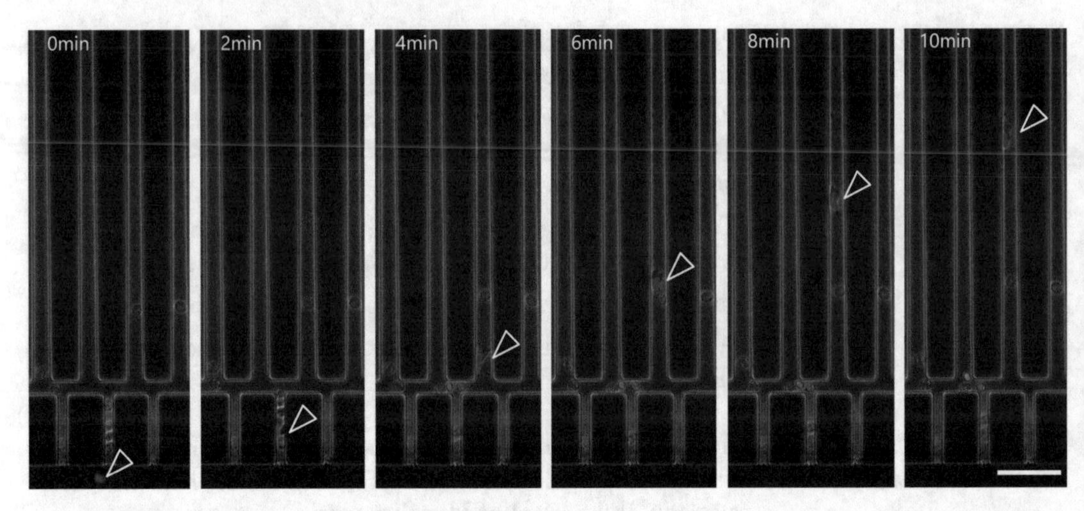

Fig. 4 Human neutrophil chemotaxis from whole blood. A montage of time-lapse images shows one neutrophil (nucleus stained in blue), which migrates from the whole blood in the central chamber (bottom) toward the LTB4-filled chemoattractant chamber (top). The scale bar is 40 μm. Time interval between successive frames is 2 min

3. When ordering a mask from a company, the resolution of the photo mask needs to be smaller than the smallest features of the design. The mask for the first layer of our design has a resolution of 3 μm. The mask for the second layer has a resolution of 20 μm.
4. SU-8 process is a standard process; thus, it is only briefly described in this chapter.
5. We recommend fabricating multiple master molds in parallel.
6. The ramping rate, spinning rate, and time for specific thickness of SU-8 film can be found from standard SU-8 protocol (http://www.microchem.com/Prod-SU8.htm).
7. The actual thickness of SU-8 after spinning may vary with temperature and humidity of the clean-room environment. These parameters are specific to each clean room and should be optimized accordingly.
8. The baking time, exposure time and development time may vary. We recommend using the parameters in the standard SU-8 protocol as a reference and then modify them.
9. The wafer should be directly dried with N_2 flow after development. Do not rinse the wafer with water or other reagent.
10. The proper alignment of the two SU-8 layers is critical. This is achieved by aligning the alignment marks in the second mask with the alignment marks on the wafer using the mask aligner.
11. We use 40 g of PDMS base and 4 g of PDMS curing agent for a 15 cm diameter Petri dish. This will yield a ~3 mm thick PDMS layer.

12. Single dishes and 6, 12, 24, and 96 glass bottom multi-well plates are available. Using a multi-well plates enables simultaneous study of neutrophil chemotaxis from multiple samples, in various experimental conditions.
13. To visualize the chemoattractant gradient, we use a fluorescent dye with similar molecular weight as the chemokine. Figure 3b demonstrates a chemoattractant gradient established between the chemoattractant chamber and central cell-loading chamber.
14. It is critical to make sure that the microscope takes images at the same focal plane during the entire imaging period. We use the Nikon perfect focus system (PFS) to automatically maintain the focal plane during the time-lapse imaging process.
15. Using the microfluidic assay, chemotaxis of single leukocytes can be investigated with high temporal and spatial resolution. The characteristic parameters of chemotaxis we investigate are the number of recruited cells, migration velocity, directionality, persistence, and trajectory.

References

1. de Oliveira S, Rosowski EE, Huttenlocher A (2016) Neutrophil migration in infection and wound repair: going forward in reverse. Nat Rev Immunol 16(6):378–391. https://doi.org/10.1038/nri.2016.49
2. Dinauer MC (2014) Disorders of neutrophil function: an overview. Methods Mol Biol 1124:501–515. https://doi.org/10.1007/978-1-62703-845-4_30
3. Butler KL, Ambravaneswaran V, Agrawal N, Bilodeau M, Toner M, Tompkins RG, Fagan S, Irimia D (2010) Burn injury reduces neutrophil directional migration speed in microfluidic devices. PLoS One 5(7):e11921. https://doi.org/10.1371/journal.pone.0011921
4. LeBlanc AK, LeBlanc CJ, Rohrbach BW, Kania SA (2015) Serial evaluation of neutrophil function in tumour-bearing dogs undergoing chemotherapy. Vet Comp Oncol 13(1):20–27. https://doi.org/10.1111/vco.12015
5. Soehnlein O, Steffens S, Hidalgo A, Weber C (2017) Neutrophils as protagonists and targets in chronic inflammation. Nat Rev Immunol 17(4):248–261. https://doi.org/10.1038/nri.2017.10
6. Brown KA, Brain SD, Pearson JD, Edgeworth JD, Lewis SM, Treacher DF (2006) Neutrophils in development of multiple organ failure in sepsis. Lancet 368(9530):157–169. https://doi.org/10.1016/S0140-6736(06)69005-3
7. Boyden S (1962) The chemotactic effect of mixtures of antibody and antigen on polymorphonuclear leucocytes. J Exp Med 115:453–466
8. Zigmond SH (1977) Ability of polymorphonuclear leukocytes to orient in gradients of chemotactic factors. J Cell Biol 75(2 Pt 1):606–616
9. Zicha D, Dunn GA, Brown AF (1991) A new direct-viewing chemotaxis chamber. J Cell Sci 99(Pt 4):769–775
10. Servant G, Weiner OD, Herzmark P, Balla T, Sedat JW, Bourne HR (2000) Polarization of chemoattractant receptor signaling during neutrophil chemotaxis. Science 287(5455):1037–1040
11. Whitesides GM (2006) The origins and the future of microfluidics. Nature 442(7101):368–373. https://doi.org/10.1038/nature05058
12. Irimia D, Ellett F (2016) Big insights from small volumes: deciphering complex leukocyte behaviors using microfluidics. J Leukoc Biol 100(2):291–304. https://doi.org/10.1189/jlb.5RU0216-056R
13. Irimia D, Geba DA, Toner M (2006) Universal microfluidic gradient generator. Anal Chem 78(10):3472–3477. https://doi.org/10.1021/ac0518710
14. Irimia D, Charras G, Agrawal N, Mitchison T, Toner M (2007) Polar stimulation and constrained cell migration in microfluidic channels. Lab Chip 7(12):1783–1790. https://doi.org/10.1039/b710524j

15. Agrawal N, Toner M, Irimia D (2008) Neutrophil migration assay from a drop of blood. Lab Chip 8(12):2054–2061. https://doi.org/10.1039/b813588f
16. Hamza B, Wong E, Patel S, Cho H, Martel J, Irimia D (2014) Retrotaxis of human neutrophils during mechanical confinement inside microfluidic channels. Integrat Biol 6(2):175–183. https://doi.org/10.1039/c3ib40175h
17. Hoang AN, Jones CN, Dimisko L, Hamza B, Martel J, Kojic N, Irimia D (2013) Measuring neutrophil speed and directionality during chemotaxis, directly from a droplet of whole blood. Technology 1(1):49. https://doi.org/10.1142/S2339547813500040
18. Jones CN, Hoang AN, Dimisko L, Hamza B, Martel J, Irimia D (2014) Microfluidic platform for measuring neutrophil chemotaxis from unprocessed whole blood. J Vis Exp (88). https://doi.org/10.3791/51215
19. Kasuga K, Yang R, Porter TF, Agrawal N, Petasis NA, Irimia D, Toner M, Serhan CN (2008) Rapid appearance of resolvin precursors in inflammatory exudates: novel mechanisms in resolution. J Immunol 181(12):8677–8687
20. Sackmann EK, Berthier E, Schwantes EA, Fichtinger PS, Evans MD, Dziadzio LL, Huttenlocher A, Mathur SK, Beebe DJ (2014) Characterizing asthma from a drop of blood using neutrophil chemotaxis. Proc Natl Acad Sci U S A 111(16):5813–5818. https://doi.org/10.1073/pnas.1324043111
21. Sackmann EK, Berthier E, Young EW, Shelef MA, Wernimont SA, Huttenlocher A, Beebe DJ (2012) Microfluidic kit-on-a-lid: a versatile platform for neutrophil chemotaxis assays. Blood 120(14):e45–e53. https://doi.org/10.1182/blood-2012-03-416453
22. Jones CN, Hoang AN, Martel JM, Dimisko L, Mikkola A, Inoue Y, Kuriyama N, Yamada M, Hamza B, Kaneki M, Warren HS, Brown DE, Irimia D (2016) Microfluidic assay for precise measurements of mouse, rat, and human neutrophil chemotaxis in whole-blood droplets. J Leukoc Biol 100(1):241–247. https://doi.org/10.1189/jlb.5TA0715-310RR
23. Mankovich AR, Lee CY, Heinrich V (2013) Differential effects of serum heat treatment on chemotaxis and phagocytosis by human neutrophils. PLoS One 8(1):e54735. https://doi.org/10.1371/journal.pone.0054735

Chapter 26

Leukocyte Migration and Deformation in Collagen Gels and Microfabricated Constrictions

Pablo J. Sáez, Lucie Barbier, Rafaele Attia, Hawa-Racine Thiam, Matthieu Piel, and Pablo Vargas

Abstract

In multicellular organisms, cell migration is a complex process. Examples of this are observed during cell motility in the interstitial space, full of extracellular matrix fibers, or when cells pass through endothelial layers to colonize or exit specific tissues. A common parameter for both situations is the fast adaptation of the cellular shape to their irregular landscape. In this chapter, we describe two methods to study cell migration in complex environments. The first one consists in a multichamber device for the visualization of cell haptotaxis toward the collagen-binding chemokine CCL21. This method is used to study cell migration as well as deformations during directed motility, as in the interstitial space. The second one consists in microfabricated channels connected to small constrictions. This procedure allows the study of cell deformations when single cells migrate through small holes and it is analogous to passage of cells through endothelial layers, resulting in a simplified system to study the mechanisms operating during transvasation. Both methods combined provide a powerful hub for the study of cell plasticity during migration in complex environments.

Key words Cell deformation, Chemokine, Chemotaxis, Collagen, Constriction, Cytoskeleton, Dendritic cell, Haptotaxis, Leukocyte, Microchannel

1 Introduction

Cell migration is a fundamental process required for immune responses in mammals. It occurs, for example, when phagocytes such as neutrophils colonize infected tissues for fast pathogen clearance [1, 2]. It also happens during the more complex adaptive immune response, in which dendritic cells (DCs) and T lymphocytes (LTs) circulate between tissues resulting in a systemic immune response [3]. Migration in any organ is a complex process that requires the physical adaptation of cells to the structural irregularities of their surroundings [2]. Cytoskeleton rearrangements have been observed during cell migration in confined environments [4–8]. In the following chapter, we describe two procedures to study cell migration in irregular landscapes, focusing our attention

Alexis Gautreau (ed.), *Cell Migration: Methods and Protocols*, Methods in Molecular Biology, vol. 1749,
https://doi.org/10.1007/978-1-4939-7701-7_26,

in cell motility and cell plasticity in collagen gels as well as in precisely designed microfabricated channels.

The first assay corresponds to a macrodevice that allows the preparation of multiple independent collagen gels, of fixed dimensions, in single microscopy dishes. This system can be used to study haptotaxis in a large number of cells as well as cell deformations imposed by the geometry of the collagen meshwork. This environment mimics some properties of the interstitial space in tissues [9].

The second assay consists in a microfabricated chip to study cellular changes in cells facing a constriction. This device corresponds to straight tubes with a drastic reduction in their diameter, forcing strong deformations in the cells moving through the gaps [5, 10]. This microfabricated system can be used to visualize cytoskeleton adaptations or cellular rearrangements in cells facing drastic spatial reductions, as when they pass through membranes during intravasation and extravasation [5, 10].

The combination of both methods provides a robust integrated system to study cell plasticity during migration in complex environments. Here we describe migration of activated DCs, but a similar procedure can be used to study other leukocytes, such as neutrophils and LTs.

2 Materials

The materials used for the protocols shown here are divided in two types: materials used as provided by the manufacturer (listed below), and those that required some preparation.

Materials for chip fabrication: glass bottom Petri dishes (35 mm), molds with customized design, oven between 65 and 75 °C, PDMS and curing agent, PDMS driller, plasma cleaner, scotch tape, surgical or razor blade, vacuum bell jar. Materials for DCs culture: bone-marrow derived dendritic cells (BMDCs), BMDC differentiation media, bovine type I collagen, standard cell culture equipment. For DC staining and fixation: antibodies, DNA staining (DAPI or Hoechst 33342), paraformaldehyde (PFA), PBS/BSA (from bovine serum albumin) 2%, and fluorescently labeled phalloidin.

1. PDMS was prepared with its cross-linker at a ratio of 10:1 by weight.
2. BMDC differentiation media (IMDM, fetal bovine serum -FBS- 10%), glutamine (20 mM), Pen-Strep (100 U/ml), and 2-mercaptoethanol (50 μM) supplemented with granulocyte-macrophage colony stimulating factor (50 ng/ml)-containing supernatant obtained from transfected J558 cells, as previously described [7].
3. CCL21 chemokine were reconstituted in sterile PBS at a concentration of 100 μg/ml.
4. PFA was dissolved in PBS to a final concentration of 4%.

3 Methods

3.1 Collagen Gels to Study Cell Haptotaxis in 3D Environments

The following method describes a device to study haptotaxis of leukocytes toward CCL21 in multiple conditions in a single classic 35 mm glass-bottom microscopy dish. Our device consists in a PDMS chip containing three parallel chambers, which function as collagen reservoirs (Fig. 1a). Each chamber can be loaded independently with collagen, allowing the study of different conditions in a single plate. The presence of multiple gels in the same chip minimizes variability due to experimental conditions (i.e., chamber preparation or chemokine exposure). Another important property of this chamber is its height (350 μm). This particular dimension allows visualization of cells located in different layers of the collagen when imaging at low-resolution (5× or 10×) using phase contrast. This method is based on the natural ability of CCL21 to bind collagen and generate gradients [9]. CCL21 binding to the collagen inside the chamber passively generates a stable gradient, as observed with a 10 kDa FITC-dextran that has a similar hydrodynamic radius (Fig. 1b). This method is complementary to other recently published procedures to study chemotaxis and haptotaxis [11–13].

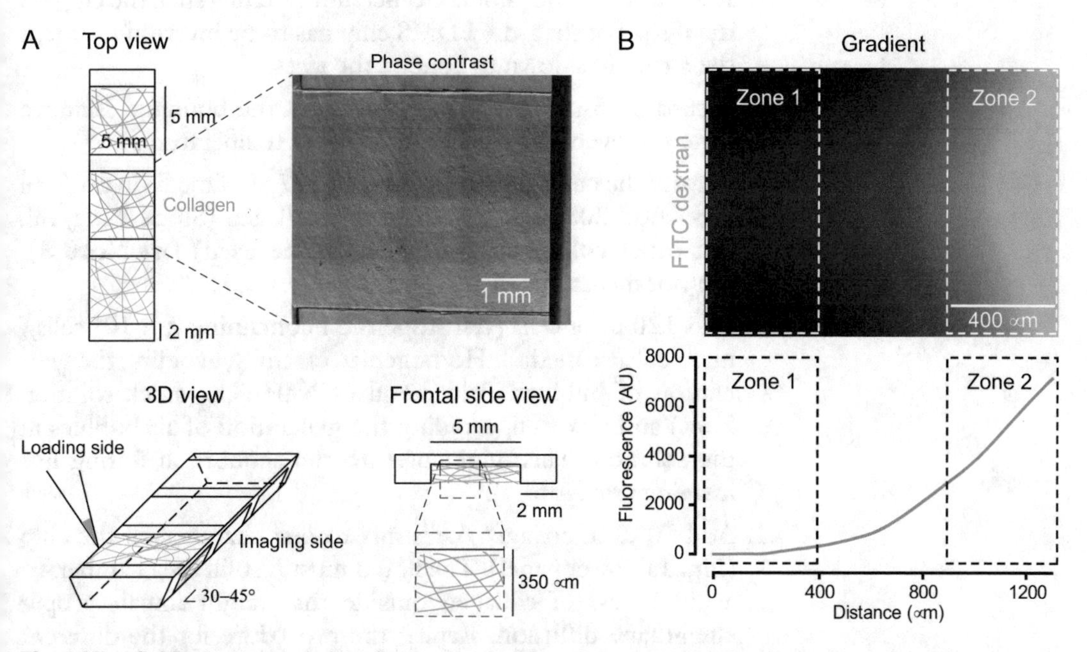

Fig. 1 Microfluidic device to study cell migration in 3D environments. (**a**) The dimensions of the chamber are depicted in the schemes and also an example of the chamber is shown. Different views and details of the custom-made PDMS chamber used to study chemotaxis are shown. (**b**) Fluorescence image evidencing the gradient of FITC dextran (50 μg/ml) in the chamber. The corresponding graph of the image (lower panel) shows the fluorescence intensity in the x axis of the chamber. Note that in the arbitrarily defined *Zone 2* there is gradient formation, while in the *Zone 1* there is not

1. To prepare the chips, mix PDMS as mentioned in Subheading 2, **item 1**.
2. Pour the mixture onto the mold forming a 2–3 mm layer. To remove bubbles, incubate 1 h in a vacuum bell jar (longer times are also acceptable) (*see* **Note 1**).
3. Once bubbles are removed, incubate the mold for 1 h in an oven at 70 °C to cure the PDMS. Then, remove and equilibrate at room temperature before proceeding to the next step.
4. Carefully remove the chips from the mold by cutting them with a surgical blade. Once removed, crop into suitable sizes to ensure fitting of the chip in the plate. For a 35 mm dish, cut the PDMS at approx. 5 mm × 20 mm (Fig. 1a). To avoid light diffraction during phase-contrast imaging, during the cutting leave a bevel (with an angle of 30–45°) in both sides of the long axis of the PDMS chip (Fig. 1a, lower panel).
5. Clean both surfaces of the PDMS chips by sticking and peeling scotch tape to remove residual dust. Cleaning can be complemented by sonicating for 30 s in ethanol 70%, and posterior drying using an air gun (*see* **Note 2**).
6. Activate both, the glass-bottom dish (without lid) and the PDMS chips (channels side up) in the plasma cleaner for 1 min.
7. Remove from the plasma cleaner and carefully stick the chip on the dish. For that, the PDMS chip has to be inverted to orient the structures down to contact the glass.
8. Incubate 15 min at 70 °C to strength the bonding. Remove from the oven and equilibrate at room temperature.
9. Prepare the collagen mix by deposing 27 μl of media in a 1.5 ml tube. Add 205 μl of Type I bovine collagen (Stock 3 mg/ml, but other collagen densities might be used) (*see* **Note 3**). Do not mix at this step.
10. Add 120 μl of cells (cell suspension containing 2×10^6 cells/ml in culture media). Homogenize carefully, avoiding the generation of bubbles. Add 13 μl of $NaHCO_3$ (stock solution 7.5%) and mix well, avoiding the generation of air bubbles in the solution. This might perturb the acquisition during live imaging (*see* **Note 4**).
11. Add 10 μl of collagen/cells mix to one chamber on the chip (Fig. 1a, lower panel). The liquid must just fill the chamber, to avoid excess of collagen outside that might impair proper chemokine diffusion. Repeat the procedure for the different samples to analyze.
12. Incubate at 37 °C for 30 min to allow collagen polymerization.
13. Add carefully 2 ml of culture medium containing 200 ng/ml of CCL21. This chemokine binds to collagen [14]. CCL21

gradient is stable for several hours. At this time, drug treatments might be added accordingly.

14. For cell tracking, perform live cell imaging at low magnification (5× or 10×) in a microscope equipped with controlled atmosphere (5% CO_2 and 37 °C) (*see* **Note 5**). Phase contrast is recommended, since it allows visualization of cells migrating in several layers of the collagen in one single acquisition (Fig. 2a).

3.2 Analysis of DC Migration in 3D Environments

1. Reconstitute the movies accordingly. To facilitate this step, a short custom script could be made using Fiji (ImageJ) software.
2. Crop the movie to analyze the first 400 μm from the border of the chamber. This zone should be defined according to gradient shape and stability (Figs. 1b and 2a).

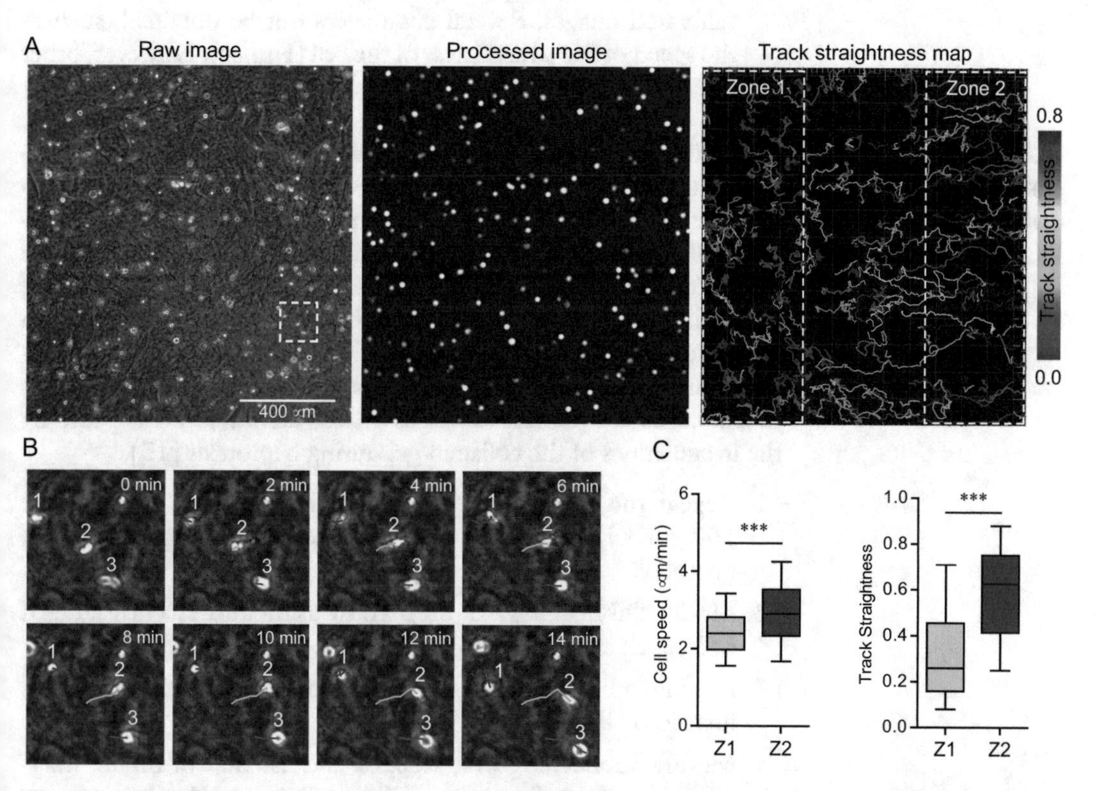

Fig. 2 Analysis of DC haptotaxis in collagen gels. (**a**) Images showing the raw and processed images of a typical example of DCs embedded in a collagen gel and exposed to CCL21 (200 ng/ml) (left and center panels). In addition, a straightness map, used as index of persistency, is shown after cell tracking using Imaris software (right panel). Every color line represents the trajectory of one cell. Regions of 400 μm without gradient (*Zone 1*) or with gradient (*Zone 2*) are depicted with a dotted line. Note than in *Zone 1* cells move randomly, whereas in *Zone 2* move persistently. (**b**) Zoom from the dotted square in (**a**) shows an image sequence of DCs migrating inside collagen gels toward a CCL21 gradient. The cell trajectories are numbered and depicted in different colors. (**c**) Graph showing the cell speed and track straightness of DCs exposed (*Zone 2*, Z2) or not (*Zone 1*, Z1) to a CCL21 gradient. ***$p < 0.001$ nonparametric Mann–Whitney test

3. To process the images and get a clean tracking it is recommended to first apply several filters as follows (all the steps are performed using Fiji (Image J) software).
4. Generate an Average time projection of the whole film, and subtract this image to the original movie. This will provide white objects in a dark background.
5. Duplicate the processed movie and apply a Mean filter (radius: 20 for DCs, but it should be adjusted when analyzing other cells). Subtract the resulting image to the movie generated in the **step 4**. This will clean the halo produced by cells' light reflection.
6. Next, apply a Gaussian Blur filter (gamma 5, but need to be adjusted accordingly to the images) to the processed movie. This will provide round white objects easily trackable using any image analysis software (Fig. 2a, b).
7. In our case, tracking was performed using Imaris software over calibrated images. Several parameters can be obtained, such as the speed or the straightness of the cell (Fig. 2c). However, other free open-source software such as Fiji (ImageJ) can be used.

3.3 Analysis of Cell Deformation in 3D Collagen Gels

In addition to the analysis of cell locomotion described in the previous sections, the assay presented here can be slightly modified to evaluate cellular deformations of the migrating cells inside the complex collagen network. To perform this analysis it is necessary to use fluorescently labeled cells and a smaller chamber of 100 μm height. This thin chamber allows the full 3D reconstruction of the cells in few z-steps, further enabling the imaging of nearly all the cells in different planes. As an example, we show the use of LifeAct-GFP DCs to analyze actin dynamics as well as the cell deformations imposed by the irregularities of the collagen gel during haptotaxis [15].

1. Repeat the steps mentioned in Subheading 3.1, but use the 100 μm height chamber and make the following updates to the protocol.
2. Modify Subheading 3.1, **step 10** by using fluorescently labeled cells, such as LifeAct-GFP DCs (Fig. 3a).
3. Modify Subheading 3.1, **step 11** loading 5 μl of collagen/cells instead of 10 μl.
4. Modify Subheading 3.1, **step 14** and use 20× or higher magnifications according to the cell size. Spinning disk microscopy is recommended because it allows fast acquisition of different planes (*see* **Note 6**).
5. Alternatively, collagen gels containing LifeAct-GFP or labeled cells might be fixed to perform further detailed microscopic analysis. Remove the medium and cover the chamber with 1 ml of 4% PFA and incubate for 30 min at room temperature.

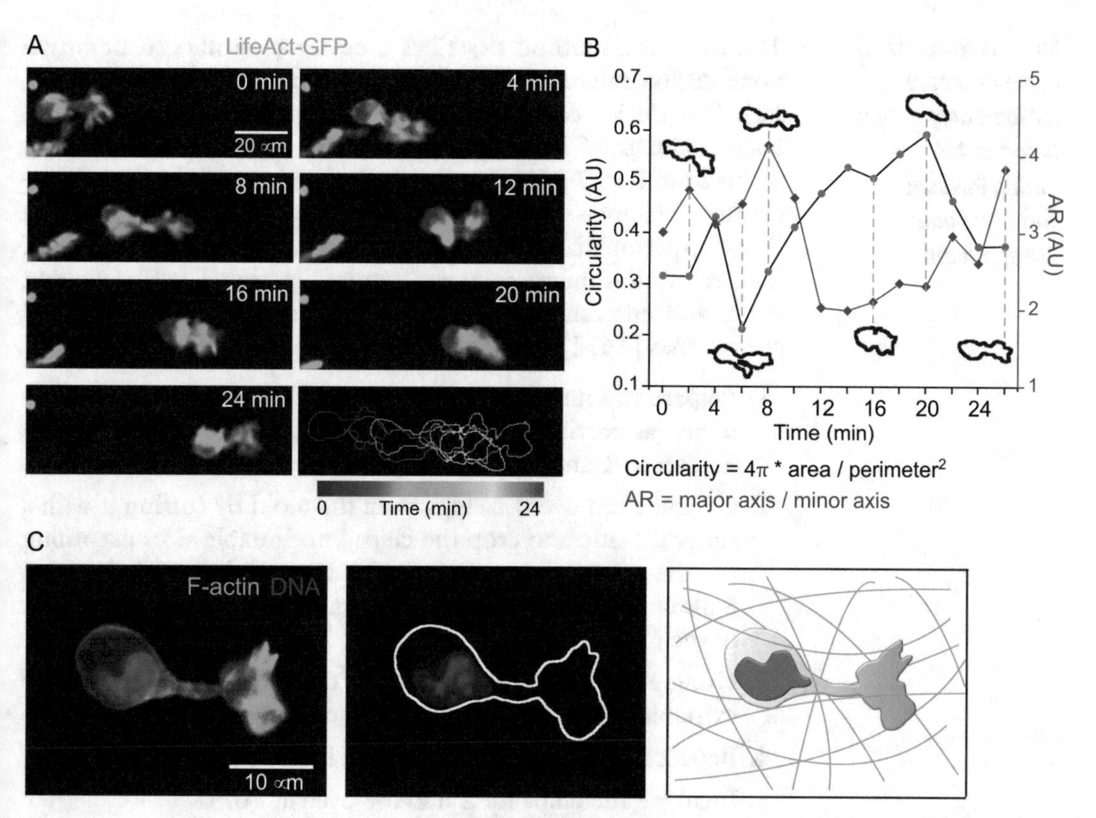

Fig. 3 F-actin dynamics during DC chemotactic migration. (**a**) Image sequence shows the z-projection of a LifeAct-GFP DC migrating inside a collagen gel toward a CCL21 gradient. The imaging was performed in a spinning disk microscope with a dry 20× objective. Time and scale bars are indicated. The last panel corresponds to the cell contour during time. Note the deformations of the cell as it progresses toward the CCL21 source. (**b**) Graph showing the dynamic changes in circularity and aspect ratio (AR) obtained with the shape descriptors tool in Fiji (Image J) software of the cell shown in (**a**). Some examples of the cell contour linked to the corresponding values in the graph are depicted. The equations to calculate each parameter are depicted. (**c**) Confocal images showing F-actin (green, phalloidin Alexa 488) and DNA (blue, Hoechst 33342) staining of DCs embedded in a collagen gel after 1 h of exposure to CCL21. The panel showing DNA staining reveals nuclear constriction during migration. Scheme representing the cell shown in the left migrating through the collagen fibers

6. Remove the PFA and wash three times with PBS.
7. Then, add 1 ml of PBS/BSA 2% (*see* **Note 7**).
8. High resolution microscopy could be performed at this stage (Fig. 3c).
9. Finally, after acquisition, the images can be analyzed using a simple tool of Fiji (ImageJ) software named "shape descriptors," which requires segmented cells in a binary mask after applying a Yen threshold over the raw stack. Circularity and AR parameters are suitable index to analyze cell shape (Fig. 3b).

3.4 Microchannels as an Approach for the Study of Cell Deformation During Passage Through Small Constrictions

The following method describes a device to study cell deformations during migration of cells through narrow constrictions [5, 10]. Our device consists in a PDMS chip containing a series of microchannels of fix height (5 μm, Fig. 4a) that decrease their width from 8 to 2 μm (Fig. 4b, c). Each chip contains many independent channels that allow the imaging of multiple cells in a single experiment (Fig. 4a). These devices are compatible with high resolution microscopy and can be functionalized by coating them with different molecules, including matrix components and chemokines [16, 17].

1. Prepare the mold using conventional photolithography techniques as previously described [5, 10], and repeat Subheading 3.1, **steps 1** and **2**.
2. Carefully remove the chips from the mold by cutting it with a surgical blade and crop the chips into suitable sizes according to the experiment. For a 35 mm dish crop them at 5 mm × 15 mm, leaving at least 2 mm space around the border of the glass, assuring the chips will fit in the bottom glass.
3. Using a PDMS driller of 3 mm of diameter, make holes at the extremity of the channels to allow cells entry (Fig. 4a).
4. Repeat Subheading Subheading 3.1, **steps 5**–**7**.
5. Incubate the chips for 2 h in the oven at 70 °C.
6. Remove the channels from the oven and put them in a bell jar for 10 min, to proceed with the coating of the PDMS chips.
7. Activate the chips in the plasma cleaner for 1 min and immediately add 5 μl of previously prepared fibronectin (or the chosen coating solution) per hole (*see* **Note 8**).
8. Verify that all the channels are properly coated by eye inspection under the microscope, and incubate them for 1 h at room temperature (*see* **Note 9**).
9. Wash three times by pouring 2 ml of sterile PBS on the top of the PDMS chips, making sure the holes contain no bubbles (*see* **Note 10**).
10. Repeat the previous step but this time using prewarm culture medium and leave the chips at 37 °C prior to load the cells.
11. Remove medium from the glass bottom dish around and on the top of the PDMS chips.
12. Prepare a cell suspension containing 20 × 10^6 cells/ml.
13. Remove the medium contained in each hole used as cell port entry.
14. Add 5 μl of cell suspension containing 1 × 10^5 cells to each hole (*see* **Note 11**).

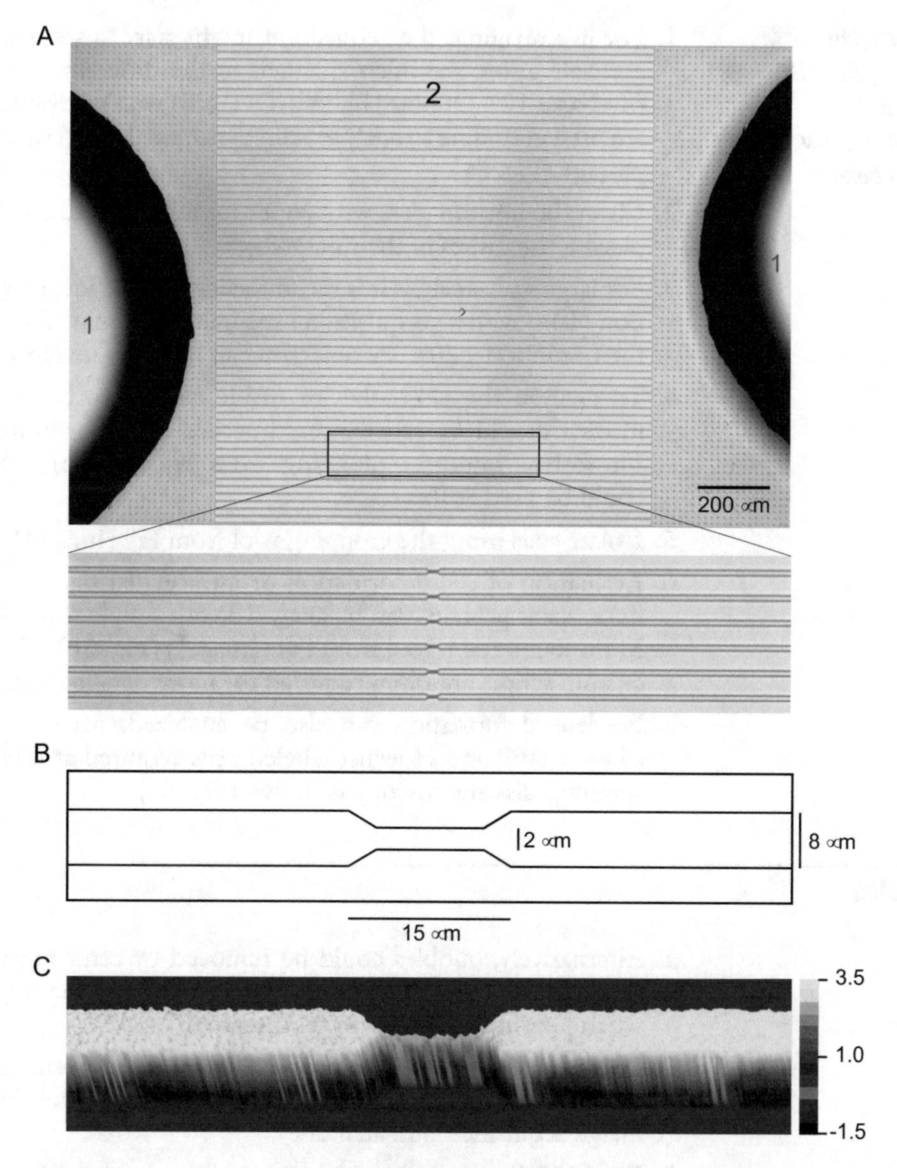

Fig. 4 Microfabricated chip with constrictions to study cell deformations. (**a**) Phase contrast picture of a whole PDMS chip, showing the two cell port entries at both extremities (1) and channels with three sequential constrictions in the middle (2). Box depicts the zoom of the region of the chip with constrictions. (**b**) Schematic drawing of a single 15 μm length and 2 μm width constriction. (**c**) Surface topography of a single constriction measured using a Wyko NT9100 Optical Profiling System

15. Incubate the loaded chips for 30–45 min at 37 °C to allow the cells to sediment.
16. Gently add 2 ml of culture medium until the chips are fully covered. Avoid pouring medium directly on top of the cell port entry holes, because this might remove the cells.

3.5 Imaging and Analysis of DC Passage Through Narrow Constrictions

1. For live imaging, the acquisition might start as soon as there are cells inside the microchannels in the different conditions (*see* **Note 12**). Of note, LifeAct-GFP cells with nuclear staining are used in routine to analyze actin dynamics during constriction passage (Fig. 5a).
2. Clean the bottom glass with paper for microscopic use, before placing the dishes in the microscope.
3. For large scale analysis it is recommended to use low magnification (10×) with an acquisition time frame of 2 min. To increase the resolution a 20× dry objective can be used (*see* **Note 13**).
4. To analyze the cell behavior inside the microchannels, the movies should be reconstituted accordingly. For example, a Bio-format importer plugging from Fiji (ImageJ) software could be used at this stage.
5. Isolate cells using the cropping tool from Fiji (ImageJ).
6. Evaluation of cell deformation or intracellular distribution of fluorescent proteins can be analyzed with simple tools, such as Multi-kymograph tool from Fiji (ImageJ) (Fig. 5b). However, custom scripts are often required for more detailed analysis.
7. Nucleus deformation can also be analyzed. An example of LifeAct-GFP and Hoechst labeled cells acquired at 40× using spinning disk microscopy is shown (Fig. 5c).

4 Notes

1. Alternatively, bubbles could be removed by centrifugation of the mixture, but still this do not prevent that some may form when pouring the solution on the mold.
2. This is a key step for this protocol because dust remaining in the chamber might disturb both chip sticking to the glass and image acquisition and analysis.
3. For BMDCs use the semiadherent cells and not the floating or adherent ones.
4. The presence of bubbles in the collagen is detrimental for the stability of the gel.
5. Using this chamber the CCL21 gradient is stable for several hours before the chemokine concentration inside the chamber is homogeneous and the gradient is lost [7].
6. To minimize phototoxicity use low laser intensity for cell visualization.
7. At this time antibody incubation and additional staining could be performed. However, the incubation time and concentration are dependent on the collagen concentration, as well as the targeted protein.

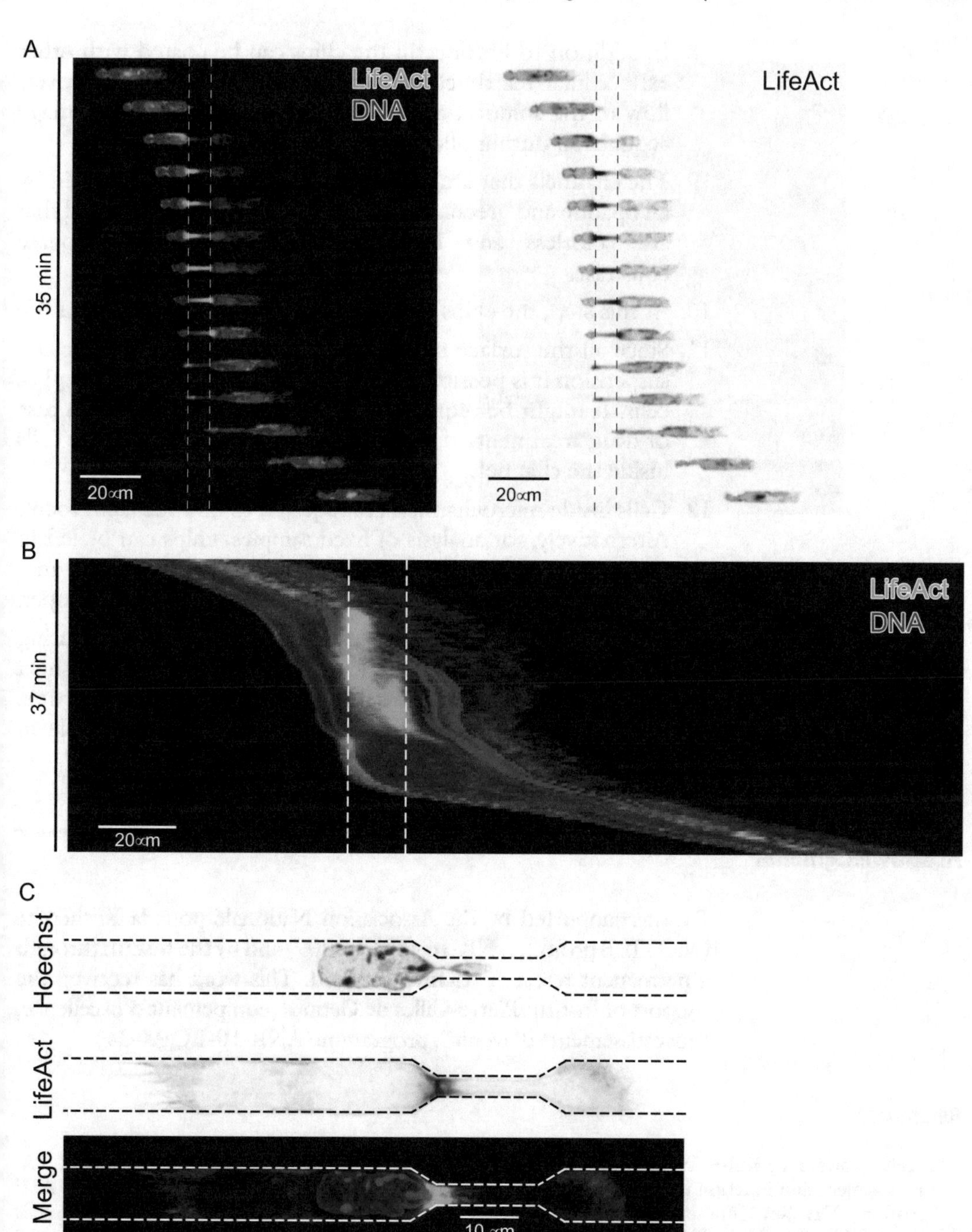

Fig. 5 F-actin and nuclear dynamics during DC passage through constrictions. (**a**) Montage showing a mature LifeAct-GFP DC stained with Hoechst 33342 passing through one constriction at a magnification of 20×. Acquisition was performed with a 2-min time frame. (**b**) Example of a kymograph using Multi-kymograph tool from Fiji (ImageJ). The duration of passage and actin accumulation in the constriction can be quantified using this tool. (**c**) Maximum z-projection showing a mature LifeAct-GFP DC stained with Hoechst 33342 inside a constriction at a magnification of 63× using spinning disk. Nuclear squeezing and F-actin accumulation around the nucleus are observed

8. In addition to fibronectin the chips can be coated with other extracellular matrix components, such as collagen. The overflow of the solution out of the well can interfere with image acquisition during phase contrast imaging.
9. The channels that are not well coated can be distinguished by an opaque and greenish color compared those well coated that are colorless and bright when observing using phase constrast.
10. At this step, the chips could be kept overnight at 4 °C in PBS.
11. Since all the surface of the hole should be filled with the cell suspension it is possible to increase the amount up to 7.5 μl of cells. It might be required to increase the cell number in case of drug treatments that might delay the entrance of the cells inside the channels.
12. Cells inside microchannels are required to find the right focus. Alternatively, for analysis of fixed samples, chips can be left in the incubator for indicated times to get enough cells to analyze, although this aspect is not discussed in the present paper.
13. If high resolution is required, spinning disk microscopy using a 63× oil objective is recommended, which is suitable for short movies with a higher time frame acquisition. Exposure time and laser power should be kept at the minimum possible to minimize cell death.

Acknowledgments

PV was supported by the Association Nationale pour la Recherche (MOTILE project, ANR-16-CE13-0009) and by the Inserm through a permanent research scientist position. This work has received the support of Institut Pierre-Gilles de Gennes (équipement d'excellence, "Investissements d'avenir", programme ANR-10-EQPX-34).

References

1. Kolaczkowska E, Kubes P (2013) Neutrophil recruitment and function in health and inflammation. Nat Rev Immunol 13(3):159–175. https://doi.org/10.1038/nri3399
2. Vargas P, Barbier L, Sáez PJ, Piel M (2017) Mechanisms for fast cell migration in complex environments. Curr Opin Cell Biol 48:72–78. https://doi.org/10.1016/j.ceb.2017.04.007
3. Worbs T, Hammerschmidt SI, Forster R (2017) Dendritic cell migration in health and disease. Nat Rev Immunol 17(1):30–48. https://doi.org/10.1038/nri.2016.116
4. Sáez PJ, Vargas P, Shoji KF, Harcha PA, Lennon-Dumenil AM, Sáez JC (2017) ATP promotes the fast migration of dendritic cells through the activity of pannexin 1 channels and P2X7 receptors. Sci Signal 10(506). https://doi.org/10.1126/scisignal.aah7107.
5. Thiam HR, Vargas P, Carpi N, Crespo CL, Raab M, Terriac E, King MC, Jacobelli J, Alberts AS, Stradal T, Lennon-Dumenil AM, Piel M (2016) Perinuclear Arp2/3-driven actin polymerization enables nuclear deformation to facilitate cell migration through complex environments. Nat

Commun 7:10997. https://doi.org/10.1038/ncomms10997

6. Ufer F, Vargas P, Engler JB, Tintelnot J, Schattling B, Winkler H, Bauer S, Kursawe N, Willing A, Keminer O, Ohana O, Salinas-Riester G, Pless O, Kuhl D, Friese MA (2016) Arc/Arg3.1 governs inflammatory dendritic cell migration from the skin and thereby controls T cell activation. Sci Immunol 1: eaaf8665–eaaf8665
7. Vargas P, Maiuri P, Bretou M, Sáez PJ, Pierobon P, Maurin M, Chabaud M, Lankar D, Obino D, Terriac E, Raab M, Thiam HR, Brocker T, Kitchen-Goosen SM, Alberts AS, Sunareni P, Xia S, Li R, Voituriez R, Piel M, Lennon-Dumenil AM (2016) Innate control of actin nucleation determines two distinct migration behaviours in dendritic cells. Nat Cell Biol 18(1):43–53. https://doi.org/10.1038/ncb3284
8. Wilson K, Lewalle A, Fritzsche M, Thorogate R, Duke T, Charras G (2013) Mechanisms of leading edge protrusion in interstitial migration. Nat Commun 4:2896. https://doi.org/10.1038/ncomms3896
9. Weber M, Hauschild R, Schwarz J, Moussion C, de Vries I, Legler DF, Luther SA, Bollenbach T, Sixt M (2013) Interstitial dendritic cell guidance by haptotactic chemokine gradients. Science 339(6117):328–332. https://doi.org/10.1126/science.1228456
10. Raab M, Gentili M, de Belly H, Thiam HR, Vargas P, Jimenez AJ, Lautenschlaeger F, Voituriez R, Lennon-Dumenil AM, Manel N, Piel M (2016) ESCRT III repairs nuclear envelope ruptures during cell migration to limit DNA damage and cell death. Science 352(6283):359–362. https://doi.org/10.1126/science.aad7611
11. Schwarz J, Bierbaum V, Merrin J, Frank T, Hauschild R, Bollenbach T, Tay S, Sixt M, Mehling M (2016) A microfluidic device for measuring cell migration towards substrate-bound and soluble chemokine gradients. Sci Rep 6:36440. https://doi.org/10.1038/srep36440
12. Schwarz J, Sixt M (2016) Quantitative analysis of dendritic cell haptotaxis. Methods Enzymol 570:567–581. https://doi.org/10.1016/bs.mie.2015.11.004
13. Haessler U, Pisano M, Wu M, Swartz MA (2011) Dendritic cell chemotaxis in 3D under defined chemokine gradients reveals differential response to ligands CCL21 and CCL19. Proc Natl Acad Sci U S A 108(14):5614–5619. https://doi.org/10.1073/pnas.1014920108
14. Yang BG, Tanaka T, Jang MH, Bai Z, Hayasaka H, Miyasaka M (2007) Binding of lymphoid chemokines to collagen IV that accumulates in the basal lamina of high endothelial venules: its implications in lymphocyte trafficking. J Immunol 179(7):4376–4382
15. Riedl J, Flynn KC, Raducanu A, Gartner F, Beck G, Bosl M, Bradke F, Massberg S, Aszodi A, Sixt M, Wedlich-Soldner R (2010) Lifeact mice for studying F-actin dynamics. Nat Methods 7(3):168–169. https://doi.org/10.1038/nmeth0310-168. nmeth0310-168 [pii]
16. Vargas P, Chabaud M, Thiam HR, Lankar D, Piel M, Lennon-Dumenil AM (2016) Study of dendritic cell migration using micro-fabrication. J Immunol Methods 432:30–34. https://doi.org/10.1016/j.jim.2015.12.005
17. Vargas P, Terriac E, Lennon-Dumenil AM, Piel M (2014) Study of cell migration in microfabricated channels. J Visual Exp 84:e51099. https://doi.org/10.3791/51099

Chapter 27

Microfluidic Devices for Examining the Physical Limits of Migration in Confined Environments

Majid Malboubi, Asier Jayo, Maddy Parsons, and Guillaume Charras

Abstract

Cell migration plays a key role in many physiological and pathological conditions during which cells migrate primarily in the 3D environments formed by tissues. Microfluidics enables the design of simple devices that can mimic in a highly controlled manner the geometry and dimensions of the interstices encountered by cells in the body. Here we describe the design, fabrication, and implementation of an array of channels with a range of cross sections to investigate migration of cells and cell clusters through confined spaces. By combining this assay with a motorized microscope stage, image data can be acquired with high throughput to determine the physical limits of migration in confined environments and their biological origin.

Key words Microfluidics, Cell deformation, Breast cancer cells, Multilayer photolithography

1 Introduction

Cell migration is involved in many aspects of development, immunity, and pathology. In embryos, neural crest cells migrate collectively from the neural tube over hundreds of microns to specific areas where they will differentiate to give rise to a number of tissues [1, 2]. In response to inflammation, white blood cells exit the blood vessels across the endothelium before migrating through the tissue to the source of inflammation [3, 4]. During cancer, cells migrate away from the primary tumor throughout the tissue either as groups or as individual cells [5]. Thus, although much has been learned from examining migration on planar substrates, there is a need to examine migration in three-dimensional environments and recent studies have made it abundantly clear that migration in 3D differs under many aspects from migration in 2D [6–10].

Microfabrication combined with soft lithography techniques allows the generation of microfluidic devices that can mimic the geometry and dimensions of the interstices encountered by cells during migration through tissues. In contrast to extracellular

Alexis Gautreau (ed.), *Cell Migration: Methods and Protocols*, Methods in Molecular Biology, vol. 1749, https://doi.org/10.1007/978-1-4939-7701-7_27, © Springer Science+Business Media, LLC 2018

matrix protein gels that form complex and poorly ordered environments, microfluidics enables the manufacture of a simplified environment that remains identical from one experiment to the next allowing straightforward quantification and comparison between experimental conditions.

Here, we demonstrate the design, manufacture, and implementation of a simple device for examining the physical limits of migration of single cells through interstices. Although the device was originally designed to examine single cell migration [11, 12], simple changes to the design also allow examination of the migration of cell clusters [13]. The device consists of a reservoir linked to a large channel via an array of smaller transverse channels (*see* Fig. 1a). After seeding the cells in the reservoir, beads soaked with chemoattractant are loaded into the large channel. Chemoattractant diffuses through the small transverse channels into the reservoir and cells migrate toward this. The device is designed such that transverse channels with many different cross sections link the reservoir to the channel. This allows comparison of the migration of cells as a function of channel cross-section in conditions that are otherwise identical. The microfluidic chip can be assembled on a

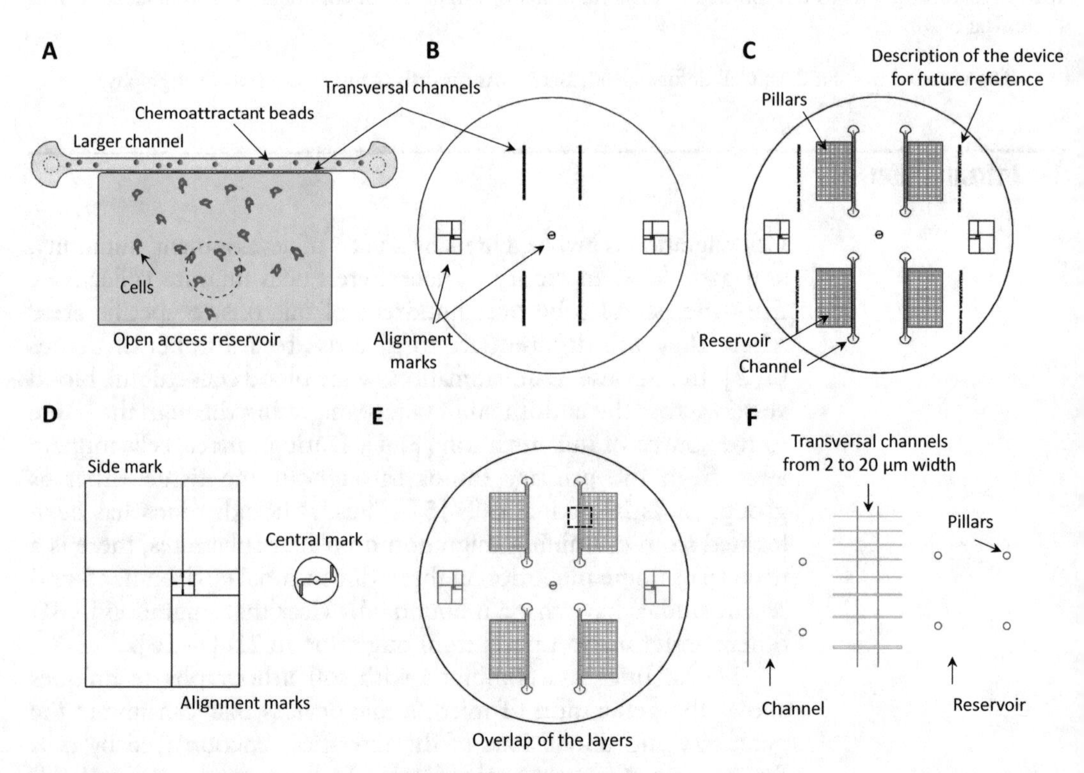

Fig. 1 Device design. (**a**) Schematic design of the device. (**b**) Layer 1, Transverse channels. (**c**) Layer 2, Open access reservoir and the channel for circulating chemoattractant. (**d**) Alignment marks. (**e**) Overlap of two layers. Layer 1 is shown in green. Layer 2 is shown in blue. Alignment marks in red. (**f**) Detailed view of transverse channels connecting the reservoir to the channel. This region is a zoomed view of the boxed region in (**e**)

coverslip or in a glass bottom petri dish for imaging. When used in combination with a motorised microscope stage, the device allows acquisition of large numbers of tracks per experiment. For use with cell clusters, the cross section of the transverse channels can be increased appropriately.

Setup of the experiment involves a microfabrication step, a soft lithography step, a cell-seeding step, and an imaging step. Once a master mold is fabricated in the microfabrication step, it can be used many times in the soft lithography step. This second step generates negative replica of the master mold using a transparent polymer to make devices for experiments. For an in depth description of the steps involved in microfluidics, we refer the interested readers to the following references [14, 15].

2 Materials

2.1 Software

1. AutoCAD for mask design.
2. ImageJ for analyzing images obtained during experiments.

2.2 Master-Mold Fabrication

1. 3-in. silicon wafers.
2. SU-8 2005 and SU-8 2050 photoresist (Microchem).
3. EC solvent (SU-8 Developer, Microchem).
4. Isopropanol.
5. Acetate and Chrome photomasks.
6. Spin coater.
7. Two hot plates set to 65 and 95 °C respectively.
8. Mask aligner with UV source.
9. Forceps for wafer manipulation.

2.3 Soft Lithography

1. PDMS prepolymer and curing agent (Sylgard 184, Dow Corning).
2. Isopropanol.
3. Acetone.
4. Compressed nitrogen cylinder and nitrogen gun.
5. Piranha solution.
6. Petri dishes.
7. Disposable plastic cups and stir rods for mixing PDMS.
8. 50 mm glass bottom dishes.
9. Scalpels.
10. Biopsy punches (1 mm diameter and 4 mm diameter).
11. Soft tubing: Tygon non-DEHP microbore tubing, 0.010″ × 0.030″ OD, 500 ft./roll.

12. Hard tubing: Cole-Parmer Microbore PTFE Tubing, 0.022″ ID × 0.042″ OD, 100 ft./roll.
13. Blunt 25 G needles.
14. Trichloro(1H 1H 2H 2H perfluorooctyl)silane.
15. Digital balance.
16. Laboratory oven.
17. Vacuum bell with pump.
18. Plasma cleaner.
19. Laminar flow hood.

2.4 Cell Culture

1. MDA MB 231 human breast carcinoma cell line (ATCC).
2. Cell culture medium: Dulbecco's modified Eagle's medium supplemented with 1% penicillin–streptomycin and glutamine and 10% v/v fetal bovine serum.
3. Trypsin 0.05% (w/v) with EDTA.
4. Pipettes/tips.
5. Fibronectin.
6. PBS.
7. 15 μm diameter polystyrene beads.
8. Ham's F12 medium without serum.
9. Humidified incubator with 5% CO_2.
10. Centrifuge.
11. Centrifuge tubes.
12. Cell culture hood.

2.5 Imaging

1. Wide-field epi-fluorescence microscope.
2. Environmental control chamber.
3. EMCCD camera.
4. Motorised stage.

3 Methods

3.1 Mask Design

The design consists of two layers (*see* **Note 1**), a first thin layer containing a series of transversal channels through which the cells will migrate (*see* Fig. 1a, b, e) and a second thicker layer consisting of the open access reservoir and the large channel (*see* Fig. 1a, c). The transverse microchannels in the first layer are 150 μm long, 5 μm high, and are arranged in groups with widths ranging from 2 to 20 μm (*see* Fig. 1e). The reservoir is 16 mm long, 10 mm wide, and 80 μm high. The large channel (chemoattractant channel) is 22 mm long, 800 μm wide, and 80 μm high (Fig. 1a). The large channel and the reservoir are separated by a gap of 100 μm.

1. Use AutoCAD to design the mask. All features must be drawn using "CLOSED POLYLINES."
2. Create a new layer, name it "Transverse microchannels," choose a colour and draw the first layer (*see* **Note 2**).
3. Create a new layer, name it "alignment marks," choose a color and draw alignment marks (*see* **Note 3**).
4. Create a new layer, call it "Reservoir," choose a color and draw the second layer (*see* **Note 4**).
5. Send the design to a mask manufacturer. The first layer needs to be printed on a chrome mask because of the small cross-section of some of the channels while the second layer can be printed on acetate because of the larger dimensions of its features (*see* **Note 5**).

3.2 Multilayer Photolithography

Photolithography must take place in a clean-room.

1. Clean a 3-in. silicon wafer first with acetone and then with isopropanol (*see* **Note 6**).
2. Dry the wafer under a stream of nitrogen (*see* **Note 7**).
3. Pour SU-8 2005 photoresist on the silicon wafer and place in spin coater. Spin following the manufacturer protocol. After holding at 500 rpm for 30s, ramp to 3000 rpm with a rate of 500 rpm/s and hold for 30 s to obtain a 5 μm thick layer of SU-8. Check that the wafer is homogenously covered with photoresist.
4. Soft-bake the wafer at 95 °C for 2 min on a hot plate.
5. Affix the chrome mask containing the transverse microchannels and the alignment marks to mask aligner.
6. Place the silicon wafer coated with resist in the mask aligner.
7. Bring the silicon wafer into contact with the mask.
8. Expose the wafer through the chrome mask for 5 s at 20 mW/cm^2 to obtain a total exposure energy of 100 mJ/cm^2 (*see* **Note 8**).
9. Release the silicon wafer.
10. Post-bake at 95 °C for 3 min on the hot plate.
11. Develop in EC solvent for 1 min in a chemical flow hood. Agitate gently.
12. Rinse with isopropanol (*see* **Note 9**).
13. Dry wafer under a stream of nitrogen.
14. Pour SU-8 2050 photoresist on the silicon wafer. Spin following the manufacturer protocol. After holding at 500 rpm for 30 s, ramp to 1900 rpm with a rate of 500 rpm/s and hold for 30 s to obtain an 80 μm thick layer of SU-8 (*see* **Note 10**).

15. Soft-bake for 3 min at 65 °C on a hot plate before moving to a 95 °C hot plate for 9 min.
16. Affix the acetate mask containing the cell-seeding reservoir, the chemoattractant channel and the alignment marks to mask aligner.
17. Place the silicon wafer coated with resist in the mask aligner.
18. Using the alignment marks, align the wafer with the acetate mask.
19. Bring the wafer into contact with the mask.
20. Expose for 11 s at 20 mW/cm^2 to obtain a total energy of 220 mJ/cm^2.
21. Release the silicon wafer.
22. Post-bake for 7 min at 95 °C on the hot plate.
23. Develop in EC solvent for 6 min in a chemical flow hood. Agitating gently throughout.
24. Rinse with isopropanol.
25. Blow nitrogen to dry the wafer (*see* **Note 11**).
26. Hard-bake at 150 °C for 6 min.

3.3 Soft Lithography

Soft lithography can be done in a normal laboratory. The soft lithography consists of a passivation step and a molding step. If a master is not used for the first time, skip passivation and go directly to **step 5**.

1. Place wafer in a vacuum bell.
2. Place a drop of Trichloro (1H 1H 2H 2H perfluorooctyl) silane in a small glass bottom petri dish in the vacuum bell.
3. Draw vacuum and leave to passivate for 3–4 h. Several wafers can be passivated simultaneously.
4. Break vacuum. Dispose of silane appropriately.
5. Place the wafer in a sufficiently large petri dish.
6. Place disposable plastic cup on balance. Tare.
7. Pour PDMS primer slowly to obtain the desired weight of PDMS primer.
8. Add curing agent such that the primer to curing agent ratio is 10:1.
9. Mix PDMS with curing agent with a stirrer (*see* **Note 12**).
10. Pour PDMS onto the mold containing the master.
11. Place in the vacuum bell. Draw vacuum and leave for 45 min to remove bubbles trapped around the features.
12. Break vacuum (*see* **Note 13**).

13. Cure at 65 °C in oven for 1 h.
14. Place the petri dish containing the master mold in a laminar flow hood to minimize dust particles adhering to PDMS.
15. Using a scalpel, cut around the device features leaving a gap of at least 5 mm (*see* **Note 14**).
16. Flip the PDMS chip over such that it is channel side up.
17. Punch holes through the PDMS at either side of the chemoattractant channel using 1 mm biopsy punches.
18. Punch a hole in the central part of the chamber using a 4 mm biopsy punch to provide an opening for cell seeding.
19. Place the PDMS chip and a 50 mm glass bottom dish in the chamber of plasma cleaner. Both surfaces to be treated should be facing up. Treat with oxygen plasma for 30 s.
20. After treatment, place the PDMS chip on the glass and very gently press the PDMS chip to encourage bonding.
21. Place the assembled device into an oven at 60 °C for 1 h to promote bonding.
22. Cut the hard tubing into 1 cm segments and push into the inlet and outlet holes of the chemoattractant channel.
23. Cut several cm of soft tubing. Push the soft tubes into the hard tubing as deep as possible. (see Fig. 2).
24. Introduce the tip of a blunt needle through the inner diameter of the soft tubing to create an interface for a luer lock syringe.
25. Sterilize the devices by immersing in 70% ethanol.
26. Dry in a sterile biosafety cabinet for 1 h before use (*see* **Note 15**).

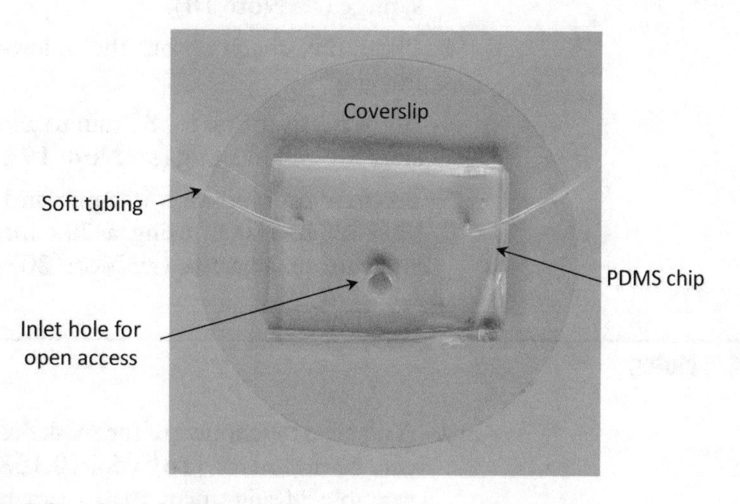

Fig. 2 Completed device

3.4 Cell Culture, Channel Preparation, and Cell Seeding

1. Passage MDA MB 231 human breast carcinoma cells (ATCC) as recommended by ATCC.
2. Circulate PBS through the chemoattractant channel using a syringe attached to soft tubing by the luer lock, the PBS will gently fill the cell-seeding reservoir.
3. Wash devices with PBS by gently circulating it through the channels and removing it from the cell-seeding reservoir.
4. Fill syringe with 10 μg/mL fibronectin in PBS and circulate through the chemoattractant channel and allow it to flow to the cell-seeding reservoir (*see* **Note 16**).
5. Incubate at 37 °C overnight in a humidified incubator (*see* **Note 17**).
6. Wash the device three times by circulating PBS through the chemoattractant channel and drawing it from the cell-seeding reservoir.
7. Fill reservoir and chemoattractant channel with Ham's F12 medium by introducing the medium through the chemoattractant channel and drawing through the cell-seeding reservoir.
8. Trypsinize growing cells, resuspend them in Ham's F12 medium at 5×10^6 cells/mL, and add up to 100 μL of the cell suspension to the reservoir chamber using a micropipette.
9. Leave cells to spread for 2 h in a humidified incubator at 37 °C.

3.5 Live Cell Imaging

1. Coat 15 μm diameter polystyrene beads with fetal bovine serum (FBS) by incubating in Ham F12 medium with 10% FBS overnight at 4 °C.
2. Introduce beads into the chemoattractant channel with a syringe (*see* **Note 18**).
3. Place the chamber on the microscope stage for live cell imaging.
4. Allow beads to rest for 30 min to allow gradient formation and start live cell imaging (*see* **Note 19**).
5. Program multiposition imaging and acquire images at 10 min intervals for 10 h, using a 20× magnification 0.4 numerical aperture air objective (*see* **Note 20**) (Fig. 3).

4 Notes

1. A detailed description of the mask design and fabrication process can be found at [16] (doi:10.1038/protex.2015.069). An example of alignment marks can be found on the Stanford

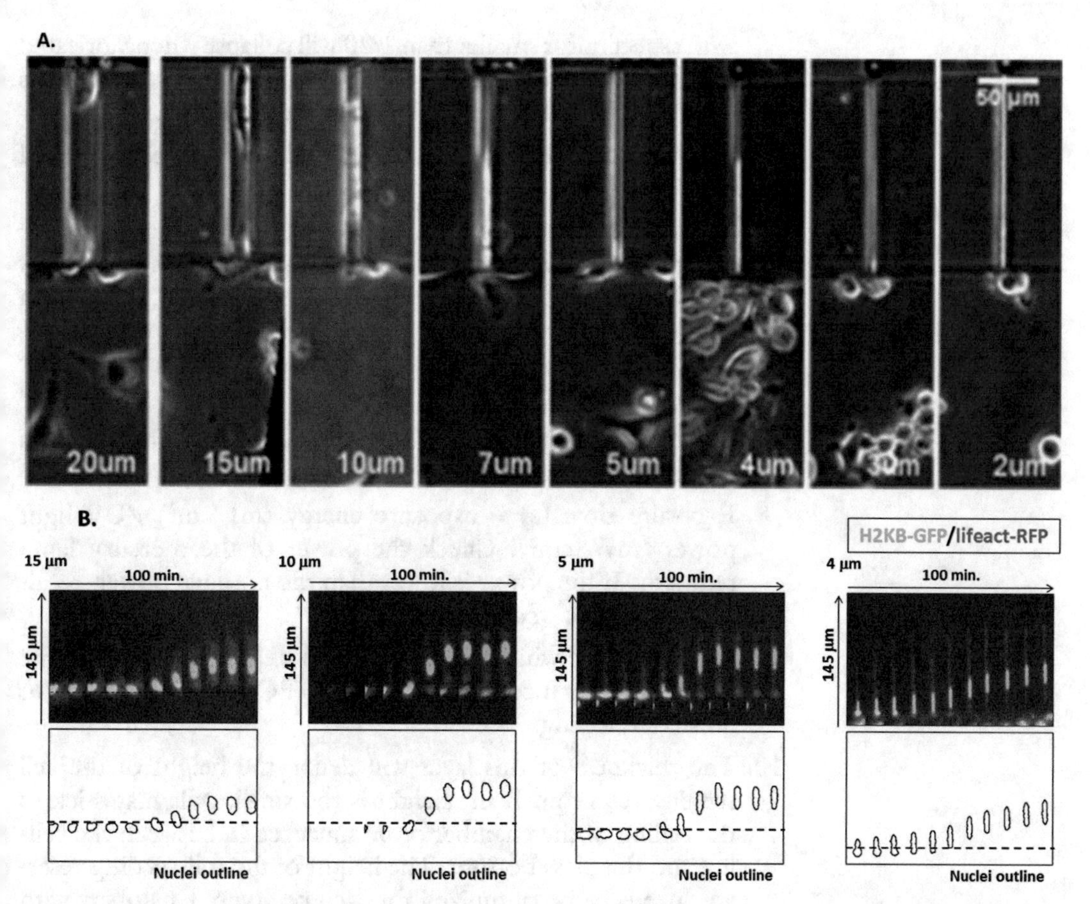

Fig. 3 (**a**) Phase contrast image of the channels from [11]. (**b**) Kymographs of representative single cells translocating their nuclei into channels of different width. The upper panel shows cells expressing GFP-H2BK (green) and RFP-lifeact (red). The lower panel shows the outline of the nuclei. Dashed lines indicate the entrance to the channels. The width of each channel is indicated in the top left above the fluorescence images

foundry website (https://web.stanford.edu/group/foundry/Mask%20Design%20Rules.html).

2. To ensure optimal bonding of the PDMS to the glass substrate, the distance between each pair of transversal microchannels should be at least ten times the width of the larger microchannel. The total width of a set of transversal channels is 1 mm enabling imaging of all channel sizes simultaneously in one field of view at 4× magnification.
3. To facilitate alignment, a large overlap region between the first and the second layer should be included.
4. Because of the large overall dimensions of the cell seeding reservoir and the chemoattractant channel, pillars were included in this layer to prevent the device from collapsing. As a rule of thumb, channels with an aspect ratio larger than 1:10 (height/width) will be structurally sound. However, channels or features

with aspect ratios smaller than 1:20 will collapse when fabricated with PDMS. To overcome this problem, regularly spaced pillars must be included in the design to prevent collapse.

5. The acetate mask should be printed with a resolution of 40,000 dpi.
6. For a stronger adhesion of SU-8 to the substrate, clean the wafers with piranha solution. Piranha solution is a mixture of concentrated sulfuric acid with hydrogen peroxide, usually in a ratio of 3:1–7:1. It is highly corrosive and an extremely powerful oxidizer and must be prepared with great care. We refer interested readers to [17] .
7. If necessary, place the wafer in an oven, or on a hot plate, for a few minutes to ensure complete dryness.
8. Exposure time (s) = exposure energy (mJ/cm^2) /UV light power (mW/cm^2). Check the power of the mercury lamp regularly. If the power is lower than the nominal power, recalculate the time of exposure.
9. If you observe white streaks after this step, it signifies that the development is incomplete. Return to EC solvent for longer to finish developing.
10. The thickness of this layer will define the height of the cell seeding reservoir. If the height is too small, cells may stick to the ceiling of the chamber or be squeezed in between the ceiling and the glass bottom. The height of the cell-seeding reservoir needs to be optimized for each cell type. Chambers with smaller heights have less nutrients available for cells. This may become an issue if long term imaging is planned.
11. Examine the wafer using a microscope to make sure that all uncross-linked SU-8 is removed, that features are sharp, and that all channels straight. Placing the beaker containing the wafer in EC solvent for a few seconds in an ultrasonic bath may help removing uncross-linked SU-8 from in between the transversal channels.
12. To introduce less dust, cover the cup with a paraffin tape and make a few holes on it.
13. **Steps 11** and **12** may need to be repeated several times until all bubbles have been removed.
14. A smaller gap may result in leakage. Do not use excessive pressure when cutting. Silicon wafers are fragile and may break. Wear gloves when handling the PDMS device and avoid touching the bottom surface containing the channels as this could compromise bonding.
15. This step must be performed in a sterile laminar flow hood to maintain sterility and minimize contamination.

16. Filling the chamber, chemoattractant channel, microchannels, and reservoir by circulating medium through the chemoattractant channels has to be a gentle process to allow the exit of any air bubbles in the device.
17. This step ensures that the PDMS becomes saturated with water vapor, which enhances cell viability within devices.
18. Beads are coated with FBS to obtain a gradual delivery of chemoattractants and stable gradient formation without flow.
19. During the experiments, petri dishes were filled with sufficient culture medium to prevent evaporation. Excess medium was also added to the petri dish outside of the device. In some experiments, to visualize gradient formation, beads were coated with rhodamine isothiocyanate and FBS, washed several times, and imaged immediately after bead injection into the upper channel.
20. Use cell lines with fluorescently tagged Histone 2BK (H2BK) for simplified visualization of the nucleus and automated cell tracking. Nuclear dyes or stains can also be used in place of H2BK. Optimize imaging interval and exposure time to minimize photodamage.

References

1. Theveneau E, Mayor R (2012) Neural crest delamination and migration: from epithelium-to-mesenchyme transition to collective cell migration. Dev Biol 366(1):34–54. https://doi.org/10.1016/j.ydbio.2011.12.041
2. Friedl P, Gilmour D (2009) Collective cell migration in morphogenesis, regeneration and cancer. Nat Rev Mol Cell Biol 10(7):445–457. https://doi.org/10.1038/nrm2720
3. Friedl P, Weigelin B (2008) Interstitial leukocyte migration and immune function. Nat Immunol 9(9):960–969. https://doi.org/10.1038/ni.f.212
4. Woodfin A, Voisin MB, Nourshargh S (2010) Recent developments and complexities in neutrophil transmigration. Curr Opin Hematol 17(1):9–17. https://doi.org/10.1097/MOH.0b013e3283333930
5. Friedl P, Alexander S (2011) Cancer invasion and the microenvironment: plasticity and reciprocity. Cell 147(5):992–1009. https://doi.org/10.1016/j.cell.2011.11.016
6. Wilson K, Lewalle A, Fritzsche M, Thorogate R, Duke T, Charras G (2013) Mechanisms of leading edge protrusion in interstitial migration. Nat Commun 4:2896. https://doi.org/10.1038/ncomms3896
7. Bergert M, Erzberger A, Desai RA, Aspalter IM, Oates AC, Charras G, Salbreux G, Paluch EK (2015) Force transmission during adhesion-independent migration. Nat Cell Biol 17(4):524–529. https://doi.org/10.1038/ncb3134
8. Ruprecht V, Wieser S, Callan-Jones A, Smutny M, Morita H, Sako K, Barone V, Ritsch-Marte M, Sixt M, Voituriez R, Heisenberg CP (2015) Cortical contractility triggers a stochastic switch to fast amoeboid cell motility. Cell 160(4):673–685. https://doi.org/10.1016/j.cell.2015.01.008
9. Liu YJ, Le Berre M, Lautenschlaeger F, Maiuri P, Callan-Jones A, Heuze M, Takaki T, Voituriez R, Piel M (2015) Confinement and low adhesion induce fast amoeboid migration of slow mesenchymal cells. Cell 160(4):659–672. https://doi.org/10.1016/j.cell.2015.01.007
10. Stroka KM, Jiang H, Chen SH, Tong Z, Wirtz D, Sun SX, Konstantopoulos K (2014) Water permeation drives tumor cell migration in confined microenvironments. Cell 157(3):611–623. https://doi.org/10.1016/j.cell.2014.02.052
11. Malboubi M, Jayo A, Parsons M, Charras G (2015) An open access microfluidic device for the study of the physical limits of cancer cell deformation during migration in confined environments. Microelectron Eng 144:42–45. https://doi.org/10.1016/j.mee.2015.02.022
12. Jayo A, Malboubi M, Antoku S, Chang W, Ortiz-Zapater E, Groen C, Pfisterer K, Tootle

T, Charras G, Gundersen GG, Parsons M (2016) Fascin regulates nuclear movement and deformation in migrating cells. Dev Cell 38(4):371–383. https://doi.org/10.1016/j.devcel.2016.07.021

13. Kuriyama S, Theveneau E, Benedetto A, Parsons M, Tanaka M, Charras G, Kabla A, Mayor R (2014) In vivo collective cell migration requires an LPAR2-dependent increase in tissue fluidity. J Cell Biol 206(1):113–127. https://doi.org/10.1083/jcb.201402093
14. Tabeling P (2010) Introduction to microfluidics. Oxford University Press, Oxford
15. Folch A (2012) Introduction to BioMEMs. CRC Press, Boca Raton, FL
16. Lake M, Narciso C, Cowdrick K, Storey T, Zhang S, Zartman J, Hoelzle D (2015) Microfluidic device design, fabrication, and testing protocols. Protocol Exchange. https://doi.org/10.1038/protex.2015.069
17. Millet LJ, Stewart ME, Sweedler JV, Nuzzo RG, Gillette MU (2007) Microfluidic devices for culturing primary mammalian neurons at low densities. Lab Chip 7(8):987–994. https://doi.org/10.1039/b705266a

Check for updates

Chapter 28

Controlling Confinement and Topology to Study Collective Cell Behaviors

Guillaume Duclos, Maxime Deforet, Hannah G. Yevick, Olivier Cochet-Escartin, Flora Ascione, Sarah Moitrier, Trinish Sarkar, Victor Yashunsky, Isabelle Bonnet, Axel Buguin, and Pascal Silberzan

Abstract

Confinement and substrate topology strongly affect the behavior of cell populations and, in particular, their collective migration. In vitro experiments dealing with these aspects require strategies of surface patterning that remain effective over long times (typically several days) and ways to control the surface topology in three dimensions. Here, we describe protocols addressing these two aspects. High-resolution patterning of a robust cell-repellent coating is achieved by etching the coating through a photoresist mask patterned directly on the coated surface. Out-of-plane curvature can be controlled using glass wires or corrugated "wavy" surfaces.

Key words Collective cell migration, Microfabrication, Surface patterning, Out-of-plane curvature, Positive curvature, Negative curvature

1 Introduction

Collective cell migration has been identified as a major mode of cell migration in vivo and can be recapitulated in vitro using for instance wound healing experiments [1–5]. In vivo, collective migrations are often accompanied by physical confinement (cells migrating between vessels or capillaries or onto such structures [5]). Therefore, in vitro experiments aimed at mimicking these biologically relevant situations often include such a confinement, usually by constraining the cells within chemically defined patterns such as stripes in a planar geometry [6] or in other topologies including the third dimension [7–9]. Two-dimensional domains, onto which cells adhere, are

Guillaume Duclos, Maxime Deforet, Hannah G. Yevick, and Olivier Cochet-Escartin contributed equally to this work.

Alexis Gautreau (ed.), *Cell Migration: Methods and Protocols*, Methods in Molecular Biology, vol. 1749, https://doi.org/10.1007/978-1-4939-7701-7_28, © Springer Science+Business Media, LLC 2018

defined within an antiadhesive substrate. Practically, requirements for the robustness of the antiadhesive coating are particularly stringent as the experiments can last days or even weeks, whereas most of the classical cell-repellent coatings are based on physisorbed polyethylene glycol (PEG)-based copolymers and are only stable for a couple of days at best.

Here, we present two technologies that combine the coating of the substrate using a robust antiadhesive polymer layer, with high-resolution photolithography. The Aam-PEG antiadhesive coating is adapted from Reference [10]. It consists in successive grafts on a glass coverslip: Silane molecules covalently bind to clean glass and form a reactive layer to which a thin layer of polyacrylamide gel can be anchored; finally, PEG-acrylamide molecules interpenetrate the gel and cross-link it to form a dense hydrophilic brush that prevents cell adhesion. The resulting coating is strongly repellent to the cells and extremely robust (up to a couple of months) [11, 12]. The Aam-PEG coating is compatible with photolithography, it can be patterned at the μm-scale resolution by plasma etching through a photoresist mask patterned directly onto the PEG coating, using standard photolithography techniques [12–14]. It can also be patterned using the deep-UV etching technique where the substrate is exposed to deep-UV radiation through a mask [15–17]. The latter method is faster than the former and avoids photolithography steps. Both methods allow for a strong confinement of cell populations for weeks, after the cells have reached confluence on the pattern.

In vivo, cells are, of course, not constrained to 2D space. Tissues routinely form complex 3D structures, many of which being highly curved. For example, cells that migrate out of tumors often wrap around vessels or extracellular matrix fibers and migrate effectively on a curved cylindrical substrate [5]. We describe a method to make model glass fibers that can be used as substrates to study collective cell migration. Their radius can be varied from ~100 μm down to submicron values [7]. Corrugated "wavy" surfaces represent an excellent assay for probing how cells migrate, orient and respond to positive and negative out-of-plane curvatures. This method requires a castellated polydimethylsiloxane (PDMS) substrate produced by thick resist photolithography that is embossed into molten polystyrene (PS). The resultant PS castellated structure is then reflowed to achieve a curved wavy substrate that can be replicated at will by molding PDMS for cell culture.

We often use glass as our final substrate for cell culture. We find that this is an adequate substrate for most purposes. Cells adhere on untreated glass probably via the adsorption of proteins present in the culture medium. However, we also provide protocols for fibronectin coating of the patterned substrates. When coated with fibronectin, cell dynamics is usually faster compared to bare glass.

After preparation, cells of interest are seeded on these various substrates and observed by optical microscopy at all relevant scales, using the most adequate modes (live imaging coupled to time-lapse or confocal imaging on fixed cells in particular when dealing with out-of-plane curvature). The role of collective migration is not only crucial for cells bearing stable cell–cell adhesions such as epithelial cells [3], but also for mesenchymal cells such as elongated fibroblasts and myoblasts [14, 18].

2 Material

Prepare all aqueous solution with deionized (DI) water. Use analytical grade reactants and solvents. Dispose of chemical waste appropriately.

2.1 Aam-PEG Coating

1. Glass coverslips (typical diameter 30 mm).
2. "Piranha" solution: 1 vol concentrated sulfuric acid–3 vol 30% hydrogen peroxide solution (*see* **Notes 1** and **14**).
3. Silane solution: 200 μL of allyltrichlorosilane (Sigma) in 15.5 mL Toluene (*see* **Note 1**).
4. Acrylamide solution: mix 3 g acrylamide (Sigma), 0.3 g *N*,*N*′-methylenebis(acrylamide) (BIS) (Sigma), 1.09 g benzophenone (Sigma) in 20 mL acetone (*see* **Notes 1** and **16**).
5. PEG solution: mix 0.6 g poly(ethylene glycol) methyl ether methacrylate (PEG) (Sigma) (*see* **Note 2**), 0.9 g BIS, 1.09 g benzophenone in 20 mL methanol (*see* **Note 16**).
6. Hot plate.
7. UV (λ = 365 nm) 25 W–Transilluminator (UVP, TFL 40 V) (*see* **Note 3**).
8. Air Plasma (Harrick Plasma cleaner).
9. Glassware (including a large glass Petri dish).
10. Common organic solvents (tolene, acetone, methanol…).

2.2 Photolithography-Based Patterning of PEG Coating

If possible, use a clean room equipped with standard photolithography equipment (necessary for small features). If high-resolution is not an issue, a clean environment may be sufficient.

1. Spin coater and mask aligner such as a MJB3 mask aligner (Karl Suss) (*see* **Note 4**). Hot plate.
2. Chrome mask with the desired pattern (glass or quartz) (Compugraphics, UK) (*see* **Note 5**).
3. S1813 Positive Photoresist (Shipley).
4. 351 Developer (Shipley) diluted in DI water (1v/4v).
5. 10 μg/mL Fibronectin solution in PBS.

6. Acetone, Isopropanol.
7. Air Plasma (Harrick plasma cleaner).
8. PBS.
9. Culture medium.

2.3 Deep-UV Patterning of PEG Coating

1. Chrome mask with the desired pattern (Synthetic quartz quality (*see* **Note 6**), Compugraphics, UK) (*see* **Note 5**).
2. Deep-UV lamp, λ = 185 nm (UVO Cleaner, Jelight, USA) (*see* **Notes 7** and **8**).
3. 10 μg/mL Fibronectin solution in PBS.
4. PBS.
5. PDMS mix: mix the two parts of the Sylgard 184 kit (Dow Corning) (1 v/10 v) (*see* **Note 1**).
6. Parafilm.
7. Teflon tweezers (Sigma).
8. Glass bottom 6-well plate (IBL) or plastic bottom 6-well plate whose polystyrene bottom has been removed by drilling.

2.4 Wire Substrates

1. Glass Pasteur pipettes (Fisher Scientific) (*see* **Note 9**) or glass rods (WPI Inc., diameter 1 mm) (*see* **Note 10**).
2. PDMS films (thickness 17 mils) (Gel-Pak, Hayward).
3. PDMS mix: mix the two parts of the Sylgard 184 kit (Dow Corning) (1 v/10 v) (*see* **Note 1**).
4. 70% Ethanol.
5. 10 μg/mL Fibronectin solution in PBS.
6. PBS.
7. 0.1 mg/mL solution of poly L-lysine-polyethylene glycol (PLL(20)-g-[3,5]-PEG (2)) (Susos, Switzerland) [16].
8. Razor blade.
9. Tweezers.
10. Glass coverslips (22 × 30 mm).
11. Glass bottom 6-well plate (IBL GmbH, Germany).
12. Air Plasma cleaner (Harrick Plasma).
13. Pipette puller (P-2000, Sutter Instruments) (*see* **Note 11**).
14. Oven.

2.5 Wavy Substrates

1. Chrome mask patterned with a periodic succession of chrome lines (width w, period p) (Compugraphics, UK) (*see* **Notes 12** and **13**).
2. SU8–2150 photoresist (Microchem).
3. SU8 developer (Microchem).

4. Fluorosilane (tridecafluoro-1,1,2,2 tetrahydrooctyl trichlorosilane) (Roth Sochiel).
5. 10 μg/mL Fibronectin solution in PBS.
6. Polystyrene polymer (granules) (Sigma).
7. PDMS mix: mix the two parts of the Sylgard 184 kit (Dow Corning) (1 v/10 v) (*see* **Note 1**).
8. Air Plasma cleaner (Harrick Plasma).
9. Heating Press (E-Z press, ICL).
10. Vacuum Oven (Precision).
11. Photolithography equipment (spin-coater, mask aligner, hot plates…).

3 Methods

3.1 Aam-PEG Coating

1. Plasma-clean the coverslips for 5 min.
2. Prepare the piranha solution (*see* **Note 14**). Soak coverslips in piranha solution and let stand for 5 min.
3. Rinse the coverslips with DI water. Dry under a jet of air. Put the dried coverslips in a clean glass Petri dish (*see* **Note 17**).
4. Prepare the allyltrichlorosilane solution. Pour the solution onto the clean coverslips and let stand for 5 min. Rinse with Toluene for 1 min. Then, rinse with Acetone followed by DI water. Dry under a jet of air (*see* **Note 15**). Bake the silanized coverslips in a glass Petri dish on a hot plate (90 °C) for 1 h.
5. Prepare the acrylamide solution (*see* **Note 16**). Put the freshly prepared acrylamide solution in a glass Petri dish and place the silanized coverslips in the dish (*see* **Note 17**). Let stand for 5 min. Place the dish on the transilluminator and expose the slides in the solution for 3 min (*see* **Note 18**) through the glass bottom of the dish (*see* **Notes 3**, **19**, and **20**). The side of the coverslips facing the illuminator is now acrylamide-coated.
6. Rinse each slide individually with Acetone and then DI water. Dry the coverslips under a jet of air (*see* **Notes 21** and **22**).
7. Prepare the PEG solution and place it in a glass bottom Petri dish. Incubate the acrylamide-coated coverslips in the solution, grafted side facing down, for 5 min (*see* **Note 17**). Expose the slides in the solution through the glass bottom of the dish with the transilluminator for 3 min (*see* **Notes 3** and **23**).
8. Rinse each slide individually with Methanol, then Acetone, and finally DI water. Dry the coverslips under a jet of air. At this point, the coated coverslips can be stored in a dust-free environment for several months.

3.2 Photolithography-Based Patterning of PEG Coating (Fig. 1a)

1. Spin-coat S1813 positive photoresist (4000 rpm, 30 s) on the coverslips prepared with the cell repellent coating as per Subheading 3.1 (*see* **Note 24**). Soft-bake the slides: 115 °C, 1 min on the hot plate (*see* **Note 25**).

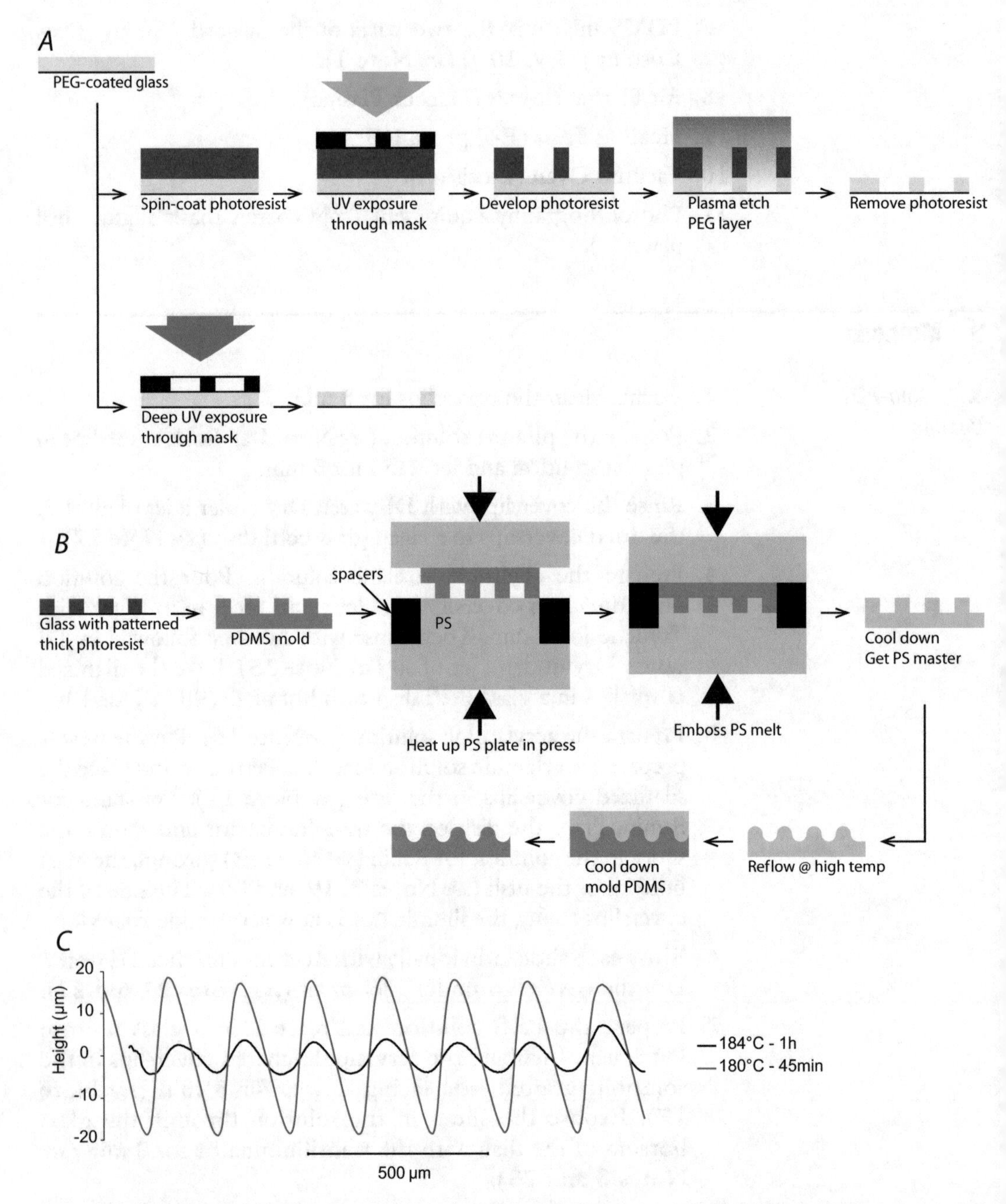

Fig. 1 (**a**) Preparation of confining patterns. The two possible methods are photolithography or deep-UV patterning. (**b**) Fabrication of wavy substrates. The castellated substrate structured with thick photoresist is duplicated in PDMS and then embossed in molten PS. The PS structure is then reflowed above glass transition. The resulting wavy pattern in PS can be duplicated indefinitely in PDMS. (**c**) Profiles of the PS mold after 45 min (red) and 1 h (black) reflow around 180 °C (the baseline has been subtracted)

2. If necessary, clean the mask with isopropanol.
3. Expose the coverslips through the mask (UV light, $\lambda = 365$ nm) with the mask aligner (*see* **Note 26**).
4. Develop the exposed slides in developer solution for 10 s.
5. Rinse with DI water. Dry under a jet of air.
6. Etch the unprotected repellent layer with four runs of 5 min each, using a low power air plasma (*see* **Note 27**).
7. Optional: Fibronectin coating: Pipette 50 μL of fibronectin solution on a Parafilm layer. Return the coverslip on to the fibronectin drop so the patterned side is in direct contact with the fibronectin solution. Incubate for 1 h.
8. Wash the coverslips with acetone to remove the protective layer of photoresist (*see* **Note 28**). Rinse with DI water and dry under a jet of air. The glass substrate is now patterned with cell-repellent areas and glass or fibronectin-coated areas.
9. Seed the cells and incubate at appropriate temperature and humidity for 2 h (*see* **Note 29**), to allow them to adhere to the adhesive parts of the substrate.
10. Aspirate medium with unattached cells. Wash with warm PBS and fill with warm culture medium. Cells can now be grown and observed (Fig. 2a).

3.3 Deep-UV Patterning of PEG Coating (Fig. 1a)

1. Prepare the coverslips with the repellant coating as in Subheading 3.1 (*see* **Note 30**).
2. Wash the quartz mask with Ethanol. Expose the mask to deep-UV light, Chrome side facing the lamp (*see* **Note 31**) for extensive cleaning.
3. Pipette 5–10 μL of DI water onto the Chrome side of the mask (*see* **Note 31**) and return the coverslip onto the drop so the Aam-PEG coating faces the quartz mask and holds to it by capillary action (*see* **Note 32**).
4. Expose the coated coverslip through the mask for 15 min with the deep-UV lamp.
5. Gently detach the coverslip from the mask using Teflon tweezers (*see* **Note 33**).
6. Rinse the coverslip with DI water and dry under a jet of air.
7. Fibronectin coating of the etched areas: Pipette 50 μL of fibronectin solution on a Parafilm layer. Return the coverslip onto the fibronectin drop so the patterned side is in direct contact with the fibronectin. Incubate for 1 h. Rinse the coverslip with PBS, DI water and dry under a jet of air. The coverslips are now patterned.
8. If needed, attach the coverslip to the bottom of a 6-well plate whose bottom has been previously removed by drilling, using

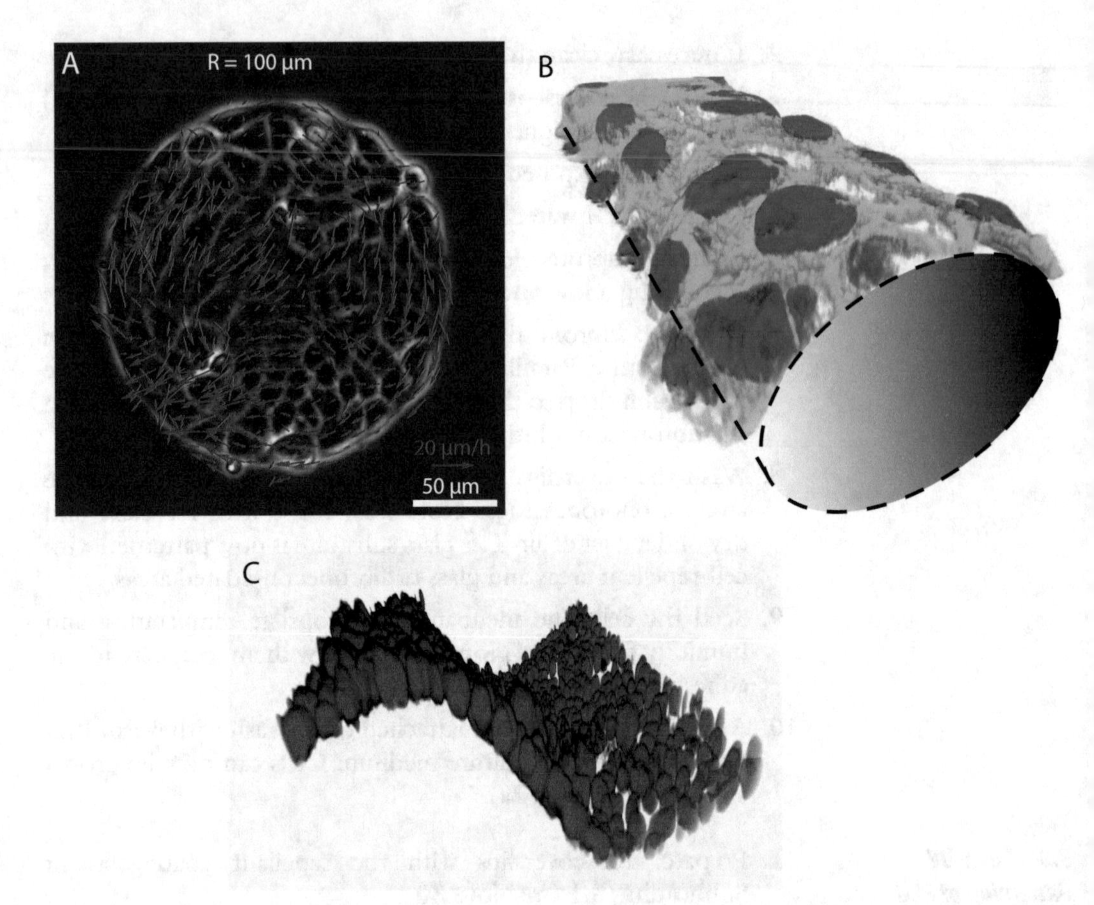

Fig. 2 (**a**) Collective displacements of MDCK cells in a confining disk. The velocity field (red) obtained by PIV is superimposed to the phase contrast image. (**b**) MDCK monolayer cultured on a 100 μm diameter cylindrical glass fiber. Only the top half is represented (Blue: nuclei; green: actin). (**c**) MDCK monolayer on a wavy surface (red nuclei). The period of the pattern is 180 μm

the PDMS mix. In that case, as fibronectin denaturates above 60 °C, cure PDMS overnight in a 37 °C incubator.

9. Seed the cells and incubate at appropriate temperature and humidity for 2 h to allow them to adhere to the substrate.
10. Aspirate medium with unattached cells. Wash with warm PBS and fill wells with warm culture medium. Cells can now be grown and observed.

3.4 Glass Wires

Glass capillaries of radius larger than 20 μm are manufactured by heating glass Pasteur pipettes over the flame of a freestanding gas torch and pulling each end (*see* **Note 34**) [7]. Glass capillaries whose radius is lower than 20 μm are manufactured from glass rods, using a pipette puller. Glass capillaries of a few fixed sizes can also be purchased (e.g., $r = 40$, 60, 85 μm, CM Scientific). This section describes how to assemble these wires into a device which can be used for migration experiments.

1. Sort wires by size under a stereomicroscope and set aside sizes of interest.
2. Cut PDMS Gel-Pak film into a rectangle (~0.5 cm × 1.5 cm) using the razor blade and peel from its polyester backing and polyethylene release liner.
3. Place the rectangular piece of Gel-Pak on a coverslip.
4. Cut glass wires to size and align them parallel on the Gel-Pak, protruding off the support on both ends.
5. Cover carefully the Gel-Pak with the freshly prepared PDMS mix to immobilize the wires. Use only a minimal amount to avoid the PDMS from coating the freely suspended wires via capillary action.
6. Cure the structure for 24 h at 65 °C.
7. Clean the structures with ethanol and DI water. Air-dry under a jet of air.
8. Expose the structures to an air plasma for 30 s to activate the PDMS and the glass wires.
9. Optional: Fibronectin coating: Incubate the wires in fibronectin solution for 1 h. Aspirate the fibronectin solution. Rinse the structures with PBS and dry under a jet of air for 5 min.
10. Pretreat the glass bottom of the 6-well plates with PLL-PEG solution (1 h) to prevent cells from growing down and off the structures onto the dish in the optical path (clouding the imaging of the wires)
11. Transfer the wire substrates to a 6-well plate with tweezers.
12. Passage cells to obtain a cell suspension of the desired concentration (*see* **Note 35**).
13. Deposit a small droplet of the cell suspension on PDMS pedestals (~10–100 μL depending on PDMS substrate area).
14. Seed the cells on the substrates and incubate at appropriate temperature and humidity for 2 h to allow the cells in suspension to adhere to the PDMS pedestal, avoiding drying of the droplet.
15. Wash with warm PBS and fill the wells with warm culture medium.
16. Once the cells have filled the support (typically after a few days of incubation) they will begin to migrate onto the wires and the 6-well plate is ready to be put under the microscope for imaging (Fig. 2b).

3.5 Wavy Substrates (Fig. 1b)

1. The pattern on the mask is transferred to a thick SU8 negative resist (the example is for SU8–2150, ending up in a castellated substrate where the patterns are about 200 μm high) (*see* **Note 36**):

(a) Spin-coat the resist on a silicon wafer (3000 rpm, 30 s).

(b) Soft-bake for 40 min, at 95 °C. Allow to cool down to room temperature.

(c) Expose through the mask with the mask aligner.

(d) Post-exposure bake: 15 min, 95 °C. Allow to cool down to room temperature.

(e) Develop in SU8 developer: 15 min. Rinse with clean developer and dry.

2. Coat with fluorinated silane: place a few drops of fluorosilane in a small polyethylene weighting boat in a vacuum desiccator. Plasma-treat the castellated SU8 structure (1 min) and put the activated pattern in the desiccator. Apply vacuum for typically 2 h (*see* **Note 37**).

3. Pour 5 g of the freshly prepared PDMS mix over the master. Degas and cure overnight at 65 °C. Allow to cool down and unmold.

4. Place 2.8 g of polystyrene granules in the heating press and inside a square metal spacer (3 cm × 3 cm; height = 3 mm). The spacer sets the thickness of the polystyrene plate and acts as a frame to confine lateral dimensions. Heat the polystyrene at 150 °C in the press for 10 min between two smooth metal plates to yield a flat polystyrene plate. Cool down to ambient temperature.

5. Melt the flat PS plate under the same experimental conditions (150 °C, 4 bars), intercalating the PDMS mold between one of the plates and the PS, to emboss the periodic castellated shape into the softened PS. Allow to cool down and unmold the PDMS master.

6. Leave the PS substrate at 80 °C (below the glass transition temperature) under vacuum overnight to pump out gas and moisture contained inside that may otherwise deform the substrate during subsequent high temperature reflow.

7. Reflow the PS plate under vacuum at 180 °C, for 50–120 min (*see* **Notes 38** and **39**) in order to modulate the wavy shape and therefore the out-of-plane curvature of the substrate. Quench to room temperature (Fig. 1c).

8. Duplicate the wavy PS substrate in PDMS (*see* **step 3** and **Note 40**).

9. Plasma-oxidize the PDMS surface for 30 s with air plasma and incubate with fibronectin solution for 1 h before culturing cells.

10. Seed the cells and incubate at appropriate temperature and humidity for 30 min to allow them to adhere to the substrate.

11. Aspirate medium with unattached cells. Wash with warm PBS and fill with warm culture medium. Cell growth and migration can now be observed (Fig. 2c).

4 Notes

1. Prepare immediately before use.
2. Use low molecular weight polymer: average M_n 950.
3. Wear UV protective glasses when using UV radiations.
4. A collimated UV lamp of sufficient power at 365 nm can also be used.
5. The Chrome features on the mask correspond to the nonadherent moiety.
6. It is mandatory that the mask is transparent for wavelengths down to 185 nm.
7. The wavelength has to be less than 200 nm.
8. These lamps generate ozone that is a toxic and irritant gas. Use under a fume hood or hook to an "ozone killer."
9. Use glass Pasteur pipettes to produce capillaries whose radius is larger than 20 μm, by heating and pulling.
10. Use these glass rods with the pipette puller. Alternatively 40, 60, and 85 μm diameter glass wires can be bought from CM Scientific and used as received.
11. If very small diameters are needed.
12. Note that we use a negative photoresist and therefore the Chrome features on the mask correspond to the valleys in the final PDMS substrate.
13. Other motifs can also be used such as arrays of disks.
14. Caution! Very exothermic when mixing. Prepare in small quantities. Add the H_2O_2 slowly in the acid. Wear reinforced gloves and face shield. Dispose of appropriately.
15. At this point, both sides of the coverslips should be hydrophobic.
16. Use a sonic bath to dissolve the products.
17. The coverslips should be well separated.
18. … or until a white halo is formed between the coverslip and the glass Petri dish.
19. The UV light comes from the bottom and illuminates the slides through the glass bottom of the Petri dish.
20. The grafting reaction occurs preferentially between the coverslips and the Petri dish bottom, presumably because of the slow diffusion of oxygen in this confined area.
21. Now, the grafted side is hydrophilic while the ungrafted silanized side is still hydrophobic. You can test which side is grafted by its wetting properties using a DI water droplet.
22. From this step on, only one side of each coverslip is coated. Keep track of which one.

23. Timing here is critical. Use the same exposure time as acrylamide step (step 3.1. 5) as a starting point.
24. Up to development, photoresist is to be handled only in inactinic light.
25. If resolution is not an issue, keep the temperature as low as possible. We had satisfactory results with $T = 80$ °C.
26. The exposure time depends on the lamp power. In doubt, follow manufacturer instructions.
27. Do not let the coverslips in the plasma for more than 5 min in one run because the heat will degrade the protective photoresist and consequently the underlying cell-repellent layer.
28. The acetone treatment does not affect the fibronectin coating as can be checked by culturing cells on these substrates.
29. 30 min on fibronectin-coated substrates.
30. Be careful not to scratch the repellent layer with the tweezers.
31. Chrome side appears darker and less shiny than the glass side.
32. Avoid bubbles as they will affect the transmission of the UV light and deform the patterns.
33. Add more DI water to facilitate the process.
34. Pull sharply when the glass becomes soft over the flame.
35. Around 2–5 10^6 cells/mL.
36. For more details, refer to manufacturer's instructions.
37. This procedure makes it easier to mold the structure with PDMS by decreasing the friction between the two polymers.
38. Vacuum is used to prevent polymer degradation.
39. Adjust reflow time and temperature to attain the exact curvature profile that is needed. Measure the profile with a stylus profilometer.
40. The PS wavy substrate is used as a master to mold as many PDMS substrates as necessary.

Acknowledgments

We gratefully acknowledge financial support from the Groupement des Entreprises Françaises dans la Lutte contre le Cancer (GEFLUC) Ile-de-France, the Région Ile-de-France Domaine d'Intérêt Majeur (DIM) Nano-K, the Association pour la Recherche sur le Cancer (ARC), the EU cofund PRESTIGE post-doc program, the EU cofund IC-3i PhD program, and the Fondation Pierre-Gilles de Gennes. The "Biology-inspired Physics at MesoScales" group is a member of the CelTisPhyBio Labex and of the Institut Pierre-Gilles de Gennes. It is a pleasure to thank Mohamed El Beheiry for his help in the 3D processing of our images.

References

1. Scarpa E, Mayor R (2016) Collective cell migration in development. J Cell Biol 212:143–155
2. Rørth P (2012) Fellow travellers: emergent properties of collective cell migration. EMBO Rep 13:984–991
3. Hakim V, Silberzan P (2017) Collective cell migration: a physics perspective. Rep Prog Phys 80(7):076601
4. Gov NS (2014) Collective cell migration. In: Kaunas R, Zemel A (eds) Cell and matrix mechanics. CRC Press, Boca Raton, FL, pp 219–238
5. Weigelin B, Bakker G-J, Friedl P (2012) Intravital third harmonic generation microscopy of collective melanoma cell invasion. IntraVital 1:32–43
6. Vedula SRK, Leong MC, Lai TL et al (2012) Emerging modes of collective cell migration induced by geometrical constraints. Proc Natl Acad Sci U S A 109:12974–12979
7. Yevick HG, Duclos G, Bonnet I et al (2015) Architecture and migration of an epithelium on a cylindrical wire. Proc Natl Acad Sci 112:5944–5949
8. Zheng Y, Chen J, Craven M et al (2012) In vitro microvessels for the study of angiogenesis and thrombosis. Proc Natl Acad Sci 109:9342–9347
9. Ye M, Sanchez HM, Hultz M et al (2014) Brain microvascular endothelial cells resist elongation due to curvature and shear stress. Sci Rep 4:4681
10. Tourovskaia A, Barber T, Wickes BT et al (2003) Micropatterns of chemisorbed cell adhesion-repellent films using oxygen plasma etching and elastomeric masks. Langmuir 19:4754–4764
11. Tourovskaia A, Figueroa-Masot X, Folch A (2006) Long-term microfluidic cultures of myotube microarrays for high-throughput focal stimulation. Nat Protoc 1:1092–1104
12. Deforet M, Hakim V, Yevick HG et al (2014) Emergence of collective modes and tri-dimensional structures from epithelial confinement. Nat Commun 5:3747
13. Nier V, Deforet M, Duclos G et al (2015) Tissue fusion over nonadhering surfaces. Proc Natl Acad Sci 112:9546–9551
14. Duclos G, Erlenkämper C, Joanny J-F et al (2017) Topological defects in confined populations of spindle-shaped cells. Nat Phys 13:58–62
15. Azioune A, Carpi N, Tseng Q et al (2010) Protein micropatterns. A direct printing protocol using deep UVs. Methods Cell Biol 97:133–146
16. Thery M, Piel M (2014) Scientific protocols – adhesive micropatterns for cells: a microcontact printing protocol. Sci Protoc. https://doi.org/10.5281/zenodo.13592
17. Azioune A, Storch M, Bornens M et al (2009) Simple and rapid process for single cell micropatterning. Lab Chip 9:1640–1642
18. Duclos G, Garcia S, Yevick HG et al (2014) Perfect nematic order in confined monolayers of spindle-shaped cells. Soft Matter 10:2346–2353

Index

Alexis Gautreau (ed.), *Cell Migration: Methods and Protocols*, Methods in Molecular Biology, vol. 1749, https://doi.org/10.1007/978-1-4939-7701-7, © Springer Science+Business Media, LLC 2018

Printed in the United States
By Bookmasters